Lecture Notes in Mathematics 2075

Editors:
J.-M. Morel, Cachan
B. Teissier, Paris

For further volumes:
http://www.springer.com/series/304

Fondazione C.I.M.E., Firenze

C.I.M.E. stands for *Centro Internazionale Matematico Estivo*, that is, International Mathematical Summer Centre. Conceived in the early fifties, it was born in 1954 in Florence, Italy, and welcomed by the world mathematical community: it continues successfully, year for year, to this day.

Many mathematicians from all over the world have been involved in a way or another in C.I.M.E.'s activities over the years. The main purpose and mode of functioning of the Centre may be summarised as follows: every year, during the summer, sessions on different themes from pure and applied mathematics are offered by application to mathematicians from all countries. A Session is generally based on three or four main courses given by specialists of international renown, plus a certain number of seminars, and is held in an attractive rural location in Italy.

The aim of a C.I.M.E. session is to bring to the attention of younger researchers the origins, development, and perspectives of some very active branch of mathematical research. The topics of the courses are generally of international resonance. The full immersion atmosphere of the courses and the daily exchange among participants are thus an initiation to international collaboration in mathematical research.

C.I.M.E. Director	C.I.M.E. Secretary
Pietro ZECCA	Elvira MASCOLO
Dipartimento di Energetica "S. Stecco"	Dipartimento di Matematica "U. Dini"
Università di Firenze	Università di Firenze
Via S. Marta, 3	viale G.B. Morgagni 67/A
50139 Florence	50134 Florence
Italy	Italy
e-mail: zecca@unifi.it	e-mail: mascolo@math.unifi.it

For more information see CIME's homepage: http://www.cime.unifi.it

CIME activity is carried out with the collaboration and financial support of:

- INdAM (Istituto Nazionale di Alta Matematica)

- MIUR (Ministero dell'Istruzione, dell'Università e della Ricerca)

Giorgio Patrizio • Zbigniew Błocki
François Berteloot • Jean-Pierre Demailly

Pluripotential Theory

Cetraro, Italy 2011

Editors:
Filippo Bracci
John Erik Fornæss

 Springer

Giorgio Patrizio
Dipartimento di Matematica "U.Dini"
Università di Firenze
Firenze, Italy

François Berteloot
Institut de Mathématiques de Toulouse
Université Paul Sabatier
Toulouse, France

Zbigniew Błocki
Institute of Mathematics
Jagiellonian University
Krakow, Poland

Jean-Pierre Demailly
Institut Fourier, Laboratoire de
  Mathématiques
University of Grenoble I
Saint-Martin d'Hères, France

ISBN 978-3-642-36420-4 ISBN 978-3-642-36421-1 (eBook)
DOI 10.1007/978-3-642-36421-1
Springer Heidelberg New York Dordrecht London

Lecture Notes in Mathematics ISSN print edition: 0075-8434
                            ISSN electronic edition: 1617-9692

Library of Congress Control Number: 2013936248

Mathematics Subject Classification (2010): 32U99, 14-XX, 32-XX, 31-XX

© Springer-Verlag Berlin Heidelberg 2013
This work is subject to copyright. All rights are reserved by the Publisher, whether the whole or part of the material is concerned, specifically the rights of translation, reprinting, reuse of illustrations, recitation, broadcasting, reproduction on microfilms or in any other physical way, and transmission or information storage and retrieval, electronic adaptation, computer software, or by similar or dissimilar methodology now known or hereafter developed. Exempted from this legal reservation are brief excerpts in connection with reviews or scholarly analysis or material supplied specifically for the purpose of being entered and executed on a computer system, for exclusive use by the purchaser of the work. Duplication of this publication or parts thereof is permitted only under the provisions of the Copyright Law of the Publisher's location, in its current version, and permission for use must always be obtained from Springer. Permissions for use may be obtained through RightsLink at the Copyright Clearance Center. Violations are liable to prosecution under the respective Copyright Law.
The use of general descriptive names, registered names, trademarks, service marks, etc. in this publication does not imply, even in the absence of a specific statement, that such names are exempt from the relevant protective laws and regulations and therefore free for general use.
While the advice and information in this book are believed to be true and accurate at the date of publication, neither the authors nor the editors nor the publisher can accept any legal responsibility for any errors or omissions that may be made. The publisher makes no warranty, express or implied, with respect to the material contained herein.

Printed on acid-free paper

Springer is part of Springer Science+Business Media (www.springer.com)

Preface

Pluripotential theory is a very powerful tool in geometry, complex analysis and dynamics. The principal subjects of investigation in pluripotential theory are plurisubharmonic functions, namely, those functions which remain subharmonic under holomorphic changes of coordinates. Plurisubharmonic functions are objects rather easy to handle and to be constructed; therefore, they are very useful and important tools in complex analysis, geometry (such as geometry of Kähler–Einstein manifolds, hyperbolicity, Green–Griffiths conjecture) and holomorphic dynamics. Among those, maximal plurisubharmonic functions and their associated Monge–Ampère equations play a fundamental role in modern mathematics. Many problems related to manifolds endowed with particular geometric structure, such as symplectic, Kählerian, iperkählerian, quaternionial-Kählerian, algebraic spinorial and Calabi–Yau manifolds and their generalizations, can be rephrased in terms of several types of complex Monge–Ampère equations about the existence of metric with constant curvature on algebraic manifolds. In particular, pluripotential theory plays a very basic role in the study of the equations associated with the existence of Einstein metrics of constant scalar curvature and extremal "a là Calabi", both in the static version and in the parabolic one (Ricci's flows and Calabi) which recently allowed to solve the Poincaré and Thurston's conjectures.

A complete and deep theory has been developed in order to characterize maximal plurisubhharmonic functions by Bedford, Taylor, Demailly, Kiselman, Siciak, Błocki and others. Indeed, maximal plurisubharmonic functions are essentially solutions of homogeneous complex Monge–Ampère equations. Special solutions to such equations are the pluricomplex Green function and the pluri-complex Poisson measures, introduced and used in reproducing formulas for plurisubharmonic functions by Klimek, Demailly, Lempert and others. Such a pluricomplex Green function turned out also to be strictly related to the Kobayashi distance and to hyperconvexity and other geometrical properties of domains in $\mathbb{C}^n$. Pluripotential theory has also a number of very important applications in algebraic geometry, in particular related to jets bundles and the solution of some of the leading conjectures in the area such as Green–Griffiths and Kobayashis conjectures. Other applications to complex dynamics in higher dimensions are also available, both in the realm of discrete dynamics and in that of holomorphic foliations. On another side, pluripotential theory and complex Monge–Ampère equations are used to characterize complex manifolds (the so-called parabolic complex manifolds and Grauert tubes).

The CIME session in Cetraro on Pluripotential theory was a great and unique occasion to present a few courses on topics of high interest in the area and to join both experts and young mathematicians in a nice environment. The school, from which these notes are taken, was aimed to provide courses on pluripotential theory and Monge–Ampère equation and applications to algebraic geometry, complex dynamics and differential geometry. The program with its wide range of topics brought together mathematicians and young researchers with different background: complex analysis and geometry, differential geometry, dynamics, and differential equations.

The courses and the notes taken from them which constitute the chapters of this volume are briefly described hereafter.

In his lectures, *François Berteloot* gives a synthetic and self-contained exposition of the theory of bifurcation currents in holomorphic families of rational maps, giving applications of pluripotential theory to complex dynamics. He constructs the Green measure (the maximal entropy measure) for a fixed rational map and discusses its dynamical properties and proves an approximation formula for its Lyapunov exponent. He presents some concrete holomorphic families of examples and proves the Branner–Hubbard result about the compactness of the connectedness locus and introduces the hypersurfaces $Per_n(w)$ whose distribution turns out to shape the bifurcation locus. He also describes the moduli space Mod_2 of degree two rational maps. Next, he studies the bifurcation current T_{bif} and gives a proof of DeMarco's fundamental results which precisely relates the Lyapunov exponent to the Green function evaluated on the critical points. He then studies how the asymptotic distribution of dynamically defined hypersurfaces is governed by the bifurcation current. Finally, he examines the higher exterior powers T_{bif}^k of the bifurcation current. He concentrates on the highest power and shows that the support of such a measure is the seat of the strongest bifurcations.

Zbigniew Błocki's lectures present two situations where the complex Monge–Ampère equation appears in Kähler geometry: the Calabi conjecture and geodesics in the space of Kähler metrics. In the first case the problem is to construct, in a given Kähler class, a metric with prescribed Ricci curvature. It turns out that this is equivalent to finding a metric with prescribed volume form and thus to solving nondegenerate complex Monge–Ampère equation on a manifold with no boundary. In the second case to find a geodesic in a Kähler class one has to solve a homogeneous complex Monge–Ampère equation on a manifold with boundary. In his self-contained lecture notes, Błocki discusses both the geometric aspects and the PDE part, mostly a priori estimates, starting from a very elementary introduction to Kähler geometry. He introduces the Calabi conjecture and its equivalence to complex Monge–Ampère equation. Later he gives basic properties of the Riemannian structure of the space of Kähler metrics, the Aubin–Yau functional and the Mabuchi K-energy as well as relation to constant scalar curvature metrics. The Lempert–Vivas example is also described. The notes contain also fundamental results on complex Monge–Ampère equations such as the basic uniqueness results as well as the comparison principle. Among other things, the continuity method, used to prove existence of solutions, is described and Yau's proof of the L^∞-estimate using Moser's iteration is presented.

Pluripotential theory is a powerful and strong tool also in algebraic geometry as shown in *Jean–Pierre Demailly*'s lectures note. In his lectures, he describes the main techniques involved in the proof of holomorphic Morse inequalities which relate certain curvature integrals to the asymptotic cohomology of large tensor powers of line or vector bundles bring a useful complement to the Riemann–Roch formula. He also describes their link with Monge–Ampère operators and intersection theory. Finally, he provides applications to the study of asymptotic cohomology functionals and the Green–Griffiths–Lang conjecture. The latter conjecture asserts that every entire curve drawn on a projective variety of general type should satisfy a global algebraic equation; via a probabilistic curvature calculation, holomorphic Morse inequalities imply that entire curves must at least satisfy a global algebraic differential equation.

Giorgio Patrizio's lectures in the CIME session were based on the lecture notes by himself and Andrea Spiro, included in this volume. In these notes the authors discuss the link between pluripotential theory and Monge–Ampère foliations. The latter turned out to have many applications in complex geometry, and the selection of a good candidate for the associated Monge–Ampère foliation is always the first step in the construction of well-behaved solutions of the complex homogeneous Monge–Ampère equation. After reviewing some basic notions on Monge–Ampère foliations, the authors concentrate on two main topics. They discuss the construction of (complete) modular data for a large family of complex manifolds, which carry regular pluricomplex Green functions. This class of manifolds naturally includes all smoothly bounded, strictly linearly convex domains and all smoothly bounded, strongly pseudoconvex circular domains of $\mathbb{C}^n$. Then they report on the problem of defining pluricomplex Green functions in the almost complex setting, providing sufficient conditions on almost complex structures, which ensure existence of almost complex Green pluripotentials and equality between the notions of stationary disks and of Kobayashi extremal disks, and allow extensions of known results to the case of non-integrable complex structures.

It is a real great pleasure to thank the speakers for their very interesting lectures and all the authors for the nice lectures notes they have carefully prepared for this volume. We also want to warmly thank all the participants to the CIME session for their enthusiasm and interest in the subject and for having created a very friendly environment which made possible to experience such a great scientific atmosphere.

Last but not least, we want to thank the CIME organization for giving us the opportunity to organize and financing this school and the GNSAGA of INDAM for support. Our special gratitude also to Pietro Zecca and Elvira Mascolo.

Also, a special thanks to Mrs. Ute McCrory at Springer for her assistance in preparing the volume.

Rome, Italy Filippo Bracci
Ann Arbor, MI John Erik Fornæss

Contents

Bifurcation Currents in Holomorphic Families of Rational Maps 1
François Berteloot

The Complex Monge–Ampère Equation in Kähler Geometry 95
Zbigniew Błocki

Applications of Pluripotential Theory to Algebraic Geometry 143
Jean-Pierre Demailly

Pluripotential Theory and Monge–Ampère Foliations 265
G. Patrizio and A. Spiro

Bifurcation Currents in Holomorphic Families of Rational Maps

François Berteloot

Je dédie ce texte à mes parents ainsi qu'à la mémoire de mon ami Giovanni Bassanelli.

Abstract These lectures are devoted to the study of bifurcations within holomorphic families of rational maps or polynomials by mean of ergodic and potential theoretic tools. After giving a general overview of the subject, we consider rational functions as ergodic dynamical systems and introduce the Green measure of a rational map and study the properties of its Lyapunov exponent. Next we consider Holomorphic families and introduce the class of hypersurfaces (Pern(w)) in the parameter space of a holomorphic family and study the connectedness locus in polynomial families. Then we introduce the bifurcation current and discuss equidistribution towards the bifurcation current and the self-intersection of the bifurcation current.

1 Introduction

In these lectures we will study bifurcations within holomorphic families of polynomials or rational maps by mean of ergodic and pluripotential theoretic tools.

A family of rational maps $(f_\lambda)_{\lambda \in M}$, whose parameter space M is a complex manifold, is called a *holomorphic family* if the map $(\lambda, z) \mapsto f_\lambda(z)$ is holomorphic on $M \times \mathbf{P}^1$ and if the degree of f_λ is constant on M. The simplest example is the quadratic polynomial family $(z^2 + \lambda)_{\lambda \in \mathbf{C}}$. The space of all rational maps of the same degree may also be considered as such a family.

F. Berteloot (✉)
C.I.M.E Course, Cetraro, Italy
e-mail: berteloo@picard.ups-tlse.fr

The interest for bifurcations within holomorphic families of rational maps started in the eighties with the seminal works of Mañé–Sad–Sullivan [42], Lyubich [41], and Douady–Hubbard [24, 25]. At the end of this decade, McMullen used Mañé–Sad–Sullivan ideas and Thurston's theory in his fundamental work on iterative root-finding algorithms [45].

In any holomorphic family, the *stability locus* is the maximal open subset of the parameter space on which the Julia set moves continuously with the parameter. Its complement is called the *bifurcation locus*. In the quadratic polynomial family, the bifurcation locus is nothing but the boundary of the Mandelbrot set.

Mañé, Sad and Sullivan have shown that the stability locus is dense and their results also enlighted the (still open) question of the density of hyperbolic parameters. McMullen proved that any stable algebraic stable family of rational maps is either trivial or affine (i.e. consists of Lattès examples), his classification of generally convergent algorithms follows from this central result.

Potential theory has been introduced in the dynamical study of polynomials by Brolin in 1965. These tools and, more precisely the pluripotential theory developed after the fundamental works of Bedford–Taylor, turned out to be extremely powerful to study holomorphic dynamical systems depending on several complex variables. In this context, the compactness properties of closed positive currents somehow supply to the lack of suitable normality criterions for holomorphic mappings. We refer to the lecture notes [23, 58] by Sibony and Dinh–Sibony for these aspects. As we shall see, these potential-theoretic tools are well adapted for studying bifurcations in one-dimensional holomorphic dynamical systems. The underlying reason for this unity of methods is certainly that holomorphic families are actually holomorphic dynamical systems of the form $M \times \mathbf{P}^1 \ni (\lambda, z) \mapsto (\lambda, f_\lambda(z)) \in M \times \mathbf{P}^1$.

The use of potential theory in the study of parameter spaces actually started rather soon. Indeed, as revealed by the work of Douady–Hubbard [26] and Sibony [57] around 1980, the Green function of $(z^2 + \lambda)$ evaluated at the critical value plays a crucial role in the study of the Mandelbrot set. The relation between this quantity and the Lyapunov exponent was also known and Przytycki explicitly raised the problem to "understand the connections between Lyapunov exponent characteristic and potential theory for rational mappings" [54]. Let us also mention that Mañé proved that the Lyapunov exponent of a rational map and the Hausdorff dimension of its maximal entropy measure are closely related: their product is equal to the logarithm of the degree of the map [43].

Around 2000, a decisive achievement was made by DeMarco. She generalized Przytycki's formula and proved that, in any holomorphic family of rational maps, the bifurcation locus is the support of a $(1, 1)$ closed positive current admitting the Lyapunov exponent function as a global potential [19, 20]. This current, denoted T_{bif}, is now called the *bifurcation current*.

In the recent years, several authors [2–4, 13, 14, 27, 29, 33, 53] have investigated the geometry of the bifurcation locus using the current T_{bif} and its exterior powers $T_{\text{bif}}^k := T_{\text{bif}} \wedge T_{\text{bif}} \wedge \cdots \wedge T_{\text{bif}}$. We shall present here most of the results obtained in these papers. They go from laminarity statements for certain regions of the bifurcation locus to Hausdorff dimension estimates and include precise density (or equidistribution) properties relative to various classes of specific parameters. Let us stress that a common feature of these papers is to use the supports of the T_{bif}^k to enlight a certain stratification of the bifurcation locus.

We have not discussed bifurcation theory for families of endomorphisms of higher dimensional complex projective spaces but have mentioned, among the techniques presented in these notes, those which also work in this more general context. These aspects have been considered by Bassanelli, Dupont, Molino and the author [2, 8], by Pham [53] and also appear in Dinh–Sibony's survey [23]. Let us mention that Deroin and Dujardin [21] have recently extended these methods for studying bifurcations in the context of Kleinian groups.

Let us now describe the contents of these notes more precisely. Section 2 mainly deals with the construction of the Green measure (the maximal entropy measure) for a fixed rational map. We show that this measure enjoys good ergodic properties and prove a very important approximation formula for its Lyapunov exponent. This formula, which roughly says that the Lyapunov exponent can be computed on cycles of big period, will be crucial for investigating the structure of the bifurcation current. The classical results presented in this section are essentially due to Lyubich. The approximation formula (and its generalization to higher dimensional rational maps) is due to Dupont, Molino and the author.

Section 3 has a double purpose. One is to present some concrete holomorphic families on which we shall often work, the other is to introduce the hypersurfaces $Per_n(w)$ whose distribution turns out to shape the bifurcation locus. We study polynomial families and prove an important result due to Branner–Hubbard about the compactness of the connectedness locus. As a by-product, we obtain some informations which will later be decisive to control the global behaviour of the bifurcation current in such families. We also study the moduli space Mod_2 of degree two rational maps, that is the space of holomorphic conjugacy classes of degree two rational maps. To some extent, this space can be considered as a holomorphic family and presents some common features with the family of cubic polynomials. Following Milnor, we show that Mod_2 can be identified in a dynamically natural way to $\mathbf{C}^2$.

In a fixed holomorphic family, $Per_n(w)$ is defined as the set of parameters for which the corresponding map has a cycle of multiplier w and exact period n. We show that the $Per_n(w)$ are complex hypersurfaces (co-dimension one complex analytic subsets) in the parameter space and give some precise global defining functions for them. This will be an important tool to study the equidistribution of the bifurcation current by the hypersurfaces $Per_n(w)$.

The core of Sect. 4 is devoted to the definition of the bifurcation current T_{bif}. It starts with a brief survey of Mañé–Sad–Sullivan work and a motivated introduction of T_{bif}. We then formally define the bifurcation current, we also introduce the activity currents of marked critical points and relate them to T_{bif}. Most of the section is devoted to DeMarco's fundamental results. We prove that T_{bif} is supported by the bifurcation locus and admits both the Lyapunov exponent function and the sum of the Green function evaluated on critical points as a global potential. We end the section with a proof of DeMarco's formula which precisely relates the Lyapunov exponent with the Green function evaluated on the critical points. Our presentation of these results only relies on computations with closed positive currents. It is due to Bassanelli and the author who also generalized it to higher dimensional rational maps.

In Sect. 5, we study how the asymptotic distribution of dynamically defined hypersurfaces is governed by the bifurcation current. Most of the results presented here are valid in polynomial families and in Mod_2 but some are actually true in any holomorphic family. We are mainly interested by the hypersurfaces $Per_n(w)$ (for $|w| \leq 1$) and the hypersurfaces $Per(c, k, n)$ defined by the pre-periodicity of a critical point c (i.e. $f^n(c) = f^k(c)$). We prove that, conveniently weighted, these hypersurfaces equidistribute the bifurcation current. The results concerning $Per(c, k, n)$ are due to Dujardin and Favre while those concerning $Per_n(w)$ are due to Bassanelli and the author, we have tried here to unify their presentation. These results yield a precise description of the laminated structure of the bifurcation locus in large regions of Mod_2. More precise laminarity results, due to Dujardin, are presented at the end of the section.

In Sect. 6, we investigate the higher exterior powers T_{bif}^k of the bifurcation current. Although we prove some general results which indicate that the supports of T_{bif}^k induce a dynamically meaningful stratification of the bifurcation locus, we concentrate our attention on the highest power, a measure, denoted μ_{bif} and called the *bifurcation measure*. Our main goal is to show that the support of μ_{bif} is the seat of the strongest bifurcations. To this purpose we first prove that this support is simultaneously approximated by extremely stable parameters (hyperbolic) and highly unstable ones (Shishikura, Misiurewicz). The approximation by hyperbolic or Shishikura parameters is due to Bassanelli and the author, the approximation by Misiurewicz parameters is due to Dujardin and Favre. We then characterize the support of μ_{bif} as being the closure of the sets of Shishikura or Misiurewicz parameters. The proof of this result, which is due to Buff and Epstein, requires to introduce new ideas based on transversality techniques. These methods have recently been extended by Gauthier for proving that the support of the bifurcation measure, in the moduli space of degree d rational maps, has full Hausdorff dimension. We end this last section with a presentation of Gauthier's proof.

We have tried to give a synthetic and self-contained presentation of the subject. In most cases, we have given complete and detailed proofs and, sometimes have substantially simplify those available in the literature. Although basics on ergodic theory are discussed in the first section, we have not included any element of pluripotential theory. For this we refer the reader to the appendix of Sibony's [58] and Dinh–Sibony's [23] lecture notes or to the book of Demailly [18]. We finally would like to recommend the recent survey of Dujardin [28] on bifurcation currents and equidistribution in parameter space.

2 Rational Maps as Ergodic Dynamical Systems

Among ergodic properties of rational maps we present those which will be used in our study. A particularly important result of this first section is an approximation formula for the Lyapunov exponent of a rational map with respect to its maximal entropy measure.

2.1 Potential Theoretic Aspects

2.1.1 The Fatou–Julia Dichotomy

A rational function f is a holomorphic map of the Riemann sphere to itself and may be represented as the ratio of two polynomials

$$f = \frac{a_0 + a_1 z + a_2 z^2 + \cdots + a_d z^d}{b_0 + b_1 z + b_2 z^2 + \cdots + b_d z^d}$$

where at least one of the coefficients a_d and b_d is not zero. The number d is the algebraic degree of f. In the sequel we shall more likely speak of *rational map*. Such a map may also be considered as a holomorphic ramified self-cover of the Riemann sphere whose topological degree is equal to d. Among these maps, polynomials are exactly those for which ∞ is totally invariant: $f^{-1}\{\infty\} = f\{\infty\} = \infty$.

It may also be convenient to identify the Riemann sphere with the one-dimensional complex projective space $\mathbf{P}^1$ that is the quotient of $\mathbf{C}^2 \setminus \{0\}$ by the action $z \mapsto u \cdot z$ of $\mathbf{C}^*$. Let us recall that the Fubini-Study form ω on $\mathbf{P}^1$ satisfies $\pi^*(\omega) = dd^c \ln \| \ \|$ where the norm is the Euclidean one on $\mathbf{C}^2$.

In this setting, the map f can be seen as induced on $\mathbf{P}^1$ by a non-degenerate and d-homogenous map of $\mathbf{C}^2$

$$F(z_1, z_2) := \left(a_0 z_2^d + a_1 z_1 z_2^{d-1} + \cdots + a_d z_1^d, b_0 z_2^d + b_1 z_1 z_2^{d-1} + \cdots + b_d z_1^d\right)$$

through the canonical projection $\pi : \mathbf{C}^2 \setminus \{0\} \to \mathbf{P}^1$.

The homogeneous map F is called a lift of f; all other lifts are proportional to F.

A point around which a rational map f does not induce a local biholomorphism is called *critical*. The image of such a point is called a *critical value*. A degree d rational map has exactly $(2d - 2)$ critical points counted with multiplicity. The *critical set* of f is the collection of all critical points and is denoted $\mathcal{C}_f$.

As for any self-map, we may study the dynamics of rational ones, that is trying to understand the behaviour of the sequence of iterates

$$f^n := f \circ \cdots \circ f.$$

The Fatou–Julia dynamical dichotomy consists in a splitting of $\mathbf{P}^1$ into two disjoint subsets on which the dynamics of f is radically different. The *Julia set* of a rational map f is the subset of $\mathbf{P}^1$ on which the dynamics of f may drastically change under a small perturbation of initial conditions while the *Fatou set* is the complement of the Julia set.

Definition 1. The Julia set $\mathcal{J}_f$ and the Fatou set $\mathcal{F}_f$ of a rational map f are respectively defined by:

$$\mathcal{J}_f := \{z \in \mathbf{P}^1 \;/\; (f^n)_n \text{ is not equicontinuous near } z\}$$
$$\mathcal{F}_f := \mathbf{P}^1 \setminus \mathcal{J}_f.$$

Both the Julia and the Fatou set are totally invariant: $\mathcal{J}_f = f(\mathcal{J}_f) = f^{-1}(\mathcal{J}_f)$ and $\mathcal{F}_f = f(\mathcal{F}_f) = f^{-1}(\mathcal{F}_f)$. In particular, f induces two distinct dynamical systems on $\mathcal{J}_f$ and $\mathcal{F}_f$.

The dynamical system $f : \mathcal{J}_f \to \mathcal{J}_f$ is chaotical. However, as it results from the Sullivan non-wandering theorem and the Fatou–Cremer classification, the dynamics of a rational map is, in some sense, totally predictable on its Fatou set.

The periodic orbits are called *cycles* and play a very important role in the understanding of the global dynamical behaviour of a map f.

Definition 2. A n-cycle is a set of n distinct points $z_0, z_1, \cdots, z_{n-1}$ such that $f(z_i) = z_{i+1}$ for $0 \le i \le n-2$ and $f(z_{n-1}) = z_0$. One says that n is the exact period of the cycle.

Each point z_i is fixed by f^n. The *multiplier* of the cycle is the derivative of f^n at some point z_i of the cycle and computed in a local chart: $\left(\chi \circ f^n \circ \chi^{-1}\right)'(\chi(z_i))$. It is easy to see that this number depends only on the cycle and neither on the point z_i or the chart χ. By abuse we shall denote it $(f^n)'(z_i)$.

The local dynamic of f near a cycle is governed by the multiplier m. This leads to the following

Definition 3. The multiplier of a n-cycle is a complex number m which is equal to the derivative of f^n computed in any local chart at any point of the cycle.

When $|m| > 1$ the cycle is called *repelling*
when $|m| < 1$ the cycle is called *attracting*
when $|m| = 1$ the cycle is called *neutral*.

Repelling cycles belongs to the Julia set and attracting ones to the Fatou set. For neutral cycles this depends in a very delicate way on the diophantine properties of the argument of m.

The first fundamental result about Julia sets is the following.

Theorem 4. *Repelling cycles are dense in the Julia set.*

It is possible to give an elementary proof of that result using the Brody–Zalcman renormalization technique (see [9]). We shall see later that repelling cycles actually equidistribute a measure whose support is exactly the Julia set.

2.1.2 The Green Measure of a Rational Map

Our goal is to endow the dynamical system $f : \mathcal{J}_f \to \mathcal{J}_f$ with an ergodic structure capturing most of its chaotical nature. This is done by exhibiting an invariant measure μ_f on $\mathcal{J}_f$ which is of constant Jacobian. Such a measure was first constructed by Lyubich [40]. For our purpose it will be extremely important to use a potential-theoretic approach which goes back to Brolin [12] for the case of polynomials. We follow here the presentation given by Dinh and Sibony in their survey [23] which also covers *mutatis mutandis* the construction of Green currents for holomorphic endomorphisms of $\mathbf{P}^k$.

The following Lemma is the key of the construction. It relies on the fundamental fact that
$$d^{-1} f^\star \omega = \omega + dd^c v$$
for some smooth function v on $\mathbf{P}^1$. This follows from a standard cohomology argument or may be seen concretely by setting $v := d^{-1} \ln \frac{\|F(z)\|}{\|z\|^d}$ for some lift F of f.

Lemma 5. *Up to some additive constant, there exists a unique continuous function g on $\mathbf{P}^1$ such that $d^{-n} f^{n\star} v \to dd^c g + \omega$ for any positive measure v which is given by $v = \omega + dd^c u$ where u is continuous.*

Proof. Let us set $g_n := v + \cdots + d^{-n+1} v \circ f^{n-1}$. One sees by induction that $d^{-n} f^{n\star} v = \omega + dd^c g_n + dd^c(d^{-n} u \circ f^n)$. As the sequence $(g_n)_n$ is clearly uniformly converging, the conclusion follows by setting $g := \lim_n g_n$. □

It might be useful to see how the function g can be obtained by using lifts.

Lemma 6. *Let f be a degree d rational map. For any lift F of f, the sequence $d^{-n} \ln \|F^n(z)\|$ converges uniformly on compact subsets of $\mathbf{C}^2 \setminus \{0\}$ to a function G_F which satisfies the following invariance and homogeneity properties:*

(i) $G_F \circ F = d G_F$
(ii) $G_F(tz) = G_F(z) + \ln|t|, \ \forall t \in \mathbf{C}.$

Moreover, $G_F - \ln \| \ \| = g \circ \pi$ where g is a function given by Lemma 5.

Proof. Let us set $G_n(z) := d^{-n} \ln \|F^n(z)\|$. As F is homogeneous and non-degenerate there exists a constant $M > 1$ such that
$$\frac{1}{M} \|z\|^d \leq \|F(z)\| \leq M \|z\|^d.$$

Thus $\frac{1}{M}\|F^n(z)\|^d \le \|F^{n+1}(z)\| \le M\|F^n(z)\|^d$ which, taking logarithms and dividing by d^{n+1} yields $|G_{n+1}(z) - G_n(z)| \le \frac{\ln M}{d^{n+1}}$. This shows that G_n is uniformly converging to G_F. The properties (i) and (ii) follows immediately from the definition of G_F.

According to the proof of Lemma 5, $g = \lim_n \left(v + \cdots + d^{-n+1} v \circ f^{n-1}\right)$ where a possible choice of v is $v \circ \pi = d^{-1} \ln \frac{\|F(z)\|}{\|z\|^d}$. To get the last assertion, it suffices to observe that $d^{-k} v \circ f^k \circ \pi = G_{k+1} - G_k$. □

The two above lemmas lead us to coin the following

Definition 7. Let F be a lift of a degree d rational map f. The Green function G_F of F on $\mathbf{C}^2$ is defined by

$$G_F := \lim_n d^{-n} \ln \|F^n(z)\|.$$

The Green function g_F of f on $\mathbf{P}^1$ is defined by

$$G_F - \ln \|\ \| = g_F \circ \pi.$$

We will often commit the abuse to denote g_f any function which is equal to g_F up to some additive constant and to call it *the* Green function of f.

The function G_F is p.s.h on $\mathbf{C}^2$ with a unique pole at the origin.

It is worth emphasize that both g_F and G_F are uniform limits of smooth functions. In particular, these functions are continuous. One may actually prove more (see [23, Proposition 1.2.3] or [2, Proposition 1.2]):

Proposition 8. *The Green functions $G_F(z)$ and $g_F(z)$ are Hölder continuous in F and z.*

We are now ready to define the Green measure μ_f and verify its first properties.

Theorem 9. *Let f be a degree $d \ge 2$ rational map and g be the Green function of f. Let $\mu_f := \omega + dd^c g$. Then μ_f is a f-invariant probability measure whose support is equal to $\mathcal{J}_f$. Moreover μ_f has constant Jacobian: $f^\star \mu_f = d\mu_f$.*

Proof. That μ_f is a probability measure follows immediately from Stokes theorem since $\int_{\mathbf{P}^1} \omega = 1$.

We shall use Lemma 5 for showing that $f^\star \mu_f = d\mu_f$. By construction $v + d^{-1} g \circ f = \lim_n v + d^{-1} g_n \circ f = \lim_n g_{n+1} = g$ and thus $d^{-1} f^\star \mu_f = \omega + dd^c v + d^{-1} dd^c (g \circ f) = \omega + dd^c (v + d^{-1} g \circ f) = \mu_f$.

The invariance property $f_\star \mu_f = \mu_f$ follows immediately from $f^\star \mu_f = d\mu_f$ by using the fact that $f_\star f^\star = d \, Id$.

Let us show that the support of μ_f is equal to $\mathcal{J}_f$. If $U \subset \mathcal{F}_f$ is open then $f^{n\star} \omega$ is uniformly bounded on U and therefore $\mu_f(U) = \lim \int_U d^{-n} f^{n\star} \omega = 0$, this shows that $\text{Supp } \mu_f \subset \mathcal{J}_f$. Conversely the identity $f^\star \mu_f = d\mu_f$ implies that $(\text{Supp } \mu_f)^c$ is invariant by f which, by Picard–Montel's theorem, implies that $(\text{Supp } \mu_f)^c \subset \mathcal{F}_f$. One should first observe that, since μ_f has continuous potentials, $\text{Supp } \mu_f$ certainly contains more than three points. □

It is sometimes useful to use the Green function G_F for defining local potentials of μ_f.

Proposition 10. *Let f be a rational map and F be a lift. For any section σ of the canonical projection π defined on some open subset U of $\mathbf{P}^1$, the function $G_F \circ \sigma$ is a potential for μ_f on U.*

Proof. On U one has $dd^c G_F \circ \sigma = dd^c g_F + dd^c \ln \|\sigma\| = dd^c g_F + \sigma^* dd^c \ln \|\cdot\| = dd^c g_F + (\pi \circ \sigma)^* \omega = dd^c g_F + \omega = \mu_f$. □

Let us underline that, by construction, the measure μ_f has continuous local potentials and in particular cannot give mass to points.

The measure μ_f can also be obtained as the image by the canonical projection $\pi : \mathbf{C}^2 \setminus \{0\} \to \mathbf{P}^1$ of a Monge–Ampère measure associated to the positive part $G_F^+ := \max(G_F, 0)$ of the Green function G_F.

Proposition 11. *Let F be a lift of a degree d rational map f and G_F be the Green function of F. The measure $\mu_F := dd^c G_F^+ \wedge dd^c G_F^+$ is supported on the compact set $\{G_F = 0\}$ and satisfies $F^\star \mu_F = d^2 \mu_F$ and $\pi_\star \mu_F = \mu_f$.*

This construction will be used only once in this text and we therefore skip its proof. Observe that the support of μ_F is contained in the boundary of the compact set $K_F := \{G_F \leq 0\}$ which is precisely the set of points z with bounded forward orbits by F.

The case of polynomials presents very interesting features, in particular the Green measure coincides with the harmonic measure of the filled-in Julia set.

Proposition 12. *Let P be a degree $d \geq 2$ polynomial on $\mathbf{C}$. The Green function $g_{\mathbf{C},P}$ of P is the subharmonic function defined on the complex plane by*

$$g_{\mathbf{C},P} := \lim_n d^{-n} \ln^+ |P^n|.$$

The Green measure of P is compactly supported in the complex plane and admits $g_{\mathbf{C},P}$ as a global potential there.

Proof. We may take $F := \left(z_2^d P(\frac{z_1}{z_2}), z_2^d\right)$ as a lift of P. Then $F^n := \left(z_2^{d^n} P^n(\frac{z_1}{z_2}), z_2^{d^n}\right)$ and

$$G_F(z_1, 1) = \frac{1}{2} \lim_n d^{-n} \ln\left(1 + |P^n(z_1)|^2\right) = \lim_n d^{-n} \ln^+ |P^n(z_1)|.$$

The conclusion then follows from Proposition 10. □

2.2 Ergodic Aspects

2.2.1 Mixing, Equidistribution Towards the Green Measure

Definition 13. Let (X, f, μ) be a dynamical system. One says that the measure μ is mixing if and only if

$$\lim_n \int_X (\varphi \circ f^n) \, \psi \, \mu = \int_X \varphi \, \mu \int_X \psi \, \mu$$

for any test functions φ and ψ.

This means that the events $\{f^n(x) \in A\}$ and $\{x \in B\}$ are asymptotically independents for any pair of Borel sets A, B.

As we shall see, the constant Jacobian property implies that Green measures are mixing and therefore ergodic.

Theorem 14. *The Green measure μ_f of any degree d rational map f is mixing.*

Proof. Let us set $c_\varphi := \int \varphi \, \mu_f$ and $c_\psi := \int \psi \, \mu_f$ where φ and ψ are two test functions. We may assume that $c_\varphi = 1$.

Since μ_f and $\varphi \mu_f$ are two probability measures, there exists a smooth function u_φ on $\mathbf{P}^1$ such that:

$$\varphi \mu_f = \mu_f + \Delta u_\varphi. \tag{1}$$

On the other hand, by the constant Jacobian property $f^* \mu_f = d \mu_f$ we have:

$$d^{-n} f^{n*} \big((\varphi - c_\varphi) \mu_f \big) = (\varphi \circ f^n - c_\varphi) \mu_f. \tag{2}$$

Now, combining (1) and (2) we get:

$$\int (\varphi \circ f^n) \psi \, \mu_f - \Big(\int \varphi \, \mu_f \Big) \Big(\int \psi \, \mu_f \Big) = \int (\varphi \circ f^n) \psi \, \mu_f - c_\varphi c_\psi =$$

$$\int \psi (\varphi \circ f^n - c_\varphi) \, \mu_f = \int \psi \, d^{-n} f^{n*}((\varphi - c_\varphi) \mu_f)$$

$$= \int (d^{-n} f_*^n \psi) \, (\varphi - 1) \, \mu_f =$$

$$\int (d^{-n} f_*^n \psi) \Delta u_\varphi = \int \psi \, d^{-n} f^{n*}(\Delta u_\varphi) = \int \psi \, d^{-n} \Delta(u_\varphi \circ f^n)$$

$$= d^{-n} \int (u_\varphi \circ f^n) \, \Delta \psi.$$

As $\int (u_\varphi \circ f^n) \, \Delta \psi$ is bounded, this leads to the desired conclusion. $\square$

It is not hard to show that a mixing measure is also ergodic.

Definition 15. Let (X, f, μ) be a dynamical system. One says that the measure μ is ergodic if and only if all integrable f-invariant functions are constants.

In particular this allows to use the classical Birkhoff ergodic theorem which says that time-averages along typical orbits coincide with the spatial-average:

Theorem 16. *Let (X, f, μ) be an ergodic dynamical system and $\varphi \in L^1(\mu)$. Then*

$$\lim_n \frac{1}{n} \sum_{k=0}^{n-1} \varphi(f^k(x)) = \int_X \varphi\, \mu$$

for μ almost every x in X.

The measure-theoretic counterpart of Fatou–Julia Theorem 4 is the following equidistribution result which has been first proved by Lyubich [40]. The content of Sect. 2.3.2 will provide another proof which exploits the mixing property.

Theorem 17. *Let f be a rational map of degree d. Let $R_n^\star$ denote the set of all n-periodic repelling points of f. Then $d^{-n} \sum_{R_n^\star} \delta_z$ is weakly converging to μ_f.*

Let us finally mention another classical equidistribution result. We refer to [23] for a potential theoretic proof.

Theorem 18. *Let f be a degree d rational map. Then*

$$\lim_n d^{-n} \sum_{\{f^n(z)=a\}} \delta_z = \mu_f$$

for any $a \in \mathbf{P}^1$ which is not exceptional for f.

We recall that a map f has *no* exceptional point unless f is a polynomial or a map of the form $z^{\pm d}$. A map of the form $z^{\pm d}$ has two exceptional points: 0 and ∞. For polynomials other than z^d, the only exceptional point is ∞.

Although this will not be used in this text, we mention that the Green measure μ_f of any degree d rational map f is the unique measure of maximal entropy for f. This means that the entropy of μ_f is maximal and, according to the variational principle, equals $\ln d$ which is the value of the topological entropy of f.

2.2.2 Natural Extension and Iterated Inverse Branches

To any ergodic dynamical system, it is possible to associate a new system which is invertible and contains all the information of the original one. It is basically obtained by considering the set of all complete orbits on which is acting a shift. This general construction is the so-called *natural extension* of a dynamical system; here is a formal definition.

Definition 19. The natural extension of a dynamical system (X, f, μ) is the dynamical system $(\hat{X}, \hat{f}, \hat{\mu})$ where

$$\hat{X} := \{\hat{x} := (x_n)_{n \in \mathbf{Z}} \ / \ x_n \in X, \ f(x_n) = x_{n+1}\}$$

$$\hat{f}(\hat{x}) := (x_{n+1})_{n \in \mathbf{Z}}$$

$$\hat{\mu}\{(x_n) \text{ s.t. } x_0 \in B\} = \mu(B).$$

The canonical projection $\pi_0 : \hat{X} \to X$ is given by $\pi_0(\hat{x}) = x_0$. One sets τ for $(\hat{f})^{-1}$.

Let us stress that $\pi_0 \circ \hat{f} = f \circ \pi_0$ and $(\pi_0)_*(\hat{\mu}) = \mu$. The measure $\hat{\mu}$ inherits most of the ergodic properties of μ.

Proposition 20. *The measure $\hat{\mu}$ is ergodic (resp. mixing) if and only if μ is ergodic (resp. mixing).*

We refer the reader to the Chap. 10 of [15] for this construction and its properties.

A powerful way to control the behaviour of inverse branches along typical orbits of the system $(\mathcal{J}_f, f, \mu_f)$ is to apply standard ergodic theory to its natural extension. This is what we shall do now. The first point is to observe that one may work with orbits avoiding the critical set of f. To this purpose one considers

$$\hat{X}_{reg} = \{\hat{x} \in \hat{\mathcal{J}}_f \ / \ x_n \notin \mathcal{C}_f \ ; \ \forall n \in \mathbf{Z}\}.$$

As $\hat{\mu}_f$ is $\hat{f}$-invariant and μ_f does not give mass to points, one sees that $\hat{\mu}_f(\hat{X}_{reg}) = 1$.

Definition 21. Let $\hat{x} \in \hat{X}_{reg}$ and $p \in \mathbf{Z}$. The injective map induced by f on some neighbourhood of x_p is denoted f_{x_p}. The inverse of f_{x_p} is defined on some neighbourhood of x_{p+1} and is denoted $f_{x_p}^{-1}$. We then set

$$f_{\hat{x}}^{-n} := f_{x_{-n}}^{-1} \circ \cdots \circ f_{x_{-1}}^{-1}.$$

The map $f_{\hat{x}}^{-n}$ is called *iterated inverse branch of f along $\hat{x}$ and of depth n*.

It will be good to keep in mind that $f_{\hat{x}}^{-n}(x_0) = x_{-n}$ and that $f_{\hat{x}}^{-1} = f_{x_{-1}}^{-1}$.

The following Proposition yields a control of the disc on which $f_{x_{-k-1}}^{-1}$ is defined.

Proposition 22. *For any sufficiently small and strictly positive ϵ, there exists a function $\alpha_\epsilon : \hat{X}_{reg} \to]0, 1[$ such that*

$$\alpha_\epsilon(\tau(\hat{x})) \geq e^{-\epsilon} \alpha_\epsilon(\hat{x}) \text{ and}$$

$$f_{x_{-k-1}}^{-1} \text{ is defined on } D(x_{-k}, \alpha_\epsilon(\tau^k(\hat{x})))$$

for $\hat{\mu}_f$-a.e. $\hat{x} \in \hat{X}_{reg}$ and every $k \in \mathbf{Z}$.

The function α_ϵ is a so-called *slow function*. The interest of such a function relies on the fact that its decreasing might be negligible with respect to other datas. For instance, in some circumstances, Proposition 22 will tell us that the local inverses $f^{-1}_{x_{-k}}$ are defined on discs whose radii may essentially be considered as constant along the orbit $\hat{x}$.

Proof. We need the following quantitative version of the inverse mapping theorem (see [11, Lemme 2]).

Lemma 23. *Let* $\rho(x) := |f'(x)|$, $r(x) := \rho(x)^2$. *There exists* $\epsilon_0 > 0$ *and, for* $\epsilon \in]0, \epsilon_0]$, $0 < C_1(\epsilon), C_2(\epsilon)$ *such that for every* $x \in \mathcal{J}_f$:

1. f is one-to-one on $D(x, C_1(\epsilon)\rho(x))$,
2. $D(f(x), C_2(\epsilon)r(x)) \subset f[D(x, C_1(\epsilon)\rho(x))]$,
3. $\text{Lip } f_x^{-1} \leq e^{\frac{\epsilon}{3}} \rho(x)^{-1}$ on $D(f(x), C_2(\epsilon)r(x))$.

Let us set $\beta_\epsilon(\hat{x}) := \text{Min}(1, C_2(\epsilon)r(x_{-1}))$. According to the two first assertions of the above Lemma, $f^{-1}_{x_{-1}} = f^{-1}_{\hat{x}}$ is defined on $D(x_0, \beta_\epsilon(\hat{x}))$ and, similarly, $f^{-1}_{x_{-k-1}} = f^{-1}_{\tau^k(\hat{x})}$ is defined on $D(x_{-k}, \beta_\epsilon(\tau^k(\hat{x})))$. All we need is to find a function α_ϵ such that $0 < \alpha_\epsilon < \beta_\epsilon$ and $\alpha_\epsilon(\tau(\hat{x})) \geq e^{-\epsilon} \alpha_\epsilon(\hat{x})$.

As μ_f admits continuous local potentials, the function $\ln \beta_\epsilon$ is $\hat{\mu}_f$-integrable. Then, by Birkhoff ergodic theorem 16, $\int_{\hat{X}} \ln \beta_\epsilon \, \hat{\mu}_f = \lim_{|n| \to +\infty} \frac{1}{|n|} \sum_{k=1}^n \ln \beta_\epsilon(\tau^k(\hat{x}))$ and, in particular

$$\lim_{|n| \to +\infty} \frac{1}{|n|} \ln \beta_\epsilon(\tau^n(\hat{x})) = 0 \text{ for } \hat{\mu}_f\text{-a.e. } \hat{x} \in \hat{X}.$$

In other words, for $\hat{\mu}_f$-a.e. $\hat{x} \in \hat{X}_{reg}$ there exists $n_0(\epsilon, \hat{x}) \in \mathbf{N}$ such that $\beta_\epsilon(\tau^n(\hat{x})) \geq e^{-|n|\epsilon}$ for $|n| \geq n_0(\epsilon, \hat{x})$. Setting then $V_\epsilon := \inf_{|n| \leq n_0(\epsilon, \hat{x})} \left(\beta_\epsilon(\tau^n(\hat{x})) e^{|n|\epsilon} \right)$ we obtain a measurable function $V_\epsilon : \hat{X}_{reg} \to]0, 1]$ such that: $\beta_\epsilon(\tau^n(\hat{x})) \geq e^{-|n|\epsilon} V_\epsilon(\hat{x})$ for $\hat{\mu}_f$-a.e. $\hat{x} \in \hat{X}_{reg}$ and every $n \in \mathbf{Z}$. It suffices to take $\alpha_\epsilon(\hat{x}) := \text{Inf}_{n \in \mathbf{Z}} \{\beta_\epsilon(\tau^n(\hat{x})) e^{|n|\epsilon}\}$. □

2.3 The Lyapunov Exponent

2.3.1 Definition, Formulas and Some Properties

Let us consider the ergodic dynamical system $(\mathcal{J}_f, f, \mu_f)$ which has been constructed in the last section. As the measure μ_f has continuous local potentials, the function $\ln |f'|$ belongs to $L^1(\mu_f)$ for any choice of a metric $|\ |$ on $\mathbf{P}^1$. We may therefore apply the Birkhoff ergodic theorem 16 to get:

$$\lim_n \frac{1}{n} \ln |(f^n)'(z)| = \lim_n \frac{1}{n} \sum_{k=0}^{n-1} \ln |f'(f^k(z))| = \int_{\mathbf{P}^1} \ln |f'| \, \mu_f, \quad \mu_f\text{-a.e.} \quad (3)$$

This identity shows that the integral $\int_{\mathbf{P}^1} \ln |f'| \, \mu_f$ does not depend on the choice of the metric $|\ |$ and leads to the following definition.

Definition 24. The Lyapunov exponent of the ergodic dynamical system $(\mathcal{J}_f, f, \mu_f)$ is the number

$$L(f) = \int_{\mathbf{P}^1} \ln |f'| \, \mu_f.$$

For simplicity we shall say that $L(f)$ is the Lyapunov exponent of f.

As the identity (3) shows, the Lyapunov exponent $L(f)$ is the exponential rate of growth of $|(f^n)'(z)|$ for a typical $z \in \mathcal{J}_f$.

Remark 25. Using the invariance property $f_* \mu_f = \mu_f$ one immediately sees that $L(f^n) = nL(f)$.

We shall need an expression of $L(f)$ which uses the formalism of line bundles. In order to prove it, we first compare the Lyapunov exponents of f with the sum of Lyapunov exponents of one of its lifts F.

Proposition 26. *Let F be a lift of some rational map f of degree d. Then the sum of Lyapunov exponents of F with respect to μ_F is given by $L(F) := \int \ln |\det F'| \, \mu_F$ and is equal to $L(f) + \ln d$.*

Proof. Let F be a polynomial lift of f. We shall compute the Lyapunov exponent using the spherical metric $|\ |_s$. Exploiting the fact that $f^*\omega = |f'|_s^2 \omega$, it is not difficult to check that

$$|f'(\xi)|_s = \frac{1}{d} \frac{\|z\|^2}{\|F(z)\|^2} \left|\det F'(z)\right|$$

for any z such that $\pi(z) = \xi$. We thus have

$$\frac{1}{n} \ln |(f^n)'(\xi)|_s + \ln d = \frac{1}{n} \ln \frac{\|z\|^2}{\|F^n(z)\|^2} + \frac{1}{n} \ln \left|\det(F^n)'(z)\right|.$$

Then the conclusion follows by Birkhoff theorem 16 since $\|F^n(z)\|$ stays away from 0 and $+\infty$ when z is in the support of μ_F and $\pi_* \mu_F = \mu_f$. $\square$

For any integer D the line bundle $\mathcal{O}_{\mathbf{P}^1}(D)$ over $\mathbf{P}^1$ is the quotient of $(\mathbf{C}^2 \setminus \{0\}) \times \mathbf{C}$ by the action of $\mathbf{C}^*$ defined by $(z, x) \mapsto (uz, u^D x)$. We denote by $[z, x]$ the elements of this quotient.

The canonical metric on $\mathcal{O}_{\mathbf{P}^1}(D)$ may be written

$$\|[z, x]\|_0 = e^{-D \ln \|z\|} |x|.$$

The homogeneity property of G_F allows us to define another metric on $\mathcal{O}_{\mathbf{P}^1}(D)$ by setting
$$\|[z,x]\|_{G_F} = e^{-DG_F(z)}|x|.$$

Let us underline that, according to Definition 7, $\|\cdot\|_{G_F} = e^{-Dg_F}\|\cdot\|_0$.

The expression given in the following Lemma will allow us to perform integration by parts and get some fundamental formulas. This will turn out to be extremely useful when we shall relate the Lyapunov exponent with bifurcations in Sect. 4.

Lemma 27. *Let f be a rational map of degree $d \geq 2$ and F be one of its lifts. Let $D := 2(d-1)$ and Jac_F be the holomorphic section of $\mathcal{O}_{\mathbf{P}^1}(D)$ induced by $\det F'$. Then*
$$L(f) + \ln d = \int_{\mathbf{P}^1} \ln \|Jac_F\|_{G_F} \, \mu_f.$$

Proof. The section Jac_F is defined by $Jac_F(\pi(z)) := [z, \det F']$ for any $z \in \mathbf{C}^2\setminus\{0\}$. Using Proposition 26, the fact that G_F vanishes on the support of μ_F and $\pi_*\mu_F = \mu_f$ we get

$$L(f) + \ln d = \int_{\mathbf{C}^2} \ln|\det F'| \, \mu_F = \int_{\{G_F=0\}} \ln|\det F'| \, \mu_F$$
$$= \int_{\{G_F=0\}} \ln\left(e^{-DG_F(z)}|\det F'|\right) \mu_F = \int_{\mathbf{C}^2} \ln\|Jac_F \circ \pi\|_{G_F} \, \mu_F$$
$$= \int_{\mathbf{P}^1} \ln\|Jac_F\|_{G_F} \, \mu_f. \qquad \square$$

It is an important and not obvious fact that $L(f)$ is strictly positive. It actually follows from the Margulis–Ruelle inequality that $L(f) \geq \frac{1}{2}\ln d$ where d is the degree of f. We will present later a simple argument which shows that this bound is equal to $\ln d$ for polynomials (see Theorem 77). Zdunik [63] and Mayer [44] have proved that the bound $\frac{1}{2}\ln d$ is taken if and only if the map f is a Lattès example. Let us summarize these results in the following statement.

Theorem 28. *The Lyapunov exponent of a degree d rational map is always greater than $\frac{1}{2}\ln d$ and the equality occurs if and only if the map is a Lattès example.*

We recall that a Lattès map is, by definition, induced on the Riemann sphere from an expanding map on a complex torus by mean of some elliptic function. We refer to the survey paper of Milnor [50] for a detailed discussion of these maps.

A remarkable consequence of the positivity of $L(f)$ is that the iterated inverse branch $f_{\hat{x}}^{-n}$ (see Definition 21) are approximately e^{-nL}-Lipschiptz and are defined on a disc whose size only depends on $\hat{x}$.

Proposition 29. *There exists $\epsilon_0 > 0$ and, for $\epsilon \in]0, \epsilon_0]$, two measurable functions $\eta_\epsilon : \hat{X}_{reg} \to]0, 1]$ and $S_\epsilon : \hat{X}_{reg} \to]1, +\infty]$ such that the maps $f_{\hat{x}}^{-n}$ are defined on*

$D(x_0, \eta_\epsilon(\hat{x}))$ and $\operatorname{Lip} f_{\hat{x}}^{-n} \leq S_\epsilon(\hat{x})e^{-n(L-\epsilon)}$ for $\hat{\mu}_f$-a.e. $\hat{x} \in \hat{X}_{reg}$ and for every $n \in \mathbf{N}$.

Proof. We may assume that $0 < \epsilon_0 < \frac{L}{3}$. Since $f_{\hat{x}}^{-n} = f_{x_{-n}}^{-1} \circ \cdots \circ f_{x_{-1}}^{-1}$, the third assertion of Lemma 23 yields $\ln \operatorname{Lip} f_{\hat{x}}^{-n} \leq n\frac{\epsilon}{3} - \sum_{k=1}^{n} \ln \rho(x_{-k})$. By Birkhoff ergodic theorem we thus have

$$\limsup \frac{1}{n} \ln \operatorname{Lip} f_{\hat{x}}^{-n} \leq -L + \frac{\epsilon}{3} \text{ for } \hat{\mu}_f\text{-a.e. } \hat{x} \in \hat{X}.$$

Then there exists $n_0(\hat{x})$ such that $\operatorname{Lip} f_{\hat{x}}^{-n} \leq e^{-n(L-\epsilon)}$ for $n \geq n_0(\hat{x})$ and it suffices to set $S_\epsilon := \max_{0 \leq n \leq n_0(\hat{x})} \left(e^{n(L-\epsilon)} \operatorname{Lip} f_{\hat{x}}^{-n} \right)$ to get the estimate

$$\operatorname{Lip} f_{\hat{x}}^{-n} \leq S_\epsilon(\hat{x}) e^{-n(L-\epsilon)} \text{ for every } n \in \mathbf{N} \text{ and } \hat{\mu}_f\text{-a.e. } \hat{x} \in \hat{X}_{reg}.$$

We now set $\eta_\epsilon := \frac{\alpha_\epsilon}{S_\epsilon}$ where α_ϵ is the given by Proposition 22. Let us check by induction on $n \in \mathbf{N}$ that $f_{\hat{x}}^{-n}$ is defined on $D(x_0, \eta_\epsilon(\hat{x}))$ for $\hat{\mu}_f$-a.e. $\hat{x} \in \hat{X}$ and every $n \in \mathbf{N}$. Here we will use the fact that the function α_ϵ is slow: $\alpha_\epsilon(\tau(\hat{x})) \geq e^{-\epsilon}\alpha_\epsilon(\hat{x})$.

Assume that $f_{\hat{x}}^{-n}$ is defined on $D(x_0, \eta_\epsilon(\hat{x}))$. Then, by our estimate on $\operatorname{Lip} f_{\hat{x}}^{-n}$, we have

$$f_{\hat{x}}^{-n}\left(D(x_0, \eta_\epsilon(\hat{x}))\right) \subset D\left(x_{-n}, e^{-n(L-\epsilon)}\alpha_\epsilon(\hat{x})\right).$$

On the other hand, by Proposition 22, the branch $f_{x_{-n-1}}^{-1}$ is defined on the disc $D(x_{-n}, \alpha_\epsilon(\tau^{n+1}(\hat{x})))$ which, as α_ϵ is slow, contains $D(x_{-n}, e^{-(n+1)\epsilon}\alpha_\epsilon(\hat{x}))$. Now, since $0 < \epsilon_0 < \frac{L}{3}$ one has $e^{-(n+1)\epsilon} \geq e^{-n(L-\epsilon)}$ and thus $f_{\hat{x}}^{-(n+1)} = f_{x_{-n-1}}^{-1} \circ f_{\hat{x}}^{-n}$ is defined on $D(x_0, \eta_\epsilon(\hat{x}))$. □

2.3.2 Lyapunov Exponent and Multipliers of Repelling Cycles

The following approximation property will play an important role in our study of bifurcation currents. We would like to mention that Deroin and Dujardin have recently used similar ideas to study the bifurcation in the context of Kleinian groups (see [21]).

Theorem 30. *Let $f : \mathbf{P}^1 \to \mathbf{P}^1$ be a rational map of degree $d \geq 2$ and L the Lyapunov exponent of f with respect to its Green measure. Then:*

$$L = \lim_n d^{-n} \sum_{p \in R_n^*} \frac{1}{n} \ln |(f^n)'(p)|$$

where $R_n^* := \{p \in \mathbf{P}^1 \ / \ p \text{ has exact period } n \text{ and } |(f^n)'(p)| > 1\}$.

Observe that the Lyapunov exponent $\lim_k \frac{1}{k} \ln|(f^k)'(p)|$ of f along the orbit of a point p is precisely equal to $\frac{1}{n} \ln|(f^n)'(p)|$ when p is n periodic. The above Theorem thus shows that the Lyapunov exponent L of f is the limit, when $n \to +\infty$, of the averages of Lyapunov exponents of repelling n-cycles.

To establish the above Theorem, we will prove that the repelling cycles equidistribute the Green measure μ_f in a somewhat constructive way and control the multipliers of the cycles which appear. For this purpose, we follow the approach used by Briend–Duval [11] where the positivity of the Lyapunov exponent plays a crucial role and which also works for endomorphisms of $\mathbf{P}^k$. This strategy actually yields to a version of Theorem 30 for endomorphisms of $\mathbf{P}^k$; this has been done in [8]. Okuyama has given a different proof of Theorem 30 in [51, 52], his proof actually does not use the positivity of the Lyapunov exponent. The proof we present here is that of [7] with a few more details.

Proof. For the simplicity of notations we consider polynomials and therefore work on $\mathbf{C}$ with the Euclidean metric. We shall denote $D(x, r)$ the open disc centered at $x \in \mathbf{C}$ and radius $r > 0$. From now on, f is a degree $d \geq 2$ polynomial whose Julia set is denoted J and whose Green measure is denoted μ. □

We shall use the natural extension (see Sect. 2.2.2) and exploit the positivity of L through Proposition 29. Let us add a few notations to those already introduced in Propositions 22 and 29. Let $0 < \epsilon_0$ be given by Proposition 22.

For $0 < \epsilon \leq \epsilon_0$ and $n, N \in \mathbf{N}$ we set:

$$\hat{X}_N^\epsilon := \{\hat{x} \in \hat{X} \ / \ \eta_\epsilon(\hat{x}) \geq \frac{1}{N} \text{ and } S_\epsilon(\hat{x}) \leq N\}$$

$$\hat{\nu}_N^\epsilon := 1_{\hat{X}_N^\epsilon} \hat{\mu}$$

$$\nu_N^\epsilon := \pi_{0\star} \hat{\nu}_N^\epsilon.$$

For $0 < \epsilon \leq L$ and $n, N \in \mathbf{N}$ we set:

$$R_n^\epsilon := \{p \in \mathbf{C} \ / \ f^n(p) = p \text{ and } |(f^n)'(p)| \geq e^{n(L-\epsilon)}\}$$

$$\mu_n^\epsilon := d^{-n} \sum_{R_n^\epsilon} \delta_p$$

$$R_n := R_n^L = \{p \in \mathbf{C} \ / \ f^n(p) = p \text{ and } |(f^n)'(p)| \geq 1\}$$

$$\mu_n := \mu_n^L = d^{-n} \sum_{R_n} \delta_p.$$

The following lemma reduces the problem to some estimates on Radon–Nikodym derivatives.

Lemma 31. *If any weak limit σ of $(\mu_n^\epsilon)_n$ for $\epsilon \in]0, \epsilon_0[$ satisfies $\frac{d\sigma}{dv_N^{\epsilon'}} \geq 1$ for some $\epsilon' > 0$ and every $N \in \mathbf{N}$ then $\mu_n^L \to \mu$ and $L = \lim_n \frac{d^{-n}}{n} \sum_{R_n^*} \ln |(f^n)'(p)|$.*

Proof. We start by showing that $\mu_n^\epsilon \to \mu$ for any $\epsilon \in]0, L]$. Let σ be a weak limit of $(\mu_n^\epsilon)_n$. Since all the μ_n^ϵ are probability measures, it suffices to show that $\sigma = \mu$.

Assume first that $0 < \epsilon < \epsilon_0$. By assumption $\frac{d\sigma}{dv_N^{\epsilon'}} \geq 1$ and therefore $\sigma \geq v_N^{\epsilon'}$ for every $N \in \mathbf{N}$. Letting $N \to +\infty$ one gets $\sigma \geq \mu$. This actually implies that $\sigma = \mu$ since

$$\sigma(J) \leq \limsup_n \mu_n^\epsilon(J) \leq \lim_n \frac{d^n + 1}{d^n} = 1 = \mu(J).$$

We have shown that $\mu_n^\epsilon \to \mu$ for $0 < \epsilon < \epsilon_0$. Let us now assume that $\frac{\epsilon_0}{2} =: \epsilon_1 \leq \epsilon$. As $\mu_n^\epsilon \geq \mu_n^{\epsilon_1}$ and $\mu_n^{\epsilon_1} \to \mu$, one gets $\sigma \geq \mu$. Just as before this implies that $\sigma = \mu$.

We now want to show that $L = \lim_n \frac{d^{-n}}{n} \sum_{R_n^*} \ln |(f^n)'(p)|$. Let us set $\varphi_n(p) := \frac{1}{n} \ln |(f^n)'(p)|$. For $M > 0$ one has

$$\mu_n^\epsilon(J)(L - \epsilon) \leq d^{-n} \sum_{R_n^\epsilon} \varphi_n(p) \leq d^{-n} \sum_{R_n} \varphi_n(p) = \int_J \ln |f'| \mu_n$$

$$\leq \int_J \text{Max}(\ln |f'|, -M) \mu_n$$

since $\mu_n^\epsilon \to \mu$ and $\mu_n = \mu_n^L \to \mu$ we get

$$(L - \epsilon) \leq \liminf d^{-n} \sum_{R_n} \varphi_n(p) \leq \limsup d^{-n} \sum_{R_n} \varphi_n(p)$$

$$\leq \int_J \text{Max}(\ln |f'|, -M) \mu.$$

To obtain $\lim d^{-n} \sum_{R_n} \varphi_n(p) = L$ it suffices to make first $M \to +\infty$ and then $\epsilon \to 0$.

Since there are less than $2nd^{\frac{n}{2}}$ periodic points whose period strictly divides n, one may replace R_n by $R_n^* := \{p \in \mathbf{P}^1 \ / \ p \text{ has exact period } n \text{ and } |(f^n)'(p)| \geq 1\}$. □

Let us now finish the proof of Theorem 30. We assume here that $0 < \epsilon < \frac{\epsilon_0}{2}$. Let $\hat{a} \in \hat{X}_N^\epsilon$ and $a := \pi(\hat{a})$. For every $r > 0$ we denote by D_r the *closed* disc centered at a of radius r. According to Lemma 31, it suffices to show that *any weak limit σ of $(\mu_n^{2\epsilon})_n$ satisfies*

$$\sigma(D_{r'}) \geq v_N^\epsilon(D_{r'}), \text{ for any integer } N \text{ and all } 0 < r' < \frac{1}{N}. \tag{4}$$

Let us pick $r' < r < \frac{1}{N}$. We set $\hat{D}_r := \pi^{-1}(D_r)$ and:

$$\hat{C}_n := \{\hat{x} \in \hat{D}_r \cap \hat{X}^\epsilon_{reg\,N} \,/\, f_{\hat{x}}^{-n}(D_r) \cap D_{r'} \neq \emptyset\}.$$

Let also consider the collection S_n of sets of the form $f_{\hat{x}}^{-n}(D_r)$ where $\hat{x}$ runs in $\hat{C}_n$. As $f_{\hat{x}}^{-n}$ is an inverse branch on D_r of the ramified cover f^n, one sees that the sets of the collection S_n are mutually disjoint.

Let us momentarily admit the two following estimates:

$$d^{-n}(\text{Card } S_n) \leq \mu_n^{2\epsilon}(D_r) \text{ for } n \text{ big enough} \tag{5}$$

$$d^{-n}(\text{Card } S_n)\,\mu(D_r) \geq \hat{\mu}\big(\hat{f}^{-n}(\hat{D}_r \cap \hat{X}^\epsilon_{reg\,N}) \cap \hat{D}_{r'}\big). \tag{6}$$

Combining (5) and (6) yields:

$$\hat{\mu}\big(\hat{f}^{-n}(\hat{D}_r \cap \hat{X}^\epsilon_{reg\,N}) \cap \hat{D}_{r'}\big) \leq \mu(D_r)\mu_n^{2\epsilon}(D_r)$$

which, by the mixing property of $\hat{\mu}$, implies

$$\nu_N^\epsilon(D_r)\mu(D_{r'}) = \hat{\mu}(\hat{D}_r \cap \hat{X}^\epsilon_{reg\,N})\hat{\mu}(\hat{D}_{r'}) \leq \mu(D_r)\sigma(D_r)$$

since $\mu(D_{r'}) > 0$, one gets (4) by making $r \to r'$.

Let us now prove the estimate (5). We have to show that D_r contains at least $(\text{Card } S_n)$ elements of $R_n^{2\epsilon}$ when n is big enough. Here we shall use Proposition 29. For every $\hat{x} \in \hat{C}_n \subset \hat{X}^\epsilon_{reg\,N}$ one has $\eta_\epsilon(\hat{x}) \geq \frac{1}{N}$ and $S_\epsilon(\hat{x}) \leq N$ and thus the map $f_{\hat{x}}^{-n}$ is defined on D_r $(r < \frac{1}{N})$ and $\text{Diam } f_{\hat{x}}^{-n}(D_r) \leq 2r\,\text{Lip } f_{\hat{x}}^{-n} \leq 2rS_\epsilon(\hat{x})e^{-n(L-\epsilon)} \leq 2rNe^{-n(L-\epsilon)}$.

As moreover $f_{\hat{x}}^{-n}(D_r)$ meets $D_{r'}$, there exists n_0, which depends only on ϵ, r and r', such that $f_{\hat{x}}^{-n}(D_r) \subset D_r$ for every $\hat{x} \in \hat{C}_n$ and $n \geq n_0$. Thus, by Brouwer theorem, $f_{\hat{x}}^{-n}$ has a fixed point $p_n \in f_{\hat{x}}^{-n}(D_r)$ for every $\hat{x} \in \hat{C}_n$ and $n \geq n_0$. Since the elements of S_n are mutually disjoint sets, we have produced $(\text{Card } S_n)$ fixed points of f^n in D_r for $n \geq n_0$. It remains to check that these fixed points belong to $R_n^{2\epsilon}$. This actually follows immediately from the estimates on $\text{Lip } f_{\hat{x}}^{-n}$. Indeed:

$$|(f^n)'(p_n)| = |(f_{\hat{x}}^{-n})'(p_n)|^{-1} \geq \big(\text{Lip } f_{\hat{x}}^{-n}\big)^{-1} \geq N^{-1}e^{n(L-\epsilon)} \geq e^{n(L-2\epsilon)}$$

for n big enough.

Finally we prove the estimate (6). Let us first observe that

$$\pi\big(\hat{f}^{-n}(\hat{D}_r \cap \hat{X}^\epsilon_{reg\,N}) \cap \hat{D}_{r'}\big) \subset \bigcup_{\hat{x} \in \hat{C}_n} f_{\hat{x}}^{-n}(D_r). \tag{7}$$

This can be easily seen: if $\hat{u} \in \hat{f}^{-n}(\hat{D}_r \cap \hat{X}^\epsilon_{reg\,N}) \cap \hat{D}_{r'}$ then $u_0 = \pi(\hat{u}) \in D_{r'} \cap f_{\hat{x}}^{-n}(D_r)$ where $\hat{x} := \hat{f}^n(\hat{u}) \in \hat{D}_r \cap \hat{X}^\epsilon_{reg\,N}$.

By the constant Jacobian property we have $\mu(f_{\hat{x}}^{-n}(D_r)) = d^{-n}\mu(D_r)$ and, since the sets $f_{\hat{x}}^{-n}(D_r)$ of the collection S_n are mutually disjoint, we obtain

$$\mu\Big(\bigcup_{\hat{x}\in\hat{C}_n} f_{\hat{x}}^{-n}(D_r)\Big) = (\text{Card } S_n)\, d^{-n}\mu(D_r). \tag{8}$$

Combining (7) with (8) yields 6:

$$\begin{aligned}(\text{Card } S_n)\, d^{-n}\mu(D_r) &\geq \mu\big[\pi\big(\hat{f}^{-n}(\hat{D}_r \cap \hat{X}^\epsilon_{reg\,N}) \cap \hat{D}_{r'}\big)\big] \\ &= \hat{\mu}\big[\pi^{-1} \circ \pi\big(\hat{f}^{-n}(\hat{D}_r \cap \hat{X}^\epsilon_{reg\,N}) \cap \hat{D}_{r'}\big)\big] \\ &\geq \hat{\mu}\big(\hat{f}^{-n}(\hat{D}_r \cap \hat{X}^\epsilon_{reg\,N}) \cap \hat{D}_{r'}\big).\end{aligned}$$

□

3 Holomorphic Families

We introduce here the main spaces in which we shall work in the next sections and present some of their structural properties.

3.1 Generalities

3.1.1 Holomorphic Families and the Space Rat_d

Let us start with a formal definition.

Definition 32. Let M be a complex manifold. A holomorphic map

$$f : M \times \mathbf{P}^1 \to \mathbf{P}^1$$

such that all rational maps $f_\lambda := f(\lambda, \cdot) : \mathbf{P}^1 \to \mathbf{P}^1$ have the same degree $d \geq 2$ is called *holomorphic family* of degree d rational maps parametrized by M. For short, any such family will be denoted $(f_\lambda)_{\lambda \in M}$.

Any degree d rational map $f := \frac{a_d z^d + \cdots + a_1 z + a_0}{b_d z^d + \cdots + b_1 z + b_0}$ is totally defined by the point $[a_d : \cdots : a_0 : b_d : \cdots : b_0]$ in the projective space $\mathbf{P}^{2d+1}$. This allows to identify the space Rat_d of degree d rational maps with a Zariski dense open subset of $\mathbf{P}^{2d+1}$.

We can be more precise by looking at the space of homogeneous polynomial maps of $\mathbf{C}^2$ which is identified to $\mathbf{C}^{2d+2}$ by the correspondence

$$(a_d, \cdots, a_0, b_d, \cdots, b_0) \mapsto \left(\sum_{i=1}^{d} a_i z_1^i z_2^{d-i}, \sum_{i=1}^{d} b_i z_1^i z_2^{d-i} \right).$$

Indeed, Rat_d is precisely the image by the canonical projection $\pi : \mathbf{C}^{2d+2} \to \mathbf{P}^{2d+1}$ of the subspace H_d of $\mathbf{C}^{2d+2}$ consisting of non-degenerate polynomials. As H_d is the complement in $\mathbf{C}^{2d+2}$ of the projective variety defined by the vanishing of the resultant $Res\left(\left(\sum_{i=1}^{d} a_i z_1^i z_2^{d-i}, \sum_{i=1}^{d} b_i z_1^i z_2^{d-i} \right) \right)$, one sees that $Rat_d = \mathbf{P}^{2d+1} \setminus \Sigma_d$ where Σ_d is an (irreducible) algebraic hypersurface of $\mathbf{P}^{2d+1}$.

From now on, we will always consider Rat_d as a quasi-projective manifold. We may therefore also see any holomorphic family of degree d rational maps with parameter space M as a holomorphic map f from M to Rat_d. In particular, we may take for M any submanifold of Rat_d; this is especially interesting when M is dynamically defined as are, for instance, the hypersurfaces $Per_n(w)$ which will be defined in the next section.

The simplest example of holomorphic family is the family of quadratic polynomials. Up to affine conjugation, any degree 2 polynomial is of the form $z^2 + a$. To understand quadratic polynomials it is therefore sufficient to work with the family $\left(z^2 + a \right)_{a \in \mathbf{C}}$.

In most cases, when considering a holomorphic family $(f_\lambda)_{\lambda \in M}$, we shall make the two following mild assumptions.

Assumptions 33. *Let $(f_\lambda)_{\lambda \in M}$ be any holomorphic family of degree d rational maps.*

A1 The marked critical points assumption *means that the critical set C_λ of f_λ is given by $2d - 2$ graphs: $C_\lambda = \cup_1^{2d-2} \{c_i(\lambda)\}$ where the maps $M \ni \lambda \mapsto c_i(\lambda) \in \mathbf{P}^1$ are holomorphic.*

A2 The no persistent neutral cycles assumption *means that if f_{λ_0} has a neutral cycle then this cycle becomes attracting or repelling under a suitable small perturbation of λ_0.*

Using a ramified cover of the parameter space M one may actually always make sure that the assumption $A2$ is satisfied.

The group of Möbius transformations, which is isomorphic to $PSL(2, \mathbf{C})$, acts by conjugation on the space Rat_d of degree d rational maps. The dynamical properties of two conjugated rational maps are clearly equivalent and it is therefore natural to work within the quotient of Rat_d resulting from this action.

The moduli space Mod_d is, by definition, the quotient of Rat_d under the action of $PSL(2, \mathbf{C})$ by conjugation. We shall denote as follows the canonical projection:

$$\Pi : Rat_d \longrightarrow Mod_d$$
$$f \mapsto \bar{f}$$

We shall usually commit the abuse of language which consists in considering an element of Mod_d as a rational map. For instance, "$\bar{f}$ has a n-cycle of multiplier w" means that every element of $\bar{f}$ possesses such a cycle. We shall also sometimes write f instead of $\bar{f}$.

Although the action of $PSL(2,\mathbf{C})$ is not free, it may be proven that Mod_d is a normal quasi-projective variety [59].

Remark 34. The following property is helpful for working in Mod_d. Every element f of Rat_d belongs to a local submanifold T_f whose dimension equals $2d - 2$ and which is transversal to the orbit of f under the action of $PSL(2,\mathbf{C})$. Moreover, T_f is invariant under the action of the stabilizer $Aut(f)$ of f which is a finite subgroup of $PSL(2,\mathbf{C})$. Finally, $\Pi(T_f)$ is a neighborhood of $\bar{f}$ in Mod_d and Π induces a biholomorphism between $T_f/Aut(f)$ and $\Pi(T_f)$.

3.1.2 The Space of Degree d Polynomials

As for quadratic polynomials, there exists a nice parametrization of the space of degree d polynomials.

Let $\mathcal{P}_d$ be the space of polynomials of degree $d \geq 2$ with $d - 1$ marked critical points up to conjugacy by affine transformations. Although this space has a natural structure of affine variety of dimension $d - 1$, we may actually work with a specific parametrization of $\mathcal{P}_d$ which we shall now present.

For every $(c,a) := (c_1, c_2, \cdots, c_{d-2}, a) \in \mathbf{C}^{d-1}$ we denote by $P_{c,a}$ the polynomial of degree d whose critical points are $(0, c_1, \cdots, c_{d-2})$ and such that $P_{c,a}(0) = a^d$. This polynomial is explicitly given by:

$$P_{c,a} := \frac{1}{d}z^d + \sum_{2}^{d-1} \frac{(-1)^{d-j}}{j}\sigma_{d-j}(c)z^j + a^d$$

where $\sigma_i(c)$ is the symmetric polynomial of degree i in $(c_1, \cdots, c_{d-2})$. For convenience we shall set $c_0 := 0$.

Thus, when considering degree d polynomials, instead of working in $\mathcal{P}_d$ we may consider the holomorphic family

$$\left(P_{c,a}\right)_{(c,a) \in \mathbf{C}^{d-1}}$$

whose parameter space M is simply $\mathbf{C}^{d-1}$. Using this parametrization, one may exhibit a finite ramified cover $\pi : \mathbf{C}^{d-1} \to \mathcal{P}_d$ (see [29, Proposition 5.1]).

It will be crucial to consider the projective compactification $\mathbf{P}^{d-1}$ of $\mathbf{C}^{d-1} = M$. This is why we wanted the expression of $P_{c,a}$ to be homogeneous in (c,a) and have used the parameter a^d instead of a. In this context, we shall denote by $\mathbf{P}_\infty$ the projective space at infinity: $\mathbf{P}_\infty := \{[c : a : 0] ; (c,a) \in \mathbf{C}^{d-1} \setminus \{0\}\}$.

3.1.3 The Moduli Space of Degree Two Rational Maps

In his paper [48], Milnor has given a particularly nice description of Mod_2 which we will now present. The reader may also consult the fourth chapter of book of Silverman [59].

Any $f \in Rat_2$ has three fixed points (counted with multiplicities) whose multipliers may be denoted μ_1, μ_2, μ_3. Let us observe that $\mu_i = 1$ if and only if one of the fixed point is not simple. The symmetric functions

$$\sigma_1 := \mu_1 + \mu_2 + \mu_3, \quad \sigma_2 := \mu_1\mu_2 + \mu_1\mu_3 + \mu_2\mu_3, \quad \sigma_3 := \mu_1\mu_2\mu_3$$

are clearly well defined on Mod_2. It follows from the holomorphic fixed point formula $\sum \frac{1}{1-\mu_i} = 1$ (see [49, Lecture 12]), applied in the generic case of three distinct fixed points, that

$$\sigma_3 - \sigma_1 + 2 = 0. \tag{9}$$

The above identity is the crucial point. It shows that the set of three multipliers $\{\mu_1, \mu_2, \mu_3\}$ is entirely determined by (σ_1, σ_2). More precisely, $\{\mu_1, \mu_2, \mu_3\}$ is the set of roots of the polynomial $X^3 - \sigma_1 X^2 + \sigma_2 X - (\sigma_1 - 2)$.

Let us mention some further useful facts.

Lemma 35. *The multipliers μ_1, μ_2, μ_3 satisfy the following identities*

$$(\mu_1 - 1)^2 = (\mu_1\mu_2 - 1)(\mu_1\mu_3 - 1)$$
$$(\mu_2 - 1)^2 = (\mu_2\mu_1 - 1)(\mu_2\mu_3 - 1)$$
$$(\mu_3 - 1)^2 = (\mu_3\mu_1 - 1)(\mu_3\mu_2 - 1).$$

In particular, if two fixed points are distinct and have multipliers μ_i, μ_j then $\mu_i\mu_j \neq 1$ and the multiplier μ_k of the remaining fixed point is given by $\mu_k = \frac{2-\mu_i-\mu_j}{1-\mu_i\mu_j}$.

Proof. The three identities are deduced from $(X - 1)^2 - (XY - 1)(XZ - 1) = X(X + Y + Z - 2 - XYZ)$ taking $\mu_1\mu_2\mu_3 - (\mu_1 + \mu_2 + \mu_3) + 2 = 0$ (see (9)) into account.

If two fixed points are distinct then one of them must be simple and thus $\mu_i \neq 1$. Then $\mu_i\mu_j \neq 1$ follows from one of the previous identities. The expression of μ_k is then obtained using $\mu_1\mu_2\mu_3 - (\mu_1 + \mu_2 + \mu_3) + 2 = 0$. □

Milnor has actually shown that (σ_1, σ_2) induces a good parametrization of Mod_2 [48].

Theorem 36. *The map $M : Mod_2 \to \mathbf{C}^2$ defined by $\bar{f} \mapsto (\sigma_1, \sigma_2)$ is one-to-one and onto.*

Proof. Let us denote by m the map which associates to any $\sigma := (\sigma_1, \sigma_2) \in \mathbf{C}^2$ the set $\{\mu_1, \mu_2, \mu_3\}$ of roots of the polynomial $X^3 - \sigma_1 X^2 + \sigma_2 X - (\sigma_1 - 2)$. It is clear that $m(\sigma) \neq \{1, 1, 1\}$ if and only if $\sigma \neq (3, 3)$.

Let $\sigma \in \mathbf{C}^2 \setminus \{(3, 3)\}$. Let us show that $M(\bar{f}) = \sigma$ uniquely determines $\bar{f} \in Mod_2$. Set $\{\mu_1, \mu_2, \mu_3\} = m(\sigma)$. At least one of the μ_i does not equal 1 and thus the map f has at least two distinct fixed points whose multipliers are, say, $\mu_1 \neq 1$ and μ_2. After conjugation we may assume that these fixed points are 0 and ∞ and that

$$f \sim z \frac{\alpha z + \beta}{\delta z + 1} \text{ with } \alpha \neq 0 \text{ and } \alpha - \beta\gamma \neq 0.$$

After conjugating by $z \mapsto \alpha^{-1} z$ one gets

$$f \sim z \frac{z + \beta'}{\delta' z + 1} \text{ with } 1 - \beta'\gamma' \neq 0$$

but, since β' and δ' are obviously respectively equal to μ_1 and μ_2 one actually has $f \sim z\frac{z+\mu_1}{\mu_2 z+1}$ with $\mu_1\mu_2 \neq 1$. We have shown that $f \sim g \sim z\frac{z+\mu_1}{\mu_2 z+1}$ as soon as $M(\bar{f}) = M(\bar{g})$.

Let us now assume that $M(\bar{f}) = (3, 3)$. In that case the multipliers of the fixed points of f are all equal to 1 and f has actually only one fixed point. After conjugation, we may assume that this fixed point is ∞ and that $f^{-1}\{\infty\} = \{0, \infty\}$. Then $f \sim \frac{p(z)}{z}$ but, since $f(z) - z = \frac{p(z)-z^2}{z}$ does not vanish on the complex plane, we must have $p(z) = z^2 + c$ and $f \sim z + \frac{c}{z}$. After conjugating by $z \mapsto \sqrt{c} z$ we get $f \sim z + \frac{1}{z}$.

We have shown that M is one-to-one. It clearly follows from the above computations that M is also onto. □

Remark 37. The proof of the above Theorem shows that the expression $z\frac{z+\mu_i}{\mu_j z+1}$ can be used as a normal form where $\mu_i \mu_j \neq 1$.

It will be extremely useful to consider the projective compactification of Mod_2 obtained through the above Theorem:

$$Mod_2 \ni \bar{f} \mapsto [\sigma_1 : \sigma_2 : 1] \in \mathbf{P}^2$$

whose corresponding line at infinity will be denoted by $\mathcal{L}$

$$\mathcal{L} := \{[\sigma_1 : \sigma_2 : 0]; \ (\sigma_1, \sigma_2) \in \mathbf{C}^2 \setminus \{0\}\}.$$

It is important to stress that this compactification is actually natural in the sense that the "behaviour near $\mathcal{L}$" captures a lot of dynamically meaningful informations. This will be discussed in Sect. 3.3.3. Let us now simply mention that the line at infinity $\mathcal{L}$ may somehow be parametrized by the limits of $\frac{\sigma_2}{\sigma_3}$.

Proposition 38. *If $[\sigma_1 : \sigma_2 : 1]$ converges to $P \in \mathcal{L}$ then the point P is of the form $[1 : \mu + \frac{1}{\mu} : 0]$ where $\mu + \frac{1}{\mu}$ is a limit of $\frac{\sigma_2}{\sigma_3} = \frac{1}{\mu_1} + \frac{1}{\mu_2} + \frac{1}{\mu_3}$ in $\mathbf{P}^1$. Moreover, $\mu \neq 0, \infty$ if and only if two of the multipliers μ_j stay bounded as $[\sigma_1 : \sigma_2 : 1]$ approaches P.*

Proof. Since (σ_1, σ_2) tends to ∞ then at least one of the μ_i's tends to ∞ too. Let us assume that $\mu_3 \to \infty$. We first look at the case where both μ_1 and μ_2 stay bounded. The third identity in Lemma 35 shows that $\mu_1 \mu_2 \to 1$. Thus $\sigma_3 \to \infty$ and $\frac{\sigma_1}{\sigma_3} = \frac{\mu_1 + \mu_2}{(\mu_1 \mu_2) \mu_3} + \frac{1}{\mu_1 \mu_2} \to 1$. As $[\sigma_1 : \sigma_2 : 1] = [\frac{\sigma_1}{\sigma_3} : \frac{\sigma_2}{\sigma_3} : \frac{1}{\sigma_3}]$ converges to $P \in \mathcal{L}$ one sees that $\frac{\sigma_2}{\sigma_3}$ must converge in $\mathbf{P}^1$ and that P is of the required form.

Let us now consider the case where $\mu_2 \to \infty$. The first identity of Lemma 35 implies that $\mu_1 \to 0$. Thus $\sigma_2 \sim \mu_2 \mu_3$ and therefore $\frac{\sigma_2}{\sigma_3} \to \infty$ and $\frac{\sigma_1}{\sigma_2} \to 1$. In that case $P = [0 : 1 : 0]$ which is again the announced form. □

3.2 The Connectedness Locus in Polynomial Families

3.2.1 Connected and Disconnected Julia Sets of Polynomials

Among rational functions, polynomials are characterized by the fact that ∞ is a totally invariant critical point. For any polynomial P the super-attractive fixed point ∞ determines a basin of attraction

$$\mathcal{B}_P(\infty) := \{z \in \mathbf{C} \text{ s.t. } \lim_n P^n(z) = \infty\}.$$

This basin is always connected and its boundary is precisely the Julia set J_P of P. The complement of $\mathcal{B}_P(\infty)$ is called the filled-in Julia set of P.

As we already saw, another nice feature of polynomials is the possibility to define a Green function $g_{\mathbf{C},P}$. The Green function $g_{\mathbf{C},P}$ is a subharmonic function on the complex plane which vanishes exactly on the filled-in Julia set of P (see Proposition 12).

Any degree d polynomial P is locally conjugated at infinity with the polynomial z^d. This means that there exists a local change of coordinates φ_P (which is called Böttcher function) such that $\varphi_P \circ P = (\varphi_P)^d$ on a neighbourhood of ∞. It is important to stress the following relation between the Böttcher and Green functions:

$$\ln |\varphi_P| = g_{\mathbf{C},P}.$$

The only obstruction to the extension of the Böttcher function φ_P to the full basin $\mathcal{B}_P(\infty)$ is the presence of other critical points than ∞ in $\mathcal{B}_P(\infty)$. This leads to the following important result:

Theorem 39. *For any polynomial P of degree $d \geq 2$ the following conditions are equivalent:*

(i) $\mathcal{B}_P(\infty)$ *is simply connected*
(ii) $\mathcal{J}_P$ *is connected*
(iii) $\mathcal{C}_P \cap \mathcal{B}_P(\infty) = \{\infty\}$
(iv) P *is conformally conjugated to* z^d *on* $\mathcal{B}_P(\infty)$.

The above theorem gives a nice characterization of polynomials having a connected Julia set. Let us apply it to the quadratic family $\left(z^2 + a\right)_{a \in \mathbf{C}}$. The Julia set $\mathcal{J}_a$ of $P_a := z^2 + a$ is connected if and only if the orbit of the critical point 0 is bounded. In other words, the set of parameters a for which $\mathcal{J}_a$ is connected is the famous Mandelbrot set.

Definition 40 (Mandelbrot set). Let P_a denote the quadratic polynomial $z^2 + a$. The Mandelbrot set $\mathcal{M}$ is defined by

$$\mathcal{M} := \{a \in \mathbf{C} \text{ s.t. } \sup_n |P_a^n(0)| < \infty\}.$$

The Mandelbrot set is thus the connectedness locus of the quadratic family. It is not difficult to show that $\mathcal{M}$ is compact. The compacity of the connectedness locus in the polynomial families of degree $d \geq 3$ is a much more delicate question which has been solved by Branner and Hubbard [10]. We shall treat it in the two next subsections and also present a somewhat more precise result which will turn out to be very useful later.

3.2.2 Polynomials with a Bounded Critical Orbit

We work here with the parametrization $\left(P_{c,a}\right)_{(c,a) \in \mathbf{C}^{d-1}}$ of $\mathcal{P}_d$ and will use the projective compactification $\mathbf{P}^{d-1}$ introduced in the Sect. 3.1.2.

We aim to show that the subset of parameters (c, a) for which the polynomial $P_{c,a}$ has at least one bounded critical orbit can only cluster on certain hypersurfaces of $\mathbf{P}_\infty$. The ideas here are essentially those used by Branner and Hubbard for proving the compactness of the connectedness locus (see [10, Chap. 1, Sect. 3]) but we also borrow from the paper [29] of Dujardin and Favre.

We shall use the following

Definition 41. The notations are those introduced in Sect. 3.1.2. For every $0 \leq i \leq d - 2$, the hypersurface Γ_i of $\mathbf{P}_\infty$ is defined by:

$$\Gamma_i := \{[c : a : 0] / \alpha_i(c, a) = 0\}$$

where α_i is the homogeneous polynomial given by:

$$\alpha_i(c, a) := P_{c,a}(c_i) = \frac{1}{d} c_i^d + \sum_{j=2}^{d-1} \frac{(-1)^{d-j}}{j} \sigma_{d-j}(c) c_i^j + a^d.$$

We denote by $\mathcal{B}_i$ the set of parameters (c,a) for which the critical point c_i of $P_{c,a}$ has a bounded forward orbit (recall that $c_0 = 0$):

$$\mathcal{B}_i := \{(c,a) \in \mathbf{C}^{d-1} \text{ s.t. } \sup_n |P_{c,a}^n(c_i)| < \infty\}.$$

A crucial observation about the intersections of hypersurfaces Γ_i is given by the next Lemma.

Lemma 42. *The intersection $\Gamma_0 \cap \Gamma_1 \cap \cdots \cap \Gamma_{d-2}$ is empty and $\Gamma_{i_1} \cap \cdots \cap \Gamma_{i_k}$ has codimension k in $\mathbf{P}_\infty$ if $0 \le i_1 < \cdots < i_k \le d-2$.*

Proof. A simple degree argument shows that $P_{c,a}(0) = P_{c,a}(c_1) = \cdots = P_{c,a}(c_{d-2}) = 0$ implies that $c_1 = \cdots = c_{d-2} = a = 0$. Thus $\Gamma_0 \cap \Gamma_1 \cap \cdots \cap \Gamma_{d-2} = \emptyset$. Then the conclusion follows from Bezout's theorem. □

Since the connectedness locus coincides with $\cap_{0 \le i \le d-2} \mathcal{B}_i$, the announced result can be stated as follows.

Theorem 43. *For every $0 \le i \le d-2$, the cluster set of $\mathcal{B}_i$ in $\mathbf{P}_\infty$ is contained in Γ_i. In particular, the connectedness locus is compact in $\mathbf{C}^{d-1}$.*

As the Green function $g_{c,a}$ of the polynomial $P_{c,a}$ is defined by

$$g_{c,a}(z) := \lim_n d^{-n} \ln^+ |P_{c,a}^n(z)|$$

one sees that

$$\mathcal{B}_i = \{(c,a) \in \mathbf{C}^{d-1} \text{ s.t. } g_{c,a}(c_i) = 0\}.$$

This is why the proof of Theorem 43 will rely on estimates on the Green functions and, more precisely, on the following result.

Proposition 44. *Let $g_{c,a}$ be the Green function of $P_{c,a}$ and G be the function defined on $\mathbf{C}^{d-1}$ by: $G(c,a) := \max\{g_{c,a}(c_k); 0 \le k \le d-2\}$. Let $\delta := \frac{\sum_{k=0}^{d-2} c_k}{d-1}$. Then the following estimates occur:*

(1) $G(c,a) \le \ln\max\{|a|, |c_k|\} + O(1)$ if $\max\{|a|, |c_k|\} \ge 1$
(2) $\max\{g_{c,a}(z), G(c,a)\} \ge \ln|z - \delta| - \ln 4$.

Let us first see how Theorem 43 may be deduced from Proposition 44.

Proof of Theorem 43. Let $\|(c,a)\|_\infty := \max\{|a|, |c_k|\}$. We simply have to check that $\alpha_i \left(\frac{(c,a)}{\|(c,a)\|_\infty} \right)$ tends to 0 when $g_{c,a}(c_i)$ stays equal to 0 and $\|(c,a)\|_\infty$ tends to $+\infty$. As $P_{c,a}(c_i) = \alpha_i(c,a)$ and $g_{c,a}(c_i) = 0$, the estimates given by Proposition 44 yield:

$$\ln \|(c,a)\|_\infty + O(1) \ge \max(dg_{c,a}(c_i), G(c,a)) = \max\{g_{c,a} \circ P_{c,a}(c_i), G(c,a)\} \ge$$

$$\ge \ln \frac{1}{4}|\alpha_i(c,a) - \delta|$$

since α_i is d-homogeneous we then get:

$$(1-d)\ln\|(c,a)\|_\infty + O(1) \geq \ln\frac{1}{4}|\alpha_i\left(\frac{(c,a)}{\|(c,a)\|_\infty}\right) - \frac{\delta}{\|(c,a)\|_\infty^d}|$$

and the conclusion follows since $\frac{\delta}{\|(c,a)\|_\infty^d}$ tends to 0 when $\|(c,a)\|_\infty$ tends to $+\infty$.

$\square$

Let us end this subsection by giving a

Proof of Proposition 44.. The first estimate is a standard consequence of the uniform growth of $P_{c,a}$ at infinity. Let us however prove it with care. We will set $A := \|(c,a)\|_\infty$ and $M_A(z) := \max\{A, |z|\}$. From

$$|P_{c,a}(z)| \leq \frac{1}{d}|z|^d\left(1 + d\max\{\frac{|\sigma_{d-j}(c)|}{j|z|^{d-j}}, \frac{|a|^d}{|z|^d}\}\right)$$

we get $|P_{c,a}(z)| \leq C_d|z|^d$ for $|z| \geq A$ where the constant C_d only depends on d. We may assume that $C_d \geq 1$. By the maximum modulus principle this yields

$$|P_{c,a}(z)| \leq C_d M_A(z)^d.$$

It is easy to check that

$$M_A(Cz) \leq CM_A(z) \text{ if } C \geq 1$$
$$M_A(M_A(z)^N) = M_A(z)^N \text{ if } A \geq 1.$$

From now on we shall assume that $A \geq 1$. By induction one gets

$$|P_{c,a}^n(z)| \leq C_d^{1+d+\cdots+d^{n-1}} M_A(z)^{d^n}$$

which implies

$$g_{c,a}(z) \leq \frac{\ln C_d}{d-1} + \ln\max\{\|(c,a)\|_\infty, |z|\} \text{ if } \|(c,a)\|_\infty \geq 1$$

and in particular

$$G(c,a) \leq \frac{\ln C_d}{d-1} + \ln\|(c,a)\|_\infty \text{ if } \|(c,a)\|_\infty \geq 1.$$

The second estimate is really more subtle. It exploits the fact that the Green function $g_{c,P}$ coincides with the log-modulus of the Böttcher coordinate function φ_P and relies on a sharp control of the distortions of this holomorphic function.

The Böttcher coordinate function $\varphi_{c,a} : \{g_{c,a} > G(c,a)\} \to \mathbf{C}$ is a univalent function such that $\varphi_{c,a} \circ P_{c,a} = \varphi_{c,a}^d$. It is easy to check that $\ln |\varphi_{c,a}| = g_{c,a}$ where it makes sense and that $\varphi_{c,a}(z) = z - \delta + O(\frac{1}{z})$ where $\delta := \frac{\sigma_1(c)}{d-1} = \frac{\sum c_k}{d-1}$.

One thus sees that $\varphi_{c,a} : \{g_{c,a} > G(c,a)\} \to \mathbf{C} \setminus \overline{D}(0, e^{G(c,a)})$ is a univalent map whose inverse $\psi_{c,a}$ satisfies $\psi_{c,a}(z) = z + \delta + O(\frac{1}{z})$ at infinity. We shall now apply the following result, which is a version of the Koebe $\frac{1}{4}$-theorem (see [10, Corollary 3.3]), to $\psi_{c,a}$.

Theorem 45. *If $F : \hat{\mathbf{C}} \setminus \overline{D}_r \to \hat{\mathbf{C}}$ is holomorphic and injective and*

$$F(z) = z + \sum_{n=1}^{\infty} \frac{a_n}{z^n}, \ z \in \mathbf{C} \setminus \overline{D}_r$$

then $\mathbf{C} \setminus \overline{D}_{2r} \subset F(\mathbf{C} \setminus \overline{D}_r)$.

Pick $z \in \mathbf{C}$ and set $r := 2\max\{e^{g_{c,a}(z)}, e^{G(c,a)}\}$. Then $z \notin \psi_{c,a}(\mathbf{C} \setminus \overline{D}_r)$ since otherwise we would have $e^{g_{c,a}}(z) = |\varphi_{c,a}(z)| > r \geq 2e^{g_{c,a}}(z)$.

Thus, according to the above distortion theorem, $z \notin \mathbf{C} \setminus \overline{D}(\delta, 2r)$. In other words $|z - \delta| \leq 2r = 4\max\{e^{g_{c,a}(z)}, e^{G(c,a)}\}$ and the desired estimate follows by taking logarithms. □

3.3 The Hypersurfaces $\text{Per}_n(w)$

We will consider here some dynamically defined subsets of the parameter space which will play a central role in our study.

3.3.1 Defining the $\text{Per}_n(w)$ Using Dynatomic Polynomials

For any holomorphic family of rational maps, the following result describes precisely the set of maps having a cycle of given period and multiplier.

Theorem 46. *Let $f : M \times \mathbf{P}^1 \to \mathbf{P}^1$ be a holomorphic family of degree $d \geq 2$ rational maps. Then for every integer $n \in \mathbf{N}^*$ there exists a holomorphic function p_n on $M \times \mathbf{C}$ which is polynomial on $\mathbf{C}$ and such that:*

1. *For any $w \in \mathbf{C} \setminus \{1\}$, the function $p_n(\lambda, w)$ vanishes if and only if f_λ has a cycle of exact period n and multiplier w.*
2. *$p_n(\lambda, 1) = 0$ if and only if f_λ has a cycle of exact period n and multiplier 1 or a cycle of exact period m whose multiplier is a primitive r^{th} root of unity with $r \geq 2$ and $n = mr$.*
3. *For every $\lambda \in M$, the degree $N_d(n)$ of $p_n(\lambda, \cdot)$ satisfies $d^{-n}N_d(n) \sim \frac{1}{n}$.*

This leads to the following

Definition 47. Under the assumptions and notations of Theorem 46, one sets

$$\text{Per}_n(w) := \{\lambda \in M / \ p_n(\lambda, w) = 0\}$$

for any integer n and any complex number w.

According to Theorem 46, $Per_n(w)$ is (at least when $w \neq 1$) the set of parameters λ for which f_λ has a cycle of exact period n and multiplier w. Moreover, $\text{Per}_n(w)$ is an hypersurface in the parameter space M or coincides with M. We also stress that the estimate on the degree $N_d(n)$ of $p_n(\lambda, \cdot)$ will be important in some of our applications.

We now start to explain the construction of the functions p_n. It clearly suffices to treat the case of the family Rat_d and then set $p_n(\lambda, w) := p_n(f_\lambda, w)$ for any holomorphic family $M \ni \lambda \mapsto f_\lambda \in \text{Rat}_d$. Our presentation is essentially based on the fourth chapter of the book of Silverman [59] and also borrows to the paper [48] of Milnor.

We will consider polynomial families; to deal with the general case one may adapt the proof by using lifts to $\mathbf{C}^2$. According to the discussion we had in Sect. 3.1.2, any degree d polynomial φ will be identified to a point in $\mathbf{C}^{d-1}$.

The key point is to associate to any integer n and any polynomial φ of degree $d \geq 2$ a polynomial $\Phi_{\varphi,n}^*$ whose roots, for a generic φ, are precisely the periodic points of φ with exact period n. Such polynomials are called *dynatomic* since they generalize cyclotomic ones, they are defined as follows.

Definition 48. For a degree d polynomial φ and an integer n one sets

$$\Phi_{\varphi,n}(z) := \varphi^n(z) - z.$$

The associated *dynatomic polynomials* are then defined by setting

$$\Phi_{\varphi,n}^*(z) := \prod_{k|n} \left(\Phi_{\varphi,k}(z)\right)^{\mu(\frac{n}{k})}$$

where $\mu : \mathbf{N}^* \to \{-1, 0, 1\}$ is the classical Möbius function.

It is clear that $\Phi_{\varphi,n}$ is a polynomial whose roots are all periodic points of φ with exact period *dividing* n, and that $\Phi_{\varphi,n}^*$ is a fraction whose roots and poles belong to the same set. Actually $\Phi_{\varphi,n}^*$ is still a polynomial but this is not at all obvious! To prove it, one will systematically exploits the fact that the sum $\sum_{k|n} \mu(\frac{k}{n})$ vanishes if $n > 1$ and computes the valuation of $\Phi_{\varphi,n}^*$ at any m-periodic point of φ for $m|n$. We then obtain a precise description of the roots of $\Phi_{\varphi,n}^*$:

Theorem 49. *Let φ be a polynomial of degree $d \geq 2$. Then $\Phi_{\varphi,n}^*$ is a polynomial whose roots are the periodic points of φ with exact period m dividing n and multiplier w satisfying $w^r = 1$ when $2 \leq r := \frac{n}{m}$. The degree $\nu_d(n)$ of $\Phi_{\varphi,n}^*$ is equivalent to d^n.*

The proof of the above Theorem will be given in the next subsection, for the moment we admit it and prove Theorem 46. Let us note that the degree $v_d(n)$ of $\Phi^*_{\varphi,n}$ is given by $v_d(n) = \sum_{k|n} d^k \mu(\frac{n}{k})$ and is clearly equivalent to d^n since $|\mu| \in \{0, 1\}$ and $\mu(1) = 1$. We may also observe that $(\varphi^n)'(z) = 1$ for any root z of $\Phi^*_{\varphi,n}$ whose period strictly divides n.

The construction of $p_n(\varphi, w)$ requires to understand the structure of the zero set Per_n of $(\varphi, z) \mapsto \Phi^*_{\varphi,n}(z)$. We recall that φ is seen as a point in $\mathbf{C}^{d-1}$. Here are the informations we need.

Proposition 50. *The set $\operatorname{Per}_n := \{(\varphi, z) \mid \Phi^*_{\varphi,n}(z) = 0\}$ is an algebraic subset of $\mathbf{C}^{d-1} \times \mathbf{C}$. The roots of $\Phi^*_{\varphi,n}$ are simple and have exact period n when $\varphi \in \mathbf{C}^{d-1} \setminus X_n$ for some proper algebraic subset X_n of $\mathbf{C}^{d-1}$.*

Proof. One sees on Definition 48 that $\Phi^*_{\varphi,n}(z)$ is rational in φ. On the other hand, $\Phi^*_{\varphi,n}(z)$ is locally bounded as it follows from the description of its roots given by Theorem 49. Thus $\Phi^*_{\varphi,n}(z)$ is actually polynomial in φ.

Let us set $\Delta(\varphi) := \prod_{i \neq j} (\alpha_i(\varphi) - \alpha_j(\varphi))$ where the α_i are the roots of $\Phi^*_{\varphi,n}$ counted with multiplicity. This is a well defined function which vanishes exactly when $\Phi^*_{\varphi,n}$ has a multiple root. This function is holomorphic outside its zero set and therefore everywhere by Rado's theorem. Then $\{\Delta = 0\}$ is an analytic subset of $\mathbf{C}^{d-1}$ which is proper since $\varphi_0 := z^d \notin \{\Delta = 0\}$.

Let Y_n be the projection of $\operatorname{Per}_n \cap \{(\varphi^n)'(z) = 1\}$ onto $\mathbf{C}^{d-1}$. By Remmert mapping theorem Y_n is an analytic subset of $\mathbf{C}^{d-1}$. Using φ_0 again one sees that $Y_n \neq \mathbf{C}^{d-1}$. Since $(\varphi^n)'(z) = 1$ when z belongs to a cycle of φ whose period strictly divides n, one may take $X_n := \{\Delta = 0\} \cup Y_n$. □

Let $Z(\Phi^*_{\varphi,n})$ be the set of roots of $\Phi^*_{\varphi,n}$ taken with multiplicity. If $z \in Z(\Phi^*_{\varphi,n})$ has exact period m with $n = mr$, we denote by $w_n(z)$ the r-th power of the multiplier of z (that is $(\varphi^n)'(z)$). As Theorem 49 tells us:

a point z is periodic of exact period n and $w_n(z) \neq 1$ if and only if

$$z \in Z(\Phi^*_{\varphi,n}) \text{ and } w_n(z) \neq 1.$$

Let us now consider the sets

$$\Lambda^*_n(\varphi) := \{w_n(z); \ z \in Z(\Phi^*_{\varphi,n})\}$$

and let us denote by $\sigma_i^{*(n)}(\varphi)$, $1 \leq i \leq v_d(n)$, the associated symmetric functions. The symmetric functions $\sigma_i^{*(n)}$ are globally defined and continuous on $\mathbf{C}^{d-1}$ and, according to Proposition 50, are holomorphic outside X_n. These functions are therefore holomorphic on $\mathbf{C}^{d-1}$. We set

$$q_n(\varphi, w) := \prod_{i=0}^{v_d(n)} \sigma_i^{*(n)}(\varphi)(-w)^{v_d(n)-i}.$$

By construction $q_n(\varphi, w) = 0$ if and only if $w \in \Lambda_n^*(\varphi)$ and q_n is holomorphic in (φ, w) and polynomial in w. As Proposition 50 shows, the elements of $Z(\Phi_{\varphi,n}^*)$ are cycles of exact period n and therefore each element of $\Lambda_n^*(\varphi)$ is repeated n times when $\varphi \notin X_n$. This means that there exists a polynomial $p_n(\varphi, \cdot)$ such that $q_n(\varphi, \cdot) = (p_n(\varphi, \cdot))^n$ when $\varphi \notin X_n$. As $p_n(\varphi, w)$ is holomorphic where it does not vanish, one sees that p_n extends holomorphically to all $\mathbf{C}^{d-1} \times \mathbf{C}$. In other words, p_n may be defined by

$$(p_n(\varphi, w))^n := q_n(\varphi, w) = \prod_{i=0}^{\nu_d(n)} \sigma_i^{*(n)}(\varphi)(-w)^{\nu_d(n)-i}.$$

The degree $N_d(n)$ of $p_n(\lambda, \cdot)$ is equal to $\frac{1}{n}\nu_d(n) = \frac{1}{n}\sum_{k|n}\mu(\frac{n}{k})d^k$. In particular $N_d(n) \sim \frac{d^n}{n}$. □

3.3.2 The Construction of Dynatomic Polynomials

We aim here to prove Theorem 49. For this purpose, let us recall that the Möbius function $\mu : \mathbf{N}^* \to \{-1, 0, 1\}$ enjoys the following fundamental property:

$$\sum_{k|n} \mu\left(\frac{n}{k}\right) = 0 \text{ for any } n \in \mathbf{N}^*. \tag{10}$$

Let us also adopt a few more notations. The valuation of $\Phi_{\varphi,n}$ (resp. $\Phi_{\varphi,n}^*$) at some point z will be denoted $a_z(\varphi, n)$ (resp. $a_z^*(\varphi, n)$). The set of m-periodic points of φ will be denoted $Per(\varphi, m)$.

The following Lemma summarizes elementary facts.

Lemma 51. *Let ψ be a polynomial and $z \in Per(\psi, 1)$. Let $\lambda := \psi'(z)$, then for $q \geq 2$ one has:*

(i) $\lambda^q \neq 1 \Rightarrow a_z(\psi, q) = a_z(\psi, 1) = 1$
(ii) $\lambda \neq 1$ and $\lambda^q = 1 \Rightarrow a_z(\psi, q) > a_z(\psi, 1) = 1$
(iii) $\lambda = 1 \Rightarrow a_z(\psi, q) = a_z(\psi, 1) \geq 2$.

Proof. We may assume $z = 0$ and set $\psi = \lambda X + \alpha X^e + o(X^e)$. Then $\Phi_{\psi,q}$ is equal to $(\lambda^q - 1)X + o(X)$ and to $q\alpha X^e + o(X^e)$ if $\lambda = 1$. The assertions (i) to (iii) then follow immediately. □

We have to compute $a_z^*(\varphi, n)$ for $z \in Per_n(\varphi, m)$ and $m|n$. We denote by λ the multiplier of z (i.e. $\lambda = (\varphi^m)'(z)$) and set $N := \frac{n}{m}$. When λ is a root of unity we denote by r its order. Clearly, Theorem 49 will be proved if we establish the following three facts:

Bifurcation Currents in Holomorphic Families of Rational Maps

F1: $N = 1 \Rightarrow a_z^*(\varphi, n) > 0$
F2: $N \geq 2$ and $\lambda^N \neq 1$ or $\lambda = 1 \Rightarrow a_z^*(\varphi, n) = 0$
F3: $N \geq 2$, $\lambda^N = 1$ and $\lambda \neq 1 \Rightarrow a_z^*(\varphi, n) \geq 0$ and $a_z^*(\varphi, n) > 0$ iff $r = N$.

Besides definitions, the following computation is based on the obvious facts that $a_z(\varphi, k) = 0$ when $k \neq qm$ for some $q \in \mathbf{N}^*$ and that $\varphi^{qm} = (\varphi^q)^m$:

$$a_z^*(\varphi, n) = \sum_{k|n} \mu\left(\frac{n}{k}\right) a_z(\varphi, k) = \sum_{qm|n} \mu\left(\frac{n}{qm}\right) a_z(\varphi, qm)$$

$$= \sum_{q|N} \mu\left(\frac{N}{q}\right) a_z(\varphi, qm) = \sum_{q|N} \mu\left(\frac{N}{q}\right) a_z(\varphi^m, q).$$

We now proceed Fact by Fact, always starting with the above identity.

(F1) $a_z^*(\varphi, n) = \mu(1) a_z(\varphi^n, 1) > 0$.

(F2) Since $\lambda^q \neq 1$ when $q|N$ although $\lambda = 1$, the assertions (i) and (iii) of Lemma 51 tell us that $a_z(\varphi^m, q) = a_z(\varphi^m, 1)$. Then $a_z^*(\varphi, n) = \left(\sum_{q|N} \mu\left(\frac{N}{q}\right)\right) a_z(\varphi^m, 1)$ which, according to (10), equals 0.

(F3) When r is not dividing q then $\lambda^q \neq 1$ and, by the assertion (i) of Lemma 51 we have $a_z(\varphi^m, q) = a_z(\varphi^m, 1)$. We may therefore write

$$a_z^*(\varphi, n) = \left(\sum_{q|N} \mu\left(\frac{N}{q}\right)\right) a_z(\varphi^m, 1) + \sum_{q|N, r|q} \mu\left(\frac{N}{q}\right) [a_z(\varphi^m, q) - a_z(\varphi^m, 1)].$$

By (10), the first term in the above expression vanishes and we get

$$a_z^*(\varphi, n) = \sum_{k|\frac{N}{r}} \mu\left(\frac{N/r}{k}\right) [a_z(\varphi^m, rk) - a_z(\varphi^m, 1)]$$

$$= \sum_{k|\frac{N}{r}} \mu\left(\frac{N/r}{k}\right) [a_z(\varphi^{mr}, k) - a_z(\varphi^m, 1)]$$

$$= a_z^*(\varphi^{mr}, \frac{N}{r}) - a_z(\varphi^m, 1) \sum_{k|\frac{N}{r}} \mu\left(\frac{N/r}{k}\right).$$

We finally consider two subcases.

If $N \neq r$ then, by (10), $\sum_{k|\frac{N}{r}} \mu\left(\frac{N/r}{k}\right) = 0$ and thus $a_z^*(\varphi, n) = a_z^*(\varphi^{mr}, \frac{N}{r})$. Since z is a fixed point of φ^{mr} whose multiplier equals $\lambda^r = 1$, Fact F2 shows that $a_z^*(\varphi^{mr}, \frac{N}{r}) = 0$.

If $N = r$ we get

$$a_z^*(\varphi, n) = a_z^*(\varphi^{mr}, 1) - a_z(\varphi^m, 1) = a_z(\varphi^{mr}, 1) - a_z(\varphi^m, 1) = a_z(\varphi^m, r) - a_z(\varphi^m, 1).$$

Since z is a fixed point of φ^m whose multiplier λ satisfies $\lambda^r = \lambda^N = 1$ and $\lambda \neq 1$, the assertion (ii) of Lemma 51 shows that this quantity is strictly positive. □

Remark 52. It follows easily from the above construction that the growth of $p_n(\lambda, w)$ is polynomial in λ when $\lambda \in \mathbf{C}^{d-1}$. This shows that $p_n(\lambda, w)$ is actually a polynomial function on $\mathbf{C}^{d-1} \times \mathbf{C}$.

3.3.3 On the Geometry of the $\text{Per}_n(w)$ in Particular Families

In this subsection we will describe some geometric properties of the hypersurfaces $\text{Per}_n(w)$ in specific families of rational maps. We are mainly interested in the behaviour at infinity in the projective compactifications of the polynomial families and the moduli space of degree two rational maps.

We start with the polynomial family of degree d polynomials (see Sect. 3.1.2). The following result is a consequence of our investigations of Sect. 3.2.2. We recall that, according to Remark 52, the sets $\text{Per}_m(\eta)$ may be seen as algebraic subsets of the projective space $\mathbf{P}^{d-1}$.

Proposition 53. *If $1 \leq k \leq d - 1$, $m_1 < m_2 < \cdots < m_k$ and $\sup_{1 \leq i \leq k} |\eta_i| < 1$ then $\text{Per}_{m_1}(\eta_1) \cap \cdots \cap \text{Per}_{m_k}(\eta_k)$ is an algebraic subset of codimension k whose intersection with $\mathbf{C}^{d-1}$ is not empty.*

Proof. By Bezout's theorem, $\text{Per}_{m_1}(\eta_1) \cap \cdots \cap \text{Per}_{m_k}(\eta_k)$ is a non-empty algebraic subset of $\mathbf{P}^{d-1}$ whose dimension is bigger than $(d - 1 - k)$.

Any cycle of attracting basins capture a critical orbit. Therefore, Theorem 43 implies that the intersection of $\mathbf{P}_\infty$ with $\text{Per}_{m_1}(\eta_1) \cap \cdots \cap \text{Per}_{m_k}(\eta_k)$ is contained in some $\Gamma_{i_1} \cap \cdots \cap \Gamma_{i_k}$ since the m_i are mutually distinct and the $|\eta_i|$ strictly smaller than 1. Then, according to Lemma 42, $\mathbf{P}_\infty \cap \text{Per}_{m_1}(\eta_1) \cap \cdots \cap \text{Per}_{m_k}(\eta_k)$ has codimension k in $\mathbf{P}_\infty$. The conclusion now follows from obvious dimension considerations. □

We now consider the space Mod_2. As it has been discussed in Sect. 3.1.3, this space can be identified to $\mathbf{C}^2$ and has a natural projective compactification which is given by

$$Mod_2 \ni \bar{f} \longmapsto [\sigma_1 : \sigma_2 : 1] \in \mathbf{P}^2$$

where $\sigma_1, \sigma_2, \sigma_3$ are the symmetric functions of the three multipliers $\{\mu_1, \mu_2, \mu_3\}$ of the fixed points (we recall that $\sigma_3 = \sigma_1 - 2$, see (9)).

The line at infinity $\mathcal{L} = \{[\sigma_1 : \sigma_2 : 0]; (\sigma_1, \sigma_2) \in \mathbf{C}^2 \setminus \{0\}\}$ enjoys an interesting dynamical parametrization (see Proposition 38).

Using this identification, the defining functions $p_n(\lambda, w)$ of $Per_n(w)$ are polynomials on $\mathbf{C}^2 \times \mathbf{C}$. Any $Per_n(w)$ may be seen as a curve in $\mathbf{P}^2$. Understanding the behaviour of these curves at infinity is crucial to investigate the structure of the bifurcation locus in Mod_2. The following facts have been proved by Milnor [48].

Proposition 54. *(1) For all $w \in \mathbf{C}$ the curve $Per_1(w)$ is actually a line whose equation in $\mathbf{C}^2$ is $(w^2 + 1)\lambda_1 - w\lambda_2 - (w^3 + 2) = 0$ and whose point at infinity is $[w : w^2 + 1 : 0]$. In particular, $Per_1(0) = \{\lambda_1 = 2\}$ is the line of quadratic polynomials, its point at infinity is $[0 : 1 : 0]$.*
(2) For $n > 1$ and $w \in \mathbf{C}$ the points at infinity of the curves $Per_n(w)$ are of the form $[u : u^2 + 1 : 0]$ with $u^q = 1$ and $q \leq n$.

Proof. (1) This is a straightforward computation using the fact that the multipliers of the fixed points $\{\mu_1, \mu_2, \mu_3\}$ are the roots of the polynomial $X^3 - \sigma_1 X^2 + \sigma_2 X - (\sigma_1 - 2)$.
(2) Let $P \in Per_n(w) \cap \mathcal{L}$. Let $\bar{\sigma} := [\sigma_1 : \sigma_2 : 1] \in Per_n(w)$ such that $\bar{\sigma} \to P$. We denote by μ_j the multipliers of the fixed points of $\bar{\sigma}$. According to Proposition 38 we may write $P = [1 : \mu + \frac{1}{\mu} : 0]$ where μ is a limit of $\frac{1}{\mu_1} + \frac{1}{\mu_2} + \frac{1}{\mu_3}$.

We first consider the case where $\mu \neq 0, \infty$. Then we may assume that $\mu_3 \to \infty$, $\mu_2 \to \mu$ and $\mu_1 \to \frac{1}{\mu}$ (recall that $\mu_1\mu_2$ must tend to 1 as it follows from the identities of Lemma 35). By Remark 37 we may use the normal form

$$f = f_{\bar{\sigma}} := z\frac{z + \mu_i}{\mu_j z + 1}.$$

Let us set $\delta := 1 - \mu_1\mu_2$ and $l(z) := \mu_2 z + 1$. An easy computation yields

$$\mu_2 \frac{f(z)}{z} = 1 - \frac{\delta}{l(z)} \tag{11}$$

and

$$\mu_2 f'(z) = 1 - \frac{\delta}{l(z)^2}. \tag{12}$$

For $|\delta| < 1$, we set:

$$D := \{|l| < |\delta|^{\frac{2}{3}}\}, \ A := \{|\delta|^{\frac{2}{3}} \leq |l| < |\delta|^{\frac{1}{3}}\}, \ C := \{|l| \geq |\delta|^{\frac{1}{3}}\}.$$

We thus have a decomposition of the Riemann sphere into three distinct regions:

$$\mathbf{P}^1 = D \cup A \cup C.$$

Observe that the disc $D \cup A$ degenerates to the point $\{\frac{-1}{\mu}\}$ as $\bar{\sigma} \to P$.

Using the identities (11) and (12) one gets the following estimates

$$\frac{f(z)}{z} = \frac{1}{\mu_2} + O\left(|\delta|^{\frac{1}{3}}\right) \text{ on } A \cup C \tag{13}$$

$$|f'(z)| \geq |\delta|^{\frac{-1}{3}} \text{ on } D \tag{14}$$

$$f'(z) = \frac{1}{\mu_2} + O\left(|\delta|^{\frac{1}{3}}\right) \text{ on } C. \tag{15}$$

Let $O := \{z_0, f(z_0), \cdots, f^{n-1}(z_0)\}$ be the orbit of some n-periodic point z_o of $f = f_{\bar{\sigma}}$ whose multiplier equals w (recall that $\bar{\sigma} \in Per_n(w)$). We assume here that $\bar{\sigma}$ is very close to P. Then it is impossible that $O \subset D \cup C$ with $O \cap D \neq \emptyset$ since otherwise, by the estimates (13) and (15), we would have $|w| = \prod_0^{n-1} |f'(f^j(z_0))| \geq C|\delta|^{-\frac{1}{3}}$. Thus, either $O \subset A \cup C$ or $O \cap D \neq \emptyset$ and $O \cap A \neq \emptyset$.

If $O \subset A \cup C$ then, using the estimate (14), we get

$$1 = \frac{f^n(z_0)}{z_0} = \prod_0^{n-1} \frac{f^{j+1}(z_0)}{f^j(z_0)} = \frac{1}{\mu_2^n}\left(1 + O\left(|\delta|^{\frac{1}{3}}\right)\right)^n$$

which implies that $\mu^n = 1$ since $\mu_2 \to \mu$ and $\delta \to 0$ when $\bar{\sigma} \to P$.

If $O \cap D \neq \emptyset$ and $0 \cap A \neq \emptyset$, we may assume that $z_0 \in A$, $f^q(z_0) \in D$ and $\{z_0, \cdots, f^{q-1}(z_0)\} \subset A \cup C$. Then

$$\frac{f^q(z_0)}{z_0} = \prod_0^{q-1} \frac{f^{j+1}(z_0)}{f^j(z_0)} = \frac{1}{\mu_2^q}\left(1 + O\left(|\delta|^{\frac{1}{3}}\right)\right)^q$$

which implies that $\mu^q = 1$ since $\mu_2 \to \mu$ and both z_0 and $f^q(z_0)$ belong to $A \cup D$ which tends to $\{\frac{-1}{\mu}\}$ when $\bar{\sigma} \to P$.

When $\mu = 0$ or $\mu = \infty$ the proof is similar but slightly more subtle. We refer the reader to the paper [48] for details. $\square$

The following Proposition, also due to Milnor (see Theorem 4.2 in [48]), implies that the curves $Per_n(w) = \{p_n(\cdot, w) = 0\}$ have no multiplicity.

Proposition 55. *Let* $N_2(n) := Card\,(Per_n(0) \cap Per_1(0))$ *be the number of hyperbolic components of period n in the Mandelbrot set. Then* $N_2(n) = \frac{\nu_2(n)}{2}$ *where $\nu_2(n)$ is defined inductively by* $\nu_2(1) = 2$ *and* $2^n = \sum_{k|n} \nu_2(k)$. *Moreover, for any $w \in \Delta$ and any $\eta \in \Delta$ we have* $Deg\, p_n(\cdot, w) = N_2(n) = Card\,(Per_n(w) \cap Per_1(\eta))$.

Epstein [31] has obtained some far advanced generalizations of the above result and, in particular, has proved the boundedness of certain hyperbolic components of Mod_2:

Theorem 56. *Let H be a hyperbolic component of Mod_2 whose elements admit two distinct attracting cycles. If neither attractor is a fixed point then H is relatively compact in Mod_2.*

4 The Bifurcation Current

In this section, we consider an arbitrary holomorphic family $(f_\lambda)_{\lambda \in M}$ of degree d rational maps with marked critical points (see (33)). Our first aim is to describe necessary and sufficient conditions for the Julia set $\mathcal{J}_\lambda$ of f_λ to move holomorphically with the parameter λ. The parameters around which such a motion does exist are called stable. We will show that the set of stable parameters is dense in the parameter space, to this purpose we will relate the stability of $\mathcal{J}_\lambda$ with the stability of the dynamics on the critical set $\mathcal{C}_\lambda$ of f_λ; this is the essence of the Mañé–Sad–Sullivan theory. We will then exhibit a closed positive $(1, 1)$-current on the parameter space whose support is precisely the complement of the stability locus; this is the bifurcation current.

4.1 Stability Versus Bifurcation

4.1.1 Motion of Repelling Cycles and Julia Sets

As Julia sets coincide with the closure of the sets of repelling cycles (see Theorems 4 and 17), it is natural to investigate how $\mathcal{J}_\lambda$ varies with λ through the parametrizations of such cycles.

Assume that f_{λ_0} has a repelling n-cycle $\{z_0, f_{\lambda_0}(z_0), \cdots, f_{\lambda_0}^{n-1}(z_0)\}$. Then, by the implicit function theorem applied to the equation $f_\lambda^n(z) - z = 0$ at (λ_0, z_0), there exists a neighbourhood U_0 of λ_0 in M and a holomorphic map

$$U_0 \ni \lambda \mapsto h_\lambda(z_0) \in \mathbf{P}^1$$

such that $h_{\lambda_0}(z_0) = z_0$ and $h_\lambda(z_0)$ is a n-periodic repelling point of f_λ for all $\lambda \in U_0$. Moreover, $h_\lambda(\cdot)$ can be extended to the full cycle so that

$$f_\lambda \circ h_\lambda = h_\lambda \circ f_{\lambda_0}.$$

This lead us to say that every repelling n-cycle of f_{λ_0} *moves holomorphically* on some neighbourhood of λ_0 and to set the following formal definition.

Definition 57. Let us denote by $\mathcal{R}_{\lambda,n}$ the set of repelling n-cycles of f_λ. Let Ω be a neighbourhood of λ_0 in M. One says that $\mathcal{R}_{\lambda_0,n}$ moves holomorphically on Ω if there exists a map

$$h : \Omega \times \mathcal{R}_{\lambda_0,n} \ni (\lambda, z) \mapsto h_\lambda(z) \in \mathcal{R}_{\lambda,n}$$

which depends holomorphically on λ and satisfies $h_{\lambda_0} = Id$, $f_\lambda \circ h_\lambda = h_\lambda \circ f_{\lambda_0}$.

More generally, the holomorphic motion of an arbitrary subset of the Riemann sphere is defined in the following way.

Definition 58. Let E be subset of the Riemann sphere and Ω be a complex manifold. Let $\lambda_0 \in \Omega$. An holomorphic motion of E over Ω and centered at λ_0 is a map

$$h : \Omega \times E \ni (\lambda, z) \mapsto h_\lambda(z) \in \hat{\mathbf{C}}$$

which satisfies the following properties:

(i) $h_{\lambda_0} = Id|_E$
(ii) $E \ni z \mapsto h_\lambda(z)$ is one-to-one for every $\lambda \in \Omega$
(iii) $\Omega \ni \lambda \mapsto h_\lambda(z)$ is holomorphic for every $z \in E$.

The interest of holomorphic motions relies on the fact that any holomorphic motion of a set E extends to the closure of E. This is a quite simple consequence of Picard–Montel theorem.

Lemma 59 (basic λ-lemma). *Let $E \subset \hat{\mathbf{C}}$ be a subset of the Riemann sphere and $\sigma : E \times \Omega \ni (z, \lambda) \mapsto \sigma(z, \lambda) \in \hat{\mathbf{C}}$ be a holomorphic motion of E over Ω. Then σ extends to a holomorphic motion $\tilde{\sigma}$ of $\overline{E}$ over Ω. Moreover $\tilde{\sigma}$ is continuous on $\overline{E} \times \Omega$.*

As $\mathcal{J}_\lambda$ is the closure of the set of repelling cycles of f_λ, this Lemma implies that the Julia set $\mathcal{J}_{\lambda_0}$ moves holomorphically over a neighbourhood V_{λ_0} of λ_0 in M as soon as *all* repelling cycles of f_{λ_0} move holomorphically on V_{λ_0}. Moreover, the holomorphic motion obtained in this way clearly conjugates the dynamics: $h_\lambda(\mathcal{J}_{\lambda_0}) = \mathcal{J}_\lambda$ and $f_\lambda \circ h_\lambda = h_\lambda \circ f_{\lambda_0}$ on $\mathcal{J}_{\lambda_0}$.

Our observations may now be gathered in the following basic Lemma.

Lemma 60. *If there exists a neighbourhood Ω of λ_0 in the parameter space M such that $\mathcal{R}_{\lambda_0,n}$ moves holomorphically on Ω for all sufficiently big n, then there exists a holomorphic motion $h_\lambda : \mathcal{J}_{\lambda_0} \to \mathcal{J}_\lambda$ which conjugates the dynamics.*

Let us mention here that there exists a much stronger version of the λ-lemma, which is due to Slodkowski and shows that the holomorphic motion actually extends as a quasi-conformal transformation of the full Riemann sphere (see Theorem 110). In particular, under the assumption of the above Lemma, f_λ and f_{λ_0} are quasi-conformally conjugated when $\lambda \in \Omega$.

We may now define the set of stable parameters and its complement; the bifurcation locus.

Definition 61. Let $(f_\lambda)_{\lambda \in M}$ be a holomorphic family of degree d rational maps.

The *stable set* $\mathcal{S}$ is the set of parameters $\lambda_0 \in M$ for which there exists a neighbourhood Ω of λ_0 and a holomorphic motion h_λ of $\mathcal{J}_{\lambda_0}$ over Ω, centered at λ_0, and such that $f_\lambda \circ h_\lambda = h_\lambda \circ f_{\lambda_0}$ on $\mathcal{J}_{\lambda_0}$.

The *bifurcation locus* B_{if} is the complement $M \setminus \mathcal{S}$.

By definition, $\mathcal{S}$ is an open subset of M but it is however not yet clear that it is not empty. We shall actually show that $\mathcal{S}$ is dense in M. To this purpose we will prove that the stability is characterized by the stability of the critical orbits. The next subsection will be devoted to this simple but remarkable fact.

4.1.2 Stability of Critical Orbits

Let us start by explaining why bifurcations are related with the instability of critical orbits.

As Lemma 60 shows, a parameter λ_0 belongs to the bifurcation locus if for any neighbourhood Ω of λ_0 in the parameter space M there exists $n_0 \geq 0$ for which $\mathcal{R}_{\lambda_0, n_0}$ does not move holomorphically on Ω.

It is not very difficult to see that this forces one of the repelling n_0-cycles of f_{λ_0}, say R_{λ_0}, to become neutral and then attracting for a certain value $\lambda_1 \in \Omega$. Now comes the crucial point. A classical result asserts that the basin of attraction of any attracting cycle of a rational map contains a critical point (see [9, Théorème II.5]). Thus, one of the critical orbits $f_\lambda^k(c_i(\lambda))$ is uniformly converging to R_λ on a neighbourhood of λ_1. Then the sequence $f_\lambda^k(c_i(\lambda))$ cannot be normal on Ω since otherwise, by Hurwitz lemma, it should converge uniformly to R_λ which is repelling for λ close to λ_0. This arguments show that the bifurcation locus is contained in the set of parameters around which the post-critical set does not move continuously.

This is an extremely important observation because it will allow us to detect bifurcations by considering only the critical orbits. It leads to the following definitions.

Definition 62. Let $(f_\lambda)_{\lambda \in M}$ be a holomorphic family of degree d rational maps. A marked critical point $c(\lambda)$ is said to be *passive* at λ_0 if the sequence $\left(f_\lambda^n(c(\lambda))\right)_n$ is normal on some neighbourhood of λ_0. If $c(\lambda)$ is not passive at λ_0 one says that it is *active*. The *activity locus* of $c(\lambda)$ is the set of parameters at which $c(\lambda)$ is active.

The key result may now be given.

Lemma 63. *In a holomorphic family with marked critical points the bifurcation locus coincides with the union of the activity loci of the critical points.*

Proof. According to our previous arguments, the bifurcation locus is contained in the union of the activity loci. It remains to show that, for any marked critical point $c(\lambda)$, the sequence $\left(f_\lambda^n(c(\lambda))\right)_n$ is normal on $\mathcal{S}$. Assume that $\lambda_0 \in \mathcal{S}$. As $\mathcal{J}_{\lambda_0}$ is a perfect compact set, we may find three distinct points a_1, a_2, a_3 on $\mathcal{J}_{\lambda_0}$ which are avoided by the orbit of $c(\lambda_0)$. Since the holomorphic motion h_λ of $\mathcal{J}_{\lambda_0}$ conjugates the dynamics, the orbit of $c(\lambda)$ avoids $\{h_\lambda(a_j);\ 1 \leq j \leq 3\}$ for all λ in a small neighbourhood of λ_0. The conclusion then follows from Picard–Montel's Theorem. □

The following Lemma is quite useful.

Lemma 64. *If λ_0 belongs to the activity locus of some marked critical point $c(\lambda)$ then there exists a sequence of parameters $\lambda_k \to \lambda_0$ such that $c(\lambda_k)$ belongs to some super-attracting cycle of f_{λ_k} or is strictly preperiodic to some repelling cycle of f_{λ_k}.*

Proof. To simplify we will assume that all critical points are marked. Since $c(\lambda)$ is active at λ_0 the critical point $c(\lambda_0)$ cannot be periodic. In particular, its pre-history cannot stay in the critical set. Then, after maybe replacing $c(\lambda_0)$ by another critical point, we may assume that there exist holomorphic maps $c_{-2}(\lambda)$, $c_{-1}(\lambda)$ near λ_0 such that $f_\lambda(c_{-2}(\lambda)) = c_{-1}(\lambda)$, $f_\lambda(c_{-1}(\lambda)) = c(\lambda)$ and Card $\{c_{-2}(\lambda), c_{-1}(\lambda), c(\lambda)\} = 3$. Now, by Picard–Montel Theorem, the sequence $\left(f_\lambda^n(c(\lambda))\right)_n$ cannot avoid the set $\{c_{-2}(\lambda), c_{-1}(\lambda), c(\lambda)\}$ on any neighbourhood of λ_0.

A similar argument, using Picard–Montel Theorem and a repelling cycle of period $n_0 \geq 3$, shows that $c(\lambda)$ becomes strictly preperiodic for λ arbitrarily close to λ_0. □

We are now ready to state and prove Mañé–Sad–Sullivan theorem.

Theorem 65. *Let $(f_\lambda)_{\lambda \in M}$ be a holomorphic family of degree d rational maps with marked critical points $\{c_1(\lambda), \cdots, c_{2d-2}(\lambda)\}$.*

A parameter λ_0 is stable if one of the following equivalent conditions is satisfied.

(1) $\mathcal{J}_{\lambda_0}$ moves holomorphically around λ_0 (see Definition 61)
(2) the critical points are passive at λ_0 (see Definition 62)
(3) f_λ has no unpersistent neutral cycles for λ sufficiently close to λ_0.

The set $\mathcal{S}$ of stable parameters is dense in M.

Proof. The equivalence between (1) and (2) is given by Lemma 63. Similar arguments may allow to show that (3) is an equivalent statement (we will give an alternative proof of that later). It remains to show that $\mathcal{S}$ is dense in M. According to Lemma 64 we may perturb λ_0 and assume that f_{λ_0} has a superattracting cycle of period bigger than 3 which persists, as an attracting cycle, for λ close enough to λ_0. If λ_0 is still active, Picard–Montel's theorem shows that a new perturbation guarantees that a critical point falls in the attracting cycle and, therefore, becomes passive. Since the number of critical points is finite, we may make all critical points passive after a finite number of perturbations. □

Example 66. In the quadratic polynomial family, the bifurcation locus is the boundary of the connectivity locus (or the Mandelbrot set). Indeed, it follows immediately from the Definition 40 of the Mandelbrot set that its boundary is the activity locus.

Example 67. The situation is more complicated in the family $(P_{c,a})_{(c,a) \in \mathbb{C}^{d-1}}$ of degree d polynomials when $d \geq 3$ (see Sect. 3.1.2). Indeed, Theorem 43 shows that the bifurcation (i.e. activity) locus is not bounded since it coincides with the boundary of $\cup_{0\leq i \leq d-2} \mathcal{B}_i$ (where $\mathcal{B}_i$ is the set of parameters for which the orbit of the critical point c_i is bounded) while the connectedness locus $\cap_{0\leq i \leq d-2} \mathcal{B}_i$ is bounded.

Although we shall not use it, we end this section by quoting a very interesting classification of the activity situations which is due to Dujardin and Favre (see [29, Theorem 4]).

Theorem 68. *Let $(f_\lambda)_{\lambda \in M}$ be a holomorphic family of degree d rational maps with a marked critical point $c(\lambda)$. If c is passive on some connected open subset U of M then exactly one of the following cases holds:*

(1) c is never preperiodic in U and the closure of its orbits move holomorphically on U.

(2) c is persistently preperiodic on U.

(3) The set of parameters for which c is preperiodic is a closed subvariety in U. Moreover, either there exists a persistently attracting cycle attracting c throughout U, or c lies in the interior of a linearization domain associated to a persistent irrationally neutral periodic point.

It is worth emphasize that the proof of that result relies on purely local, and subtle, arguments.

4.1.3 Some Remarkable Parameters

- As we already mentioned, the basin of any attracting cycle of a rational map contains a critical point. As a consequence, any rational map of degree d has at most $2d - 2$ attracting cycles. Then, using perturbation arguments, Fatou and Julia proved that the number of non-repelling cycles is bounded by $6d - 6$. The precise bound has been obtained by Shishikura [55] using quasiconformal surgery, Epstein has given a more algebraic proof based on quadratic differentials (see [30]).

Theorem 69. *A rational map of degree d has at most $2d - 2$ non-repelling cycles.*

In particular, any degree d rational map cannot have more than $2d - 2$ neutral cycles. Shishikura has also shown that the bound $2d - 2$ is sharp (we will give another proof, using bifurcation currents, in Sect. 6.2.2). Let us mention that the Julia set of a degree d map having $2d - 2$ Cremer cycles coincides with the full Riemann sphere. These results motivate the following definition.

Definition 70. *The set S_{hi} of degree d Shishikura rational maps is defined by*

$$S_{\text{hi}} = \{f \in Rat_d \ / \ f \text{ has } 2d - 2 \text{ neutral cycles}\}.$$

In a holomorphic family $(f_\lambda)_{\lambda \in M}$ we shall denote by $S_{\text{hi}}(M)$ the set of parameters λ for which f_λ is Shishikura.

According to Theorem 65, one has $S_{\text{hi}} \subset B_{\text{if}}$. In the last section, we will obtain some informations on the geometry of the set S_{hi} and, in particular, reprove that it is not empty.

- Any repelling cycle of $f \in Rat_d$ is an invariant (compact) set on which f is uniformly expanding. Some rational map may be uniformly expanding on much bigger compact sets. Such sets are called *hyperbolic* and are necessarily contained in the Julia set. A rational map which is uniformly expanding on its Julia set is said to be hyperbolic. Let us give a precise definition.

Definition 71. Let f be a rational map. A compact set K of the Riemann sphere is said to be *hyperbolic* for f if it is invariant ($f(K) \subset K$) and if there exists $C > 0$ and $M > 1$ such that

$$|(f^n)'(z)|_s \geq CM^n \; ; \; \forall z \in K, \; \forall n \geq 0.$$

One says that f is uniformly expanding on K (recall that $|\;|_s$ is the spherical metric).

It may happen that a critical orbit is captured by some hyperbolic set and, in particular, by a repelling cycle. Such rational maps play a very important role since they allow to define transfer maps carrying informations from the dynamical plane to the parameter space. A particular attention will be devoted to those having all critical orbits captured by a hyperbolic set.

Definition 72. The set M_{is} of degree d *Misiurewicz rational maps* is defined by

$$M_{\text{is}} = \{f \in Rat_d \;/\; \text{all critical orbits of } f \text{ are captured by a compact hyperbolic set}\}.$$

When the hyperbolic set is an union of repelling cycles the map is said to be strongly Misiurewicz and the set of such maps is denoted M_{iss}. In a holomorphic family $(f_\lambda)_{\lambda \in M}$ we shall denote by $M_{\text{is}}(M)$ (resp. $M_{\text{iss}}(M)$) the set of parameters λ for which f_λ is Misiurewicz (resp. strongly Misiurewicz).

Within a holomorphic family $(f_\lambda)_{\lambda \in M}$, one may show that a critical point whose orbit is captured by a hyperbolic set and leaves this set under a small perturbation is active. In particular, we have the following inclusion: $M_{\text{is}} \subset B_{\text{if}}$. To prove this, one first has to construct a holomorphic motion of the hyperbolic set and then linearize along its orbits (see [33]). When the hyperbolic set is a cycle, the motion is given by the implicit function theorem and the linearizability is a well known fact.

Lemma 73. *Let $(f_\lambda)_{\lambda \in M}$ be a holomorphic family of degree d rational maps with a marked critical point $c(\lambda)$. Assume that f_λ has a repelling n-cycle $\mathcal{R}(\lambda) := \{z_\lambda, f_\lambda(z_\lambda), \cdots, f_\lambda^{n-1}(z_\lambda)\}$ for $\lambda \in U$. If $f_\lambda^k(c(\lambda)) \in \mathcal{R}(\lambda)$ for $\lambda = \lambda_0 \in U$ but not for all $\lambda \in U$ then $c(\lambda)$ is active at λ_0.*

Proof. We may assume that $n = 1$ which means that z_λ is fixed by f_λ. Shrinking U and linearizing we get a a family of local biholomorphisms ϕ_λ which depends holomorphically on λ and such that $\phi_\lambda(0) = z_\lambda$ and $f_\lambda \circ \phi_\lambda(u) = \phi_\lambda(m_\lambda u)$ (see [9, Théorème II.1 and Remarque II.2]). As z_λ is repelling, one has $|m_\lambda| > 1$. Let us set $u(\lambda) := \phi_\lambda^{-1}(f_\lambda^k(c(\lambda)))$, then $f_\lambda^{p+k}(c_\lambda) = f_\lambda^p(\phi_\lambda(u(\lambda))) = \phi_\lambda((m_\lambda)^p u(\lambda))$ which shows that $f_\lambda^{p+k}(c_\lambda)$ is not normal at λ_0 since, by assumption, $u(\lambda_0) = 0$ but u does not vanish identically on U. □

- As we already mentioned, a rational map which is uniformly expanding on its Julia set is called hyperbolic. The dynamical study of such maps turns out to be much easier.

Definition 74. The set H_{yp} of degree d *hyperbolic rational maps* is defined by

$$H_{yp} = \{f \in Rat_d \mid f \text{ is uniformly expanding on its Julia set}\}.$$

In a holomorphic family $(f_\lambda)_{\lambda \in M}$ we shall denote by $H_{yp}(M)$ the set of parameters λ for which f_λ is hyperbolic.

There are several characterizations of hyperbolicity. One may show that a rational map f is hyperbolic if and only if its postcritical set does not contaminate its Julia set:

$$f \text{ is hyperbolic} \Leftrightarrow \overline{\cup_{n \geq 0} f^n(\mathcal{C}_f)} \cap \mathcal{J}_f = \emptyset.$$

As a consequence, in any holomorphic family $(f_\lambda)_{\lambda \in M}$, hyperbolic parameters are stable: $H_{yp}(M) \subset \mathcal{S}$. This characterization also implies that a hyperbolic map has only attracting or repelling cycles.

The Fatou's conjecture asserts that the hyperbolic parameters are dense in any holomorphic family, it is an open problem even for quadratic polynomials. According to Mañé–Sad–Sullivan Theorem it can be rephrased as follows

Fatou's conjecture 75. $H_{yp}(M) = \mathcal{S}$ *for any holomorphic family* $(f_\lambda)_{\lambda \in M}$.

Let us also mention that Mañé, Sad and Sullivan have shown that hyperbolic and non-hyperbolic parameters cannot coexist in the same stable component (i.e. connected component of $\mathcal{S}$). A given stable component is therefore either *hyperbolic* (if all parameters in the component are hyperbolic) or *non-hyperbolic* (when the component does not contain any hyperbolic parameter). Fatou's conjecture thus claims that non-hyperbolic components do not exist.

Although the bifurcation locus of the quadratic polynomial family is clearly accumulated by hyperbolic parameters, this is far from being clear in other families and seems to be an interesting question. In the last section we will show that parameters which, in some sense, produce the strongest bifurcations, are accumulated by hyperbolic parameters.

4.2 Potential Theoretic Approach

4.2.1 Lyapunov Exponent and Bifurcations

We wish here to explain why the current $dd^c L(\lambda)$, where $L(\lambda)$ is the Lyapunov exponent of f_λ, is a natural candidate for being a bifurcation current in any holomorphic family $(f_\lambda)_{\lambda \in M}$.

To support this idea we will first use the material discussed in Sects. 2.3.2 and 3.3.1 and relate the Lyapunov exponent with the distribution of hypersurfaces $\operatorname{Per}_n(e^{i\theta})$ in M.

According to the Mañé–Sad–Sullivan Theorem 65, the bifurcation locus is the closure of the union of all such hypersurfaces:

$$B_{\text{if}} = \overline{\cup_n \cup_\theta \operatorname{Per}_n(e^{i\theta})}.$$

One may thus expect that the sequence of currents $\frac{1}{2\pi}\int_0^{2\pi}[\operatorname{Per}_n(e^{i\theta})]$, correctly weighted, converges to some current supported by B_{if}. As we shall see, the approximation formula 30 actually implies that the following convergence occurs in the sense of currents:

$$\frac{d^{-n}}{2\pi}\int_0^{2\pi}[\operatorname{Per}_n(e^{i\theta})]\,d\theta \to T_{\text{bif}}.$$

We will now briefly prove this, more details and a generalization will be provided by Theorem 101.

Let us denote $p_n(\cdot, e^{i\theta})$ the canonical defining functions for the hypersurfaces $\operatorname{Per}_n(e^{i\theta})$ given by Theorem 46. We have to show that the sequence of p.s.h functions

$$L_n(\lambda) := \frac{d^{-n}}{2\pi}\int_0^{2\pi} \ln|p_n(\lambda, e^{i\theta})|\,d\theta.$$

converges to L in $L^1_{loc}(M)$.

Writing

$$p_n(\lambda, e^{i\theta}) =: \prod_{i=1}^{N_d(n)}\left(e^{i\theta} - w_{n,j}(\lambda)\right)$$

and using the fact that $\ln^+|a| = \frac{1}{2\pi}\int_0^{2\pi}\ln|a - e^{i\theta}|\,d\theta$ yields

$$L_n(\lambda) = \frac{1}{2\pi d^n}\int_0^{2\pi}\ln\prod_j|e^{i\theta} - w_{n,j}(\lambda)|d\theta = d^{-n}\sum_j \ln^+|w_{n,j}(\lambda)|. \quad (16)$$

Using the fact that $d^{-n}N_d(n) \sim \frac{1}{n}$ (see Theorem 46) one sees that the sequence L_n is locally bounded from above and it thus remains to see that it converges pointwise to L.

According to Theorem 46, the set $\{w_{n,j}(\lambda) \ / \ w_{n,j}(\lambda) \neq 1\}$ coincides with the set of multipliers of cycles of exact period n (counted with multiplicity) from which the cycles of multiplier 1 are deleted.

Since f_λ has a finite number of non-repelling cycles (Fatou's theorem), all cycles appearing in (16) are repelling for n big enough. The conclusion then follows from Theorem 30. □

As a second evidence, we will relate $dd^c L$ with the instability of the critical dynamics. Mañé–Sad–Sullivan Theorem 65 also revealed that the bifurcations are due to the activity of one (or more) critical point. It is not very difficult to check that the activity locus of a marked critical point $c(\lambda)$ is exactly supported by the current $dd^c g_\lambda (c(\lambda))$ where g_λ is the Green function of f_λ. A bifurcation current could therefore also be defined by

$$dd^c \sum_j g_\lambda (c(\lambda))$$

where the sum is taken over the critical points.

Remark 76. It seems that the quantity $g_\lambda (c(\lambda))$ appeared for the first time in [57] where it was used to study the bifurcations in the quadratic polynomial family.

It turns out that the currents $dd^c \sum_j g_\lambda (c(\lambda))$ and $dd^c L$ do coincide. We will justify this for a polynomial family and, to this purpose, will establish a fundamental formula which relates the Lyapunov exponent (see Definition 24) of a polynomial with its critical points.

Theorem 77 (Przytycki's formula). *Let P be a unitary degree d polynomial, $L(P)$ its Lyapunov exponent and $g_{C,P}$ its Green function. Then*

$$L(P) = \ln d + \sum g_{C,P}(c)$$

where the sum is taken over the critical points of P counted with multiplicity.

Proof. Let us write $c_1, c_2, \cdots, c_{d-1}$ the critical points of P. Then

$$L(P) = \int_C \ln |P'| \, \mu_P = \int_C \ln |d \prod_{j=1}^{d-1} (z - c_j)| \, \mu_P = \ln d + \sum_{j=1}^{d-1} \int_C \ln |z - c_j| \, \mu_P.$$

Now, since $\mu_P = dd^c g_{C,P}$, the formula immediately follows by an integration by parts. □

Remark 78. As the Green function $g_{C,P}$ of a polynomial P is positive (see Proposition 12) the above formula implies that $L(P) \geq \ln d$.

Most of the remaining of this section will be devoted to extend these ideas to families of rational maps.

4.2.2 Przytycki's Generalized Formula

The proof of Przytycki's formula relies on a simple integration by parts. It is more delicate to perform such an integration by part in the case of a rational map f. To this purpose we will work in the line bundle $\mathcal{O}_{\mathbf{P}^1}(D)$ for $D := 2(d-1)$ which we endow with two metrics, the flat one $\|[z,x]\|_0$ and the Green metric $\|[z,x]\|_{G_F}$ whose potential is the Green function G_F of some lift F of f (see Sect. 2.3.1).

In this general situation the integration by part yields the following formula. We refer to Definition 7 for the definitions and notations related to the Green functions.

Proposition 79. *Let f be a rational map of degree $d \geq 2$ and F be one of its lifts. Let $D := 2(d-1)$ and Jac_F be the holomorphic section of $\mathcal{O}_{\mathbf{P}^1}(D)$ induced by $\det F'$. Let g_F be the Green function of f on $\mathbf{P}^1$ and μ_f be the Green measure of f. Then*

$$L(f) + \ln d = \int_{\mathbf{P}^1} g_F[\mathcal{C}_f] - 2(d-1) \int_{\mathbf{P}^1} g_F(\mu_f + \omega) + \int_{\mathbf{P}^1} \ln \|Jac_F\|_0\, \omega.$$

Proof. Let us recall that $\|\cdot\|_G = e^{-Dg_F}\|\cdot\|_0$. According to Lemma 27 we have:

$$L(f) + \ln d = \int_{\mathbf{P}^1} \ln \|Jac_F\|_{G_F}\, \mu_f = \int_{\mathbf{P}^1} \ln \|Jac_F\|_{G_F}\, dd^c g_F + \int_{\mathbf{P}^1} \ln \|Jac_F\|_{G_F}\, \omega$$

which, after integrating by parts, yields

$$L(f) + \ln d = \int_{\mathbf{P}^1} g_F\, (dd^c \ln \|Jac_F\|_{G_F}) + \int_{\mathbf{P}^1} \ln \|Jac_F\|_{G_F}\, \omega$$

and by Poincaré–Lelong formula:

$$L(f) + \ln d = \int_{\mathbf{P}^1} g_F\, ([\mathcal{C}_f] - D\mu_f) + \int_{\mathbf{P}^1} (\ln \|Jac_F\|_0 - Dg_F)\, \omega. \tag{17}$$

$\square$

When working with a holomorphic family of polynomials $(P_\lambda)_M$, Przytycki's formula says that the Lyapunov function $L(P_\lambda)$ and the sum of values of the Green function on critical points differ from a constant. In particular, $L(P_\lambda)$ is a p.s.h function on M and these two functions induce the same $(1,1)$ current on M. It is then rather clear, using Mañé–Sad–Sullivan Theorem 65, that this current is exactly supported by the bifurcation locus.

We aim to generalize this to holomorphic families of rational maps and will therefore compute the dd^c of the left part of formula (17). The following formula has been established in [2].

Theorem 80. *Let $(f_\lambda)_M$ be a holomorphic family of degree d rational maps which admits a holomorphic family of lifts $(F_\lambda)_M$. Let p_M (resp $p_{\mathbf{P}^1}$) be the canonical projection from $M \times \mathbf{P}^1$ onto M (resp. $\mathbf{P}^1$). Then*

$$dd^c L(\lambda) = (p_M)_\star \left((dd^c_{\lambda,z} g_\lambda(z) + \hat{\omega}) \wedge [C] \right) \qquad (18)$$

where $C := \{(\lambda,z) \in M \times \mathbf{P}^1 \ / \ z \in \mathcal{C}_\lambda\}$, $g_\lambda := g_{F_\lambda}$ is the Green function of f_λ on $\mathbf{P}^1$, $L(\lambda)$ the Lyapunov exponent of f_λ and $\hat{\omega} := p_{\mathbf{P}^1}^\star \omega$.
Moreover, the function $\lambda \mapsto \int_{\mathbf{P}^1} g_\lambda(\mu_\lambda + \omega)$ is pluriharmonic on M.

Proof. We may work locally and define a holomorphic section Jac_λ of $\mathcal{O}_{\mathbf{P}^1}(D)$ induced by $\det F_\lambda'$. Then $\widetilde{Jac}(\lambda, [z]) := (\lambda, Jac_\lambda([z]))$ is a holomorphic section of the line bundle $M \times \mathcal{O}_{\mathbf{P}^1}(D)$ over $M \times \mathbf{P}^1$.

Let us rewrite Proposition 79 on the form $L(\lambda) + \ln d = H(\lambda) - D\ B(\lambda)$ where $H(\lambda) := \int_{\mathbf{P}^1} g_\lambda [\mathcal{C}_\lambda] + \int_{\mathbf{P}^1} \ln \|Jac_\lambda\|_0 \ \omega$, $B(\lambda) := \int_{\mathbf{P}^1} g_\lambda(\mu_\lambda + \omega)$ and $D := 2(d-1)$.

We first compute $dd^c H$. Let Φ denote a $(m-1, m-1)$ test form where m is the dimension of M. Then $\langle dd^c H, \Phi \rangle = I_1 + I_2$ where $I_1 := \int_M dd^c \Phi \int_{\mathbf{P}^1} g_\lambda [\mathcal{C}_\lambda]$ and $I_2 := \int_M dd^c \Phi \int_{\mathbf{P}^1} \ln \|Jac_\lambda\|_0 \ \omega$.

Slicing and then integrating by parts, we get

$$I_1 = \int_M dd^c \Phi \int_{\mathbf{P}^1} (g[C])_\lambda = \int_{M \times \mathbf{P}^1} (p_M)^\star (dd^c \Phi) \wedge (g[C])$$
$$= \int_{M \times \mathbf{P}^1} (p_M)^\star \Phi \wedge dd^c g \wedge [C].$$

Performing the same computation and then using the Poincaré–Lelong identity $dd^c \ln \|\widetilde{Jac}\|_0 = [C] - D\hat{\omega}$ one gets

$$I_2 = \int_M dd^c \Phi \int_{\mathbf{P}^1} \left(\ln \|\widetilde{Jac}\|_0 \ \hat{\omega} \right)_\lambda$$
$$= \int_{M \times \mathbf{P}^1} (p_M)^\star (dd^c \Phi) \wedge \ln \|\widetilde{Jac}\|_0 \ \hat{\omega} =$$
$$= \int_{M \times \mathbf{P}^1} (p_M)^\star \Phi \wedge dd^c \ln \|\widetilde{Jac}\|_0 \wedge \hat{\omega} = \int_{M \times \mathbf{P}^1} (p_M)^\star \Phi \wedge \hat{\omega} \wedge [C].$$

This shows that $dd^c H = (p_M)_\star \left((dd^c_{\lambda,z} g_\lambda(z) + \hat{\omega}) \wedge [C] \right)$.

It remains to show that $dd^c B = 0$. It seems very difficult to prove this by calculus, we will use a trick which exploits the dynamical situation. If one replace the family $(f_\lambda)_M$ by the family $(f_\lambda^2)_M$ then L becomes $2L$ and $dd^c H$ becomes $2dd^c H$ while B is unchanged. Thus, applying Proposition 79 to $(f_\lambda)_M$ and taking dd^c yields $dd^c L = dd^c H - 2(d-1) dd^c B$ but, with the family $(f_\lambda^2)_M$, this yields $2dd^c L = 2dd^c H - 2(d^2 - 1) dd^c B$. By comparison one obtains $dd^c B = 0$. □

Remark 81. The paper [2] actually covers the case of holomorphic families of endomorphisms of $\mathbf{P}^k$. In this setting, the formula (18) becomes

$$dd^c L(\lambda) = (p_M)_\star \big((dd^c_{\lambda,z} g_\lambda(z) + \hat{\omega})^k \wedge [C]\big)$$

where $L(\lambda)$ is now the sum of Lyapunov exponents of f_λ with respect to the Green measure μ_λ.

In [22], Dinh and Sibony established that $L(\lambda)$ is a *p.s.h* function in the more general setting of polynomial-like mappings.

4.2.3 Activity Currents and the Bifurcation Current

We have seen with Mañé–Sad–Sullivan Theorem 65 that the bifurcations, within a holomorphic family of degree d rational maps $(f_\lambda)_{\lambda \in M}$, are due to the activity of the critical points (see in particular Lemma 63). We will use this and the formula given by Theorem 80, to define a closed positive $(1, 1)$-current T_{bif} on M whose support is the bifurcation locus and which admits the Lyapunov function as global potential. Besides, we will also introduce a collection of $2d - 2$ closed positive $(1, 1)$-currents which detect the activity of each critical point and see that the bifurcation current T_{bif} is the sum of these currents. Here are the formal definitions.

Definition 82. Let $(f_\lambda)_{\lambda \in M}$ be any holomorphic family of degree d rational maps with marked critical points $\{c_i(\lambda); 1 \le i \le 2d - 2\}$. The *bifurcation current* T_{bif} is defined by

$$T_{\text{bif}} := dd^c L(\lambda)$$

where $L(\lambda)$ is the Lyapunov exponent of f_λ with respect to its Green measure.

The *activity current* T_i of the marked critical point c_i is defined by

$$T_i := (p_M)_\star \big((dd^c_{\lambda,z} g_\lambda(z) + \hat{\omega}) \wedge [C_i]\big)$$

where C_i is the graph $\{(\lambda, c_i(\lambda); \lambda \in M\}$ in $M \times \mathbf{P}^1$ and $g_\lambda := g_{F_\lambda}$ is the Green function of f_λ on $\mathbf{P}^1$ for some (local) holomorphic family of lifts (F_λ).

Let us observe that the current $dd^c_{\lambda,z} g_\lambda(z) + \hat{\omega}$ somehow interpolates the Green measures μ_λ of f_λ and is actually dynamically obtained. Indeed, one may consider the map $\hat{f}: M \times \mathbf{P}^1 \to M \times \mathbf{P}^1$ defined by $\hat{f}(\lambda, z) := (\lambda, f_\lambda(z))$ and then get it by taking the limit of the sequence of currents $d^{-n}(\hat{f})^n(\hat{\omega})$ in $M \times \mathbf{P}^1$. To establish the convergence one proceeds like in Lemma 5. The activity current T_i is the projection on M of the restriction of this current to the hypersurface C_i.

We now give local potentials for the activity currents.

Lemma 83. *Let F_λ be a local holomorphic family of lifts of f_λ and G_λ be the Green function of F_λ on $\mathbf{C}^2$. Let $\hat{c}_i(\lambda)$ be a local lift of $c_i(\lambda)$. Then $G_\lambda(\hat{c}_i(\lambda))$ is a local potential of T_i.*

Proof. This is a straightforward computation using the fact that, for any local section σ of the canonical projection $\pi : \mathbf{C}^2 \setminus \{0\} \to \mathbf{P}^1$, the function $G_\lambda(\sigma(z))$ is a local potential of $dd^c_{\lambda,z} g_\lambda(z) + \hat{\omega}$ (see Proposition 6). □

The following result has been originally proved by DeMarco in [19, 20].

Theorem 84. *Let $(f_\lambda)_{\lambda \in M}$ be any holomorphic family of degree d rational maps with marked critical points $\{c_i(\lambda); 1 \le i \le 2d - 2\}$. The support of the activity current T_i is the activity locus of the marked critical point c_i.*

The support of the bifurcation current T_{bif} is the bifurcation locus of the family $(f_\lambda)_{\lambda \in M}$ and $T_{bif} = \sum_1^{2d-2} T_i$.

Proof. Let us first show that c_i is passive on the complement of $Supp\, T_i$. If $T_i = 0$ on a small ball $B \subset M$ then, by Lemma 83, $G_\lambda(\hat{c}_i(\lambda))$ is pluriharmonic and therefore equal to $\ln|h_i(\lambda)|$ for some non-vanishing holomorphic function h_i on B. Replacing $\hat{c}_i(\lambda)$ by $\frac{\hat{c}_i(\lambda)}{h_i(\lambda)}$ one gets, thanks to the homogeneity property of G_λ (see Proposition 6), $G_\lambda(\hat{c}_i(\lambda)) = 0$. This implies that

$$\{F_\lambda^n(\hat{c}_i(\lambda)) \,/\, n \ge 1, \lambda \in B\} \subset \cup_{\lambda \in B} G_\lambda^{-1}\{0\}$$

where, after reducing B, the set $\cup_{\lambda \in B} G_\lambda^{-1}\{0\}$ is a relatively compact in $\mathbf{C}^2$. Montel's theorem then tells us that $\left(F_\lambda^n(\hat{c}_i(\lambda))\right)_n$ and, thus, $\left(f_\lambda^n(c_i(\lambda))\right)_n$ are normal on B.

Let us now show that T_i vanishes where c_i is passive. Assume that a subsequence $\left(f_\lambda^{n_k}(c_i(\lambda))\right)_k$ is uniformly converging on a small ball $B \subset M$. Then we may find a local section σ of $\pi : \mathbf{C}^2 \setminus \{0\} \to \mathbf{P}^1$ such that $F_\lambda^{n_k}(\hat{c}_i(\lambda)) = h_{n_k}(\lambda) \cdot \sigma \circ f_\lambda^{n_k}(c_i(\lambda))$ where h_{n_k} is a non-vanishing holomorphic function on B. As $G_\lambda \circ F_\lambda = dG_\lambda$ (see Proposition 6), this yields

$$G_\lambda(\hat{c}_i(\lambda)) = d^{-n_k}\left(\ln|h_{n_k}(\lambda)| + G_\lambda \circ \sigma \circ f_\lambda^{n_k}(c_i(\lambda))\right)$$

which, after taking dd^c and making $k \to +\infty$, implies that T_i vanishes on B.

That $dd^c L = T_{\mathrm{bif}} = \sum_1^{2d-2} T_i$ follows immediately from Theorem 80. Then, Mañé–Sad–Sullivan Theorem 65 implies that the support of T_{bif} is the bifurcation locus. □

Remark 85. As proved by Mañé, the product of the Lyapunov exponent of a degree d rational map and the Hausdorff dimension of its Green measure is equal to $\ln d$ [43]. This suggest that for bifurcation investigations it could be more natural to study the Borelian support of the measure than its topological support.

It is important to stress here that the identity $dd^c L = T_{\mathrm{bif}} = \sum_1^{2d-2} T_i$ may be seen as a potential-theoretic expression of Mañé–Sad–Sullivan theory. In practice, the expression $T_{\mathrm{bif}} = dd^c L$ will be used to investigate the set of parameters S_{hi} while the expression $T_{\mathrm{bif}} = \sum_1^{2d-2} T_i$ will be used for the parameters M_{is} (see Sect. 4.1.3).

In the next sections, we will deeply use the fact that the Lyapunov function is a potential of T_{bif}, together with the approximation formula 30, to analyse how the hypersurfaces $Per_n(w)$ may shape the bifurcation locus.

The continuity of the Lyapunov function L will turn out to be decisive for this study. Mañé was the first to establish the continuity of L (see [43]), using Theorem 80 and Lemma 83 one may see that this function is actually Hölder continuous.

Theorem 86. *The Lyapunov function of any holomorphic family of degree d rational maps is p.s.h and Hölder continuous.*

Proof. According to Theorem 80 and Lemma 83 the functions L and $\sum G_\lambda(\hat{c}_i(\lambda))$ differ from a pluriharmonic function. The conclusion follows from the fact that $G_\lambda(z)$ is Hölder continuous in (λ, z) (see [2, Proposition 1.2]). □

4.2.4 DeMarco's Formula

Using Theorem 80 we will get an explicit version of the formula given by Proposition 79. This result was first obtained by DeMarco who used a completely different method. We refer to the paper of Okuyama [51] for yet another proof.

The key will be to compute the integral $\int_{\mathbf{P}^1} g_F (\mu_f + \omega)$ which appears in the formula given by Proposition 79. To this purpose we shall use the *resultant* of a homogeneous polynomial map F of degree d on $\mathbf{C}^2$. The space of such maps can be identified with $\mathbf{C}^{2d+2}$. The resultant $Res\ F$ of F polynomially depends on F and vanishes if and only if F is degenerate. Moreover $Res\ (z_1^d, z_2^d) = 1$ and Res is $2d$-homogeneous: $Res\ aF = a^{2d}\ Res\ F$.

Lemma 87. $\int_{\mathbf{P}^1} g_F (\mu_f + \omega) = \frac{1}{d(d-1)} \ln |Res\ F| - \frac{1}{2}$.

Proof. The function $B(F) := \int_{\mathbf{P}^1} g_F (\mu_f + \omega)$ is well defined on $\mathbf{C}^{2d+2} \setminus \Sigma$ where Σ is the hypersurface where Res vanishes. Moreover, according to Theorem 80, B is pluriharmonic. As B is locally bounded from above, it extends to some $p.s.h$ function through Σ. Then, by Siu's theorem, there exists some positive constant c such that $dd^c B(F) = c\ dd^c \ln |Res F|$ which means that $B - c \ln |Res F|$ is pluriharmonic on $\mathbf{C}^{2d+2}$.

Let φ be a non-vanishing holomorphic function on $\mathbf{C}^{2d+2}$ such that

$$B - c \ln |Res F| = \ln |\varphi|.$$

Using the homogeneity of Res and the fact that $B(aF) = \frac{2}{d-1} \ln |a| + B(F)$ (one easily checks that $g_{aF} = \frac{1}{d-1} \ln |a| + g_F$) one gets:

$$|\varphi(aF)| = |a|^{\frac{2}{d-1} - 2cd}.$$

Making $a \to 0$ one sees that $c = \frac{1}{d(d-1)}$ and φ is constant. To compute this constant one essentially tests the formula on $F_0 := (z_1^d, z_2^d)$ (see [2, Proposition 4.10]). □

We are now ready to prove the main result of this subsection.

Theorem 88. *Let f be a rational map of degree $d \geq 2$ and F be one of its lifts. Let G_F be the Green function of F on $\mathbf{C}^2$ and Res F be the resultant of F. Then*

$$L(f) + \ln d = \sum_{j=1}^{2d-2} G_F(\hat{c}_j) - \frac{2}{d} \ln |Res\ F|$$

where $\hat{c}_1, \hat{c}_2, \cdots, \hat{c}_{2d-2}$ are chosen so that $det F'(z) = \prod_{j=1}^{2d-2} \hat{c}_j \wedge z$.

Proof. Taking Lemma 87 into account, the formula given by Proposition 79 becomes

$$L(f) + \ln d = \int_{\mathbf{P}^1} g_F[\mathcal{C}_f] - 2(d-1)\left(\frac{1}{d(d-1)} \ln |Res\ F| - \frac{1}{2}\right)$$
$$+ \int_{\mathbf{P}^1} \ln \|Jac_F\|_0\ \omega.$$

Observe that $\omega = \pi_* m$ where $m := (dd^c \ln^+ \|\cdot\|)^2$ is the normalized Lebesgue measure on the Euclidean unit sphere of $\mathbf{C}^2$. Then

$$\int_{\mathbf{C}^2} \ln |det F'|\ m = \int_{\mathbf{C}^2} \ln\left(e^{-D\|\cdot\|} |det F'|\right)\ m$$
$$= \int_{\mathbf{C}^2} \ln \|J_F \circ \pi\|_0\ m = \int_{\mathbf{P}^1} \ln \|Jac_F\|_0\ \omega.$$

Let us pick U_j in the unitary group of $\mathbf{C}^2$ such that $U_j^{-1}(\hat{c}_j) = (\|\hat{c}_j\|, 0)$. Then $U_j(z) \wedge \hat{c}_j = -z_2 \|\hat{c}_j\|$ and, since $\int_{\mathbf{C}^2} \ln |z_2|\ m = -\frac{1}{2}$ one gets

$$\int_{\mathbf{P}^1} \ln \|Jac_F\|_0\ \omega = \int_{\mathbf{C}^2} \ln |det F'|\ m = \sum_j \int_{\mathbf{C}^2} \ln |U_j(z) \wedge \hat{c}_j| = \sum_j \|\hat{c}_j\| - (d-1).$$

On the other hand, $\int_{\mathbf{P}^1} g_F[\mathcal{C}_f] = \sum_j g_F \circ \pi(\hat{c}_i) = \sum_j G_F(\hat{c}_j) - \sum_j \ln \|\hat{c}_j\|$ and the conclusion follows. □

5 Equidistribution Towards the Bifurcation Current

In this section we study how the bifurcation current may be approximated by various weighted hypersurfaces which are dynamically defined. As an application, we construct holomorphic motions and describe the laminated structure of the bifurcation locus in certain regions of the moduli space Mod_2.

5.1 A General Method

We will present a general method which may be used to prove that a sequence of currents $T_n := dd^c h_n$, which admit global p.s.h potentials h_n on the parameter space $\mathbf{C}^k$ of some holomorphic family $(f_\lambda)_{\mathbf{C}^k}$, is converging to the bifurcation current T_{bif}.

The method itself only relies on standard potential-theoretic arguments but dynamical informations are then necessary to apply it. More specifically, one needs to control the bifurcation current near infinity, this is why the method works well in polynomial families but seems difficult to handle in more general families.

In Sect. 5.2 we shall use it to study the distribution of critically periodic parameters and in Sect. 5.3.3 to study the distribution of the hypersurfaces $Per_n(e^{i\theta})$.

We have to prove that $(h_n)_n$ converges in L^1_{loc} to L and, for this, will use a well known *compactness principle* for subharmonic functions:

Theorem 89. *Let (φ_j) be a sequence of subharmonic functions which is locally uniformly bounded from above on some domain $\Omega \subset \mathbf{R}^n$. If (φ_j) does not converge to $-\infty$ then a subsequence (φ_{j_k}) converges in $L^1_{loc}(\Omega)$ to some subharmonic function φ. In particular, (φ_j) converges in $L^1_{loc}(\Omega)$ to some subharmonic function φ if it converges pointwise to φ.*

We thus first need to check that $(h_n)_n$ is locally bounded from above and does not converge to $-\infty$, the following result may be used to check this last property.

Lemma 90 (Hartogs). *Let (φ_j) be a sequence of subharmonic functions and g be a continuous function defined on some domain $\Omega \subset \mathbf{R}^n$. If $\limsup_j \varphi_j(x) \le g(x)$ for every $x \in \Omega$ then, for any compact $K \subset \Omega$ and every $\epsilon > 0$ one has $\varphi_j(x) \le g(x) + \epsilon$ on K for j big enough.*

Then one has to show that L is the unique limit value of $(h_n)_n$ for the L^1_{loc} topology. To this purpose, we shall apply the following generalized maximum principle to some limit value φ of $(h_n)_n$ and to $\psi = L$.

Lemma 91. *Two p.s.h functions φ, ψ on $\mathbf{C}^k$ coincide if the following conditions are satisfied:*

(i) ψ is continuous
(ii) $\varphi \le \psi$
(iii) $Supp (dd^c \varphi) \subset Supp (dd^c \psi)$
(iv) $\varphi = \psi$ on $Supp (dd^c \psi)$
(v) for any $\lambda_0 \in \mathbf{C}^k$ there exists a complex line $\mathcal{L}$ through λ_0 such that $\varphi = \psi$ on the unbounded component of $\mathcal{L} \setminus (\mathcal{L} \cap Supp (dd^c \psi))$.

Proof. Because of (iv), we only have to show that $\varphi(\lambda_0) = \psi(\lambda_0)$ when λ_0 lies in the complement of $Supp (dd^c \psi)$. According to (v), we may find a complex line $\mathcal{L}$ in $\mathbf{C}^k$ containing λ_0 and such that ψ and φ coincide on the unbounded component Ω_∞ of $\mathcal{L} \setminus (\mathcal{L} \cap Supp (dd^c \psi))$. We may therefore assume that $\lambda_0 \notin \Omega_\infty$. By (i), (iii) and (iv) $\varphi|_\mathcal{L}$ coincides with the continuous function $\psi|_\mathcal{L}$ on $Supp \Delta \varphi|_\mathcal{L}$

which, by the continuity principle, implies that $\varphi|_{\mathcal{L}}$ is continuous. Let Ω_0 be the (bounded) component of $\mathcal{L} \setminus (\mathcal{L} \cap \mathrm{Supp}\,(dd^c\psi))$ containing λ_0. The continuous function $(\varphi - \psi)|_{\mathcal{L}}$ vanishes on $b\Omega_0$ (see (iv)), is harmonic on Ω_0 (see (iii)) and negative (see (ii)). The maximum principle now implies that $\varphi(\lambda_0) = \psi(\lambda_0)$. □

It is worth emphasize that quite precise informations about the behaviour of the bifurcation locus at infinity are required to apply the above Lemma to our problems.

5.2 Distribution of Critically Periodic Parameters in Polynomial Families

The aim of this section is to present a result due to Dujardin and Favre (see [29]) concerning the asymptotic distribution of degree d polynomials which have a pre-periodic critical point. We will work in the context of polynomial families, this will allow us to modify the original proof and significantly simplify it. We refer to the paper of Dujardin–Favre for results dealing with general holomorphic families of rational maps.

We work here in the family $(P_{c,a})_{(c,a) \in \mathbf{C}^{d-1}}$ of degree d polynomials which has been introduced in the Sect. 3.1.2. Let us recall that $P_{c,a}$ is the polynomial of degree d whose critical set is $\{0 = c_0, c_1, \cdots, c_{d-2}\}$ and such that $P_{c,a}(0) = a^d$.

For $0 \le i \le d-2$ and $0 \le k < n$, we denote by $\mathrm{Per}(i, n, k)$ the hypersurface of $\mathbf{C}^{d-1}$ defined by

$$\mathrm{Per}(i,n,k) := \{(c,a) \in \mathbf{C}^{d-1} \,/\, P_{c,a}^n(c_i) = P_{c,a}^k(c_i)\}.$$

The result we want to establish, which has been first proved by Dujardin and Favre in [29], is the following.

Theorem 92. *In the family of degree d polynomials, for any sequence of integers $(k_n)_n$ such that $0 \le k_n < n$ one has $\sum_{i=0}^{d-2} \lim_n d^{-n}[\mathrm{Per}(i,n,k_n)] = T_{\mathrm{bif}}$.*

To simplify, we shall write λ the parameters $(c,a) \in \mathbf{C}^{d-1}$. We follow the strategy described in the previous subsection.

The bifurcation current T_{bif} is given by $T_{\mathrm{bif}} = \sum_{i=0}^{d-2} dd^c g_\lambda(c_i)$ (see Lemma 83) where g_λ is the Green function of P_λ (see the Sect. 3.2.1). It thus suffices to show that for any fixed $0 \le i \le d-2$ the following sequence of potentials

$$h_n(\lambda) := d^{-n} \ln |P_\lambda^n(c_i) - P_\lambda^{k_n}(c_i)|$$

converges in L^1_{loc} to $g_\lambda(c_i)$.

To this purpose we shall compare these potentials with the functions

$$g_n(\lambda) := d^{-n} \ln \max\left(1, |P_\lambda^n(c_i)|\right)$$

which do converge locally uniformly to $g_\lambda(c_i)$.

The first point is to check that the sequence $(h_n)_n$ is locally uniformly bounded from above. We shall actually prove a little bit more.

Lemma 93. *For any compact $K \subset \mathbf{C}^{d-1}$ and any $\epsilon > 0$ there exists an integer n_0 such that $h_n|_K \leq g_n|_K + \epsilon$ for $n \geq n_0$.*

Proof. It is not difficult to see that there exists $R \geq 1$ such that

$$(1-\epsilon)|z|^{d^n} \leq |P_\lambda^n(z)| \leq (1+\epsilon)|z|^{d^n} \qquad (19)$$

for every $\lambda \in K$, every $n \in \mathbf{N}$ and every $|z| \geq R$.

We now proceed by contradiction and assume that there exists $\lambda_p \in K$ and $n_p \to +\infty$ such that $h_{n_p}(\lambda_p) \geq g_{n_p}(\lambda_p) + \epsilon$. This means that

$$|P_{\lambda_p}^{n_p}(c_i) - P_{\lambda_p}^{k_{n_p}}(c_i)| \geq e^{\epsilon d^{n_p}} \max\left(1, |P_{\lambda_p}^{n_p}(c_i)|\right). \qquad (20)$$

Let us set $B_p := P_{\lambda_p}^{k_{n_p}}(c_i)$. By (20) we have $\lim_p |B_p| = +\infty$ and thus $|B_p| \geq R$ for p big enough. Then, using (19), one may write

$$P_{\lambda_p}^{n_p}(c_i) = P_{\lambda_p}^{n_p - k_{n_p}}(B_p) = (u_p B_p)^{d^{n_p - k_{n_p}}}$$

where $(1-\epsilon) \leq |u_p| \leq (1+\epsilon)$ and the estimate (20) becomes

$$|(u_p B_p)^{d^{n_p - k_{n_p}}} - B_p| \geq e^{\epsilon d^{n_p}} |u_p B_p|^{d^{n_p - k_{n_p}}}.$$

This is clearly impossible when $p \to +\infty$. □

We now have to check that $(h_n)_n$ does not converge to $-\infty$. The following technical Lemma deals with that and will also play an important role in the remaining of the proof.

Lemma 94. *If c_i belongs to some attracting basin of P_{λ_0} then there exists a neighbourhood V_0 of λ_0 such that $\sup_n \sup_{V_0} (h_n - g_n) \geq 0$.*

Proof. If V_0 is a sufficiently small neighbourhood of λ_0 then $P_\lambda^n(c_i) \to a_\lambda$ where a_λ is an attracting cycle of P_λ for every $\lambda \in V_0$. We will assume that $P_\lambda(a_\lambda) = a_\lambda$.

Let us now proceed by contradiction and suppose that there exists $\epsilon > 0$ such that

$$|P_\lambda^n(c_i) - P_\lambda^{k_n}(c_i)| \leq e^{-\epsilon d^n} \max\left(1, |P_\lambda^n(c_i)|\right), \quad \forall \lambda \in V_0, \forall n. \qquad (21)$$

Since $P_\lambda^n(c_i) \to a_\lambda$, (21) would imply

$$|P_\lambda^n(c_i) - P_\lambda^{k_n}(c_i)| \leq C e^{-\epsilon d^n}, \quad \forall \lambda \in V_0, \forall n. \qquad (22)$$

Bifurcation Currents in Holomorphic Families of Rational Maps   55

The estimate (22) implies that a_λ is a super-attracting fixed point for any $\lambda \in V_0$ which, in turn, implies that $a_\lambda = \infty$ for all $\lambda \in V_0$. But in that case we would have $|P_\lambda^{k_n}(c_i)| \leq \frac{1}{2}|P_\lambda^n(c_i)|$ for n big enough and the estimate (21) would be violated. □

We finally have to show that $g := g_\lambda(c_i) = \lim g_n$ is the only limit value of the sequence $(h_n)_n$ for the L^1_{loc} convergence. Assume that (after taking a subsequence!) h_n is converging in L^1_{loc} to h. To prove that the functions h and g coincide we shall check that they satisfy the assumptions of Lemma 91.

Our modification of the original proof essentially stays in the third step.

First step: $h \leq g$.

Let B_0 be a ball of radius r centered at λ_0 and let $\epsilon > 0$. By the mean value property we have

$$h(\lambda_0) \leq \frac{1}{|B_0|} \int_{B_0} h = \lim_n \frac{1}{|B_0|} \int_{B_0} h_n$$

but, according to Lemma 93, $h_n \leq g_n + \epsilon$ on B_0 for n big enough and thus

$$h(\lambda_0) \leq \epsilon + \lim_n \frac{1}{|B_0|} \int_{B_0} g_n = \epsilon + \frac{1}{|B_0|} \int_{B_0} g.$$

As g is continuous, the conclusion follows by making $r \to 0$ and then $\epsilon \to 0$.

Second step: $h = g$ on $\mathrm{Supp}\, dd^c g$.

Combining Lemma 94 and the result of step one we will first establish the following

Fact $(h - g)$ vanishes when c_i is captured by an attracting basin.

Suppose to the contrary that c_i is captured by an attracting cycle of P_{λ_0} and $(h - g)(\lambda_0) < 0$. As the function $(h - g)$ is upper semi-continuous, we may shrink V_0 so that $(h - g) \leq -\epsilon < 0$ on V_0. Now, as c_i is passive on V_0, the function $(h - g)$ is p.s.h on V_0 and, after shrinking V_0 again, Hartogs Lemma 90 implies that $(h_n - g_n) \leq -\frac{\epsilon}{2}$ on V_0 for n big enough. This contradicts Lemma 94.

Now, if $\lambda_0 \in \mathrm{Supp}\, dd^c g$ then $\lambda_0 = \lim_k \lambda_k$ where λ_k is a parameter for which c_i is captured by some attracting cycle (see Lemma 64). As $(h - g)$ is upper semi-continuous we get $(h - g)(\lambda_0) \geq \limsup(h - g)(\lambda_k) = 0$ and, by the first step, $(h - g)(\lambda_0) = 0$.

Third step: $\mathrm{Supp}\, dd^c h \subset \mathrm{Supp}\, dd^c g$.

Let Ω be a connected component of $\mathbf{C}^{d-1} \setminus \mathrm{Supp}\, dd^c g$. We have to show that h is pluriharmonic on Ω. We proceed by contradiction. If $dd^c h$ does not vanish on Ω then there exists some n_0 for which some irreducible component $\mathcal{H}$ of

$$\{P_\lambda^{n_0}(c_i) - P_\lambda^{k_{n_0}}(c_i) = 0\}$$

meets Ω. When $\lambda \in \mathcal{H}$ then c_i is captured by a cycle since $P_\lambda^{k_{n_0}}(c_i) =: z(\lambda)$ satisfies $P_\lambda^{m_0}(z(\lambda)) = P_\lambda^{m_0} \circ P_\lambda^{k_{n_0}}(c_i) = P_\lambda^{k_{n_0}}(c_i) = z(\lambda)$ for $m_0 := n_0 - k_{n_0} > 0$.

Let us show that $z(\lambda)$ is a neutral periodic point. We first observe that the vanishing of $dd^c g$ on Ω forces $z(\lambda)$ to be non-repelling and thus $|(P_\lambda^{m_0})'(z(\lambda))| \leq 1$ on $\mathcal{H} \cap \Omega$.

Let us now see why $z(\lambda)$ cannot be attracting. If this would be the case then, by the above Fact, we would have $h(\lambda_0) = g(\lambda_0)$ for a certain $\lambda_0 \in \mathcal{H} \cap \Omega$. As $(h-g)$ is negative and p.s.h on Ω this implies, via the maximum principle, that $h = g$ on Ω. This is impossible since $dd^c h$ is supposed to be non vanishing on Ω.

We thus have $(P_\lambda^{m_0})'(z(\lambda)) = e^{iv_0}$ on $\mathcal{H} \cap \Omega$ and therefore $z(\lambda)$ belongs to a neutral cycle whose period p_0 divides m_0 and whose multiplier is a q_0-root of e^{iv_0} where $m_0 = p_0 q_0$. In other words, $\mathcal{H} \cap \Omega$ is contained in a finite union of hypersurfaces of the form $\mathrm{Per}_n(e^{i\theta})$. This implies that

$$\mathcal{H} \subset \mathrm{Per}_{n_0}(e^{i\theta_0}).$$

for some integer n_0 and some real number θ_0.

Finally, using a global argument, we will see that this is impossible. Let us recall the following dynamical fact.

Lemma 95. *Every polynomial which has a neutral cycle also has a bounded, non-preperiodic, critical orbit.*

Thus, when $\lambda \in \mathcal{H}$, the polynomial P_λ has two distinct bounded critical orbits; the orbit of c_i which is preperiodic and the orbit of some other critical point which is given by the above Lemma. This shows that $\mathcal{H}$ cannot meet the line $\{c_0 = c_1 = \cdots = c_{d-2} = 0\} := M_d$ since the corresponding polynomials (which are given by $\frac{1}{d}z^d + a^d$ were $a \in \mathbf{C}$) have only one critical orbit. We will now work in the projective compactification of $\mathbf{C}^{d-1}$ introduced in Sect. 3.1.2. By Theorem 43, $\mathcal{H}$ and M_d cannot meet at infinity. This contradicts Bezout's theorem.

Fourth step: for any $\lambda_0 \in \mathbf{C}^{d-1}$ there exists a complex line $\mathcal{L}$ through λ_0 such that $h = g$ on the unbounded component of $\mathcal{L} \setminus (\mathcal{L} \cap \mathrm{Supp}(dd^c g))$.

Here one uses again Theorem 43 to pick a line $\mathcal{L}$ through λ_0 which meets infinity at some point $\xi_0 \notin \overline{\mathrm{Supp}\, dd^c g}$. Then, for any λ in the unbounded component of $\mathcal{L} \setminus (\mathcal{L} \cap \mathrm{Supp}\,(dd^c g))$ the critical point c_i belongs to the super-attracting basin of ∞ and thus, as we saw in second step, $h(\lambda) = g(\lambda)$. □

The general result obtained by Dujardin and Favre may be stated as follows.

Theorem 96. *Let $(f_\lambda)_M$ be a holomorphic family of degree d rational maps with a marked critical point c_λ which is not stably preperiodic. Let $\mathrm{Per}(n,k)$ be the hypersurface in M defined by:*

$$\mathrm{Per}(n,k) := \{\lambda \in M \;/\; f_\lambda^n(c_\lambda) = f_\lambda^k(c_\lambda)\}.$$

Assume that the following assumption is satisfied:

(H) *For every $\lambda_0 \in M$ there exists a curve $\Gamma \subset M$ passing through λ_0 such that $\{\lambda \in \Gamma \;/\; c_\lambda \text{ is attracted by a cycle}\}$ has a relatively compact complement in Γ.*

Then, for any sequence of integers $(k_n)_n$ *such that* $0 \le k_n < n$ *one has*

$$\lim_n d^{-n-(1-e)k_n}[Per(n,k_n)] = T_c$$

where T_c *is the activity current of the marked critical point* c_λ *and* e *the cardinal of the exceptional set of* f_λ *for a generic* λ.

It would be interesting to remove the assumption (H) which seems to be a technical one. Observe however that the above Theorem covers the case of the moduli space Mod_2.

5.3 Distribution of Rational Maps with Cycles of a Given Multiplier

Let $f : M \times \mathbf{P}^1 \to \mathbf{P}^1$ be an arbitrary holomorphic family of degree $d \ge 2$ rational maps.

We want to investigate the asymptotic distribution of the hypersurfaces $Per_n(w)$ in M when $|w| < 1$. Concretely, we will consider the current of integration $[Per_n(w)]$ or, more precisely, the currents

$$[Per_n(w)] := dd^c \ln|p_n(\lambda, w)|$$

where $p_n(\cdot, w)$ are the canonical defining functions for the hypersurfaces $Per_n(w)$ constructed in Sect. 3.3 by mean of dynatomic polynomials. We ask if the following convergence occurs:

$$\lim_n \frac{1}{d^n}[Per_n(w)] = T_{\text{bif}}.$$

The question is easy to handle when $|w| < 1$, more delicate when $|w| = 1$ and widely open when $|w| > 1$.

5.3.1 The Case of Attracting Cycles

We aim to prove the following general result (see [3]).

Theorem 97. *For any holomorphic family of degree* d *rational maps* $(f_\lambda)_{\lambda \in M}$ *one has* $d^{-n}[Per_n(w)] \to T_{bif}$ *when* $|w| < 1$.

Let us have a look to the case $w = 0$. Comparing $Per_n(0)$ with the hypersurfaces $Per(n, 0)$ considered in the the last section, one sees that the above result may be derived from the Theorem 96 of Dujardin and Favre but that we do not need any special assumption. Moreover, for the quadratic polynomial family one obtains the equidistribution of centers of hyperbolic components of the Mandelbrot set. This was first proved by Levin [39].

Using arithmetical methods, Favre and Rivera-Letelier [32] have estimated the equidistribution speed of d^{-n} [Per$_n(0)$] for unicritical families $(z^d + c)$.

Proof. Let us set
$$L_n(\lambda, w) := d^{-n} \ln |p_n(\lambda, w)|.$$

Since, by definition, $T_{\text{bif}} = dd^c \; L(\lambda)$ where $L(\lambda)$ be the Lyapunov exponent of $(\mathbf{P}^1, f_\lambda, \mu_\lambda)$ and μ_λ is the Green measure of f_λ, all we have to show is that L_n converges to L in $L^1_{\text{loc}}(M)$. Here again we shall use the compactness principle for subharmonic functions (see Theorem 89).

The situation is purely local and therefore, taking charts, we may assume that $M = \mathbf{C}^k$. We write the polynomials p_n as follows:

$$p_n(\lambda, w) =: \prod_{i=1}^{N_d(n)} \left(w - w_{n,j}(\lambda)\right).$$

Using the fact that $d^{-n} N_d(n) \sim \frac{1}{n}$ (see Theorem 46) one sees that the sequence L_n is locally uniformly bounded from above.

According to Theorem 46, the set $\{w_{n,j}(\lambda) \; / \; w_{n,j}(\lambda) \neq 1\}$ coincides with the set of multipliers of cycles of exact period n (counted with multiplicity) from which the cycles of multiplier 1 are deleted. We thus have

$$\sum_{j=1}^{N_d(n)} \ln^+ |w_{n,j}(\lambda)| = \frac{1}{n} \sum_{p \in R_n^*(\lambda)} \ln |(f_\lambda^n)'(p)| \qquad (23)$$

where $R_n^*(\lambda) := \{p \in \mathbf{P}^1 \; / \; p \text{ has exact period } n \text{ and } |(f_\lambda^n)'(p)| > 1\}$. Since f_λ has a finite number of non-repelling cycles (Fatou's theorem), one sees that there exists $n(\lambda) \in \mathbf{N}$ such that

$$n \geq n(\lambda) \Rightarrow |w_{n,j}(\lambda)| > 1, \text{ for any } 1 \leq j \leq N_d(n). \qquad (24)$$

By (23) and (24), one gets

$$L_n(\lambda, 0) = d^{-n} \sum_{j=1}^{N_d(n)} \ln |w_{n,j}(\lambda)| = d^{-n} \sum_{j=1}^{N_d(n)} \ln^+ |w_{n,j}(\lambda)| = \frac{d^{-n}}{n} \sum_{R_n^*(\lambda)} \ln |(f_\lambda^n)'(p)|$$

for $n \geq n(\lambda)$ which, by Theorem 30, yields:

$$\lim_n L_n(\lambda, 0) = L(\lambda), \; \forall \lambda \in M. \qquad (25)$$

If now $|w| < 1$, it follows from (24) that $L_n(\lambda, w) - L_n(\lambda, 0) = d^{-n} \sum_j \ln \frac{|w_{n,j}(\lambda) - w|}{|w_{n,j}(\lambda)|}$ and $\ln(1 - |w|) \leq \ln \frac{|w_{n,j}(\lambda) - w|}{|w_{n,j}(\lambda)|} \leq \ln(1 + |w|)$ for $1 \leq j \leq N_d(n)$ and $n \geq n(\lambda)$. We thus get

$$d^{-n}N_d(n)\ln(1-|w|) \le |L_n(\lambda,w) - L_n(\lambda,0)| \le d^{-n}N_d(n)\ln(1+|w|)$$

for $n \ge n(\lambda)$. Using (25) and the fact that $d^{-n}N_d(n) \sim \frac{1}{n}$ we obtain $\lim_n L_n(\lambda,w) = L(\lambda)$ for any $(\lambda,w) \in M \times \Delta$. The L^1_{loc} convergence of $L_n(\cdot,w)$ now follows immediately from Theorem 89. □

Remark 98. We have proved that $L_n(\lambda,w) := d^{-n}\ln|p_n(\lambda,w)|$ converges pointwise to $L(\lambda)$ on M when $|w| < 1$.

The above discussion shows that the pointwise convergence of $L_n(\lambda,w)$ to L (and therefore the convergence $d^{-n}[\text{Per}_n(w)] \to T_{\text{bif}}$) is quite a straightforward consequence of Theorem 30 when $|w| < 1$. However, when $|w| \ge 1$ and λ is a non-hyperbolic parameter, the control of $L_n(\lambda,w) = d^{-n}\sum \ln|w - w_{n,j}(\lambda)|$ is very delicate because f_λ may have many cycles whose multipliers are close to w. This is why we introduce the *p.s.h* functions L_n^+ which both coincide with L_n on the hyperbolic components and are rather easily seen to converge nicely. These functions will be extremely helpful later.

Definition 99. The *p.s.h* functions L_n^+ are defined by:

$$L_n^+(\lambda,w) := d^{-n} \sum_{j=1}^{N_d(n)} \ln^+|w - w_{n,j}(\lambda)|$$

where $p_n(\lambda,w) =: \prod_{j=1}^{N_d(n)}(w - w_{n,j}(\lambda))$ are the polynomials associated to the family $(f_\lambda)_{\lambda \in M}$ by Theorem 46.

The interest of considering these functions stays in the next Lemma.

Lemma 100. *The sequence L_n^+ converges pointwise and in L^1_{loc} to L on $M \times \mathbf{C}$. For every $w \in \mathbf{C}$ the sequence $L_n^+(\cdot,w)$ converges in L^1_{loc} to L on M.*

Proof. We will show that $L_n^+(\cdot,w)$ converges pointwise to L on M for every $w \in \mathbf{C}$. As $(L_n^+)_n$ is locally uniformly bounded, this implies the convergence of $L_n^+(\cdot,w)$ in $L^1_{loc}(M)$ (Theorem 89) and the convergence of L_n^+ in $L^1_{loc}(M \times \mathbf{C})$ then follows by Lebesgue's theorem.

As $L_n(\lambda,0) \to L(\lambda)$ (see Remark 98), we have to estimate $L_n^+(\lambda,w) - L_n(\lambda,0) =: \epsilon_n(\lambda,w)$ on M. Let us fix $\lambda \in M$, $w \in \mathbf{C}$ and pick $R > |w|$. Since f_λ has a finite number of non-repelling cycles (Fatou's theorem), one sees that there exists $n(\lambda) \in \mathbf{N}$ such that

$$n \ge n(\lambda) \Rightarrow |w_{n,j}(\lambda)| > 1, \text{ for any } 1 \le j \le N_d(n).$$

We may then decompose $\epsilon_n(\lambda,w)$ in the following way:

$$\epsilon_n(\lambda, w) = d^{-n} \sum_{1 \le |w_{n,j}(\lambda)| < R+1} \ln^+ |w_{n,j}(\lambda) - w| + d^{-n} \sum_{|w_{n,j}(\lambda)| \ge R+1} \ln \frac{|w_{n,j}(\lambda) - w|}{|w_{n,j}(\lambda)|}$$
$$- d^{-n} \sum_{1 \le |w_{n,j}(\lambda)| < R+1} \ln |w_{n,j}(\lambda)|.$$

We may write this decomposition as $\epsilon_n(\lambda, w) =: \epsilon_{n,1}(\lambda, w) + \epsilon_{n,2}(\lambda, w) - \epsilon_{n,1}(\lambda, 0)$.

Clearly, $0 \le \epsilon_{n,1}(\lambda, w) \le d^{-n} N_d(n) \ln(2R+1)$ and thus $\lim_n \epsilon_{n,1}(\lambda, w) = 0$. Similarly, $\lim_n \epsilon_{n,2}(\lambda, w) = 0$ follows from the fact that one has:

$$\ln(1 - \frac{R}{R+1}) \le \ln \frac{|w_{n,j}(\lambda)| - R}{|w_{n,j}(\lambda)|} \le \ln \frac{|w_{n,j}(\lambda) - w|}{|w_{n,j}(\lambda)|}$$
$$\le \ln \frac{|w_{n,j}(\lambda)| + R}{|w_{n,j}(\lambda)|} \le \ln(1 + \frac{R}{R+1}).$$

for $|w_{n,j}(\lambda)| > R + 1 > |w| + 1$. $\square$

As the functions L_n^+ and L_n coincide on hyperbolic components, the above Lemma would easily yield the convergence of $d^{-n}[\text{Per}_n(w)]$ towards T_{bif} *for any* $w \in \mathbf{C}$ if the density of hyperbolic parameters in M was known. The remaining of this section is, in some sense, devoted to overcome this difficulty. We shall first do this in a general setting by averaging the multipliers. Then we will restrict ourself to polynomial families and, using the nice distribution of hyperbolic parameters near infinity, will show that $d^{-n}[\text{Per}_n(e^{i\theta})]$ converges towards T_{bif}.

5.3.2 Averaging the Multipliers

Although the convergence of $\lim_n \frac{1}{d^n}[\text{Per}_n(w)]$ to T_{bif} is not clear when $|w| \ge 1$, one easily obtains the convergence by averaging over the argument of the multiplier w. The following result is due to Bassanelli and the author [3].

Theorem 101. *For any holomorphic family of degree d rational maps $(f_\lambda)_{\lambda \in M}$ one has $\frac{d^{-n}}{2\pi} \int_0^{2\pi} [\text{Per}_n(re^{i\theta})] \, d\theta \to T_{\text{bif}}$, when $r \ge 0$.*

Proof. One essentially has to investigate the following sequences of p.s.h functions

$$L_n^r(\lambda) := \frac{d^{-n}}{2\pi} \int_0^{2\pi} \ln |p_n(\lambda, re^{i\theta})| \, d\theta.$$

We will see that $L_n^r(\lambda) \ge C \frac{\ln r}{n}$ where C only depends on the family and that L_n^r converges to L in $L_{loc}^1(M)$.

For that, we essentially will compare L_n^r with $L_n(\lambda, 0) = L_n^0$ by using the formula $\ln \max(|a|, r) = \frac{1}{2\pi} \int_0^{2\pi} \ln|a - re^{i\theta}| d\theta$. Indeed, writing

$$p_n(\lambda, w) =: \prod_{i=1}^{N_d(n)} (w - w_{n,j}(\lambda))$$

this formula yields

$$L_n^r(\lambda) = \frac{1}{2\pi d^n} \int_0^{2\pi} \ln \prod_j |re^{i\theta} - w_{n,j}(\lambda)| d\theta = d^{-n} \sum_j \ln \max(|w_{n,j}(\lambda)|, r). \tag{26}$$

According to Theorem 46, the set $\{w_{n,j}(\lambda) \ / \ w_{n,j}(\lambda) \neq 1\}$ coincides with the set of multipliers of cycles of exact period n (counted with multiplicity) from which the cycles of multiplier 1 are deleted. Using the fact that $d^{-n} N_d(n) \sim \frac{1}{n}$ (see Theorem 46) one sees that the sequence $L_n^r(\lambda)$ is locally bounded from above and is uniformly bounded from below by $C \frac{\ln r}{n}$. Since f_λ has a finite number of non-repelling cycles (Fatou's theorem), there exists $n(\lambda) \in \mathbf{N}$ such that

$$n \geq n(\lambda) \Rightarrow |w_{n,j}(\lambda)| > 1, \text{ for any } 1 \leq j \leq N_d(n).$$

Now we deduce from (26) that for $n \geq n(\lambda)$:

$$L_n^r(\lambda) = d^{-n} \sum_j \ln |w_{n,j}(\lambda)| + d^{-n} \sum_{1 \leq |w_{n,j}(\lambda)| < r} \ln \frac{r}{|w_{n,j}(\lambda)|}$$

$$= L_n(\lambda, 0) + d^{-n} \sum_{1 \leq |w_{n,j}(\lambda)| < r} \ln \frac{r}{|w_{n,j}(\lambda)|}$$

and thus

$$0 \leq L_n^r(\lambda) - L_n(\lambda, 0) = d^{-n} \sum_{1 \leq |w_{n,j}(\lambda)| < r} \ln \frac{r}{|w_{n,j}(\lambda)|} \leq d^{-n} N_d(n) \ln^+ r.$$

Recalling that $d^{-n} N_d(n) \sim \frac{1}{n}$ and $L_n(\lambda, 0) \to L(\lambda)$ (see Remark 98), this implies that L_n^r converges pointwise to L and, by Theorem 89, that $(L_n^r)_n$ converges to L in $L_{loc}^1(M)$.

Now, to get the conclusion, one has to justify the following identity:

$$dd^c L_n^r = \frac{d^{-n}}{2\pi} \int_0^{2\pi} [Per_n(re^{i\theta})] d\theta.$$

Going back to definitions and taking $dd^c \ln|p_n(\lambda, re^{i\theta})| = [Per_n(re^{i\theta})]$ into account, one sees that this is a consequence of Fubini's theorem if one checks that $\ln|p_n(\lambda, re^{i\theta})|$ is locally integrable. Let K be a compact subset of M and c_n be an upper bound for $\ln|p_n(\lambda, re^{i\theta})|$ on $K \times [0, 2\pi]$. Then, the negative function $\ln|p_n(\lambda, e^{i\theta})| - c_n$ is indeed integrable on $K \times [0, 2\pi]$ as it follows from the fact that $L_n^r(\lambda) \geq C \frac{\ln r}{n}$:

$$\int_K \left(\int_0^{2\pi} \left(\ln|p_n(\lambda, re^{i\theta})| - c_n \right) d\theta \right) dV = 2\pi d^n \int_K L_n^r dV - 2\pi c_n \int_K dV$$

$$\geq \left(d^n C \frac{\ln r}{n} - c_n \right) 2\pi \int_K dV.$$

□

Remark 102. We have proved that $L_n^r(\lambda) := \dfrac{d^{-n}}{2\pi} \int_0^{2\pi} \ln|p_n(\lambda, re^{i\theta})| \, d\theta$ is pointwise converging to $L(\lambda)$ on M.

The following result is essentially a potential-theoretic consequence of the former one. It implicitly contains some information about the convergence of $d^{-n} [Per_n(w)]$ for arbitrary choices of w but seems hard to improve without furtherly use dynamical properties (see [4]).

Theorem 103. *For any family of degree d rational maps $(f_\lambda)_{\lambda \in M}$ one has*

$$d^{-n} \, dd^c_{(\lambda, w)} \ln|p_n(\lambda, w)| \to dd^c L(\lambda)$$

where $p_n(\cdot, w)$ are the canonical defining functions for the hypersurfaces $Per_n(w)$ given by Theorem 46.

Proof. Let us set
$$L_n(\lambda, w) := d^{-n} \ln|p_n(\lambda, w)|.$$

As we have seen in the two last subsections (see Remarks 98 and 102)

$$L_n(\lambda, 0) \to L(\lambda)$$

$$L_n^r(\lambda) := \frac{d^{-n}}{2\pi} \int_0^{2\pi} \ln|p_n(\lambda, re^{i\theta})| \, d\theta \to L(\lambda) \text{ for any } r \geq 0.$$

Let us also recall that the function L is continuous on M (see Theorem 86).

As the functions L_n are p.s.h and the sequence $(L_n)_n$ is locally uniformly bounded from above, we shall again use the compacity properties of p.s.h functions given by Theorem 89. Since $L_n(\lambda, 0)$ converges to $L(\lambda)$, the sequence $(L_n)_n$ does not converge to $-\infty$ and it therefore suffices to show that, among p.s.h functions on $M \times \mathbf{C}$, the function L is the only possible limit for $(L_n)_n$ in $L^1_{loc}(M \times \mathbf{C})$.

Let φ be a *p.s.h* function on $M \times \mathbf{C}$ and $(L_{n_j})_j$ a subsequence of $(L_n)_n$ which converges to φ in $L^1_{loc}(M \times \mathbf{C})$. Pick $(\lambda_0, w_0) \in M \times \mathbf{C}$. We have to prove that $\varphi(\lambda_0, w_0) = L(\lambda_0)$.

Let us first observe that $\varphi(\lambda_0, w_0) \leq L(\lambda_0)$. Take a ball B_ϵ of radius ϵ and centered at $(\lambda_0, w_0) \in M \times \mathbf{C}$. By the submean value property and the L^1_{loc}-convergence of L_n^+ (see Lemma 100) we have:

$$\varphi(\lambda_0, w_0) \leq \frac{1}{|B_\epsilon|} \int_{B_\epsilon} \varphi \, dm = \lim_j \frac{1}{|B_\epsilon|} \int_{B_\epsilon} L_{n_j} \, dm$$

$$\leq \lim_j \frac{1}{|B_\epsilon|} \int_{B_\epsilon} L_{n_j}^+ \, dm = \frac{1}{|B_\epsilon|} \int_{B_\epsilon} L \, dm$$

making then $\epsilon \to 0$, one obtains $\varphi(\lambda_0, w_0) \leq L(\lambda_0)$ since L is continuous.

Let us now check that $\limsup_j L_{n_j}(\lambda_0, w_0 e^{i\theta}) = L(\lambda_0)$ for almost all $\theta \in [0, 2\pi]$. Let $r_0 := |w_0|$. By Lemma 100, the sequence L_n^+ converges pointwise to L and therefore:

$$\limsup_j L_{n_j}(\lambda_0, w_0 e^{i\theta}) \leq \limsup_j L_{n_j}^+(\lambda_0, w_0 e^{i\theta}) = L(\lambda_0).$$

On the other hand, by pointwise convergence of $L_n^{r_0}$ to L and Fatou's lemma we have:

$$L(\lambda_0) = \lim_n L_n^{r_0}(\lambda_0) = \limsup_j \frac{1}{2\pi} \int_0^{2\pi} L_{n_j}(\lambda_0, r_0 e^{i\theta}) d\theta$$

$$\leq \frac{1}{2\pi} \int_0^{2\pi} \limsup_j L_{n_j}(\lambda_0, r_0 e^{i\theta}) d\theta$$

and the desired property follows immediately.

To end the proof we argue by contradiction and assume that $\varphi(\lambda_0, w_0) < L(\lambda_0)$. As φ is upper semi-continuous and L continuous, there exists a neighbourhood V_0 of (λ_0, w_0) and $\epsilon > 0$ such that

$$\varphi - L \leq -\epsilon \text{ on } V_0.$$

Pick a small ball B_{λ_0} centered at λ_0 and a small disc Δ_{w_0} centered at w_0 such that $B_0 := B_{\lambda_0} \times \Delta_{w_0}$ is relatively compact in V_0. Then, according to Hartogs Lemma 90, we have:

$$\limsup_j \left(\sup_{B_0} (L_{n_j} - L) \right) \leq \sup_{B_0} (\varphi - L) \leq -\epsilon.$$

This is impossible since, as we have seen before, we may find $(\lambda_0, r_0 e^{i\theta_0}) \in B_0$ such that $\limsup_j \left(L_{n_j}(\lambda_0, r_0 e^{i\theta_0}) - L(\lambda_0) \right) = 0$. □

Remark 104. Using standard techniques, one may deduce from the above Theorem that the set of multipliers w for which the bifurcation current T_{bif} is not a limit of the sequence $d^{-n}[\text{Per}_n(w)]$ is contained in a polar subset of the complex plane.

5.3.3 The Case of Neutral Cycles in Polynomial Families

We return to the family $(P_{c,a})_{(c,a) \in \mathbf{C}^{d-1}}$ of degree d polynomials. Recall that $P_{c,a}$ is the polynomial of degree d whose critical set is $\{0 = c_0, c_1, \cdots, c_{d-2}\}$ and such that $P_{c,a}(0) = a^d$ (see Sect. 3.1.2).

We want to prove that, in this family, $\lim_n d^{-n}[\text{Per}_n(w)] = T_{\text{bif}}$ for $|w| \leq 1$. Taking the results of the previous subsection into account (see Theorem 97), it remains to treat the case $|w| = 1$ and prove the following result due to Bassanelli and the author (see [4]).

Theorem 105. *In the family of degree d polynomials* $\lim_n d^{-n}[\text{Per}_n(e^{i\theta})] = T_{\text{bif}}$ *for any $\theta \in [0, 2\pi]$.*

We will follow the strategy described in Sect. 5.1. As we shall see, the proof would be rather simple if we would know that the bifurcation locus is accumulated by hyperbolic parameters. This is however unknown when $d \geq 3$ and is a source of technical difficulties (see the fourth step).

Proof. We denote by λ the parameter in $\mathbf{C}^{d-1}$ (i.e. $\lambda := (c, a)$) and set

$$L_n(\lambda) := d^{-n} \ln |p_n(\lambda, e^{i\theta})|$$

where the polynomials $p_n(\lambda, w)$ are those given by Theorem 46. We have to show that the sequence $(L_n)_n$ converges to L in L^1_{loc}.

We have already seen that $(L_n)_n$ is a uniformly locally bounded sequence of p.s.h functions on $\mathbf{C}^{d-1}$. Since the family $\{P_{c,a}\}_{(c,a) \in \mathbf{C}^{d-1}}$ contains hyperbolic parameters, on which the $L_n(\lambda) = L_n^+(\lambda, e^{i\theta})$, it follows from Lemma 100 that the sequence $(L_n)_n$ does not converge to $-\infty$. Thus, according to Theorem 89, we have to show that L is the only limit value of the sequence $(L_n)_n$ for the L^1_{loc} convergence.

Assume that (after taking a subsequence!) $(L_n)_n$ is converging in L^1_{loc} to φ. To prove that the p.s.h functions φ and L coincide we shall check that they satisfy the assumptions of Lemma 91.

First step: $\varphi \leq L$.

Since $L_n^+(\lambda, e^{i\theta})$ converges to L in L^1_{loc} (see Lemma 100) and $L_n(\lambda) \leq L_n^+(\lambda, e^{i\theta})$ we get

$$\varphi(\lambda_0) \leq \frac{1}{|B_\epsilon|} \int_{B_\epsilon} \varphi \, dm \leq \frac{1}{|B_\epsilon|} \int_{B_\epsilon} L \, dm$$

for any small ball B_ϵ centered at λ_0. The desired inequality then follows by making $\epsilon \to 0$ since the function L is continuous (see Theorem 86).

Second step: $Supp \, dd^c \varphi \subset Supp \, dd^c L$.

Since there are no persistent neutral cycles in the family $(P_{c,a})_{(c,a)\in \mathbf{C}^{d-1}}$, the hypersurfaces $Per_n(e^{i\theta})$ are contained in the bifurcation locus. This means that the functions L_n are pluriharmonic on $\mathbf{C}^{d-1} \setminus Supp \, dd^c L$. The same is thus true for the limit φ.

Third step: for any $\lambda_0 \in \mathbf{C}^{d-1}$ there exists a complex line $\mathcal{L}$ through λ_0 such that $\varphi = L$ on the unbounded component of $\mathcal{L} \setminus (\mathcal{L} \cap Supp(dd^c L))$.

By Theorem 43 we may pick a line $\mathcal{L}$ through λ_0 which meets infinity far from the cluster set of $\cup_i \mathcal{B}_i$ in $\mathbf{P}_\infty$. This means that for any λ in the unbounded component of $\mathcal{L} \setminus (\mathcal{L} \cap Supp \, (dd^c L))$ all critical points c_i belong to the super-attracting basin of ∞ and thus, λ is a hyperbolic parameter. This implies that $L_n(\lambda) = L_n^+(\lambda, e^{i\theta})$ and that, by Lemma 100, $\varphi(\lambda) = L(\lambda)$.

Fourth step: $\varphi = L$ on $Supp \, dd^c L$.

This is the most delicate part of the proof, it somehow proceeds by induction on d. To simplify the exposition, we will only treat the cases $d = 2$ and $d = 3$.

When $d = 2$ the parameter space is $\mathbf{C}$ and the bifurcation locus is the boundary $b\mathcal{M}$ of the Mandelbrot set. The unbounded stable component $(\mathcal{M})^c$ is hyperbolic and thus, as we saw in the last step, $(\varphi - L) = 0$ there. Since $(\varphi - L)$ is negative and u.s.c, this implies that $\varphi = L$ on $b\mathcal{M} = Supp \, dd^c L$.

Let us stress that this ends the proof when $d = 2$; the complex line $\mathcal{L}$ of the third step is the parameter space itself in that case !

We now assume that $d = 3$, the parameter space is then $\mathbf{C}^2$. Let us consider the sets U_k of parameters which do admit an attracting k-cycle:

$$U_k := \bigcup_{|w|<1} Per_k(w).$$

We have to show that $(\varphi - L)$ vanishes on the bifurcation locus. Since the bifurcation locus is accumulated by curves of the form $Per_k(0)$ (by Theorem 97 $\lim_k d^{-k}[Per_k(0)] = T_{\text{bif}}$), and the function $(\varphi - L)$ is negative and upper semi-continuous, it suffices to prove that

$$(\varphi - L) = 0 \text{ on all sets } U_k.$$

Let us first treat the problem on a curve $\mathcal{C} := Per_k(\eta)$ for $|\eta| < 1$ and show that

$(\star)$ the sequence $L_n|_\mathcal{C}$ converges uniformly to $L|_\mathcal{C}$ on the stable components.

We may assume that $\mathcal{C}$ is irreducible and desingularize it. This gives a one-dimensional holomorphic family $(P_{\pi(u)})_{u \in M}$. Keeping in mind that the elements of this family are degree 3 polynomials which do admit an attracting basin of period k and using the fact that the connectedness locus in $\mathbf{C}^2$ is compact (see Theorem 43), one sees that the family $(P_{\pi(u)})_{u \in M}$ enjoys the same properties than the quadratic polynomial family:

1. The bifurcation locus is contained in the closure of hyperbolic parameters.
2. The set of non-hyperbolic parameters is compact in M.

Exactly as for the quadratic polynomial family this implies that the sequence $L_n|_\mathcal{C}$ converges in L^1_{loc} to $L|_\mathcal{C}$ and the convergence is locally uniform on stable components since, as we already observed, the functions L_n are pluriharmonic there.

We now want to show that $\varphi = L$ on any open subset U_k. Again, as the stable parameters are dense and $(\varphi - L)$ is u.s.c and negative, it suffices to show that $\varphi = L$ on any stable component of U_k. On such a component the functions L_n are pluriharmonic and thus actually converge locally uniformly to φ. Then, $(\star)$ clearly implies that $\varphi = L$ on Ω. □

Remark 106. The relatively simple behaviour of the bifurcation locus near infinity within polynomials families is crucial in the above proof. It is an open problem to show that $\lim_n d^{-n}[\text{Per}_n(e^{i\theta})] = T_{\text{bif}}$ in general families of rational maps. The first reasonable case to study would be that of the moduli space Mod_2 for which precise informations concerning the bifurcation locus at infinity have been obtained by Epstein [31].

5.4 Laminated Structures in Bifurcation Loci

5.4.1 Holomorphic Motion of the Mandelbrot Set in Mod_2

We work here in the moduli space Mod_2 of degree two rational maps which, as we saw in Sect. 3.1.3, can be identified to $\mathbf{C}^2$. Our aim is to show that the bifurcation locus in the region

$$U_1 := \{\lambda \in \mathbf{C}^2 / f_\lambda \text{ has an attracting fixed point}\}$$

can be obtained by holomorphically moving the boundary $b\mathcal{M}$ of the Mandelbrot set. We remind that $b\mathcal{M}$ is the bifurcation locus of $Per_1(0)$ which is a complex line contained in U_1 and can be identified to the family of quadratic polynomials.

We will see simultaneously that the bifurcation current is uniformly laminar in the region U_1. Let us first recall some basic facts about holomorphic motions.

Definition 107. *Let M be a complex manifold and $E \subset M$ be any subset. A holomorphic motion of E in M is a map*

$$\sigma : E \times \Delta \ni (z, u) \mapsto \sigma(z, u) =: \sigma_u(z) \in M$$

which satisfies the following properties:

(i) $\sigma_0 = Id|_E$
(ii) $E \ni z \mapsto \sigma_u(z) \in M$ is one-to-one for every $u \in \Delta$
(iii) $\Delta \ni u \mapsto \sigma_u(z) \in M$ is holomorphic for every $z \in E$.

When the family of holomorphic discs in M enjoys good compactness properties, any holomorphic motion extends to the closure. In particular, when M is the Riemann sphere $\hat{\mathbf{C}}$, the Picard–Montel theorem combined with Hurwitz lemma easily leads to some famous extension statement which is usually called λ-lemma since the "time" parameter is denoted λ rather than u (see Lemma 59).

The main result of this subsection is the following. It was first proved by Goldberg and Keene (see [34]). The proof we present here exploits the formalism of bifurcation currents and is due to Bassanelli and the author (see [3]), this approach turns out to be much simpler and also provides some information on the laminarity of the bifurcation currents.

Theorem 108. *Let Ω_{hyp} be the union of all hyperbolic components of the Mandelbrot set $\mathcal{M}$ and $\heartsuit$ the main cardioid. Let B_{if_1} be the bifurcation locus in U_1 and $T_{bif}|_{U_1}$ be the associated bifurcation current. Let μ_1 be the harmonic measure of $\mathcal{M}$.*

There exists a continuous holomorphic motion

$$\sigma : \left((\Omega_{hyp} \setminus \heartsuit) \cup b\mathcal{M} \right) \times \Delta \to U_1$$

such that

$$\sigma (b\mathcal{M} \times \Delta) = B_{if_1} \text{ and } T_{bif}|_{U_1} = \int_{Per_1(0)} [\sigma(z, \Delta)] \, \mu_1.$$

In particular, B_{if_1} is a lamination with μ_1 as transverse measure. Moreover, the map σ is holomorphic on $(\Omega_{hyp} \setminus \heartsuit) \times \Delta$ and preserves the curves $Per_n(w)$ for $n \geq 2$ and $|w| \leq 1$.

Proof. First step: Holomorphic motion of $(\Omega_{hyp} \setminus \heartsuit)$.

The curve $Per_1(0)$ is actually the complex line $\lambda_1 = 2$. We will write $(2, \lambda_2) =: z$ the points of this line.

Let us consider $U_n := \{\lambda \in \mathbf{C}^2 / f_\lambda$ has an attracting cycle of period $n\}$ and $\Omega_n := U_n \cap Per_1(0)$. We recall that $\Omega_{hyp} := \heartsuit \cup \bigcup_{n \geq 2} \Omega_n$.

Let us also set $U_{n,1} := U_n \cap U_1$. By the Fatou–Shishikura inequality, a quadratic rational map has at most two non-repelling cycles. Thus $U_{n,1} \cap U_{m,1} = \emptyset$ when $n \neq m$ and there exists a well defined holomorphic map

$$\psi_n : U_{n,1} \to \Delta \times \Delta$$

which associates to every $\lambda \in U_{n,1}$ the pair $(w_n(\lambda), w_1(\lambda))$ where $w_n(\lambda)$ is the multiplier of the attracting n-cycle of λ and $w_1(\lambda)$ the multiplier of its attracting fixed point.

The cornerstone is the following transversality statement due to Douady and Hubbard (see also [2]):

Lemma 109. *The map ψ induces a biholomorphism*

$$\psi_{n,j} : U_{n,1,j} \to \Delta \times \Delta$$

on each connected component $U_{n,1,j}$ of $U_{n,1}$.

By the above Lemma, the connected components $\Omega_{n,j}$ of Ω_n coincides with $U_{n,1,j} \cap Per_1(0)$ and one clearly obtains a holomorphic motion $\sigma : (\Omega_{hyp} \setminus \heartsuit) \times \Delta \to U_1$ by setting:

$$\sigma(z,t) := (\psi_{n,j})^{-1}(w_n(z), t)$$

for any $z \in \Omega_{n,j}$.

Second step: extension of σ to $b\mathcal{M}$.

The key point here is that $\sigma(z,t) =: (\alpha(z,t), \beta(z,t))$ belongs to the complex line $Per_1(t)$ which, according to Proposition 54, is given by the equation

$$(t^2 + 1)\lambda_1 - t\lambda_2 - (t^3 + 2) = 0.$$

Thus $\sigma(z,t)$ is completely determined by $\beta(z,t)$:

$$\alpha(z,t) = \frac{1}{1+t^2}\left(t\beta(z,t) + t^3 + 2\right), \quad \forall t \in \Delta. \tag{27}$$

We will now identify $Per_1(0)$ with the deleted Riemann sphere $\hat{\mathbf{C}} \setminus \{\infty\}$ and set $\beta(\infty, t) = \infty$ for all $t \in \Delta$. Then, the map $\beta : (\{\infty\} \cup (\Omega_{hyp} \setminus \heartsuit)) \times \Delta \to \hat{\mathbf{C}}$ is clearly a holomorphic motion which, by Lemma 59, extends to the closure of $(\Omega_{hyp} \setminus \heartsuit)$. We thus obtain a continuous holomorphic motion

$$\beta : (\{\infty\} \cup (\Omega_{hyp} \setminus \heartsuit) \cup b\mathcal{M}) \times \Delta \to \hat{\mathbf{C}}.$$

As, by construction, $\beta(z,t) \neq \infty$ when $z \neq \infty$, the identity (27) shows that $\sigma(z,t) = (\alpha(z,t), \beta(z,t))$ extends to a continuous holomorphic motion of $((\Omega_{hyp} \setminus \heartsuit) \cup b\mathcal{M})$.

Third step: laminarity properties.

Let us show that $T_{bif}|_{U_1} = \int_{Per_1(0)} [\sigma(z, \Delta)] \, \mu_1$. According to the approximation formula given by Theorem 97 and applied on $Per_1(0)$ we have:

$$\mu_1 = \lim_m 2^{-m} \sum_{z \in Per_1(0) \cap Per_m(0)} \delta_{\sigma(z,0)}. \tag{28}$$

Let us set $T := \int_{Per_1(0)} [\sigma(z, \Delta)] \, \mu_1$. We have to check that $T = T_{bif}|_{U_1}$. Let ϕ be a $(1,1)$-test form in U_1. As the holomorphic motion σ is continuous, the function $z \mapsto \langle [\sigma(z,\Delta)], \phi \rangle$ is continuous as well. Then, using (28) one gets

$$\langle T, \phi \rangle = \lim_m 2^{-m} \sum_{z \in Per_1(0) \cap Per_m(0)} \langle [\sigma(z, \Delta)], \phi \rangle = \lim_m 2^{-m} \langle [Per_m(0)], \phi \rangle \quad (29)$$

where the last equality uses the fact that, according to Proposition 55, the curves $Per_m(0)$ have no multiplicity in U_1. Now the conclusion follows by using (29) and the approximation formula of Theorem 97 in U_1.

By construction, the map σ is holomorphic on $(\Omega_{hyp} \setminus \heartsuit) \times \Delta$ and preserves the curves $Per_n(w)$ for $n \geq 2$ and $|w| < 1$. This extends to $|w| = 1$ by continuity.

Using the continuity of σ, one easily sees that $\sigma(b\mathcal{M}, \Delta)$ is closed in U_1 and therefore contains the support of $\int_{Per_1(0)} [\sigma(z, \Delta)] \, \mu_1$. By the above formula we thus have

$$B_{\mathrm{if}_1} = Supp(T_{\mathrm{bif}}|_{U_1}) \subset \sigma(b\mathcal{M}, \Delta).$$

The opposite inclusion easily follows from the construction of σ: any point in $\sigma(b\mathcal{M}, \Delta)$ is a limit of $z_m \in Per_1(0) \cap Per_m(0)$ where $m \to +\infty$. □

Instead of using the basic λ-Lemma we could have use its far advanced generalization due to Slodkowski and get a motion on the full line $Per_1(0)$.

Theorem 110 (Slodkowski λ-lemma). *Let $E \subset \hat{\mathbf{C}}$ be a subset of the Riemann sphere and $\sigma : E \times \Delta \ni (z, t) \mapsto \sigma(z, t) \in \hat{\mathbf{C}}$ be a holomorphic motion. Then σ extends to a holomorphic motion $\tilde{\sigma}$ of $\hat{\mathbf{C}}$. Moreover $\tilde{\sigma}$ is continuous on $\overline{E} \times \Delta$ and $z \mapsto \tilde{\sigma}(z, t)$ is a K-quasi-conformal homeomorphism for $K := \frac{1+|t|}{1-|t|}$.*

Our reference for quasi-conformal maps is the book [36] where one can also find a nice proof of Slodkowski theorem due to Chirka and Rosay.

Using Slodkowski Theorem one may obtain further informations on the motion given by Theorem 108. We refer to our paper [3] for a proof.

Theorem 111. *Let $\sigma : ((\Omega_{hyp} \setminus \heartsuit) \cup b\mathcal{M}) \times \Delta \to U_1$ be the holomorphic motion given by Theorem 108. Then σ extends to a continuous holomorphic motion $\tilde{\sigma} : Per_1(0) \times \Delta \longrightarrow U_1$ which is onto. All stable components in U_1 are of the form $\tilde{\sigma}(\omega \times \Delta)$ for some stable component ω in $Per_1(0)$. Moreover, the map $z \mapsto \tilde{\sigma}(z, t)$ is a quasi-conformal homeomorphism for each t and $\tilde{\sigma}$ is one-to-one on $(Per_1(0) \setminus \overline{\heartsuit}) \times \Delta$ where $\heartsuit$ is the main cardioid.*

Theorem 111 shows that non-hyperbolic components exist in U_1 if and only if such components exist within the quadratic polynomial family $Per_1(0)$. Let us underline that, in relation with Fatou's problem on the density of hyperbolic rational maps, it is conjectured that such components do not exist.

It might be useful to note that holomorphic motions enjoy good Hölder regularity properties. We will end this subsection by giving a basic result in this direction.

Lemma 112. *Let $h : B(0, r) \times E \longrightarrow \mathbf{P}^1$ be a holomorphic motion of some $E \subset \mathbf{P}^1$ parametrized by a ball in $\mathbf{C}^k$. Then, for any $z_0 \in E$ we may find $0 < r_1 \leq r$ and $\eta > 0$ such that the following estimates hold for $0 < r' < r_1$, $\lambda \in B(0, r')$ and $z, z' \in D(z_0, \eta) \cap E$:*

$$C'(r')|z-z'|^{\frac{r_1+\|\lambda\|}{r_1-\|\lambda\|}} \leq |h(\lambda,z) - h(\lambda,z')| \leq C(r')|z-z'|^{\frac{r_1-\|\lambda\|}{r_1+\|\lambda\|}}$$

where $C(r'), C'(r')$ are strictly positive constants.

Proof. Let $z_0 \in E$. As h is continuous, one finds $0 < \eta < 1$ and $0 < r_1 \leq r$ such that

$$|h(\lambda,z) - h(\lambda,z_0)| < 1$$

for any $\lambda \in B(0,r_1)$ and any $z \in D(z_0,\eta) \cap E$. Let us now pick two distinct points z,z' in $D(z_0,\eta) \cap E$, a parameter λ in $B(0,r_1) \setminus \{0\}$ and set

$$g_{z,z'}^\lambda(t) := -\log \frac{|h(r_1 t \frac{\lambda}{\|\lambda\|},z) - h(r_1 t \frac{\lambda}{\|\lambda\|},z')|}{2}, \ t \in D.$$

Since $z \neq z'$ and $z,z' \in D(z_0,\eta) \cap E$ one has

$$0 < |h(r_1 t \frac{\lambda}{\|\lambda\|},z) - h(r_1 t \frac{\lambda}{\|\lambda\|},z')| < 2$$

for any $t \in \Delta$ and therefore $g_{z,z'}^\lambda$ is a positive harmonic function on the unit disc Δ. Harnack inequalities yield

$$\frac{1-|t|}{1+|t|} g_{z,z'}^\lambda(0) \leq g_{z,z'}^\lambda(t) \leq \frac{1+|t|}{1-|t|} g_{z,z'}^\lambda(0)$$

for any $t \in D$, which means that

$$\left(\frac{|z-z'|}{2}\right)^{\frac{1+|t|}{1-|t|}} \leq \frac{|h(rt\frac{\lambda}{\|\lambda\|},z) - h(rt\frac{\lambda}{\|\lambda\|},z')|}{2} \leq \left(\frac{|z-z'|}{2}\right)^{\frac{1-|t|}{1+|t|}}$$

for $t \in D$. Taking $t = \|\lambda\|/r_1$ we get

$$\left(\frac{|z-z'|}{2}\right)^{\frac{r_1+\|\lambda\|}{r_1-\|\lambda\|}} \leq \frac{|h(\lambda,z) - h(\lambda,z')|}{2} \leq \left(\frac{|z-z'|}{2}\right)^{\frac{r_1-\|\lambda\|}{r_1+\|\lambda\|}}.$$

The conclusion follows by setting $C(r') := 2^{\frac{2r'}{r_1+r'}}$ and $C'(r') := 2^{\frac{-2r'}{r_1-r'}}$ for $0 < r' < r_1$. □

5.4.2 Further Laminarity Statements for T_{bif}

The following result is an analogue of Theorem 108 in the regions

$$U_n := \{\lambda \in Mod_2/f_\lambda \text{ has an attracting cycle of period } n\}.$$

It shows, in particular, that the bifurcation current in Mod_2 is uniformly laminar in the regions U_n. It has been established by Bassanelli and the author in [3].

Theorem 113. *Let B_{if_n} be the bifurcation locus in U_n and $T_{bif}|_{U_n}$ be the associated bifurcation current. Let $B_{if_n^c}$ be the bifurcation locus in the central curve $Per_n(0)$ and μ_n^c be the associated bifurcation measure. Then, there exists a map*

$$\sigma : B_{if_n^c} \times \Delta \longrightarrow B_{if_n}$$
$$(\lambda, t) \longmapsto \sigma(\lambda, t)$$

such that:

(1) $\sigma\left(B_{if_n^c} \times \Delta\right) = B_{if_n}$,
(2) σ is continuous, $\sigma(\lambda, \cdot)$ is one-to-one and holomorphic for each $\lambda \in B_{if_n^c}$,
(3) $p_n(\sigma(\lambda, t), t) = 0; \forall \lambda \in B_{if_n^c}, \forall t \in \Delta$,
(4) the discs $(\sigma(\lambda, \Delta))_{\lambda \in B_{if_n^c}}$ are mutually disjoint.

Moreover the bifurcation current in U_n is given by

$$T_{bif}|_{U_n} = \int_{B_{if_n^c}} [\sigma(\lambda, \Delta)] \, \mu_n^c$$

and, in particular, B_{if_n} is a lamination with μ_n^c as transverse measure.

The proof is similar to that of Theorem 113 but requires a special treatment for the extension problems since there is no λ-lemma available. The key is to use the fact that, by construction, the starting motion $\sigma : \left(\bigcup_{m \neq n}(U_m \cap U_n)\right) \times \Delta \longrightarrow U_n$ satisfies the following property:

$$p_n(\sigma(\lambda, t), t) = 0; \quad \forall \lambda \in B_{if_n^c}, \quad \forall t \in \Delta.$$

Such a motion is what we call a p_n-guided holomorphic motion. Using Zalcman rescaling lemma, one proves the following compactness property for guided holomorphic motions. We stress that here, an holomorphic motion $\mathcal{G}$ is seen as a family of disjoints holomorphic discs σ and $\mathcal{G}_{t_0}$ is the set of points $\sigma(t_0)$.

Theorem 114. *Let $p(\lambda, w)$ be a polynomial on $\mathbf{C}^2 \times \mathbf{C}$ such that the degree of $p(\cdot, w)$ does not depend on $w \in \Delta$. Let $\mathcal{G}$ be a p-guided holomorphic motion in $\mathbf{C}^2$ such that any component of the algebraic curve $\{p(\cdot, t) = 0\}$ contains at least three points of $\mathcal{G}_t$ for every $t \in \Delta$. Then, for any $\mathcal{F} \subset \mathcal{G}$ such that $\mathcal{F}_{t_0}$ is relatively compact in $\mathbf{C}^2$ for some $t_0 \in \Delta$, there exists a continuous p-guided holomorphic motion $\hat{\mathcal{F}}$ in $\mathbf{C}^2$ such that $\mathcal{F} \subset \hat{\mathcal{F}}$ and $\hat{\mathcal{F}}_{t_0} = \overline{\mathcal{F}_{t_0}}$.*

The above Theorem plays the role of the λ-lemma in the proof of Theorem 113. We refer to the paper [3] for details.

We will now end this section by presenting some more precise laminarity result for T_{bif} which is due to Dujardin (see [27])

Theorem 115. *Within the polynomial family of degree* 3, *the bifurcation current is laminar on every open set where one critical point is passive and, in particular, outside the connectedness locus.*

We refer to Sect. 3.2 for the discussion of the connectedness locus within polynomial families.

A few comments on the concept of laminar currents are necessary here. We restrict ourself to the case of a positive $(1, 1)$-current T on a complex manifold M. One says that T is *locally uniformly laminar* on M if any point of the support of T admits a neighbourhood on which T is of the form

$$T = \int_\tau [\Delta_t]\, \mu(t)$$

where Δ_t is a lamination by holomorphic discs with a transverse measure μ.

With this terminology, Theorem 113 says that T_{bif} is uniformly laminar on the open sets of Mod_2 where an attracting basin of given period exists. In particular, T_{bif} is locally uniformly laminar on any open set of Mod_2 where a critical point is attracted by a cycle.

One says that T is *laminar* on M if there exists a sequence of open sets $\Omega_i \subset M$ and a sequence of currents T_i which are locally uniformly laminar on Ω_i and such that the sequence $(T_i)_i$ increasingly converges to T. This is actually equivalent to say that

$$T = \int_A [\Delta_a]\, \mu(a)$$

where $(\Delta_a)_{a \in A}$ is a family of compatible (in the sense of analytic continuation) holomorphic discs parametrized by an abstract set A and μ is a measure on A.

The methods used in Mod_2 for proving Theorem 113 work also in the cubic polynomial family (the situation is actually technically simpler) and thus, for this family too, the bifurcation current is locally uniformly laminar on the open subsets where a critical point is attracted by a cycle. As it is not hard to see, this would imply Theorem 115 if the hyperbolicity conjecture was known to hold in the cubic polynomial family. To overcome this difficulty, one uses more sophisticated tools on laminar currents. Precisely, the proof of theorem 115 is based on a laminarity criterion due to De Thelin [17] which says that a current $T := \lim_n \frac{1}{d_n}[C_n]$, where $[C_n]$ is a sequence of integration currents on curves in $\mathbf{C}^2$, is laminar if *genus* $(C_n) = O(d_n)$.

6 The Bifurcation Measure

The powers $(T_{\text{bif}})^k$ of the bifurcation current detect stronger bifurcations and allow to define a very interesting stratification of the bifurcation locus. Among them, the highest power is a measure whose support should be the seat of the strongest bifurcations. In this section, we will survey the basic properties of this measure and then describe its support as the closure of some sets of remarkable parameters. Although some results about the intermediate powers of T_{bif} can be easily deduced from our exposition, we will not discuss explicitly these currents here and refer the reader to the survey of Dujardin [28] for more details on this topic.

6.1 A Monge–Ampère Mass Related with Strong Bifurcations

6.1.1 Basic Properties

Since the bifurcation current T_{bif} of any holomorphic family $(f_\lambda)_{\lambda \in M}$ of degree d rational maps has a continuous potential L (see Definition 82 and Theorem 86), one may define the powers $(T_{\text{bif}})^k := T_{\text{bif}} \wedge T_{\text{bif}} \wedge \cdots \wedge T_{\text{bif}}$ for any $k \leq m := dim M$. We recall that for any closed positive current T, the product $dd^c L \wedge T$ is defined by $dd^c L \wedge T := dd^c(LT)$. In particular, $(T_{\text{bif}})^m$ is a positive measure on M which is equal to the Monge–Ampère mass of the Lyapunov function L.

Definition 116. Let $(f_\lambda)_{\lambda \in M}$ be a holomorphic family of degree d rational maps parametrized by a complex manifold M of dimension m. The *bifurcation measure* μ_{bif} of the family is the positive measure on M defined by

$$\mu_{\text{bif}} = \frac{1}{m!}(T_{\text{bif}})^m = \frac{1}{m!}(dd^c L)^m$$

where T_{bif} is the bifurcation current and L the Lyapunov function of the family.

The following proposition is a direct consequence of the definition and the fact that μ_{bif} has locally bounded potentials.

Proposition 117. *The support of μ_{bif} is contained in the bifurcation locus and μ_{bif} does not charge pluripolar sets.*

It is actually possible to define the bifurcation measure in the moduli space Mod_d of degree d rational maps and show that this measure has strictly positive and finite mass (see [2, Proposition 6.6]). Although all the results we will present here are true in Mod_d, we will restrict ourself to the technically simpler situation of holomorphic families. The example we have in mind are the polynomial families and the moduli space Mod_2 which, in some sense, can be treated as a holomorphic family (see Theorem 36).

In arbitrary holomorphic families, the measure μ_{bif} can identically vanish. Moreover, when $\mu_{\text{bif}} > 0$, it is usually quite involved to prove it. Note however that this will follow from standard arguments in polynomial families. The following simple observation already shows that $\mu_{\text{bif}} > 0$ in Mod_2 (and more generally Mod_d), it has also its own interest.

Proposition 118. *In any holomorphic family, all rigid Lattès examples belong to the support of μ_{bif}.*

Proof. The parameters corresponding to rigid Lattès examples are isolated. Thus, as Theorem 28 shows, the Lyapunov function L takes the value $\frac{\ln d}{2}$ and has a strict minimum at any such parameter. Applying the comparison principle to L and some constant function $\frac{\ln d}{2} + \epsilon$, one sees that the Monge–Ampère measure $(dd^c L)^m$ cannot vanish around a strict minimum of L. This yields the conclusion. $\square$

Using pluripotential theory, it is possible to show that any Lattès example in Mod_d lies in the support of $(T_{bif})^{2d-3}$. Buff and Gauthier have recently shown that such maps actually belong to the support of μ_{bif} (see [14]). Their proof requires transversality statements and uses the quadratic differentials techniques.

Theorem 119. *In Mod_d, Lattès examples belong to the support of the bifurcation measure.*

We end this subsection by showing that the activity currents have no self-intersection. This is a useful geometric information which, in particular, shows that the activity of all critical points is a necessary condition for a parameter to be in the support of the bifurcation measure. It was first proved by Dujardin–Favre in the context of polynomial families (see [29, Proposition 6.9]), we present here a general argument due to Gauthier [33].

Theorem 120. *Let $(f_\lambda)_{\lambda \in M}$ be any holomorphic family of degree d rational maps with marked critical points. The activity currents T_i satisfy $T_i \wedge T_i = 0$. In particular, when $m := \dim M = 2d - 2$ (or $m = d - 1$ for polynomial families) then*

$$\mu_{bif} = T_1 \wedge T_2 \wedge \cdots \wedge T_m$$

and the support of μ_{bif} is contained in the intersection of the activity loci of the critical points.

The proof is very close to that of a density statement which will be presented in the next section, it combines the following potential-theoretic Lemma with a dynamical observation.

Lemma 121. *Let u be a continuous p.s.h function on some open subset Ω in $\mathbf{C}^2$. Let Γ be the union of all analytic subsets of Ω on which u is harmonic. If the support of $dd^c u$ is contained in $\overline{\Gamma}$ then $dd^c u \wedge dd^c u$ vanishes on Ω.*

Proof. Let us set $\mu := dd^c u \wedge dd^c u$. Let B_r be an open ball of radius r whose closure is contained in Ω, we have to show that $\mu\left(B_{\frac{r}{2}}\right) = 0$.

Denote by h the solution of the Dirichlet–Monge–Ampère problem with data u on bB_r:

$$h = u \text{ on the boundary of } B_r$$

$$dd^c h \wedge dd^c h = 0 \text{ on } B_r \text{ (i.e. } h \text{ is maximal on } B_r\text{).}$$

The function h is $p.s.h$ and continuous on $\overline{B_r}$ (see [5]). As h is $p.s.h$ maximal and coincides with the $p.s.h$ function u on $b\overline{B_r}$, we have $u \leq h$ on $\overline{B_r}$. For any $\epsilon > 0$ we define

$$D_\epsilon := \{\lambda \in B_{\frac{r}{2}} \; / \; 0 \leq h(\lambda) - u(\lambda) \leq \epsilon\}.$$

We will see that our assumption implies that

$$\text{Supp } \mu \cap B_{\frac{r}{2}} \subset D_\epsilon \text{ for all } \epsilon > 0. \tag{30}$$

Indeed, if γ is a complex curve in Ω on which u is harmonic then, the maximum modulus principle, applied to $(h - u)$ on $\gamma \cap B_r$ implies that $h = u$ on γ. Then, as $(h - u)$ is continuous on B_r and $\text{Supp } \mu \subset \text{Supp } dd^c u \subset \overline{\Gamma}$, we get $\text{Supp } \mu \cap B_r \subset \{h = u\}$.

Now, a result due to Briend–Duval (see [11] or [23, Théorème A.10.2]) says that

$$\mu(D_\epsilon) \leq C\epsilon \tag{31}$$

where C only depends on u and B_r. From (30) and (31) we deduce that $\mu\left(B_{\frac{r}{2}}\right) = 0$. □

We may now end the proof of Theorem 120.

Proof. We only have to show that $T_i \wedge T_i = 0$, the remaining then follows from the identity $T_{\text{bif}} = \sum_i T_i$ (see Theorem 84).

The statement is local and we may therefore assume that $M = \mathbf{C}^k$. Moreover, an elementary slicing argument allows to reduce the dimension to $k = 2$. We apply the above Lemma with $u = G_\lambda(\hat{c}_i(\lambda))$ (see Lemma 83). We have to show that the support of $dd^c u = T_i$ is accumulated by curves on which the critical point c_i is passive. These curves are of the form $\{f_\lambda^n(c_i(\lambda)) = c_i(\lambda)\}$ and their existence follows from Lemma 64. □

Remark 122. An example, due to A. Douady, shows that the activity of all critical points is not sufficient for a parameter to be in the support of the bifurcation measure. We will present this example in the next subsection (see Example 125).

6.1.2 Some Concrete Families

We first discuss the case of the polynomial families introduced in Sect. 3.1.2. We follow here the paper [29] by Dujardin and Favre.

Proposition 123. *The bifurcation measure μ_{bif} of the degree d polynomial family $(P_{c,a})_{(c,a) \in \mathbf{C}^{d-1}}$ is a probability measure supported on the connectedness locus $\mathcal{C}$. It coincides with the pluricomplex equilibrium measure of the compact set $\mathcal{C}$ and its support is the Shilov boundary of $\mathcal{C}$.*

Proof. Let us recall that the Green function of the polynomial $P_{c,a}$ is denoted $g_{c,a}$. The connectedness locus $\mathcal{C}$ is a compact subset of $\mathbf{C}^{d-1}$ which coincides with the intersection $\cap_{0 \le i \le d-2} \mathcal{B}_i$ where $\mathcal{B}_i$ is the set of parameters for which the orbit of the critical point c_i is bounded (see Theorem 43). As the support of the activity current T_i is contained in $b\mathcal{B}_i$ we deduce that $Supp\ \mu_{\mathrm{bif}} \subset \mathcal{C}$ from Theorem 120. All the remaining follows from the fact that

$$\mu_{\mathrm{bif}} = (dd^c \mathcal{G})^{d-1}$$

where

$$\mathcal{G} := sup\{u\ p.s.h\ /\ u - \ln^+ max\{|a|, |c_k|\} \le O(1), u \le 0\ \text{on}\ \mathcal{C}\}.$$

is the pluricomplex Green function of $\mathcal{C}$ with pole at infinity. An identity which we shall now prove.

Let us first establish that $\mu_{\mathrm{bif}} = (dd^c G)^{d-1}$ where $G := max\{g_0, g_1, \cdots, g_{d-2}\}$ and $g_i := g_{c,a}(c_i)$. We show by induction that $T_0 \wedge T_1 \wedge \cdots \wedge T_l = (dd^c G_l)^{l+1}$ for $0 \le l \le d - 2$ where $G_l := max\{g_0, g_1, \cdots, g_l\}$. This comes from the following computation:

$$\begin{aligned} T_0 \wedge T_1 \wedge \cdots \wedge T_{l-1} \wedge T_l &= dd^c \left(g_l (dd^c G_{l-1})^l\right) = dd^c \left(G_l (dd^c G_{l-1})^l\right) \\ &= dd^c \left(G_{l-1} (dd^c G_{l-1})^{l-1} \wedge dd^c G_l\right) \\ &= dd^c \left(G_l (dd^c G_{l-1})^{l-1} \wedge dd^c G_l\right) = (dd^c G_l)^{l+1} \end{aligned}$$

the second equality follows from $g_l = G_l$ on the support of $(dd^c G_{l-1})^l$ and the fourth from $G_l = G_{l-1}$ on the support of $dd^c G_l$, the last equality is obtained by repeating the same arguments $l - 1$ times.

It remains to show that $G = \mathcal{G}$. The proof is standard and relies on the estimate given by Proposition 44 and the fact that G is maximal outside $\mathcal{C}$. We refer to the paper [29, Proposition 6.14] for more details. □

Remark 124. The above result shows that the support of the bifurcation measure is topologically much smaller than the bifurcation locus. Indeed, the Shilov boundary is usually a tiny part of the full boundary. For instance, the boundary of the bidisc $\Delta \times \Delta$ is $(\Delta \times S^1) \cup (S^1 \times \Delta)$ while its Shilov boundary is the real torus $S^1 \times S^1$.

We will now present the example mentioned at the end of last subsection.

Example 125. In the holomorphic family of degree 3 polynomials

$$\left((1 + \alpha_1)z + (\frac{1}{2} + \alpha_2)z^2 + z^3\right)_{\alpha \in V_0}$$

where V_0 is a neighbourhood of the origin in $\mathbf{C}^2$ the critical points are both active at the origin $(0, 0)$ but $(0, 0) \notin Supp\ \mu_{\mathrm{bif}}$.

This family is a deformation of the polynomial $P_0 := z + \frac{1}{2}z^2 + z^3$. If V_0 is small enough we have two marked critical points $c_1(\alpha)$ and $c_2(\alpha)$. The origin 0 is a parabolic fixed point for P_0 and we may assume that P_α has two fixed points counted with multiplicity near 0 for all $\alpha \in V_0$. As P is real, its critical points are complex conjugate and both of them are attracted by the parabolic fixed point at 0, moreover their orbits are not stationary.

We first show that $(0,0) \notin Supp\ \mu_{\text{bif}}$. When the fixed points of P_α are distinct, we denote by $m_1(\alpha)$ and $m_2(\alpha)$ their multipliers. When this is the case, it turns out that either $|m_1(\alpha)| < 1$ or $|m_2(\alpha)| < 1$ and thus one of the fixed points attracts a critical point. This can be seen by using the holomorphic fixed point formula (see [49, Lecture 12]).

By Theorem 120, this implies that $\alpha \notin Supp\ \mu_{\text{bif}}$. We thus see that μ_{bif} is supported on the subvariety of parameters α for which the fixed point is double. By Proposition 117 this implies that μ_{bif} vanishes near $(0,0)$.

Let us now see that both critical points are active. We may assume that the family is parametrized by a disc D in $\mathbf{C}$ such that P_α has two distinct fixed points when $\alpha \neq 0$. Assume to the contrary that a critical point $c(\alpha)$ is passive. Then, after taking a subsequence, the sequence $u_n(\alpha) := P_\alpha^n(c(\alpha))$ is uniformly converging to $u(\alpha)$. Since the polynomial P_0 is real, its critical points are complex conjugate and must therefore both be attracted by the parabolic fixed point 0. Thus $u_n(0)$ converges to 0 and never belongs to the analytic set $Z := \{(\alpha, z) \in D \times \mathbf{P}^1\ /\ P_\alpha(z) = z\}$. If $u_n(\alpha) \in Z$ for $\alpha \neq 0$ then the orbit of $c(\alpha)$ is captured by a fixed point which, since $c(\alpha)$ is passive, must be attracting. In that case the curve $(\alpha, u(\alpha))$ is contained in Z. If $u_n(\alpha)$ never belongs to Z then, by Hurwitz lemma, the curve $(\alpha, u(\alpha))$ is also contained in Z. This is impossible since, for some α close to 0, the critical orbit should be attracted by a repelling fixed point. □

The situation in the moduli space Mod_2 is more complicated. We recall that Mod_2 can be identified to $\mathbf{C}^2$. Using the results which will be obtained in the last section of this section and the holomorphic motions constructed in Sect. 5.4, it is possible to show that the support of the bifurcation locus is not bounded.

6.2 Density Statements

Our aim here is to explain why the support of the bifurcation measure may be considered as the locus of the strongest bifurcations. To this purpose we will show that the remarkable parameters introduced in Sect. 4.1.3 accumulate the support of the bifurcation measure. These informations will be obtained through equidistribution arguments for the bifurcation current and its powers.

Let us mention that, using further techniques, some of the above mentioned remarkable parameters will be shown to belong to the support of the bifurcation measure in the next section.

6.2.1 Strongly Misiurewicz Parameters

The results given in this subsection are essentially due to Dujardin and Favre. We present them in the setting of polynomial families and refer the reader to the original paper [29] for a greater generality.

Theorem 126. *In the degree d polynomial family $(P_{c,a})_{(c,a)\in \mathbf{C}^{d-1}}$ let us define a sequence of analytic sets by:*

$$W_{n_0,\cdots,n_l} := \cap_{j=0}^{l}\{P_{c,a}^{n_j}(c_j) = P_{c,a}^{k(n_j)}(c_j)\}$$

where $l \leq d-2$ and $k(n_j) < n_j$. Then

$$\lim_{n_{d-2}\to\infty}\cdots\lim_{n_0\to\infty}\frac{1}{d^{n_{d-2}}+\cdots+d^{n_0}}[W_{n_0,\cdots,n_{d-2}}] = \mu_{bif}$$

and $W_{n_0,\cdots,n_{d-2}}$ is finite.

Proof. We treat the case $d = 3$ which is actually not very different from the general case.

Let us first observe that W_{n_0,n_1} has codimension at least two and is contained in the connectedness locus which is compact (see Theorem 43). Thus W_{n_0,n_1} is a finite set.

Applying a version of Theorem 92 suitably adapted to the family W_{n_0} yields

$$\lim_{n_1\to\infty} d^{n_1}[W_{n_0,n_1}] = T_1 \wedge [W_{n_0}]$$

where T_1 is the activity current of the critical point c_1. By the same Theorem one has $\lim_{n_0\to\infty} d^{n_0}[W_{n_0}] = T_0$ where T_0 is the activity current of c_0 and this, since T_1 has continuous potentials, gives

$$\lim_{n_0\to\infty} T_1 \wedge d^{n_0}[W_{n_0}] = T_1 \wedge T_0.$$

The conclusion follows immediately since, according to Theorem 120, $\mu_{bif} = T_0 \wedge T_1$. □

An important consequence of the above result is that the support of the bifurcation measure is accumulated by strongly Misiurewicz polynomials. An alternative proof of that fact will be given in the next subsection for arbitrary families. We refer to Definition 72 for a definition of Misiurewicz parameters.

Corollary 127. *In polynomial families, the support of the bifurcation measure is contained in the closure of strongly Misiurewicz parameters: $\mathrm{Supp}\,\mu_{bif} \subset \overline{M_{iss}}$.*

Proof. By the above Theorem

$$\lim_{n_{d-2}\to\infty}\cdots\lim_{n_0\to\infty}\frac{1}{d^{n_{d-2}}+\cdots+d^{n_0}}[\cap_0^{d-2}\{P_{c,a}^{n_j}(c_j)=P_{c,a}^{n_j-1}(c_j)\}]=\mu_{\text{bif}}. \quad (32)$$

Let us observe that

$$H_j := \{P_{c,a}^{n_j}(c_j)=P_{c,a}^{n_j-1}(c_j)\} = Preper_{n_j} \cup Fix_j$$

where, for parameters in $Preper_{n_j}$ the critical point c_j is strictly preperiodic to a (necessarily) repelling fixed point while c_j is fixed for parameters in Fix_j.

Now Theorem 92 may be rewritten as

$$\lim_{n_j\to\infty}\frac{1}{d^{n_j}}[Preper_{n_j}] + \frac{\alpha_{n_j}}{d^{n_j}}[Fix_j] = T_j \quad (33)$$

but, as T_j cannot charge the hypersurface Fix_j, we must have $\frac{\alpha_{n_j}}{d^{n_j}} \to 0$. Thus $\lim_{n_j\to\infty}\frac{1}{d^{n_j}}[Preper_j]=T_j$ and (32) yields

$$\lim_{n_{d-2}\to\infty}\cdots\lim_{n_0\to\infty}\frac{1}{d^{n_{d-2}}+\cdots+d^{n_0}}[\cap_{j=0}^{d-2}Preper_{n_j}]=\mu_{\text{bif}}.$$

The conclusion follows immediately since $\cap_{j=0}^{d-2}Preper_{n_j} \subset M$ iss. □

6.2.2 Shishikura or Hyperbolic Parameters

We aim here to show that the support of the bifurcation measure in Mod_d is simultaneously accumulated by Shishikura and hyperbolic parameters (see Sect. 4.1.3 for definitions):

$$Supp\,\mu_{\text{bif}} \subset \overline{S\text{hi}} \cap \overline{H\text{yp}}.$$

It is worth emphasize that both statements will be deduced in the same way from the following generalized version of Theorem 101. The results discussed in this subsection and the next one are due to Bassanelli and the author [2, 3].

Theorem 128. *Let μ_{bif} be the bifurcation measure of a holomorphic family $(f_\lambda)_{\lambda\in M}$ of rational maps. Let m denote the complex dimension of M. Let $0 < r \leq 1$. Then there exists increasing sequences of integers $k_2(n),\ldots,k_m(n)$ such that:*

$$\mu_{bif} = \lim_n \frac{d^{-(n+k_2(n)+\cdots+k_m(n))}}{m!(2\pi)^m}$$

$$\times \int_{[0,2\pi]^m} [Per_n(re^{i\theta_1})] \wedge \bigwedge_{j=2}^m [Per_{k_j(n)}(re^{i\theta_j})]\,d\theta_1\cdots d\theta_m.$$

Moreover, we may assume that $k_j(n) \neq k_i(n)$ when $i \neq j$.

We will derive that result from Theorem 101 by simple calculus arguments with currents.

Proof. For any fixed variety $Per_p(re^{i\theta_p})$, the set of $\theta \in [0, 2\pi]$ for which $Per_p(re^{i\theta_p})$ shares a non trivial component with $Per_m(re^{i\theta})$ for some $m \in \mathbf{N}^*$ is at most countable. This follows from Fatou's theorem on the finiteness of the set of non-repelling cycles. Thus, the wedge products $[Per_n(re^{i\theta_1})] \wedge [Per_{k_2(n)}(re^{i\theta_2})] \cdots \wedge [Per_{k_m(n)}(re^{i\theta_m})]$ make sense for almost every $(\theta_1, \cdots, \theta_m) \in [0, 2\pi]^m$ and the integrals

$$\int_{[0,2\pi]^m} [Per_n(re^{i\theta_1})] \wedge \bigwedge_{j=2}^{m} [Per_{k_j(n)}(re^{i\theta_j})] \, d\theta_1 \cdots d\theta_m$$

are well defined.

Next, we need the following formula which has been justified for $q = 1$ at the end of the proof of Theorem 101. The proof is similar for $q > 1$ and we shall omit it. Recall that $L_n^r(\lambda) := \frac{d^{-n}}{2\pi} \int_0^{2\pi} \ln |p_n(\lambda, re^{i\theta})| \, d\theta$.

$$dd^c L_{n_1}^r \wedge \cdots \wedge dd^c L_{n_q}^r = \frac{d^{-(n_1 + \cdots + n_q)}}{(2\pi)^q} \int_{[0,2\pi]^q} \bigwedge_{k=1}^{q} [Per_{n_k}(re^{i\theta_k})] d\theta_1 \cdots d\theta_q.$$

To prove the convergence, we may replace M by $\mathbf{C}^m$ since the problem is local. The conclusion is obtained by using Theorem 101, the above formula and the next Lemma inductively.

Lemma 129. *If $S_n \to (dd^c L)^p$ for some sequence $(S_n)_n$ of closed, positive (p, p)-currents on M then $dd^c L_{k(n)}^r \wedge S_n \to (dd^c L)^{p+1}$ for some increasing sequence of integers $k(n)$.*

Let us briefly justify Lemma 129. Let us denote by s_n the trace measure of S_n, as M has been identified with $\mathbf{C}^m$ this measure is given by $s_n := S_n \wedge (dd^c |z|^2)^{m-p}$. Since S_n is positive, s_n is positive as well. Let us consider the sequence $(u_k)_k$ defined by $u_k := L_k^r - L$. We know that $(u_k)_k$ converges pointwise to 0 (see Remark 102) and is locally uniformly bounded (the function L is continuous). The positive current S_n may be considered as a (p, p) form whose coefficients are measures which are dominated by the trace measure s_n. Thus, by the dominated convergence theorem, $(L_k^r - L)S_n = u_k S_n$ tends to 0 as $k \to \infty$ and n is fixed. On the other hand, LS_n converges to LS because L is continuous. It follows that some subsequence $L_{k(n)}^r S_n$ converges to LS. □

Corollary 130. *In the moduli space $\overline{Mod_d}$ the support of the bifurcation measure μ_{bif} is contained in $\overline{S}_{hi} \cap \overline{H}_{yp}$.*

Proof. Use Remark 34 to work with families and then apply Theorem 128. For $0 < r < 1$ one gets $Supp \, \mu_{bif} \subset \overline{H}_{yp}$ and for $r = 1$ $Supp \, \mu_{bif} \subset \overline{S}_{hi}$. □

We also stress that the above result yields a rather simple proof of the existence of Shishikura maps, the original one was based on quasi-conformal surgery.

Corollary 131. *In the moduli space Mod_d one may find maps having $2d - 2$ neutral cycles.*

Proof. Combine Corollary 130 with the fact that $\mu_{bif} > 0$ (see Proposition 118). □

Let us end this subsection with a final remark.

Theorem 128 remains true, with the same proof, if one replace the integrals by

$$\int_{[0,2\pi]^m} [Per_n(r_1 e^{i\theta_1})] \wedge \bigwedge_{j=2}^{m} [Per_{k_j(n)}(r_j e^{i\theta_j})] \, d\theta_1 \cdots d\theta_m$$

where $0 < r_j \leq 1$ for $1 \leq j \leq m$. As a consequence, if $\alpha + \nu = m$ and $\mathcal{P}_{\alpha,\nu}$ is the set of parameters λ such that f_λ has α distinct attracting cycles and ν distinct neutral cycles then $Supp \, \mu_{bif}$ is contained in the closure of $\mathcal{P}_{\alpha,\nu}$.

6.2.3 Shishikura or Hyperbolic Parameters with Chosen Multipliers

As Theorem 105 shows, polynomial with a neutral cycle of a given multiplier are dense in the support of the bifurcation current. We believe that a similar property is still true for the bifurcation measure which means that Shishikura parameters with arbitrarily fixed multipliers should be dense in the support μ_{bif}. The following result goes in this direction.

Theorem 132. *Denote by $p(f)$ (resp. $s(f)$, $c(f)$) the number of distinct parabolic (resp. Siegel, Cremer) cycles of $f \in Mod_d$. Then*

$$Supp \, \mu_{bif} \subset \overline{\{f \in Mod_d \,/\, p(f) = p, s(f) = s \text{ and } c(f) = c\}}$$

for any triple of integers p, s and c such that $p + s + c = 2d - 2$.

The proof is essentially based on Lemma 121 and Mañé–Sad–Sullivan theorem. More precisely we will use the following

Lemma 133. *Let E be a dense subset of $[0, 2\pi]$. Then for any holomorphic family of degree d rational map $(f_\lambda)_M$ the set*

$$\cup_n \cup_{\theta \in E} Per_n(e^{i\theta})$$

is dense in the bifurcation locus.

Proof. Use Mañé–Sad–Sullivan theorem or Theorem 101 with $r = 1$. □

Let us now prove Theorem 132. We restrict ourself to Mod_2. The general case requires to use a slicing argument, we refer to [2] for details.

Proof. Let E_1 and E_2 be two dense subsets of $[0, 2\pi]$. Let λ_0 be a point in the support of μ_{bif} and U_0 be an arbitrarily small neighbourhood of λ_0. By Lemma 133, the support of the bifurcation current $dd^c L$ is accumulated by holomorphic discs contained in $\cup_n \cup_{\theta \in E_1} Per_n(e^{i\theta})$. Among such discs, let us consider those which go through U_0 and pick one disc Γ_1 on which the Lyapunov function L is not harmonic. Such a disc exists since otherwise, according to Lemma 121, the measure μ_{bif} would vanish on U_0. The bifurcation locus of $(f_\lambda)_{\Gamma_1}$ is not empty and thus, to get a Shishikura parameter in U_0 with multipliers $e^{i\theta_1}$ and $e^{i\theta_2}$ where $\theta_j \in E_j$, it suffices to apply again Lemma 133 with the dense set E_2 to the family $(f_\lambda)_{\Gamma_1}$. □

Using Theorem 97 instead Theorem 101 one may prove, with exactly the same arguments as above, that the support of μ_{bif} is accumulated by hyperbolic parameters with attracting cycles of given multipliers.

Theorem 134. *Let $w_1, w_2, \cdots, w_{2d-2}$ be complex numbers belonging to the open unit disc. Then any $\lambda_0 \in \text{Supp } \mu_{\text{bif}}$ is accumulated by maps $f \in Mod_d$ having $2d - 2$ attracting cycles whose respective multipliers are the w_i.*

6.3 The Support of the Bifurcation Measure

We will establish that the inclusions obtained in the former section are actually equalities. This will give a precise meaning to our interpretation of the support of the bifurcation measure in Mod_d as a strong-bifurcation locus.

6.3.1 A Transversality Result

Transversality statements play a very important role for understanding the structure of parameter spaces. We have already encounter such results like for instance Lemma 109 or the Fatou–Shishikura inequality. We refer to the fundamental work of Epstein [30] for a general and synthetic treatment of transversality problems in holomorphic dynamics.

All the results presented here are true in Rat_d or in the moduli spaces Mod_d. For simplicity we shall restrict ourself to the moduli space Mod_2 which will be treated as a holomorphic family $(f_\lambda)_{\lambda \in \mathbb{C}^2}$ (see Theorem 36). We shall also assume, to simplify, to have two marked critical points $c_j(\lambda)$ $j = 1, 2$. Our exposition also covers the case of the family of cubic polynomials.

Assume that $f_0 \in Mod_2$ is strongly Misiurewicz. This means that there exists two repelling cycles

$$\mathcal{C}_j(0) := \{z_j(0), \cdots, f_0^{n_j-1}(z_j(0))\}$$

and an integer $k_0 \geq 1$ such that

$$f_0^{k_0}(c_j(0)) = z_j(0) \text{ but } c_j(0) \notin \mathcal{C}_j(0)$$

for $j = 1, 2$.

By the implicit function theorem, we may follow the cycles $\mathcal{C}_j$ on a small ball $B(0, r)$ centered at the origin. Writing $\mathcal{C}_j(\lambda) := \{z_j(\lambda), \cdots, f_\lambda^{n_j-1}(z_j(\lambda))\}$ the cycles corresponding to the parameter $\lambda \in B(0, r)$, we may define an important tool for studying the parameter space near f_0.

Definition 135. The map $\chi : B(0, r) \to \mathbf{C}^2$ defined by

$$\lambda \mapsto \left(f_\lambda^{k_0}(c_j(\lambda)) - z_j(\lambda) \right)_{j=1,2}$$

is called *activity map* near the strongly Misiurewicz parameter f_0.

The activity map χ measures the difference between two natural holomorphic motions of the point $z_j(0)$. One as a repelling periodic point and the other as a post-critical point. The definition of χ implicitly uses local charts of $\mathbf{P}^1$ near $z_j(0)$. As we shall see later, the activity map will allow to transfer informations from the dynamical space of f_0 to the parameter space. For this, the next result will be essential.

Theorem 136. *The activity map χ near a strongly Misiurewicz parameter f_0 is locally invertible.*

This Theorem was proved by Buff and Epstein (see [13]) in the general setting of Rat_d. In that case one has to assume that f_0 is not a flexible Lattès map (such maps do not exist in degree two). The proof of Buff and Epstein uses quadratic differentials techniques. We shall prove here a weaker statement which is due to Gauthier [33] and is sufficient for the applications we have in mind.

Theorem 137. *The activity map χ near a strongly Misiurewicz parameter f_0 is locally proper.*

The proof of that result is based on more classical arguments going back to Sullivan (see also [62] or [1]). The key point relies on the following Lemma.

Lemma 138. *Any holomorphic curve contained in M_{iss} must consist of flexible Lattès maps.*

Proof. Assume that $(f_\lambda)_{\lambda \in D}$ is a holomorphic family parametrized by some one-dimensional disc D and that all f_λ are strongly Misiurewicz parameters when $\lambda \in D$. It actually suffices to show that f_{λ_1} is Lattès for some $\lambda_1 \in D$.

Since all f_λ are strongly Misiurewicz, the Julia set of f_λ coincides with $\mathbf{P}^1$ for all $\lambda \in D$. Moreover, by Lemma 64, the family $(f_\lambda)_{\lambda \in D}$ is stable. According to a Theorem of Mañé–Sad–Sullivan (see [42, Theorem B]), the stability of $(f_\lambda)_{\lambda \in D}$ implies the existence of a quasiconformal holomorphic motion $\Phi : D \times \mathbf{P}^1 \to \mathbf{P}^1$ which conjugates f_λ to f_0 on $\mathbf{P}^1$. Let us denote by η^λ the Beltrami form satisfying

$$\frac{\partial \Phi_\lambda}{\partial \bar{z}} = \eta^\lambda \frac{\partial \Phi_\lambda}{\partial z}.$$

There exists $\lambda_1 \in D \setminus \{0\}$ for which the support of η^{λ_1} has strictly positive Lebesgue measure. Indeed, if this would not be the case, f_λ would be holomorphically conjugated to f_0 for all $\lambda \in D$. Then the Julia set of f_{λ_1} carries an invariant line field and thus f_{λ_1} is a flexible Lattès map. This last argument uses the fact that the conical set of a strongly Misiurewicz map coincides with its Julia set (see [9, Theorem VII.22] and [46, Corollary 3.18]). □

We may now easily prove Theorem 137.

Proof. Let us first establish that at least one critical point must be active. Assume to the contrary that both critical points are passive around f_0. Then, according to Lemma 73, f_λ is strongly Misiurewicz for all $\lambda \in B(0, r)$ after maybe reducing r. Cutting $B(0, r)$ by a disc D passing through the origin we obtain, by Lemma 138, a disc of flexible Lattès maps. Since this is impossible in Mod_2 (and by assumption in other cases) we have reached a contradiction and proved that at least one critical point, say c_1, is active at f_0.

The activity of c_1 means that $\chi_1^{-1}(0)$ has codimension one. The conclusion is obtained by repeating the argument on the hypersurface $\chi_1^{-1}(0)$. □

6.3.2 The Bifurcation Measure and Strong-Bifurcation Loci

We want to establish that the inclusion $Supp\ \mu_{bif} \subset \overline{S_{hi}} \cap \overline{M_{iss}}$ obtained in Sect. 6.2.2 is actually an equality. This is the reason for which we shall consider the support of the bifurcation measure as a strong-bifurcation locus. This is essentially a consequence of the following result due to Buff and Epstein [13].

Theorem 139. *In the moduli space Mod_d the set of strongly Misiurewicz parameters is contained in the support of the bifurcation measure: $M_{iss} \subset Supp\ \mu_{bif}$.*

Corollary 140. *In the moduli space Mod_d one has $Supp\ \mu_{bif} = \overline{S_{hi}} = \overline{M_{iss}}$.*

Proof. To simplify the presentation we will work in the degree 3 polynomial family $(P_{c,a})_{(c,a) \in \mathbb{C}^2}$. As usual we write λ the parameter (c, a) and $c_1(\lambda), c_2(\lambda)$ the marked critical points, the fact that in this setting $c_1 = 0$ does not play any role here.

Assume that p_0 is a strongly Misiurewicz polynomial. By definition, there exists an integer k_0 such that $p_0^{k_0}(c_j(0)) =: z_j(0)$ is a repelling periodic point for $j = 1, 2$. To get lighter notations we shall assume that the $z_j(0)$ are fixed repelling points.

We denote by $z_j(\lambda)$ the repelling fixed points which are obtained by holomorphically moving $z_j(0)$ on some neighbourhood of the origin and by $w_j(\lambda)$ the corresponding multipliers. Observe that $|w_j(\lambda)| \geq a > 1$ on a sufficiently small neighbourhood of 0.

The activity map χ (see Definition 135) may be written: $\chi = (\chi_1, \chi_2)$ where

$$\chi_j(\lambda) = p_\lambda^{k_0}(c_j(\lambda)) - z_j(\lambda).$$

We will use here Theorem 136 and assume that χ is locally invertible at the origin. It is possible to adapt the proof for using the weaker transversality statement given by Theorem 137. For this one uses the fact that the sets obtained by rescaling the ramification locus of χ are not charged by the measure μ_{bif} and, thanks to some Besicovitch covering argument, reduces the problem to some estimate similar to those which we will now perform in the invertible case. We refer to the papers [13, 33] for details.

Let us denote by $D^2(0, \epsilon)$ the bidisc centered at the origin and of multiradius (ϵ, ϵ) in $\mathbf{C}^2$. For ϵ small enough we may define a sequence of rescaling

$$\delta_n : D^2(0, \epsilon) \to \Omega_n$$

by setting $\delta_n(x) := \chi^{-1}\left(\frac{x_1}{(w_1(0))^n}, \frac{x_2}{(w_2(0))^n}\right)$. To prove the Theorem, it suffices to show that $\mu_{\text{bif}}(\Omega_n) > 0$ for all n. The crucial point of the proof is revealed by the following computations.

$$2\mu_{\text{bif}}(\Omega_n) = \int_{\Omega_n} T_{\text{bif}}^2 = \int_{D^2(0,\epsilon)} \delta_n^\star (T_1 + T_2)^2 \geq \int_{D^2(0,\epsilon)} \delta_n^\star (T_1 \wedge T_2) =$$

$$= \int_{D^2(0,\epsilon)} \delta_n^\star [dd^c g_\lambda(c_1(\lambda)) \wedge dd^c g_\lambda(c_2(\lambda))]$$

using the homogeneity property of the Green function

$$g_\lambda(c_j(\lambda)) = 3^{-(k_0+n)} g_\lambda\left(p_\lambda^{k_0+n}(c_j(\lambda))\right)$$

one thus gets

$$2 \cdot 3^{(k_0+n)} \mu_{\text{bif}}(\Omega_n)$$
$$\geq \int_{D^2(0,\epsilon)} \delta_n^\star \left[dd^c g_\lambda \circ p_\lambda^{k_0+n}(c_1(\lambda)) \wedge dd^c g_\lambda \circ p_\lambda^{k_0+n}(c_2(\lambda))\right] =$$
$$= \int_{D^2(0,\epsilon)} dd^c g_{\delta_n(x)} \circ p_{\delta_n(x)}^{k_0+n}(c_1(\delta_n(x))) \wedge dd^c g_{\delta_n(x)} \circ p_{\delta_n(x)}^{k_0+n}(c_2(\delta_n(x))).$$

Let us express the quantities $p_{\delta_n(x)}^{k_0+n}(c_j(\delta_n(x)))$ by using the activity map χ. By definition we have $p_\lambda^{k_0+n}(c_j(\lambda)) = p_\lambda^n(z_j(\lambda) + \chi_j(\lambda))$ and thus

$$p_{\delta_n(x)}^{k_0+n}(c_j(\delta_n(x))) = p_{\delta_n(x)}^n \left(z_j(\delta_n(x)) + \frac{x_j}{w_j(0)^n}\right).$$

To conclude, we momentarily admit the following

Claim. $p_{\delta_n(x)}^n \left(z_j(\delta_n(x)) + \frac{x_j}{w_j(0)^n} \right)$ is uniformly converging to some local biholomorphism $\psi_j : \mathbf{C}, 0 \to \mathbf{P}^1_{,z_j(0)}$.

As the Green function $g_\lambda(z)$ is continuous in (λ, z), the Claim clearly implies that $g_{\delta_n(x)} \circ p_{\delta_n(x)}^{k_0+n} (c_2(\delta_n(x)))$ uniformly converges towards $g_0(\psi_0(x_j))$ and our estimate yields

$$\liminf_n \mu_{\mathrm{bif}}(\Omega_n) \geq \int_{D^2(0,\epsilon)} dd^c g_0(\psi_1(x_1)) \wedge dd^c g_0(\psi_2(x_2)) =$$

$$= \left(\int_{\psi_1(D(0,\epsilon))} dd^c g_0 \right) \left(\int_{\psi_2(D(0,\epsilon))} dd^c g_0 \right) > 0$$

where the positivity of the last term follows from the fact that the repelling fixed points $z_j(0) = \psi_j(0)$ belong to the Julia set of p_0.

It remains to justify the Claim. Let us write $w_j(\lambda)$ on the form $w_j(0)(1 + \epsilon_j(\lambda))$. As $\|\delta_n(x)\| \leq C\frac{1}{a^n}$ one sees that $(1 + \epsilon_j(\delta_n(x)))^n$ is uniformly converging to 1. Then

$$p_{\delta_n(x)}^n \left(z_j(\delta_n(x)) + \frac{x_j}{w_j(0)^n} \right) = p_{\delta_n(x)}^n \left[z_j(\delta_n(x)) + \frac{x_j}{w_j(\delta_n(x))^n} (1 + \epsilon_j(\delta_n(x)))^n \right]$$

behaves like $p_{\delta_n(x)}^n \left[z_j(\delta_n(x)) + \frac{x_j}{w_j(\delta_n(x))^n} \right]$.

Now let us linearize p_λ near the repelling fixed points $z_j(\lambda)$. The linearization holomorphically depends on the parameter λ and one gets local biholomorphisms $\psi_{j,\lambda}$ such that $\psi_{j,\lambda}(0) = z_j(\lambda)$ and

$$p_\lambda \circ \psi_{j,\lambda}(z) = \psi_{j,\lambda}(w(\lambda)z) \text{ on } B(0, \frac{\epsilon}{|w(\lambda)|}).$$

Then the local biholomorphism ψ_j of the Claim is simply $\psi_{j,0}$. □

Remark 141. By Theorem 120, $(T_1 + T_2)^2 = 2T_1 \wedge T_2$ and therefore the estimate in the proof of Theorem 139 becomes an equality:

$$2 \cdot 3^{(k_0+n)} \mu_{\mathrm{bif}}(\Omega_n)$$

$$= \int_{D^2(0,\epsilon)} dd^c g_{\delta_n(x)} \circ p_{\delta_n(x)}^{k_0+n}(c_1(\delta_n(x))) \wedge dd^c g_{\delta_n(x)} \circ p_{\delta_n(x)}^{k_0+n}(c_2(\delta_n(x))).$$

This allows to estimate the pointwise Hausdorff dimension of μ_{bif} at strongly Misiurewicz parameters.

Let us recall that, for polynomial families, Theorem 139 was first proved by Dujardin and Favre in [29]. Their proof is based on a totally different approach, which consists in comparing the measure μ_{bif} with a landing measure for a family of external rays and uses some deep result due to Bielefeld et al. [37] and Kiwi [38].

This approach has the advantage to yield to a generalization in any degree of some results due to Graczyk and Swiatek [35] and Smirnov [60]. They prove the following.

Theorem 142. *A μ_{bif} generic polynomial has the topological Collet–Eckman property. In particular the following properties are satisfied for μ_{bif}-almost all parameters:*

(i) All cycles are repelling.
(ii) The Julia set and the filled-in Julia set coincide and their Hausdorff dimension is strictly less than 2.
(iii) Each critical point has a dense forward orbit in the Julia set.

Let us end this subsection with a generalization of Buff–Epstein's theorem. It is actually true that all Misiurewicz parameters and not only strongly ones (see Definition 72 for definitions) belong to the support of the bifurcation measure. This has been proved by Gauthier in [33], his result can be stated as follows:

Theorem 143. *In the moduli space Mod_d one has $\overline{M_{is}} = Supp \, \mu_{bif}$.*

The proof is conceptually similar to that of Theorem 139 but is technically more involved. Let us stress the main new difficulties. To define the activity map χ one has to construct a natural holomorphic motion for hyperbolic invariant set, this has been done by de Melo-van Strien [16, Theorem 2.3, page 225]. The transversality statement concerning χ is a weaker one but, as we explained in the last subsection, one may overcome this by using the flexibility of positive currents. The final argument requires to linearize the maps along repelling orbits contained in hyperbolic sets.

Using the transversality map χ associated to Misiurewicz parameters it is possible to construct a "transfer map" which copies hyperbolic sets in the parameter space with a precise control of distortion. This allows to estimate the dimension of *Supp* μ_{bif} from below and is particularly useful when dealing with hyperbolic sets of big Hausdorf dimension. This is what we shall discuss in the last subsection.

6.3.3 Hausdorff Dimension Estimates

Our aim here is to present the following result obtained by T. Gauthier in his thesis (see [33]).

Theorem 144. *The strong-bifurcation locus in Mod_d is homogeneous and has full Hausdorff dimension.*

Let us recall that a subset E of any metric space is said to be *homogeneous* if its Hausdorff dimension coincides with that of any non-empty intersection $U \cap E$ with an open set U.

For the quadratic polynomial family, the strong-bifurcation locus coincides with the usual one. In that case, Gauthier's theorem restates a celebrated result of Shishikura [56] which says that the boundary of the Mandelbrot set is of Hausdorff dimension 2. More generally, Tan Lei [61] has proved that the boundary of the connectedness locus of the degree d polynomial family has maximal Hausdorff dimension. As Remark 124 shows, Gauthiers's result asserts that, for polynomial families, the full dimension is actually concentrated on a tiny part of the boundary of the connectedness locus.

The approach of Gauthier exploits his former result (Theorem 143) and the transfer techniques elaborated by Shishikura and Tan-Lei which he slightly simplifies. We shall explain this below.

Combining Theorems 144 and 132 yields the

Corollary 145. *Denote by $p(f)$ (resp. $s(f)$, $c(f)$) the number of distinct parabolic (resp. Siegel, Cremer) cycles of $f \in Rat_d$. Let p, s and c be three integers such that $p + s + c = 2d - 2$. Then the set*

$$\overline{\{f \in Rat_d \; / \; p(f) = p, s(f) = s \text{ and } c(f) = c\}}$$

is homogeneous and has maximal Hausdorff dimension $2(2d + 1)$.

Let us now sketch the proof of Theorem 144 in the moduli space Mod_2 of quadratic rational maps which we again will consider as a holomorphic family parametrized by $\mathbf{C}^2$ with marked critical points.

We recall that the hyperbolic dimension $\dim_{\text{hyp}}(f)$ of a map f is the supremum of the dimensions of all compact f-invariant hyperbolic homogeneous sets. We shall actually prove the following basic dimension estimate:

Proposition 146. *For $f \in M_{\text{is}}$ one has $\dim_H(M_{\text{is}} \cap \Omega) \geq 2 \dim_{\text{hyp}}(f)$ for any neighborhood Ω of $[f]$ in Mod_2.*

It is absolutely not trivial to deduce Theorem 144 from the above proposition. Adapting some ideas of McMullen [47], one may show that the hyperbolic dimension of a Misiurewicz map equals the Hausdorff dimension of its Julia set which is the Riemann sphere. Then Proposition 146 actually gives $\dim_H(M_{is}) = 4$ and the Theorem 144 follows since, by Theorem 143, $M_{\text{is}} \subset Supp \, \mu_{\text{bif}}$.

Let us now prove Proposition 146. Let f be Misiurewicz (and not a flexible Lattès map) and Ω be a neighbourhood of f in Mod_2. Let E_f be a compact f-hyperbolic set such that

$$\dim_H(E_f \cap U) \geq \dim_{\text{hyp}}(f) - \epsilon$$

for any open set U intersecting E_f. Note that the critical orbits of f are captured by some hyperbolic set which, a priori, is different from that set E_f. The proof uses two technical tools which are certainly of independent interest. The first one allows to make all critical orbits fall into some *big* hyperbolic set, which is a deformation of E_f, by an arbitrarily small perturbation of f. More specifically one shows the following:

Fact 147. *One may find a Misiurewicz map $g \in \Omega$ whose critical points eventually fall into E_g (i.e. $g^N(c_i) \in E_g$ for $i = 1, 2$) where E_g is a compact g-hyperbolic set such that*
$$\dim_H(E_g \cap U_i) \geq \dim_{\text{hyp}}(f) - 3\epsilon$$
for any open neighbourhood U_i of $g^N(c_i)$.

The second is a construction of a transfer map with good regularity properties:

Fact 148. *Let g be a Misiurewicz map with associated g-hyperbolic set E_g (i.e. $g^N(c_i) \in E_g$ for $i = 1, 2$). For any neighbourhood Ω of g in Mod_2 there exists a neighbourhood $U_1 \times U_2$ of $(g^N(c_1), g^N(c_2))$ in $\mathbf{P}^1 \times \mathbf{P}^1$ and a transfer map $\mathcal{T}$*
$$\mathcal{T} : \big((E_g \cap U_1) \times (E_g \cap U_2)\big) \to M_{is} \cap \Omega$$
such that:
$$\|\mathcal{T}(z_1, z_2) - \mathcal{T}(z'_1, z'_2)\| \geq C_\epsilon \|(z_1, z_2) - (z'_1, z'_2)\|^{\frac{1+\epsilon}{1-\epsilon}}$$
for $0 < \epsilon < \epsilon_0$ and $(z_1, z_2), (z'_1, z'_2) \in U_1 \times U_2$.

Let us stress that the proofs of both Facts require transversality (Theorem 137) and uses the Hölder continuity properties of holomorphic motions (Lemma 112). We shall prove them later and right now see how they easily lead to the desired dimension estimate.

Let $g \in \Omega$ be the Misiurewicz map given by Fact 147. Consider the transfer map $\mathcal{T}$ associated to g and given by Fact 148. Then, by construction, we have:
$$\dim_H(M_{is} \cap \Omega) \geq \dim_H \left[\mathcal{T}\big((E_g \cap U_1) \times (E_g \cap U_2)\big)\right].$$

The estimate of Fact 148 then gives:
$$\dim_H(M_{is} \cap \Omega) \geq \tfrac{1-\epsilon}{1+\epsilon} \dim_H \left[(E_g \cap U_1) \times (E_g \cap U_2)\right]$$
$$\geq \tfrac{1-\epsilon}{1+\epsilon}\big(\dim_H(E_g \cap U_1) + \dim_H(E_g \cap U_2)\big).$$

But the choice of the map g (see Fact 147) gives
$$\dim_H(M_{is} \cap W) \geq \tfrac{1-\epsilon}{1+\epsilon} 2(\dim_{\text{hyp}}(f) - 3\epsilon).$$

and the conclusion follows making $\epsilon \to 0$.

We now prove the two Facts.

Proof of Fact 147. One may construct a holomorphic motion
$$h : B(0, \rho) \times E_f \cup \overline{\bigcup_{n \geq k_0} f^n(C_f)} \to \mathbf{P}^1$$

such that $f_\lambda \circ h_\lambda = h_\lambda \circ f_0$ on E_f. Here $B(0,\rho)$ is a ball in $\mathbf{C}^2$ centered at $f_0 := f$. As in Definition 135, one may associate to this motion an activity map χ defined on a neighbourhood of 0 by

$$\chi : \lambda \mapsto \left(f_\lambda^{k_0}(c_j(\lambda)) - h_\lambda(f_0^{k_0}(c_j(0))) \right)_{j=1,2}.$$

The transversality Theorem 137 being also valid in this context, the hypersurfaces $\chi_1^{-1}\{0\}$ and $\chi_2^{-1}\{0\}$ intersect properly and, therefore, the critical point c_2 is active in $\chi_1^{-1}\{0\}$. This will allow us to drastically change the orbit of c_2 by a small perturbation of λ in $\chi_1^{-1}\{0\}$.

Let z_1, z_2 and z_3 be three distinct points of E_f. By Montel's Theorem, there exists $1 \leq i \leq 3$, a parameter $\lambda_1 \in \chi_1^{-1}\{0\}$ which is arbitrarily close to 0 and an integer $N_1 \geq 1$ such that $f_{\lambda_1}^{N_1}(c_2(\lambda_1)) = h_{\lambda_1}(z_i)$. When f_{λ_1} is sufficiently close to f (i.e. λ_1 close to 0) one has

$$\dim_H(h_{\lambda_1}(E_f) \cap U) \geq \dim_{\mathrm{hyp}}(f) - 2\epsilon$$

for any small enough neighborhood U of z_i. This follows from the Hölder regularity of holomorphic motions.

It turns out, since λ_1 has been taken in $\chi_1^{-1}\{0\}$, that the map f_{λ_1} is again Misiurewicz with critical orbits captured by $E := h_{\lambda_1}(E_f)$. We may therefore define a new activity map (still denoted χ) near λ_1 and repeat the above procedure working this time in the family $\chi_2^{-1}\{0\}$. This yields a map $g := f_{\lambda_2}$ which is arbitrarily close to f and an integer $N \geq 1$ such that

$$f_{\lambda_2}^N(C_{f_{\lambda_2}}) \subset h_{\lambda_2}(E) \text{ and } \dim_H(h_{\lambda_2}(E) \cap U) \geq \dim_{\mathrm{hyp}}(f) - 3\epsilon$$

for any neighborhood $U \subset \mathbf{P}^1$ of any $c_i(\lambda_2)$.

Proof of Fact 148. As in the proof of Fact 147, one may construct a holomorphic motion

$$h : B(0,\rho) \times E_g \longrightarrow \mathbf{P}^1$$

such that $g_\lambda \circ h_\lambda = h_\lambda \circ g_0$ on E_g. Here $B(0,\rho)$ is a ball in $\mathbf{C}^2$ centered at $g_0 := g$.

By a Theorem of Bers and Royden (see [6]), this motion extends to a holomorphic motion

$$h : B(0,\rho/3) \times \mathbf{P}^1 \longrightarrow \mathbf{P}^1.$$

Set

$$X : B(0,\rho/3) \times (\mathbf{P}^1)^2 \longrightarrow (\mathbf{P}^1)^2$$
$$(\lambda,(z_1,z_2)) \mapsto \left(g_\lambda^N(c_1(\lambda)) - h_\lambda(z_1), g_\lambda^N(c_2(\lambda)) - h_\lambda(z_2) \right).$$

For the sake of simplicity, we assume that the activity map of g, which coincides with $X(\cdot,(g^N(c_1), g^N(c_2)))$, is locally invertible at g. Then, since X is continuous, a Rouché-like Theorem shows that the map $X\big(\cdot,(z_1,z_2)\big)$ is locally invertible at g for (z_1,z_2) close enough to $(g^N(c_1), g^N(c_2))$.

This allows to define the transfer map $\mathcal{T}$ by

$$\mathcal{T}(z_1,z_2) := X(\cdot,(z_1,z_2))^{-1}\{(0,0)\},$$

for $(z_1,z_2) \in U_1 \times U_2$, where U_i are small enough neighborhood of $g^N(c_i)$. By construction $\mathcal{T}(z_1,z_2)$ is a Misiurewicz map when $(z_1,z_2) \in (E_g \cap U_1) \times (E_g \cap U_2)$ and thus, shrinking U_i one has

$$\mathcal{T}\big((E_g \cap U_1) \times (E_g \cap U_2)\big) \subset M_{\text{is}} \cap \Omega.$$

The estimate is obtained by exploiting the fact that the holomorphic motion $h(\lambda,z)$ is Hölder continuous in z and holomorphic in λ. □

Acknowledgements It is my pleasure to thank my colleagues Charles Favre, Thomas Gauthier and Nessim Sibony for their useful comments on the first draft of these notes. I also would like to thank the anonymous referee for having carefully read the manuscript and having helped me to improve it.

References

1. M. Aspenberg, Perturbations of rational misiurewicz maps (2008). Preprint math.DS/0804.1106
2. G. Bassanelli, F. Berteloot, Bifurcation currents in holomorphic dynamics on P^k. J. Reine Angew. Math. **608**, 201–235 (2007)
3. G. Bassanelli, F. Berteloot, Lyapunov exponents, bifurcation currents and laminations in bifurcation loci. Math. Ann. **345**(1), 1–23 (2009)
4. G. Bassanelli, F. Berteloot, Distribution of polynomials with cycles of given mutiplier. Nagoya Math. J. **201**, 23–43 (2011)
5. E. Bedford, B.A. Taylor, The Dirichlet problem for a complex Monge-Ampere equation. Bull. Am. Math. Soc. **82**(1), 102–104 (1976)
6. L. Bers, H.L. Royden, Holomorphic families of injections. Acta Math. **157**(3–4), 259–286 (1986)
7. F. Berteloot, Lyapunov exponent of a rational map and multipliers of repelling cycles. Riv. Math. Univ. Parma **1**(2), 263–269 (2010)
8. F. Berteloot, C. Dupont, L. Molino, Normalization of bundle holomorphic contractions and applications to dynamics. Ann. Inst. Fourier (Grenoble) **58**(6), 2137–2168 (2008)
9. F. Berteloot, V. Mayer, in *Rudiments de dynamique holomorphe*, Cours Spécialisés, vol. 7 (Société Mathématique de France, Paris, 2001)
10. B. Branner, J.H. Hubbard, The iteration of cubic polynomials. I. The global topology of parameter space. Acta Math. **160**(3–4), 143–206 (1988)
11. J.-Y. Briend, J. Duval, Exposants de Liapounoff et distribution des points périodiques d'un endomorphisme de $\mathbf{P}^k$. Acta Math. **182**(2), 143–157 (1999)
12. H. Brolin, Invariant sets under iteration of rational functions. Ark. Mat. **6**, 103–144 (1965)
13. X. Buff, A.L. Epstein, in *Complex Dynamics: Families and Friends* ed. by D. Schleicher. Bifurcation Measure and Postcritically Finite Rational Maps. (A K Peters, Ltd., Wellesley, 2009), pp. 491–512

14. X. Buff, T. Gauthier, Perturbations of flexible Lattès maps. Bull. Soc. Math. Fr. (to appear)
15. I.P. Cornfeld, S.V. Fomin, Ya.G. Sinaĭ, in *Ergodic Theory*. Grundlehren der Mathematischen Wissenschaften [Fundamental Principles of Mathematical Sciences], vol. 245 (Springer, New York, 1982). Translated from the Russian by A.B. Sosinskiĭ
16. W. de Melo, S. van Strien, in *One-Dimensional Dynamics*. Ergebnisse der Mathematik und ihrer Grenzgebiete (3) [Results in Mathematics and Related Areas (3)] vol. 25. (Springer, Berlin, 1993)
17. H. De Thelin, Sur la laminarité de certains courants. Ann. Sci. École Norm. Sup. **4**(37), 304–311 (2004)
18. J.-P. Demailly, Complex analytic and differential geometry. Free accessible book (http://www-fourier.ujf-grenoble.fr/~demailly/manuscripts/agbook.pdf)
19. L. DeMarco, Dynamics of rational maps: a current on the bifurcation locus. Math. Res. Lett. **8**(1–2), 57–66 (2001)
20. L. DeMarco, Dynamics of rational maps: Lyapunov exponents, bifurcations, and capacity. Math. Ann. **326**(1), 43–73 (2003)
21. B. Deroin, R. Dujardin, Random walks, kleinian groups, and bifurcation currents (2010) Preprint arXiv:math.GT/1011.1365v2
22. T.-C. Dinh, N. Sibony, Dynamique des applications d'allure polynomiale. J. Math. Pures Appl. **82**(2), 367–423 (2003)
23. T.-C. Dinh, N. Sibony, in *Holomorphic Dynamical Systems*. Dynamics in Several Complex Variables: Endomorphisms of Projective Spaces and Polynomial-Like Mappings. Lecture Notes in Mathematics, vol. 1998 (Springer, Berlin, 2010), pp. 165–294
24. A. Douady, J.H. Hubbard, in *Étude Dynamique des Polynômes Complexes. Partie I*. Publications Mathématiques d'Orsay [Mathematical Publications of Orsay], vol. 84 (Université de Paris-Sud, Département de Mathématiques, Orsay, 1984)
25. A. Douady, J.H. Hubbard, in *Étude Dynamique des Polynômes Complexes. Partie II*, Publications Mathématiques d'Orsay [Mathematical Publications of Orsay], vol. 85 (Université de Paris-Sud, Département de Mathématiques, Orsay, 1985). With the collaboration of P. Lavaurs, Tan Lei, and P. Sentenac
26. A. Douady, J.H. Hubbard, Itération des polynômes quadratiques complexes. C. R. Acad. Sci. Paris Sér. I Math. **294**(3), 123–126 (1982)
27. R. Dujardin, in *Complex Dynamics: Families and Friends* ed. by D. Schleicher. Cubic polynomials: A Measurable View on Parameter Space. (A K Peters, Ltd., Wellesley, 2009), pp. 451–490
28. R. Dujardin, Bifurcation currents and equidistribution on parameter space (2011). Preprint math.DS/1111.3989
29. R. Dujardin, C. Favre, Distribution of rational maps with a preperiodic critical point. Am. J. Math. **130**(4), 979–1032 (2008)
30. A. Epstein, Transversality in holomorphic dynamics (2009). Preprint
31. A.L. Epstein, Bounded hyperbolic components of quadratic rational maps. Ergodic Theory Dyn. Syst. **20**(3), 727–748 (2000)
32. C. Favre, J. Rivera-Letelier, équidistribution quantitative des points de petite hauteur sur la droite projective. Math. Ann. **335**, 311–361 (2006)
33. T. Gauthier, Strong bifurcation loci of full Hausdorff dimension. Ann. Sci. Ecole Norm. Sup. 4°série, **45**(6), 947–984 (2012)
34. L.R. Goldberg, L. Keen, The mapping class group of a generic quadratic rational map and automorphisms of the 2-shift. Invent. Math. **101**(2), 335–372 (1990)
35. J. Graczyk, G. Świątek, Harmonic measure and expansion on the boundary of the connectedness locus. Invent. Math. **142**(3), 605–629 (2000)
36. J.H. Hubbard, in *Teichmüller theory and applications to geometry, topology, and dynamics*, vol. 1 (Matrix Editions, Ithaca, 2006) Teichmüller theory, With contributions by Adrien Douady, William Dunbar, Roland Roeder, Sylvain Bonnot, David Brown, Allen Hatcher, Chris Hruska and Sudeb Mitra, With forewords by William Thurston and Clifford Earle

37. J.H. Hubbard, B. Bielefeld, Y. Fisher, The classification of critically preperiodic polynomials as dynamical systems. J. Am. Math. Soc. **5**(4), 721–762 (1992)
38. K. Jan, Combinatorial continuity in complex polynomial dynamics. Proc. Lond. Math. Soc. **3**(1), 215–248 (1994)
39. G.M. Levin, On the theory of iterations of polynomial families in the complex plane. J. Sov. Math. **52**(6), 3512–3522 (1990)
40. M.Ju. Ljubich, Entropy properties of rational endomorphisms of the Riemann sphere. Ergodic Theory Dyn. Syst. **3**(3), 351–385 (1983)
41. M.Ju. Ljubich, Some typical properties of the dynamics of rational mappings. Russ. Math. Surv. **38**(5), 154–155 (1983)
42. R. Mañé, P. Sad, D. Sullivan, On the dynamics of rational maps. Ann. Sci. École Norm. Sup. (4), **16**(2), 193–217 (1983)
43. R. Mañé, in *Dynamical Systems*, Valparaiso 1986. The Hausdorff Dimension of Invariant Probabilities of Rational Maps. Lecture Notes in Mathematics, vol. 1331 (Springer, Berlin, 1988), pp. 86–117
44. V. Mayer, Comparing measures and invariant line fields. Ergodic Theory Dyn. Syst. **22**(2), 555–570 (2002)
45. C.T. McMullen, Families of rational maps and iterative root-findings algorithms. Ann. Math. **125**(4), 467–493 (1987)
46. C.T. McMullen, in *Complex Dynamics and Renormalization*. Annals of Mathematics Studies, vol. 135 (Princeton University Press, Princeton, 1994)
47. C.T. McMullen, Hausdorff dimension and conformal dynamics. II. Geometrically finite rational maps. Comment. Math. Helv. **75**(4), 535–593 (2000)
48. J. Milnor, Geometry and dynamics of quadratic rational maps. Exp. Math. **2**(1), 37–83 (1993). With an appendix by the author and Lei Tan
49. J. Milnor, in *Dynamics in One Complex Variable*, 3rd edn. Annals of Mathematics Studies, vol. 160 (Princeton University Press, Princeton, 2006)
50. J. Milnor, in *Dynamics on the Riemann Sphere*. On Lattès Maps (European Mathematical Society, Zürich, 2006), pp. 9–43
51. Y. Okuyama, Lyapunov exponents in complex dynamics and potential theory (2010). Preprint arXiv math.CV/1008.1445v1
52. Y. Okuyama, Repelling periodic points and logarithmic equidistribution in non-archimedean dynamics. Acta Arith. **153**(3), 267–277 (2012)
53. N.-m. Pham, Lyapunov exponents and bifurcation current for polynomial-like maps (2005). Preprint, arXiv:math.DS/0512557v1
54. F. Przytycki, Hausdorff dimension of harmonic measure on the boundary of an attractive basin for a holomorphic map. Invent. Math. **80**(1), 161–179 (1985)
55. M. Shishikura, On the quasiconformal surgery of rational functions. Ann. Sci. École Norm. Sup. (4), **20**(1), 1–29 (1987)
56. M. Shishikura, The Hausdorff dimension of thref/ e boundary of the Mandelbrot set and Julia sets. Ann. Math. (2), **147**(2), 225–267 (1998)
57. N. Sibony, Itérées de polynômes et fonction de Green (1982) Exposé d'Orsay
58. N. Sibony, in *Dynamique et Géométrie Complexes*, Lyon, 1997. Dynamique des Applications Rationnelles de P^k. Panoramic Synthèses, vol. 8 (Société Mathématique de France, Paris, 1999), pp. ix–x, xi–xii, 97–185
59. J.H. Silverman, in *The Arithmetic of Dynamical Systems*. Graduate Texts in Mathematics, vol. 241 (Springer, New York, 2007)
60. S. Smirnov, Symbolic dynamics and Collet-Eckmann conditions. Int. Math. Res. Not. (7), 333–351 (2000)
61. L. Tan, Hausdorff dimension of subsets of the parameter space for families of rational maps. (A generalization of Shishikura's result). Nonlinearity **11**(2), 233–246 (1998)
62. S. van Strien, Misiurewicz maps unfold generically (even if they are critically non-finite). Fund. Math. **163**(1), 39–54 (2000)
63. A. Zdunik, Parabolic orbifolds and the dimension of the maximal measure for rational maps. Invent. Math. **99**(3), 627–649 (1990)

The Complex Monge–Ampère Equation in Kähler Geometry

Zbigniew Błocki

Course given at CIME Summer School in Pluripotential Theory
Cetraro, Italy, July 11-16, 2011

Abstract We will discuss two main cases where the complex Monge–Ampère equation (CMA) is used in Käehler geometry: the Calabi–Yau theorem which boils down to solving nondegenerate CMA on a compact manifold without boundary and Donaldson's problem of existence of geodesics in Mabuchi's space of Käehler metrics which is equivalent to solving homogeneous CMA on a manifold with boundary. At first, we will introduce basic notions of Käehler geometry, then derive the equations corresponding to geometric problems, discuss the continuity method which reduces solving such an equation to a priori estimates, and present some of those estimates. We shall also briefly discuss such geometric problems as Käehler–Einstein metrics and more general metrics of constant scalar curvature.

1 Introduction

We present two situations where the complex Monge–Ampère equation (CMA) appears in Kähler geometry: the Calabi conjecture and geodesics in the space of Kähler metrics. In the first case the problem is to construct, in a given Kähler class, a metric with prescribed Ricci curvature. It turns out that this is equivalent to finding a metric with prescribed volume form, and thus to solving nondegenerate CMA on a manifold with no boundary. This was eventually done by Yau [47], building up on earlier work by Calabi, Nirenberg and Aubin. On the other hand, to find a geodesic in a Kähler class (the problem was posed by Donaldson [20]) one has to solve a homogeneous CMA on a manifold with boundary (this was observed independently by Semmes [39] and Donaldson [20]). Existence of weak geodesics was proved

---

Z. Błocki (✉)
Jagiellonian University, Institute of Mathematics, Łojasiewicza 6, 30-348 Kraków, Poland
e-mail: Zbigniew.Blocki@im.uj.edu.pl; umblocki@cyf-kr.edu.pl

by Chen [18] but Lempert and Vivas [33] showed recently that these geodesics do not have to be smooth. Their partial regularity is nevertheless of interest from the geometric point of view.

In Sects. 2–6 we discuss mostly geometric aspects, whereas Sects. 7–13 concentrate on the PDE part, mostly a priori estimates. We start with a very elementary introduction to Kähler geometry in Sect. 2, assuming the reader is familiar with Riemannian geometry. The Calabi conjecture and its equivalence to CMA are presented in Sect. 3, where the problem of extremal metrics is also briefly discussed. Basic properties of the Riemannian structure of the space of Kähler metrics (introduced independently by Mabuchi [35] and Donaldson [20]) are presented in Sect. 4. The Aubin–Yau functional and the Mabuchi K-energy as well as relation to constant scalar curvature metrics are discussed there as well. The Lempert–Vivas example is described in Sect. 5. Assuming Sects. 7–13, where appropriate results on CMA are shown, in Sect. 6 we present a theorem due to Chen [18] that a Kähler class with the distance defined by this Riemannian structure is a metric space.

The fundamental results on CMA are formulated in Sect. 7, where also basic uniqueness results as well as the comparison principle are showed. The continuity method, used to prove existence of solutions, is described in Sect. 8. It reduces the problem to a priori estimates. Yau's proof of the L^∞-estimate using Moser's iteration is presented in Sect. 9, whereas Sects. 10–12 deal with the first and second order estimates (Sects. 11–12 are not needed in the empty boundary case, that is in the proof of the Calabi conjecture). Higher order estimates then follow from the general, completely *real* Evans–Krylov theory, this is explained in Sect. 13. A slight novelty of this approach in the proof of Yau's theorem is the use of Theorem 25 below which enables us to use directly this real Evans–Krylov theory, instead of proving its complex version (compare with [10, 40] or [13]).

The author would like to thank the organizers of the CIME school in Pluripotential Theory, Filippo Bracci and John Erik Fornæss, for the invitation and a very good time he had in Cetraro in July 2011.

2 Basic Notions of Kähler Geometry

Let M be a complex manifold of dimension n and by $J : TM \to TM$ denote its complex structure. We start with a Hermitian metric h on M and set

$$\langle X, Y \rangle := \operatorname{Re} h(X, Y), \quad \omega(X, Y) := -\operatorname{Im} h(X, Y), \quad X, Y \in TM.$$

Then $\langle \cdot, \cdot \rangle$ is a Riemannian metric on M, ω a real 2-form on M and

$$\langle JX, Y \rangle = \omega(X, Y), \quad \langle JX, JY \rangle = \langle X, Y \rangle. \tag{1}$$

The Riemannian metric $\langle \cdot, \cdot \rangle$ determines unique Levi–Civita connection ∇.

By $T_{\mathbb{C}}M$ denote the complexification of TM (treated as a real space) and extend J, $\langle\cdot,\cdot\rangle$, ω, and ∇ to $T_{\mathbb{C}}M$ in a $\mathbb{C}$-linear way. In local coordinates $z^j = x^j + iy^j$ the vector fields $\partial/\partial x^j$, $\partial/\partial y^j$ span TM over $\mathbb{R}$. We also have

$$J(\partial/\partial x_j) = \partial/\partial y_j, \quad J(\partial/\partial y_j) = -\partial/\partial x_j.$$

The vector fields

$$\partial_j := \frac{\partial}{\partial z^j}, \quad \partial_{\bar{j}} := \frac{\partial}{\partial \bar{z}^j},$$

span $T_{\mathbb{C}}M$ over $\mathbb{C}$ and

$$J(\partial_j) = i\partial_j, \quad J(\partial_{\bar{j}}) = -i\partial_{\bar{j}}.$$

Set

$$g_{j\bar{k}} := \langle \partial_j, \partial_{\bar{k}} \rangle \left(= \langle \partial_{\bar{k}}, \partial_j \rangle \right).$$

Then $\overline{g_{j\bar{k}}} = g_{k\bar{j}}$ and by (1)

$$\langle \partial_j, \partial_k \rangle = \langle \partial_{\bar{j}}, \partial_{\bar{k}} \rangle = 0.$$

If $X = X^j \partial_j + \bar{X}^j \partial_{\bar{j}}$ then $X \in TM$ and it follows that

$$|X|^2 = 2g_{j\bar{k}} X^j \bar{X}^k,$$

thus $(g_{j\bar{k}}) > 0$. By (1)

$$\omega = ig_{j\bar{k}} dz^j \wedge d\bar{z}^k \tag{2}$$

(we see in particular that ω is a form of type $(1,1)$).

Proposition 1. *For a Hermitian metric h the following are equivalent*

(i) $\nabla J = 0$;
(ii) $d\omega = 0$;
(iii) $\omega = i\partial\bar{\partial}g$ *locally for some smooth real-valued function g.*

Proof. (i)$\Rightarrow$(ii) By (1)

$$3\,d\omega(X,Y,Z) = X\omega(Y,Z) + Y\omega(Z,X) + Z\omega(X,Y)$$
$$- \omega([X,Y],Z) - \omega([Y,Z],X) - \omega([Z,X],Y)$$
$$= \langle (\nabla_X J)Y, Z \rangle + \langle (\nabla_Y J)Z, X \rangle + \langle (\nabla_Z J)X, Y \rangle.$$

(ii)$\Rightarrow$(i) Similarly one can show that

$$3\,d\omega(X,Y,Z) - 3\,d\omega(X,JY,JZ) = 2\,\langle (\nabla_X J)Y, Z \rangle + \langle X, N(Y, JZ) \rangle,$$

where
$$N(X,Y) = [X,Y] + J[JX,Y] + J[X,JY] - [JX,JY]$$
is the Nijenhuis tensor (in our case it vanishes, because J is integrable).

(ii)$\Rightarrow$(iii) Locally we can find a real 1-form γ such that $\omega = d\gamma$. We may write $\gamma = \bar{\beta} + \beta$, where β is a $(0,1)$-form. Then, since $d = \partial + \bar{\partial}$,
$$\omega = \partial\bar{\beta} + \partial\beta + \bar{\partial}\bar{\beta} + \bar{\partial}\beta.$$
It follows that $\bar{\partial}\beta = 0$, because ω is a $(1,1)$-form. Therefore we can find (locally) a complex-valued, smooth function f with $\beta = \bar{\partial}f$ and
$$\omega = \partial\beta + \bar{\partial}\bar{\beta} = 2i\partial\bar{\partial}(\mathrm{Im}\, f).$$
We can thus take $g = 2\mathrm{Im}\, f$.

(iii)$\Rightarrow$(ii) is obvious. $\square$

The metric satisfying equivalent conditions in Proposition 1 is called *Kähler*. It is thus a Hermitian metric on a complex manifold for which the Riemannian structure is compatible with the complex structure. The corresponding form ω is also called *Kähler*, it is characterized by the following properties: ω is a smooth, real, positive, closed $(1,1)$-form.

From now on we will use the lower indices to denote partial differentiation w.r.t. z^j and $\bar{z}^k$, so that for example $\partial^2 g/\partial z^j \partial\bar{z}^k = g_{j\bar{k}}$ and (2) is compatible with $\omega = i\partial\bar{\partial}g$.

Volume form. Since $\langle \partial_j, \partial_{\bar{k}} \rangle = g_{j\bar{k}}$ and $\langle \partial_j, \partial_k \rangle = 0$, we can easily deduce that
$$\langle \frac{\partial}{\partial x^j}, \frac{\partial}{\partial x^k} \rangle = \langle \frac{\partial}{\partial y^j}, \frac{\partial}{\partial y^k} \rangle = 2\mathrm{Re}\, g_{j\bar{k}}, \quad \langle \frac{\partial}{\partial x^j}, \frac{\partial}{\partial y^k} \rangle = -\langle \frac{\partial}{\partial x^k}, \frac{\partial}{\partial y^j} \rangle = 2\mathrm{Im}\, g_{j\bar{k}}.$$
From this, using the notation $x^{j+n} = y^n$,
$$\sqrt{\det \left(\langle \frac{\partial}{\partial x^j}, \frac{\partial}{\partial x^k} \rangle \right)_{1 \le j,k \le 2n}} = 2^n \det(g_{j\bar{k}}).$$
It follows that the volume form on M is given by
$$2^n \det(g_{j\bar{k}})\, d\lambda = \frac{\omega^n}{n!},$$
where $d\lambda$ is the Euclidean volume form and $\omega^n = \omega \wedge \cdots \wedge \omega$. In the Kähler case it will be however convenient to get rid of the constant and define the volume as
$$dV := \omega^n.$$

Christoffel symbols. From now on we assume that ω is a Kähler form on M and $\langle \cdot, \cdot \rangle$ is the associated metric. Write

$$\nabla_{\partial_j} \partial_k = \Gamma^l_{jk} \partial_l + \Gamma^{\bar{l}}_{jk} \partial_{\bar{l}}, \quad \nabla_{\partial_{\bar{j}}} \partial_k = \Gamma^l_{\bar{j}k} \partial_l + \Gamma^{\bar{l}}_{\bar{j}k} \partial_{\bar{l}}.$$

Since $\nabla J = 0$, we have for example $i \nabla_{\partial_j} \partial_k = \nabla_{\partial_j}(J\partial_k) = J \nabla_{\partial_j} \partial_k$ and it follows that $\Gamma^{\bar{l}}_{jk} = 0$. Similarly we show that $\Gamma^l_{\bar{j}k} = \Gamma^{\bar{l}}_{\bar{j}k} = 0$, so the only non-vanishing Christoffel symbols are $\Gamma^l_{jk} = \overline{\Gamma^{\bar{l}}_{\bar{j}\bar{k}}}$. Denoting further $g_j = \partial g / \partial z^j$, $g_{\bar{k}} = \partial g / \partial \bar{z}^k$ (which by Proposition 1(iii) is consistent with the previous notation) we have

$$g_{j\bar{k}l} = \partial_l \langle \partial_l, \partial_{\bar{k}} \rangle = \Gamma^p_{lj} g_{p\bar{q}},$$

which means that

$$\Gamma^l_{jk} = \overline{\Gamma^{\bar{l}}_{\bar{j}\bar{k}}} = g^{l\bar{q}} g_{j\bar{q}k}, \tag{3}$$

where $g^{p\bar{q}}$ is determined by

$$g^{j\bar{q}} g_{k\bar{q}} = \delta_{jk}. \tag{4}$$

Riemannian curvature. Recall that it is defined by

$$R(X, Y) = \nabla_X \nabla_Y - \nabla_Y \nabla_X - \nabla_{[X,Y]}$$

(we extend it to $T_{\mathbb{C}} M$) and

$$R(X, Y, Z, W) = \langle R(X, Y)Z, W \rangle. \tag{5}$$

The classical properties are

$$R(Y, X) = -R(X, Y),$$
$$R(X, Y, Z, W) = -R(Y, X, Z, W) = -R(X, Y, W, Z) = R(Z, W, X, Y), \tag{6}$$
$$R(X, Y)Z + R(Y, Z)X + R(Z, X)Y = 0$$

(the latter is the first Bianchi identity). From $\nabla J = 0$ it follows that

$$R(X, Y)J = JR(X, Y)$$

and from (6) we infer

$$R(X, Y, Z, W) = R(X, Y, JZ, JW) = R(JX, JY, Z, W).$$

It follows that

$$R(JX, JY) = R(X, Y),$$

thus
$$R(\partial_j, \partial_k) = R(\partial_{\bar{j}}, \partial_{\bar{k}}) = 0.$$

We have
$$R(\partial_j, \partial_{\bar{k}})\partial_p = -\nabla_{\partial_{\bar{k}}}\nabla_{\partial_j}\partial_p = -\partial_{\bar{k}}(\Gamma^l_{jp})\partial_l$$

and
$$R(\partial_j, \partial_{\bar{k}})\partial_{\bar{q}} = \nabla_{\partial_j}\nabla_{\partial_{\bar{k}}}\partial_{\bar{q}} = \partial_j(\overline{\Gamma^l_{kq}})\partial_{\bar{l}}.$$

Therefore, if we write
$$R(\partial_j, \partial_{\bar{k}})\partial_p = R^l_{j\bar{k}p}\partial_l, \quad R(\partial_j, \partial_{\bar{k}})\partial_{\bar{q}} = R^{\bar{l}}_{j\bar{k}\bar{q}}\partial_{\bar{l}},$$

then
$$R^l_{j\bar{k}p} = -\overline{R^{\bar{l}}_{k\bar{j}\bar{p}}} = -(g^{l\bar{i}}g_{j\bar{i}p})_{\bar{k}}.$$

The relevant coefficients for (5) are
$$R_{j\bar{k}p\bar{q}} := R(\partial_j, \partial_{\bar{k}}, \partial_p, \partial_{\bar{q}}) = g_{l\bar{q}}R^l_{j\bar{k}p}$$

by (3). Applying a first-order differential operator (with constant coefficients) D to both sides of (4) we get
$$Dg^{p\bar{q}} = -g^{p\bar{i}}g^{s\bar{q}}Dg_{s\bar{i}} \qquad (7)$$

and thus
$$R_{j\bar{k}p\bar{q}} = -g_{j\bar{k}p\bar{q}} + g^{s\bar{t}}g_{j\bar{t}p}g_{s\bar{k}\bar{q}}.$$

Ricci curvature. Recall that the Ricci curvature is defined by
$$Ric(X, Y) := tr(Z \mapsto R(Z, X)Y).$$

We extend it to $T_{\mathbb{C}}M$. If we write $Z = Z^p\partial_p + \tilde{Z}^q\partial_{\bar{q}}$ then
$$R(Z, \partial_j)\partial_k = -\tilde{Z}^q R^l_{j\bar{q}k}\partial_l, \quad R(Z, \partial_{\bar{j}})\partial_{\bar{k}} = Z^p R^{\bar{l}}_{p\bar{j}\bar{k}}\partial_{\bar{l}},$$
$$R(Z, \partial_j)\partial_{\bar{k}} = -\tilde{Z}^q R^{\bar{l}}_{j\bar{q}k}\partial_{\bar{l}}, \quad R(Z, \partial_{\bar{k}})\partial_j = Z^p R^l_{p\bar{k}j}\partial_l.$$

It follows that
$$Ric(\partial_j, \partial_k) = Ric(\partial_{\bar{j}}, \partial_{\bar{k}}) = 0$$

and
$$Ric_{j\bar{k}} := Ric(\partial_j, \partial_{\bar{k}}) = \overline{R^p_{p\bar{j}k}} = -(g^{p\bar{q}}g_{p\bar{q}\bar{k}})_j.$$

Since
$$D\det(g_{p\bar{q}}) = M^{p\bar{q}}Dg_{p\bar{q}},$$

where $(M^{p\bar{q}}) = \det(g_{s\bar{t}})(g^{p\bar{q}})$ is the adjoint matrix to $(g_{p\bar{q}})$, we have

$$D\bigl(\log \det(g_{p\bar{q}})\bigr) = g^{p\bar{q}} D g_{p\bar{q}}. \tag{8}$$

Therefore
$$Ric_{j\bar{k}} = -(\log \det(g_{p\bar{q}}))_{j\bar{k}}. \tag{9}$$

From the proceeding calculations we infer in particular
$$Ric(JX, JY) = Ric(X, Y).$$

The associated Ricci 2-form is defined by
$$Ric_\omega(X, Y) := Ric(JX, Y)$$

(since Ric is symmetric, Ric_ω is antisymmetric). We then have
$$Ric_\omega = iRic_{j\bar{k}} dz^j \wedge d\bar{z}^k = -i\partial\bar{\partial}(\log \det(g_{p\bar{q}})).$$

An important consequence of this formula is the following: if $\tilde{\omega}$ is another Kähler form on M then
$$Ric_\omega - Ric_{\tilde{\omega}} = i\partial\bar{\partial} \log \frac{\tilde{\omega}^n}{\omega^n}. \tag{10}$$

In particular, Ric_ω and $Ric_{\tilde{\omega}}$ are $\partial\bar{\partial}$-cohomologous.

Scalar curvature. It is the trace of the mapping $Ric : T_{\mathbb{C}} M \to T_{\mathbb{C}} M$ defined by the relation
$$\langle Ric\, X, Y \rangle = Ric(X, Y).$$

Since
$$Ric\, \partial_j = g^{p\bar{q}} Ric_{j\bar{q}} \partial_p,$$

we will obtain
$$S = 2g^{p\bar{q}} Ric_{p\bar{q}} = 2n \frac{Ric_\omega \wedge \omega^{n-1}}{\omega^n}.$$

Bisectional curvature. It is defined by
$$\sigma(X, Y) = R(X, JX, Y, JY) = R(X, Y, X, Y) + R(X, JY, X, JY),$$

where the last equality is a consequence of the first Bianchi identity. If we write $X = X^j \partial_j + \tilde{X}^k \partial_{\bar{k}}$, $Y = Y^p \partial_p + \tilde{Y}^q \partial_{\bar{q}}$, then
$$\sigma(X, Y) = -2i X^j \tilde{X}^k R(\partial_j, \partial_{\bar{k}}, Y, JY) = 4 R_{j\bar{k}p\bar{q}} X^j \tilde{X}^k Y^p \tilde{Y}^q.$$

An upper bound for the bisectional curvature is a positive constant $C > 0$ satisfying
$$\sigma(X, Y) \leq C |X|^2 |Y|^2, \quad X, Y \in TM. \tag{11}$$

Since $|X^j\partial_j + \bar{X}^k\partial_{\bar{k}}|^2 = 2g_{j\bar{k}}X^j\bar{X}^k$, it follows that (11) is equivalent to

$$R_{j\bar{k}p\bar{q}}a^j\bar{a}^k b^p\bar{b}^q \leq C g_{j\bar{k}}a^j\bar{a}^k g_{p\bar{q}}b^p\bar{b}^q, \quad a,b \in \mathbb{C}^n. \tag{12}$$

Similarly we can define a lower bound.

Gradient. For a real-valued function φ on M its gradient $\nabla\varphi$ is defined by the relation

$$\langle \nabla\varphi, X \rangle = X\varphi.$$

Therefore

$$\nabla\varphi = g^{j\bar{k}}(\varphi_{\bar{k}}\partial_j + \varphi_j\partial_{\bar{k}})$$

and

$$|\nabla\varphi|^2 = 2g^{j\bar{k}}\varphi_j\varphi_{\bar{k}}.$$

Laplacian. It is given by

$$\Delta\varphi := tr(X \mapsto \nabla_X \nabla\varphi).$$

For $X = X^j\partial_j + \tilde{X}^k\partial_{\bar{k}}$ we have

$$\nabla_X \nabla\varphi = X^j \big[(g^{p\bar{q}}\varphi_{\bar{q}})_j \partial_p + g^{p\bar{q}}\varphi_{\bar{q}}\Gamma^k_{jp}\partial_k + (g^{p\bar{q}}\varphi_p)_j \partial_{\bar{q}}\big]$$
$$+ \tilde{X}^k \big[(g^{p\bar{q}}\varphi_{\bar{q}})_{\bar{k}} \partial_p + (g^{p\bar{q}}\varphi_p)_{\bar{k}} \partial_{\bar{q}} + g^{p\bar{q}}\varphi_p \overline{\Gamma^j_{kq}}\partial_{\bar{j}}\big].$$

From (3) and (7) we will get

$$\Delta\varphi = 2g^{j\bar{k}}\varphi_{j\bar{k}}.$$

Lichnerowicz operator. For a real-valued function φ we can write

$$\nabla\varphi = \nabla'\varphi + \overline{\nabla'\varphi},$$

where

$$\nabla'\varphi = g^{j\bar{k}}\varphi_{\bar{k}}\partial_j \in T^{1,0}M.$$

The Lichnerowicz operator is defined by

$$\mathcal{L}\varphi := \bar{\partial}\nabla'\varphi = \big(g^{j\bar{k}}\varphi_{\bar{k}}\big)_{\bar{q}}\partial_j \otimes d\bar{z}^q,$$

so that $\nabla\varphi$ is a holomorphic vector field iff $\mathcal{L}\varphi = 0$.

Proposition 2. $\mathcal{L}^*\mathcal{L}\varphi = \Delta^2\varphi + \langle Ric_\omega, i\partial\bar{\partial}\varphi\rangle + \langle \nabla S, \nabla\varphi\rangle.$

Proof. Since

$$|\mathcal{L}\varphi|^2 = 4g^{p\bar{q}}g_{j\bar{t}}\big(g^{j\bar{k}}\varphi_{\bar{k}}\big)_{\bar{q}}\big(g^{s\bar{t}}\varphi_s\big)_p$$

and
$$(g^{p\bar{q}} \det(g_{j\bar{k}}))_p = 0$$
for every q, it follows that
$$\mathcal{L}^*\mathcal{L}\varphi = 4\mathrm{Re}\left[g^{s\bar{t}}\left(g^{p\bar{q}}\left(g_{j\bar{t}}(g^{j\bar{k}}\varphi_{\bar{k}})_{\bar{q}}\right)_p\right)_s\right].$$

We can compute that
$$g^{p\bar{q}}\left(g_{j\bar{t}}(g^{j\bar{k}}\varphi_{\bar{k}})_{\bar{q}}\right)_p = (g^{p\bar{q}}\varphi_{p\bar{q}})_{\bar{t}} - g^{p\bar{q}}(g^{a\bar{k}}g_{a\bar{t}\bar{q}})_p\varphi_{\bar{k}}$$

and thus
$$\frac{1}{4}\mathcal{L}^*\mathcal{L}\varphi = g^{s\bar{t}}(g^{p\bar{q}}\varphi_{p\bar{q}})_{s\bar{t}} - g^{s\bar{t}}g^{p\bar{q}}(g^{a\bar{k}}g_{a\bar{t}\bar{q}})_p\varphi_{s\bar{k}} - \mathrm{Re}\left[g^{s\bar{t}}\left(g^{p\bar{q}}(g^{a\bar{k}}g_{a\bar{t}\bar{q}})_p\right)_s\varphi_{\bar{k}}\right].$$

One can check that
$$-g^{s\bar{t}}g^{p\bar{q}}(g^{a\bar{k}}g_{a\bar{t}\bar{q}})_p = g^{s\bar{q}}g^{p\bar{k}}Ric_{p\bar{q}}$$
$$-g^{s\bar{t}}\left(g^{p\bar{q}}(g^{a\bar{k}}g_{a\bar{t}\bar{q}})_p\right)_s = \frac{1}{2}g^{j\bar{k}}S_j$$

and the result follows. $\square$

Poisson bracket. It is defined by the relation
$$\{\varphi,\psi\}\omega^n = n\,d\varphi\wedge d\psi\wedge\omega^{n-1}$$
or, in local coordinates,
$$\{\varphi,\psi\} = ig^{j\bar{k}}(\varphi_{\bar{k}}\psi_j - \varphi_j\psi_{\bar{k}}).$$

If one of φ,ψ,η has a compact support then
$$\int_M \{\varphi,\psi\}\eta\,\omega^n = \int_M \varphi\{\psi,\eta\}\omega^n.$$

d^c**-operator.** It is useful to introduce the operator $d^c := \frac{i}{2}(\bar{\partial}-\partial)$. It is real (in the sense that it maps real forms to real forms) and $dd^c = i\partial\bar{\partial}$. One can easily show that
$$dd^c\varphi\wedge\omega^{n-1} = \frac{1}{2n}\Delta\varphi\,\omega^n$$
and
$$d\varphi\wedge d^c\psi\wedge\omega^{n-1} = \frac{1}{2n}\langle\nabla\varphi,\nabla\psi\rangle\omega^n.$$

The operator d^c clearly depends only on the complex structure. In the Kähler case we have however the formula

$$d^c \varphi = -\frac{1}{2} i_{\nabla\varphi} \omega \qquad (13)$$

(where $i_X \omega(Y) = \omega(X, Y)$).

Normal coordinates. Near a fixed point we can holomorphically change coordinates in such a way that $g_{j\bar{p}} = \delta_{jk}$ and $g_{j\bar{k}l} = g_{j\bar{k}lm} = 0$. By a linear transformation we can obtain the first condition. Then consider the mapping

$$F^m(z) := z^m + \frac{1}{2} a_{jk}^m z^j z^k + \frac{1}{6} b_{jkl}^m z^j z^k z^l$$

(the origin being our fixed point), where a_{jk}^m is symmetric in j, k and b_{jkl}^m symmetric in j, k, l. Then for $\tilde{g} = g \circ F$ we have

$$\tilde{g}_{j\bar{k}l}(0) = g_{j\bar{k}l}(0) + a_{jl}^k$$

$$\tilde{g}_{j\bar{k}lm}(0) = g_{j\bar{k}lm}(0) + 3 g_{j\bar{k}p}(0) a_{lm}^p + b_{jlm}^k$$

and we can choose the coefficients of F in such a way that the left-hand sides vanish.

3 Calabi Conjecture and Extremal Metrics

A complex manifold is called *Kähler* if it admits a Kähler metric. We will be particularly interested in compact Kähler manifolds. If ω is a Kähler form on a compact complex manifold M then the (p, p)-form ω^p is not exact, because if $\omega^p = d\alpha$ for some α, then

$$\int_M \omega^n = \int_M d(\alpha \wedge \omega^p) = 0$$

which is a contradiction. Since ω^p is a real closed $2p$-form, it follows that for compact Kähler manifolds $H^{2p}(M, \mathbb{R}) \neq 0$.

Example. Hopf surface $M := (\mathbb{C}^2 \setminus \{0\})/\{2^n : n \in \mathbb{Z}\}$ is a compact complex surface, topologically equivalent to $S^1 \times S^3$. Therefore $H^2(M, \mathbb{R}) = 0$ and thus M is not Kähler. □

dd^c**-lemma.** It follows from (10) that for two Kähler forms $\omega, \tilde{\omega}$ on M the $(1, 1)$-forms $Ric_\omega, Ric_{\tilde{\omega}}$ are dd^c-cohomologous, in particular d-cohomologous. The following result, called a dd^c-lemma, shows that these two notions are in fact equivalent for $(1, 1)$-forms on a compact manifold:

Theorem 3. *If a (1,1)-form on a compact Kähler manifold is d-exact then it is dd^c-exact.*

We will follow the proof from [44]. Theorem 3 will be an easy consequence of the following:

Lemma 4. *Assume that β is a (0,1)-form on a compact Kähler manifold such that $\bar\partial \beta = 0$. Then $\partial \beta = \partial \bar\partial f$ for some $f \in C^\infty(M, \mathbb{C})$.*

Proof. Let ω be a Kähler form on M. Since
$$\int_M \partial \beta \wedge \omega^{n-1} = \int_M d\beta \wedge \omega^{n-1} = \int_M d(\beta \wedge \omega^{n-1}) = 0,$$
we can find $f \in C^\infty(M, \mathbb{C})$ solving
$$\partial \bar\partial f \wedge \omega^{n-1} = \partial \beta \wedge \omega^{n-1}.$$
Set $\gamma := \beta - \bar\partial f$, we have to show that $\partial \gamma = 0$. Since $\bar\partial \gamma = 0$,
$$\int_M \partial \gamma \wedge \overline{\partial \gamma} \wedge \omega^{n-2} = \int_M d(\gamma \wedge \overline{d\gamma} \wedge \omega^{n-2}) = 0.$$

Locally we may write
$$\gamma = \gamma_{\bar k} d\bar z^k$$
and
$$\partial \gamma = \gamma_{\bar k j} dz^j \wedge d\bar z^k.$$
One can then show that
$$\partial \gamma \wedge \overline{\partial \gamma} \wedge \omega^{n-2} = \frac{1}{n(n-1)}\big(g^{j\bar k}g^{p\bar q}\gamma_{\bar k p}\overline{\gamma_{\bar j q}} - |g^{j\bar k}\gamma_{\bar k j}|^2\big)\omega^n.$$

Now $\partial \gamma \wedge \omega^{n-1} = 0$ means that $g^{j\bar k}\gamma_{\bar k j} = 0$ and it follows that $\gamma_{k\bar j} = 0$. □

Proof of Theorem 3. Write $\alpha = \beta_1 + \beta_2$, where β_1 is a $(1,0)$, and β_2 a $(0,1)$-form. Then
$$d\alpha = \partial \beta_1 + \partial \beta_2 + \bar\partial \beta_1 + \bar\partial \beta_2.$$
Since $d\alpha$ is of type $(1,1)$, it follows that $\bar\partial \beta_1 = \partial \beta_2 = 0$. By Lemma 4 we have $\partial \beta_1 = \partial \bar\partial f_1$ and $\partial \beta_2 = \partial \bar\partial f_2$ for some $f_1, f_2 \in C^\infty(M, \mathbb{C})$. Therefore
$$d\alpha = \partial \beta_2 + \bar\partial \beta_1 = \partial \bar\partial (f_2 - \bar f_1).$$
□

From now on we assume that M is a compact Kähler manifold. For a Kähler form ω on M by $c_1(M)$ we denote the cohomology class $\{Ric_\omega\}$. By (10) it is independent of the choice of ω; in fact $c_1(M) = c_1(M)_\mathbb{R}/2\pi$, where $c_1(M)_\mathbb{R}$ is the first Chern class.

Calabi conjecture ([17]). Let $\tilde{R}$ be a (1,1)-form on M cohomologous to Ric_ω (we write $R \sim Ric_\omega$). Then we ask whether there exists another, unique Kähler form $\tilde{\omega} \sim \omega$ on M such that $\tilde{R} = Ric_{\tilde{\omega}}$. In other words, the question is whether the mapping

$$\{\omega\} \ni \tilde{\omega} \longmapsto Ric_{\tilde{\omega}} \in c_1(M)$$

is bijective.

Derivation of the Monge–Ampère equation. By dd^c-lemma we have $Ric_\omega = \tilde{R} + dd^c\eta$ for some $\eta \in C^\infty(M)$. We are thus looking for $\varphi \in C^\infty(M)$ such that $\omega_\varphi := \omega + dd^c\varphi > 0$ and

$$dd^c\Big(\log \frac{\omega_\varphi^n}{\omega^n} - \eta\Big) = 0,$$

that is

$$\log \frac{\omega_\varphi^n}{\omega^n} - \eta = c,$$

a constant. This means that

$$\omega_\varphi^n = e^{c+\eta}\omega^n.$$

Since $\omega_\varphi^n - \omega^n$ is exact, from the Stokes theorem we infer

$$\int_M \omega_\varphi^n = \int_M \omega^n =: V.$$

Therefore the constant c is uniquely determined. It follows that to solve the Calabi conjecture is equivalent to solve the following Dirichlet problem for the complex Monge–Ampère operator on M: for $f \in C^\infty(M)$, $f > 0$, satisfying $\int_M f\omega^n = V$, there exists, unique up to an additive constant, $\varphi \in C^\infty(M)$ such that $\omega + dd^c\varphi > 0$ and

$$(\omega + dd^c\varphi)^n = f\omega^n. \tag{14}$$

This problem was solved by Yau [47], the proof will be given in Sects. 7–13. The solution of Calabi conjecture has many important consequences (see e.g. [48]). The one which is particularly interesting in algebraic geometry is that for a compact Kähler manifold M with $c_1(M) = 0$ there exists a Kähler metric with vanishing Ricci curvature. Except for the torus $\mathbb{C}^n/\Lambda$ such a metric can never be written down explicitly.

Kähler–Einstein metrics. A Kähler form ω is called *Kähler–Einstein* if $Ric_\omega = \lambda\omega$ for some $\lambda \in \mathbb{R}$. A necessary condition for M is thus that $c_1(M)$ is definite which means that it contains a definite representative. There are three possibilities: $c_1(M) = 0$, $c_1(M) < 0$ and $c_1(M) > 0$. Assume that it is the case, we can then find a Kähler metric ω with $\lambda\omega \in c_1(M)$, that is $Ric_\omega = \lambda\omega + dd^c\eta$ for some $\eta \in C^\infty(M)$. We are looking for $\varphi \in C^\infty(M)$ such that $Ric_{\omega_\varphi} = \lambda\omega_\varphi$ which, similarly as before, is equivalent to

$$(\omega + dd^c\varphi)^n = e^{-\lambda\varphi+\eta+c}\omega^n. \tag{15}$$

If $c_1(M) = 0$ then $\lambda = 0$ and (15) is covered by (14). If $c_1(M) < 0$ one can solve (15) in a similar way as (14). It was done by Aubin [1] and Yau [47], in fact, the L^∞-estimate in this case is very simple.

The case $c_1(M) > 0$ (such manifolds are called Fano) is the most difficult. The first obstruction to the existence of Kähler–Einstein metrics is a result of Matsushima [36] which says that in this case the Lie algebra of holomorphic vector fields must be reductive (that is it must be a complexification of a compact real subalgebra). By the result of Tian [41] this is the only obstruction in dimension 2 but in [43] he constructed a 3-dimensional Fano manifold with no holomorphic vector fields and no Kähler–Einstein metric. In fact, the Fano surfaces can be classified: they are exactly $\mathbb{P}^2$, $\mathbb{P}^1 \times \mathbb{P}^1$ or $\mathbb{P}^2$ blown up at k points in general position, where $1 \leq k \leq 8$. Among those only $\mathbb{P}^2$ blown up at one or two points have non-reductive algebras of holomorphic vector fields, and thus all the other surfaces admit Kähler–Einstein metrics—see [43] or a recent exposition of Tosatti [45].

Uniqueness of Kähler–Einstein metrics in a given Kähler class $\{\omega\}$ (satisfying the necessary condition $\lambda\{\omega\} \subset c_1(M)$) for $c_1(M) = 0$ and $c_1(M) < 0$ follows quite easily from the equation (15). In the Fano case $c_1(M) > 0$ it holds up to a biholomorphism—it was proved by Bando and Mabuchi [3] (see also [6,7]).

Constant scalar curvature metrics. Given a compact Kähler manifold (M, ω) we are interested in a metric in $\{\omega\}$ with constant scalar curvature (csc). With the notation $S_\varphi = S_{\omega_\varphi}$ we are thus looking for φ satisfying $S_\varphi = \bar{S}$, where $\bar{S}$ is a constant. First of all we note that $\bar{S}$ is uniquely determined by the Kähler class:

$$\bar{S}\int_M \omega^n = \int_M S_\varphi \omega_\varphi^n = 2n\int_M Ric_\varphi \wedge \omega_\varphi^{n-1} = 2n\int_M Ric_\omega \wedge \omega^{n-1}. \tag{16}$$

Secondly, the csc problem is more general than the Kähler–Einstein problem. For if $\lambda\{\omega\} \subset c_1(M)$, that is $Ric_\omega = \lambda\omega + dd^c\eta$ for some $\eta \in C^\infty(M)$, and ω_φ is a csc metric then $\bar{S} = 2n\lambda$ and $Ric_\varphi \wedge \omega_\varphi^{n-1} = \lambda\omega_\varphi^n$. But since

$$Ric_\varphi - \lambda\omega_\varphi = dd^c[\eta - \log\frac{\omega_\varphi^n}{\omega^n} - \lambda\varphi],$$

it follows that $Ric_\varphi = \lambda\omega_\varphi$.

The equation $S_\varphi = \bar{S}$ is of order 4 and therefore very difficult to handle directly. The question of uniqueness of csc metrics was treated in [19]. A general conjecture links existence of csc metrics with stability in the sense of geometric invariant theory. So far it has been fully answered only in the case of toric surfaces (Donaldson [21]). See [37] for an extensive survey on csc metrics.

4 The Space of Kähler Metrics

We consider the class of Kähler potentials w.r.t. a Kähler form ω:

$$\mathcal{H} := \{\varphi \in C^\infty(M) : \omega_\varphi > 0\}.$$

It is an open subset of $C^\infty(M)$ and thus has a structure of an infinite dimensional differential manifold (its differential structure is determined by the relation

$$C^\infty(U, C^\infty(M)) = C^\infty(M \times U)$$

for any region U in $\mathbb{R}^m$). For $\varphi \in \mathcal{H}$ the tangent space $T_\varphi \mathcal{H}$ may be thus identified with $C^\infty(M)$. On $T_\varphi \mathcal{H}$, following Mabuchi [M], we define a scalar product:

$$\langle\langle \psi, \eta \rangle\rangle := \frac{1}{V} \int_M \psi \eta \, \omega_\varphi^n, \quad \psi, \eta \in T_\varphi \mathcal{H}.$$

Also by $\varphi = \varphi(t)$ denote a smooth curve $[a,b] \to \mathcal{H}$ (which is an element of $C^\infty(M \times [a,b])$). For a vector field ψ on φ (which we may also treat as an element of $C^\infty(M \times [a,b])$) we want to define a connection $\nabla_{\dot\varphi} \psi$ (where we denote $\dot\varphi = d\varphi/dt$), so that

$$\frac{d}{dt} \langle\langle \psi, \eta \rangle\rangle = \langle\langle \nabla_{\dot\varphi} \psi, \eta \rangle\rangle + \langle\langle \psi, \nabla_{\dot\varphi} \eta \rangle\rangle \tag{17}$$

(where η is another vector field on φ). Since

$$\frac{d}{dt} \omega_\varphi^n = n \, dd^c \dot\varphi \wedge \omega_\varphi^{n-1} = \frac{1}{2} \Delta \dot\varphi \, \omega_\varphi^n, \tag{18}$$

where Δ denotes the Laplacian w.r.t. ω_φ, we have

$$\frac{d}{dt} \langle\langle \psi, \eta \rangle\rangle = \frac{1}{V} \int_M \left(\dot\psi \eta + \psi \dot\eta + \frac{1}{2} \psi \eta \Delta \dot\varphi \right) \omega_\varphi^n$$

$$= \frac{1}{V} \int_M \left(\dot\psi \eta + \psi \dot\eta - \frac{1}{2} \langle \nabla(\psi\eta), \nabla\dot\varphi \rangle \right) \omega_\varphi^n$$

$$= \frac{1}{V} \int_M \left[\left(\dot\psi - \frac{1}{2} \langle \nabla\psi, \nabla\dot\varphi \rangle \right) \eta + \psi \left(\dot\eta - \frac{1}{2} \langle \nabla\eta, \nabla\dot\varphi \rangle \right) \right] \omega_\varphi^n.$$

This shows that the right way to define a connection on $\mathcal{H}$ is

$$\nabla_{\dot\varphi} \psi := \dot\psi - \frac{1}{2} \langle \nabla\psi, \nabla\dot\varphi \rangle,$$

The Complex Monge–Ampère Equation in Kähler Geometry

where ∇ on the right-hand side denotes the gradient w.r.t. ω_φ. A curve φ in $\mathcal{H}$ is therefore a *geodesic* if $\nabla_{\dot\varphi}\dot\varphi = 0$, that is

$$\ddot\varphi - \frac{1}{2}|\nabla\dot\varphi|^2 = 0. \tag{19}$$

Curvature.

Theorem 5 (Mabuchi [35], Donaldson [20]). *We have the following formula for the curvature of $\langle\langle\cdot,\cdot\rangle\rangle$*

$$R(\psi,\eta)\gamma = -\frac{1}{4}\{\{\psi,\eta\},\gamma\}, \quad \psi,\eta,\gamma \in T_\varphi\mathcal{H}, \; \varphi \in \mathcal{H}.$$

In particular, the sectional curvature is given by

$$K(\psi,\eta) = -\frac{1}{4}\|\{\psi,\eta\}\|^2 \leq 0.$$

Proof. Without loss of generality we may evaluate the curvature at $0 \in \mathcal{H}$. Let $\varphi \in C^\infty([0,1] \times [0,1], \mathcal{H})$ be such that $\varphi(0,0) = 0$ and at $s = t = 0$ we have $\varphi_s(= d\varphi/ds) = \psi$, $\varphi_t = \eta$. Take $\gamma \in C^\infty([0,1]^2, C^\infty(M)) = C^\infty(M \times [0,1]^2)$. We have

$$\nabla_{\varphi_s}\nabla_{\varphi_t}\gamma - \nabla_{\varphi_t}\nabla_{\varphi_s}\gamma = \nabla_{\varphi_s}\left(\gamma_t - \frac{1}{2}\langle\nabla\varphi_t,\nabla\gamma\rangle\right) - \nabla_{\varphi_t}\left(\gamma_s - \frac{1}{2}\langle\nabla\varphi_s,\nabla\gamma\rangle\right)$$

$$= -\frac{1}{2}\frac{d}{ds}\langle\nabla\varphi_t,\nabla\gamma\rangle - \frac{1}{2}\langle\nabla\varphi_s,\nabla\gamma_t\rangle + \frac{1}{4}\langle\nabla\varphi_s,\nabla\langle\nabla\varphi_t,\nabla\gamma\rangle\rangle$$

$$+ \frac{1}{2}\frac{d}{dt}\langle\nabla\varphi_s,\nabla\gamma\rangle + \frac{1}{2}\langle\nabla\varphi_t,\nabla\gamma_s\rangle - \frac{1}{4}\langle\nabla\varphi_t,\nabla\langle\nabla\varphi_s,\nabla\gamma\rangle\rangle.$$

Denoting $u = g + \varphi$ we get

$$\frac{d}{dt}\langle\nabla\varphi_s,\nabla\gamma\rangle = \frac{d}{dt}\left[u^{j\bar k}\left(\gamma_j\varphi_{s\bar k} + \gamma_{\bar k}\varphi_{sj}\right)\right]$$

$$= -\frac{1}{4}\langle i\partial\bar\partial\eta, i\partial\psi \wedge \bar\partial\gamma + i\partial\gamma \wedge \bar\partial\psi\rangle + \langle\nabla\varphi_s,\nabla\gamma_t\rangle + \langle\nabla\varphi_{st},\nabla\gamma\rangle$$

where in the last line we have evaluated at $s = t = 0$. Therefore at $s = t = 0$ we have

$$\nabla_{\varphi_s}\nabla_{\varphi_t}\gamma - \nabla_{\varphi_t}\nabla_{\varphi_s}\gamma = \frac{1}{4}\langle\nabla\psi,\nabla\langle\nabla\eta,\nabla\gamma\rangle\rangle - \frac{1}{4}\langle\nabla\eta,\nabla\langle\nabla\psi,\nabla\gamma\rangle\rangle$$

$$+ \frac{1}{8}\langle i\partial\bar\partial\psi, i\partial\eta \wedge \bar\partial\gamma + i\partial\gamma \wedge \bar\partial\eta\rangle$$

$$- \frac{1}{8}\langle i\partial\bar\partial\eta, i\partial\psi \wedge \bar\partial\gamma + i\partial\gamma \wedge \bar\partial\psi\rangle.$$

We can now show, using for example normal coordinates, that the right-hand side is equal to $-\frac{1}{4}\{\{\psi,\eta\},\gamma\}$. □

Derivation of the homogeneous complex Monge–Ampère equation. Writing locally $u = g + \varphi$, since g is independent of t, we can rewrite (19) as

$$\ddot{u} - u^{p\bar{q}}\dot{u}_p\dot{u}_{\bar{q}} = 0.$$

Multiplying both sides by $\det(u_{j\bar{k}})$ (which is non-vanishing) we arrive at the equation

$$\det\begin{pmatrix} & & u_{1t} \\ (u_{j\bar{k}}) & & \vdots \\ & & u_{nt} \\ u_{t\bar{1}} & \cdots & u_{t\bar{n}} & u_{tt} \end{pmatrix} = 0.$$

This suggests to complexify the variable t, either simply by adding an imaginary variable, or introducing the new one $\zeta(=z^{n+1})\in\mathbb{C}_*$, so that $t = \log|\zeta|$. Then for $v(\zeta) = u(\log|\zeta|)$ we have $v_\zeta = \dot{u}/2\zeta$ and $v_{\zeta\bar{\zeta}} = \ddot{u}/4|\zeta|^2$. We have thus obtained the following characterization of geodesics in $\mathcal{H}$:

Proposition 6 ((Semmes [39], Donaldson [20])). *For $\varphi_0, \varphi_1 \in \mathcal{H}$ existence of a geodesic in $\mathcal{H}$ joining φ_0 and φ_1 is equivalent to solving the following Dirichlet problem for the homogeneous complex Monge–Ampère equation:*

$$\begin{cases} \varphi \in C^\infty(M\times\{e^0 \leq |\zeta| \leq e^1\}) \\ \omega + dd^c\varphi(\cdot,\zeta) > 0, \quad e^0 \leq |\zeta| \leq e^1 \\ (\omega + dd^c\varphi)^{n+1} = 0 \\ \varphi(\cdot,\zeta) = \varphi_j, \quad |\zeta| = e^j, \ j = 0,1. \end{cases}$$

□

Although ω is a degenerate form on $M\times\mathbb{C}$, it is not a problem: write

$$\omega + dd^c\varphi = \tilde{\omega} + dd^c(\varphi - |\zeta|^2), \tag{20}$$

where $\tilde{\omega} = \omega + dd^c|\zeta|^2$ is a Kähler form on $M\times\mathbb{C}$, and consider the related problem.

The existence of geodesic is thus equivalent to solving the homogeneous Monge–Ampère equation on a compact Kähler manifold with boundary. From the uniqueness of this equation (see e.g. the next section) it follows in particular that given two potentials in $\mathcal{H}$ there exists at most one geodesic joining them.

As shown recently by Lempert and Vivas [33], it is not always possible to join two metrics by a smooth geodesic (see Sect. 6). However, for $\varepsilon > 0$ we can introduce a notion of an ε-*geodesic*: instead of (19) it solves

$$(\ddot{\varphi} - \frac{1}{2}|\nabla\dot{\varphi}|^2)\omega_\varphi^n = \varepsilon\omega^n$$

which is equivalent to the following non-degenerate complex Monge–Ampère equation:

$$(\omega + dd^c\varphi)^{n+1} = \frac{\varepsilon}{4|\zeta|^2}(\omega + dd^c|\zeta|^2)^{n+1}. \tag{21}$$

As shown by Chen [18] (see also [15]), smooth ε-geodesics always exist (see Theorem 19 below) and they approximate weak geodesics. Existence of ε-geodesics will be used in Sect. 6 below to show that $\mathcal{H}$ with a distance defined by its Riemannian structure is a metric space (this result is due to Chen [18], see also [15]).

Normalization, Aubin–Yau functional. The Riemannian structure on $\mathcal{H}$ will induce such a structure on the Kähler class $\{\omega\} = \mathcal{H}/\mathbb{R}$, which is independent of the choice of ω. For this we need a good normalization on $\mathcal{H}$. The right tool for this purpose is the Aubin–Yau functional (see e.g. [2])

$$I : \mathcal{H} \to \mathbb{R}$$

which is characterized by the following properties

$$I(0) = 0, \quad d_\varphi I.\psi = \frac{1}{V}\int_M \psi\, \omega_\varphi^n, \quad \varphi \in \mathcal{H},\ \psi \in C^\infty(M). \tag{22}$$

This means that we are looking for I with $dI = \alpha$, where the 1-form α is given by

$$\alpha(\varphi).\psi = \frac{1}{V}\int_M \psi\, \omega_\varphi^n. \tag{23}$$

Such an I exists provided that α is closed. But by (18)

$$d\alpha(\varphi).(\psi,\tilde{\psi}) = d_\varphi(\alpha(\varphi).\psi).\tilde{\psi} - d_\varphi(\alpha(\varphi).\tilde{\psi}).\psi = \frac{n}{V}\int_M (\psi\Delta\tilde{\psi} - \tilde{\psi}\Delta\psi)\,\omega_\varphi^n = 0$$

and it follows that there is I satisfying (22).

For any curve $\tilde{\varphi}$ in $\mathcal{H}$ joining 0 with φ we have

$$I(\varphi) = \int_0^1 \frac{1}{V}\int_M \dot{\tilde{\varphi}}\,\omega_{\tilde{\varphi}}^n\, dt.$$

Taking $\tilde{\varphi}(t) = t\varphi$, since (with some abuse of notation)

$$\frac{d}{dt}\frac{(\omega + t\, dd^c\varphi)^{n+1} - \omega^{n+1}}{(n+1)\, dd^c\varphi} = (\omega + t\, dd^c\varphi)^n = \omega_{t\varphi}^n,$$

we obtain the formula

$$I(\varphi) = \frac{1}{n+1} \sum_{p=0}^{n} \frac{1}{V} \int_M \varphi \, \omega_\varphi^p \wedge \omega^{n-p}.$$

We also get
$$I(\varphi + c) = I(\varphi) + c$$
for any constant c.

Now for any curve φ in $\mathcal{H}$ by (22) and (17) we have
$$\left(\frac{d}{dt}\right)^2 I(\varphi) = \frac{d}{dt} \langle\langle \dot\varphi, 1\rangle\rangle = \langle\langle \nabla_{\dot\varphi} \dot\varphi, 1\rangle\rangle$$

and it follows that I is affine along geodesics. Moreover, if φ is a geodesic then so is $\varphi - I(\varphi)$. Therefore, by uniqueness of geodesics, $\mathcal{H}_0 := I^{-1}(0)$ is a totally geodesic subspace of $\mathcal{H}$. The bijective mapping
$$\mathcal{H}_0 \ni \varphi \longmapsto \omega_\varphi \in \{\omega\}$$
induces the Riemannian structure on $\{\omega\}$. By (22) we have
$$T_\varphi \mathcal{H}_0 = \{\psi \in C^\infty(M) : \int_M \psi \omega_\varphi^n = 0\}.$$

One can easily show that this Riemannian structure on $\{\omega\}$ is independent of the choice of ω.

Mabuchi K-energy [34]. It is defined by the condition
$$K(0) = 0, \quad d_\varphi K.\psi = -\frac{1}{V}\int_M \psi(S_\varphi - \bar S)\omega_\varphi^n, \tag{24}$$

where $\bar S$ is the average of scalar curvature S_φ (it is given by (16)). We are thus looking for K satisfying $dK = \beta + \bar S \alpha$, where α is given by (23) and
$$\beta(\varphi).\psi = -\frac{1}{V}\int_M \psi S_\varphi \omega_\varphi^n = -\frac{2n}{V}\int_M \psi \, Ric_\varphi \wedge \omega_\varphi^{n-1},$$

we have to show that $d\beta = 0$. We compute
$$d_\varphi(Ric_\varphi).\psi = d_\varphi(Ric_\omega - dd^c \log \frac{\omega_\varphi^n}{\omega^n}).\psi = -\frac{1}{2}dd^c \Delta\psi \tag{25}$$

and thus
$$d_\varphi(\beta(\varphi).\psi).\tilde\psi = \frac{1}{V}\int_M \tilde\psi\Big(\frac{1}{2}\Delta^2\tilde\psi \, \omega_\varphi^n - 2n(n-1)dd^c\tilde\psi \wedge Ric_\varphi \wedge \omega_\varphi^{n-2}\Big).$$

It is clear that the latter expression is symmetric in ψ and $\tilde{\psi}$ and therefore $d\beta = 0$.
To get a precise formula for K take as before $\tilde{\varphi} = t\varphi$. Similarly we have

$$\frac{d}{dt} \frac{\omega_{t\varphi}^n - \omega^n}{n\, dd^c \varphi} = \omega_{t\varphi}^{n-1}$$

and

$$\frac{d}{dt}\left[\left(\log \frac{\omega_{\tilde{\varphi}}^n}{\omega^n}\right)\omega_{\tilde{\varphi}}^n\right] = n\left(1 + \log \frac{\omega_{\tilde{\varphi}}^n}{\omega^n}\right) dd^c \dot{\tilde{\varphi}} \wedge \omega_{\tilde{\varphi}}^{n-1}.$$

Using this we will easily get (see also [37, 42])

$$K(\varphi) = \frac{2}{V} \int_M \left[\left(\log \frac{\omega_\varphi^n}{\omega^n}\right)\omega_\varphi^n - \varphi \sum_{p=0}^{n-1} Ric_\omega \wedge \omega_\varphi^p \wedge \omega^{n-p-1}\right] + \bar{S}\, I(\varphi).$$

The usefulness of the K-energy in some geometric problems becomes clear in view of the following two results:

Proposition 7 (Mabuchi [35], Donaldson [20]). *For any smooth curve φ in $\mathcal{H}$ we have*

$$\left(\frac{d}{dt}\right)^2 K(\varphi) = -\frac{1}{V} \int_M \nabla_{\dot{\varphi}} \dot{\varphi} (S_\varphi - \bar{S}) \omega_\varphi^n + \frac{1}{2V} \int_M |\mathcal{L}\dot{\varphi}|^2 \omega_\varphi^n.$$

In particular, the K-energy is convex along geodesics.

Proof. We have

$$\frac{d}{dt} K(\varphi) = -\langle\langle \dot{\varphi}, S_\varphi - \bar{S}\rangle\rangle,$$

therefore

$$\left(\frac{d}{dt}\right)^2 K(\varphi) = -\langle\langle \nabla_{\dot{\varphi}} \dot{\varphi}, S_\varphi - \bar{S}\rangle\rangle - \langle\langle \dot{\varphi}, \nabla_{\dot{\varphi}} S_\varphi\rangle\rangle.$$

Moreover

$$-\langle\dot{\varphi}, \nabla_{\dot{\varphi}} S_\varphi\rangle = \frac{1}{V} \int_M \dot{\varphi}\left(\frac{1}{2}\langle\nabla S_\varphi, \nabla \dot{\varphi}\rangle - \frac{d}{dt} S_\varphi\right) \omega_\varphi^n.$$

Write $u = g + \varphi$. Then

$$S_\varphi = -2u^{p\bar{q}}\left(\log \det(u_{j\bar{k}})\right)_{p\bar{q}}$$

and, since $\dot{g} = 0$,

$$\frac{d}{dt} S_\varphi = 2u^{p\bar{t}} u^{s\bar{q}} \big(\log\det(u_{j\bar{k}})\big)_{p\bar{q}} \dot{\varphi}_{s\bar{t}} - 2u^{p\bar{q}} \big(u^{j\bar{k}} \dot{\varphi}_{j\bar{k}}\big)_{p\bar{q}}$$
$$= -\frac{1}{2} \langle Ric_\varphi, \nabla^2 \dot{\varphi}\rangle - \frac{1}{2} \Delta^2 \dot{\varphi}.$$

The result now follows from Proposition 2. □

Proposition 8 (Donaldson [20]). *Let ω_{φ_0} and ω_{φ_1} be csc metrics. Assume moreover that φ_0 and φ_1 can be joined by a smooth geodesic. Then there exists a biholomorphism F of M such that $\omega_{\varphi_0} = F^* \omega_{\varphi_1}$.*

Proof. Let φ be this geodesic and set $h := K(\varphi)$. Then, since $S_{\varphi_0} = S_{\varphi_1} = \bar{S}$, we have $\dot{h}(0) = \dot{h}(1) = 0$ and by Proposition 7 h is convex. Therefore $\ddot{h} = 0$ and, again by Proposition 7, $\mathcal{L}\dot{\varphi} = 0$, that is $\nabla\dot{\varphi}$ is a flow of holomorphic vector fields. By F denote the flow of biholomorphisms generated by $\frac{1}{2}\nabla\dot{\varphi}$ (so that $\dot{F} = \frac{1}{2}\nabla\dot{\varphi} \circ F$, $F|_{t=0} = id$).

We have to check that $\omega_{\varphi_0} = F^* \omega_\varphi$, it will be enough to show that $\frac{d}{dt} F^* \omega_\varphi = 0$. We compute

$$\frac{d}{dt} F^* \omega_\varphi = F^* \big(L_{\frac{1}{2}\nabla\dot{\varphi}}\omega_\varphi + dd^c \dot{\varphi}\big) = F^* d\big(\frac{1}{2} i_{\nabla_f}\omega_\varphi + d^c \dot{\varphi}\big) = 0$$

by (13) (where $L_X = i_X \circ d + d \circ i_X$ is the Lie derivative). (This argument from symplectic geometry is called a Moser's trick.) □

In view of the Lempert–Vivas counterexample Proposition 8 is not sufficient to prove the uniqueness of csc metrics. For a more direct approach to this problem see [19].

5 Lempert–Vivas Example

It is well known that in general one cannot expect C^∞-regularity of solutions of the homogeneous Monge–Ampère equation. The simplest example is due to Gamelin and Sibony [25]: the function

$$u(z,w) := \big(\max\{0, |z|^2 - 1/2, |w|^2 - 1/2\}\big)^2$$

satisfies $dd^c u \geq 0$, $(dd^c u)^2 = 0$ in the unit ball $\mathbb{B}$ of $\mathbb{C}^2$,

$$u(z,w) = \big(|z|^2 - 1/2\big)^2 = \big(|w|^2 - 1/2\big)^2 \in C^\infty(\partial\mathbb{B}),$$

but $u \notin C^2(\mathbb{B})$.

For some time it was however an open problem whether there exists a smooth geodesic connecting arbitrary two elements in $\mathcal{H}$. In the special case of toric Kähler manifolds it was in fact shown in [28] that it is indeed the case. This suggests that a possible counterexample would have to be more complicated, as the Gamelin–Sibony example from the flat case is *toric*.

The counterexample to the geodesic problem was found recently by Lempert and Vivas [33]. It works on Kähler manifolds with a holomorphic isometry $h : M \to M$ satisfying $h^2 = id$ and having an isolated fixed point. We will consider the simplest situation, that is the Riemann sphere $\mathbb{P}$ with the Fubini-Study metric $\omega = dd^c(\log(1 + |z|^2))$ and $h(z) = -z$. The key is the following result:

Lemma 9 (Lempert–Vivas [33]). *Take $\varphi \in \mathcal{H}$ with*

$$\varphi(-z) = \varphi(z) \tag{26}$$

Assume that there is a geodesic of class C^3 joining 0 with φ. Then either $1+\varphi_{z\bar{z}}(0) = |1 - \varphi_{zz}(0)|$ or $|\varphi_{zz}(0)| \leq |\varphi_{z\bar{z}}(0)|$, in particular

$$|\varphi_{zz}(0)| \leq 2 + \varphi_{z\bar{z}}(0). \tag{27}$$

Proof. By $\tilde{\varphi}$ denote the geodesic joining 0 with φ. We can assume that it is a C^3 function defined on $\mathbb{P} \times \bar{S}$, where $S = \{0 < \mathrm{Im}\, w < 1\}$, and such that

$$\tilde{\varphi}(z, w + \sigma) = \tilde{\varphi}(z, w), \quad \sigma \in \mathbb{R}.$$

Moreover, by uniqueness of the Dirichlet problem (see Theorem 21 below) by (26) we have

$$\tilde{\varphi}(-z, w) = \tilde{\varphi}(z, w).$$

On $\mathbb{C} \times \bar{S}$ set $u := g + \tilde{\varphi}$. Then $u \in C^3(\mathbb{C} \times \bar{S})$,

$$u_{z\bar{z}} u_{w\bar{w}} - |u_{z\bar{w}}|^2 = 0,$$

$u_{z\bar{z}} > 0$, u is independent of $\sigma = \mathrm{Re}\, w$, $u(\cdot, 0) = g$, $u(\cdot, i) = g + \varphi$.

Since $(u_{j\bar{k}})$ is of maximal rank, it is well known (see e.g. [4]) that there is a C^1 foliation of $\mathbb{C} \times \bar{S}$ by holomorphic discs (with boundary) which are tangent to $dd^c u$. This foliation is also invariant under the mapping $(z, w) \mapsto (-z, w)$ and thus $\{0\} \times \bar{S}$ is one of the leaves. The neighboring leaves are graphs of functions defined on $\bar{S}$: there exists $f \in C^1(U \times \bar{S})$, where U is a neighborhood of 0, $f(z, \cdot)$ holomorphic in S and $\{(f(z, w), w) : w \in \bar{S}\}$ is the leaf passing through $(z, 0)$. Since this leaf is tangent to $dd^c u$, it follows that

$$u_{z\bar{z}}(f(z,w), w)\overline{f_w(z,w)} + u_{z\bar{w}}(f(z,w), w) = 0$$

which is equivalent to the fact that $u_z(f(z, w), w)$ is holomorphic in w. Set

$$\Phi(w) := \frac{d}{dt}\bigg|_{t=0} f(t, w)$$

and

$$\Psi(w) := \frac{d}{dt}\bigg|_{t=0} u_z(f(t, w), w) = u_{zz}(0, w)\Phi(w) + u_{z\bar{z}}(0, w)\overline{\Phi(w)}.$$

Then Φ, Ψ are holomorphic in S, Φ is C^1 on $\bar{S}$, and Ψ is continuous on $\bar{S}$.
Since u is independent of $\operatorname{Re} w$, we can write

$$\Psi(w) = \begin{cases} \overline{\Phi(w)}, & \operatorname{Im} w = 0 \\ P\,\overline{\Phi(w)} + Q\,\Phi(w), & \operatorname{Im} w = 1, \end{cases}$$

where $P = g_{z\bar{z}}(0) + \varphi_{z\bar{z}}(0) = 1 + \varphi_{z\bar{z}}(0) > 0$ and $Q = g_{zz}(0) + \varphi_{zz}(0) = \varphi_{zz}(0)$.
Since $\Psi_{\bar{w}} = 0$,

$$u_{zz\bar{w}}(0, w)\Phi(w) + u_{z\bar{z}\bar{w}}(0, w)\overline{\Phi(w)} + u_{z\bar{z}}(0, w)\overline{\Phi'(w)} = 0.$$

On $\{\operatorname{Im} w = 0\}$ we thus have

$$\begin{cases} \Phi'(\sigma) = A\Phi(\sigma) + B\overline{\Phi(\sigma)} \\ \Phi(0) = 1, \end{cases}$$

where

$$A = -u_{z\bar{z}w}(0, 0), \quad B = -u_{\bar{z}\bar{z}w}(0, 0).$$

Therefore Φ on $\{\operatorname{Im} w = 0\}$ is of the form

$$\Phi(\sigma) = xe^{\lambda\sigma} + \bar{y}e^{\bar{\lambda}\sigma},$$

where

$$\begin{pmatrix} A & B \\ \bar{B} & \bar{A} \end{pmatrix} \begin{pmatrix} x \\ y \end{pmatrix} = \lambda \begin{pmatrix} x \\ y \end{pmatrix}$$

and $x + \bar{y} = 1$. Note that $A \in i\mathbb{R}$ (because $u_\sigma = 0$), and thus either $\lambda \in \mathbb{R}$ or $\lambda \in i\mathbb{R}$.

By the Schwarz reflection principle and analytic continuation we obtain

$$\Phi(w) = xe^{\lambda w} + \bar{y}e^{\bar{\lambda}w}, \quad w \in \bar{S}.$$

Similarly, since

$$\Psi(\sigma) = \overline{\Phi(\sigma)} = \bar{x}e^{\bar{\lambda}\sigma} + ye^{\lambda\sigma},$$

we infer

$$\Psi(w) = \bar{x}e^{\bar{\lambda}w} + ye^{\lambda w}, \quad w \in \bar{S}.$$

Therefore, using the fact that $\Psi(w) = P\,\overline{\Phi(w)} + Q\,\Phi(w)$ on $\{\operatorname{Im} w = 1\}$, we get

$$\bar{x}e^{\bar{\lambda}(\sigma+i)} + ye^{\lambda(\sigma+i)} = P\left(\bar{x}e^{\bar{\lambda}(\sigma-i)} + ye^{\lambda(\sigma-i)}\right) + Q\left(xe^{\lambda(\sigma+i)} + \bar{y}e^{\bar{\lambda}(\sigma+i)}\right).$$

We have to consider two cases. If $\lambda \in \mathbb{R}$ then

$$e^{\lambda i} = Pe^{-\lambda i} + Qe^{\lambda i}.$$

This means that

$$P = e^{2\lambda i}(1 - Q),$$

in particular

$$P = |1 - Q|.$$

If $\lambda = i\mu \in i\mathbb{R}$ then we will get

$$\begin{cases} \bar{x}e^{\mu} = P\bar{x}e^{-\mu} + Q\bar{y}e^{\mu} \\ ye^{-\mu} = Pye^{\mu} + Qxe^{-\mu}. \end{cases}$$

Rewrite this as

$$\begin{cases} \bar{x}(e^{2\mu} - P) = \bar{y}Qe^{2\mu} \\ \bar{y}(e^{-2\mu} - P) = \bar{x}\bar{Q}e^{-2\mu}. \end{cases}$$

Since at least one of x, y does not vanish, we will obtain

$$|Q|^2 = (e^{2\mu} - P)(e^{-2\mu} - P) \le (1 - P)^2. \qquad \square$$

If φ is a smooth compactly supported function in $\mathbb{C}$ then $\varphi \in \mathcal{H}$ provided that $g_{z\bar{z}} + \varphi_{z\bar{z}} > 0$. The following lemma shows that there are such functions satisfying (26) but not (27):

Lemma 10. *For every real a and $\varepsilon > 0$ there exists smooth φ with support in the unit disc, satisfying (26), and such that $\varphi_{zz}(0) = a$, $\varphi_{z\bar{z}}(0) = 0$, $|\varphi_{z\bar{z}}| \le \varepsilon$ in $\mathbb{C}$.*

Proof. We may assume that $a > 0$. The function we seek will be of the form

$$\varphi(z) = \operatorname{Re}(z^2)\chi(|z|^2),$$

where $\chi \in C^{\infty}(\mathbb{R}_+)$ is supported in the interval $(0, 1)$ and constant near 0. Then $\varphi_{zz}(0) = \chi(0)$, $\varphi_{z\bar{z}}(0) = 0$ and

$$\varphi_{z\bar{z}} = \operatorname{Re}(z^2)\left(3\chi'(|z|^2) + |z|^2\chi''(|z|^2)\right).$$

We are looking for χ of the form

$$\chi(t) = f(-\log t),$$

where $f \in C^\infty(\mathbb{R}_+)$ is supported in $(1, \infty)$ and equal to a near ∞. We have, with $t = |z|^2$,

$$|\varphi_{z\bar z}| \leq t\big|3\chi'(t) + t\chi''(t)\big| = \big|-2f'(-\log t) + f''(-\log t)\big|.$$

We can now easily arrange f in such a way that $|f'|$ and $|f''|$ are arbitrarily small. $\square$

6 Metric Structure of $\mathcal{H}$

Although smooth geodesics in $\mathcal{H}$ do not always exist, one can make a geometric use of existence of ε-geodesics. The Riemannian structure gives a distance on $\mathcal{H}$:

$$d(\varphi_0, \varphi_1) = \inf\{l(\varphi) : \varphi \in C^\infty([0,1], \mathcal{H}), \varphi(0) = \varphi_0, \varphi(1) = \varphi_1\}, \quad \varphi_0, \varphi_1 \in \mathcal{H},$$

where

$$l(\varphi) = \int_0^1 |\dot\varphi| dt = \int_0^1 \sqrt{\frac{1}{V} \int_M \dot\varphi^2 \omega_\varphi^n}\, dt$$

(note that the family in the definition of d is always nonempty, for example $\varphi(t) = (1-t)\varphi_0 + t\varphi_1$ is a smooth curve in $\mathcal{H}$ connecting φ_0 with φ_1). We will show the following result of Chen [18] (see also [15]):

Theorem 11. $(\mathcal{H}, d)$ *is a metric space.*

The only problem with this result is to show that $d(\varphi_0, \varphi_1) > 0$ if $\varphi_0 \neq \varphi_1$. The main tool in the proof will be existence of ε-geodesics. In fact, making use of results proved in Sects. 7–13 and the standard elliptic theory, we have the following existence result for ε-geodesics:

Theorem 12. *For $\varphi_0, \varphi_1 \in \mathcal{H}$ and $\varepsilon > 0$ there exists a unique ε-geodesic φ connecting φ_0 with φ_1. Moreover, it depends smoothly on φ_0, φ_1, i.e. if $\varphi_0, \varphi_1 \in C^\infty([0,1], \mathcal{H})$ then there exists unique $\varphi \in C^\infty([0,1] \times [0,1], \mathcal{H})$ such that $\varphi(0, \cdot) = \varphi_0$, $\varphi(1, \cdot) = \varphi_1$, and $\varphi(\cdot, t)$ is an ε-geodesic for every $t \in [0,1]$. In addition,*

$$\Delta\varphi, |\nabla\dot\varphi|, \ddot\varphi \leq C, \tag{28}$$

(here Δ and ∇ are taken w.r.t. ω) where C is independent of ε (if ε is small).

We start with the following lemma:

Lemma 13. *For an ε-geodesic φ connecting $\varphi_0, \varphi_1 \in \mathcal{H}$ we have*

$$\frac{1}{V}\int_M \dot\varphi^2 \omega_\varphi^n \geq \mathcal{E}(\varphi_0,\varphi_1) - 2\varepsilon \sup_{[0,1]} ||\dot\varphi||,$$

where

$$\mathcal{E}(\varphi_0,\varphi_1) := \max\left\{\frac{1}{V}\int_{\{\varphi_0>\varphi_1\}} (\varphi_0-\varphi_1)^2 \omega_{\varphi_0}^n, \frac{1}{V}\int_{\{\varphi_1>\varphi_0\}} (\varphi_1-\varphi_0)^2 \omega_{\varphi_1}^n\right\}.$$

In particular,

$$l(\varphi)^2 \geq \mathcal{E}(\varphi_0,\varphi_1) - 2\varepsilon \sup_{[0,1]} ||\dot\varphi||.$$

Proof. For

$$E := \frac{1}{V}\int_M \dot\varphi^2 \omega_\varphi^n$$

we have

$$\dot E = \frac{1}{V}\int_M (2\dot\varphi\ddot\varphi + \frac{1}{2}\dot\varphi^2 \Delta\dot\varphi)\omega_\varphi^n = \frac{2}{V}\int_M \dot\varphi(\ddot\varphi - \frac{1}{2}|\nabla\dot\varphi|^2)\omega_\varphi^n = \frac{2\varepsilon}{V}\int_M \dot\varphi \omega^n.$$

Thus $|\dot E| \leq 2\varepsilon \sup_{[0,1]} ||\dot\varphi||$ which implies that

$$E(t) \geq \max\{E(0), E(1)\} - 2\varepsilon \sup_{[0,1]} ||\dot\varphi||.$$

Since $\ddot\varphi \geq 0$,

$$\dot\varphi(0) \leq \varphi(1) - \varphi(0) \leq \dot\varphi(1).$$

For $z \in M$ with $\varphi_1(z) > \varphi_0(z)$ we thus have $\dot\varphi(z,1)^2 \geq (\varphi_1(z) - \varphi_0(z))^2$. Therefore

$$E(1) \geq \frac{1}{V}\int_{\{\varphi_1>\varphi_0\}} (\varphi_1-\varphi_0)^2 \omega_{\varphi_1}^n.$$

Similarly

$$E(0) \geq \frac{1}{V}\int_{\{\varphi_0>\varphi_1\}} (\varphi_0-\varphi_1)^2 \omega_{\varphi_0}^n$$

and the desired estimate follows. □

Theorem 14. *Suppose $\psi \in C^\infty([0,1],\mathcal{H})$ and $\tilde\psi \in \mathcal{H} \setminus \psi([0,1])$. For $\varepsilon > 0$ by φ denote an element of $C^\infty([0,1]\times[0,1],\mathcal{H})$ uniquely determined by the following property: $\varphi(\cdot,t)$ is an ε-geodesic connecting $\tilde\psi$ with $\psi(t)$. Then for ε sufficiently small*

$$l(\varphi(\cdot,0)) \leq l(\psi) + l(\varphi(\cdot,1)) + C\varepsilon,$$

where $C > 0$ is independent of ε.

Proof. Without loss of generality we may assume that $V = 1$. Set

$$l_1(t) := \int_0^t ||\dot\psi|| d\tilde t, \quad l_2(t) := l(\varphi(\cdot,t)).$$

It is enough to show that $l_1' + l_2' \geq -C\varepsilon$ on $[0,1]$. We clearly have

$$l_1' = ||\dot\psi|| = \sqrt{\int_M \dot\psi^2 \omega_\psi^n}.$$

On the other hand,

$$l_2(t) = \int_0^1 \sqrt{E(s,t)}\, ds,$$

where

$$E = \int_M \varphi_s^2 \omega_\varphi^n$$

(using the notation $\varphi_s = \partial\varphi/\partial s$). We have

$$E_s = 2\int_M \varphi_s \nabla_{\varphi_s}\varphi_s \omega_\varphi^n = 2\varepsilon \int_M \varphi_s \omega^n$$

and

$$\begin{aligned}
E_t &= \int_M \left(2\varphi_s\varphi_{st} + \frac{1}{2}\varphi_s^2 \Delta\varphi_t\right)\omega_\varphi^n \\
&= 2\int_M \varphi_s\left(\varphi_{st} - \frac{1}{2}\langle\nabla\varphi_s, \nabla\varphi_t\rangle\right)\omega_\varphi^n \\
&= 2\int_M \varphi_s \nabla_{\varphi_s}\varphi_t \omega_\varphi^n \\
&= 2\frac{\partial}{\partial s}\langle\!\langle\varphi_s,\varphi_t\rangle\!\rangle - 2\int_M \varphi_t \nabla_{\varphi_s}\varphi_s \omega_\varphi^n \\
&= 2\frac{\partial}{\partial s}\int_M \varphi_s\varphi_t \omega_\varphi^n - 2\varepsilon\int_M \varphi_t \omega^n.
\end{aligned}$$

Therefore

$$\begin{aligned}
l_2' &= \frac{1}{2}\int_0^1 E^{-1/2} E_t\, ds \\
&= \int_0^1 E^{-1/2} \frac{\partial}{\partial s}\int_M \varphi_s\varphi_t \omega_\varphi^n\, ds - \varepsilon \int_0^1 E^{-1/2}\int_M \varphi_t \omega^n\, ds,
\end{aligned}$$

and the first term is equal to

The Complex Monge–Ampère Equation in Kähler Geometry

$$\left[E^{-1/2}\int_M \varphi_s\varphi_t\omega_\varphi^n\right]_{s=0}^{s=1} + \frac{1}{2}\int_0^1 E^{-3/2}E_s\int_M \varphi_s\varphi_t\omega_\varphi^n\,ds$$

$$= \left(\int_M \eta^2\omega_\varphi^n\right)^{-1/2}\int_M \eta\dot\psi\omega_\varphi^n - \varepsilon\int_0^1 E^{-3/2}\int_M \varphi_s\omega^n\int_M \varphi_s\varphi_t\omega_\varphi^n\,ds,$$

where $\eta = \varphi_s(1,\cdot)$; we have used that $\varphi_t(0,\cdot) = 0$, $\varphi_t(1,\cdot) = \dot\psi$, and

$$E(1,\cdot) = \int_M \eta^2\omega_\varphi^n.$$

From the Schwarz inequality it now follows that $l'_1 + l'_2 \geq -R$, where

$$R = \varepsilon\int_0^1 E^{-1/2}\int_M \varphi_t\omega^n\,ds + \varepsilon\int_0^1 E^{-3/2}\int_M \varphi_s\omega^n\int_M \varphi_s\varphi_t\omega_\varphi^n\,ds.$$

By Lemma 13

$$E(s,t) \geq \mathcal{E}(\tilde\psi,\psi(t)) - 2\varepsilon\sup_{[0,1]}||\varphi_s(\cdot,t)||.$$

Since $\mathcal{E}(\tilde\psi,\psi(t))$ is continuous and positive for $t \in [0,1]$, it follows that for ε sufficiently small

$$E \geq c > 0$$

and thus $R \leq C\varepsilon$. □

We are now in position to show that the geodesic distance is the same as d:

Theorem 15. *Let φ^ε be an ε-geodesic connecting $\varphi_0, \varphi_1 \in \mathcal{H}$. Then*

$$d(\varphi_0,\varphi_1) = \lim_{\varepsilon\to 0^+} l(\varphi^\varepsilon).$$

Proof. Let $\psi \in C^\infty([0,1],\mathcal{H})$ be an arbitrary curve connecting $\varphi_0, \varphi_1 \in \mathcal{H}$. We have to show that

$$\overline{\lim}_{\varepsilon\to 0^+} l(\varphi^\varepsilon) \leq l(\psi).$$

Without loss of generality we may assume that $\varphi_1 \notin \psi([0,1))$. Extend φ^ε to a function from $C^\infty([0,1] \times [0,1),\mathcal{H})$ in such a way that $\varphi^\varepsilon(0,\cdot) \equiv \varphi_1$, $\varphi^\varepsilon(1,\cdot) \equiv \psi$ on $[0,1)$ and $\varphi^\varepsilon(\cdot,t)$ is an ε-geodesic for $t \in [0,1)$. By Theorem 14 for $t \in [0,1)$ we have

$$l(\varphi^\varepsilon(\cdot,0)) \leq l(\psi|_{[0,t]}) + l(\varphi^\varepsilon(\cdot,t)) + C(t)\varepsilon.$$

Since clearly

$$\lim_{t\to 1^-} l(\psi|_{[0,t]}) = l(\psi),$$

it remains to show that

$$\varlimsup_{t \to 1^-} \varlimsup_{\varepsilon \to 0^+} l(\varphi^\varepsilon(\cdot, t)) = 0.$$

But it follows immediately from the following:

Lemma 16. *For an ε-geodesic φ connecting $\varphi_0, \varphi_1 \in \mathcal{H}$ we have*

$$l(\varphi) \leq \sqrt{V} \left(||\varphi_0 - \varphi_1||_{L^\infty(M)} + \frac{\varepsilon}{2\lambda^n} \right),$$

where $\lambda > 0$ is such that $\omega_{\varphi_0} \geq \lambda \omega$, $\omega_{\varphi_1} \geq \lambda \omega$.

Proof. Since $\ddot{\varphi} \geq 0$,

$$\dot{\varphi}(0) \leq \dot{\varphi} \leq \dot{\varphi}(1).$$

So to estimate $|\dot{\varphi}|$ we need to bound $\dot{\varphi}(0)$ from below and $\dot{\varphi}(1)$ from above. The function

$$v(\zeta) = 2b \log^2 |\zeta| + (a - 2b) \log |\zeta| - a$$

satisfies $v_{\zeta\bar{\zeta}} = b|\zeta|^{-2}$, $v = -a$ on $|\zeta| = 1$, and $v = 0$ on $|\zeta| = e$. We want to choose a, b so that $\varphi_1 + v \leq \varphi$ on $\tilde{M} := M \times \{1 \leq |\zeta| \leq e\}$.

On one hand, if $a := ||\varphi_0 - \varphi_1||_{L^\infty(M)}$ then $\varphi_1 + v \leq \varphi$ on $\partial \tilde{M}$. On the other one we have (if $b > 0$)

$$(\omega + dd^c(\varphi_1 + v))^{n+1} \geq \left(\lambda \omega + \frac{b}{|\zeta|^2} dd^c |\zeta|^2\right)^{n+1} = \frac{b\lambda^n}{|\zeta|^2} (\omega + dd^c |\zeta|^2)^{n+1}.$$

Therefore, by (21) if $b := \varepsilon/4\lambda^n$ we will get $\omega_{\varphi_1+v}^{n+1} \geq \omega_\varphi^{n+1}$ and $\varphi_1 + v \leq \varphi$ on $\tilde{M}$ by comparison principle. We will obtain

$$\dot{\varphi}(1) \leq \frac{d}{dt}\left(2bt^2 + (a-2b)t - a\right)\Big|_{t=1} = ||\varphi_0 - \varphi_1||_{L^\infty(M)} + \frac{\varepsilon}{2\lambda^n}.$$

Similarly we can show the lower bound for $\dot{\varphi}(0)$ and the estimate follows from the definition of $l(\varphi)$. $\square$

Combining Theorem 14 with Lemma 13 we get the following quantitative estimate from which Theorem 11 follows:

Theorem 17. *For $\varphi_0, \varphi_1 \in \mathcal{H}$ we have*

$$d(\varphi_0, \varphi_1) \geq \sqrt{\max\left\{ \int_{\{\varphi_0 > \varphi_1\}} (\varphi_0 - \varphi_1)^2 \omega_{\varphi_0}^n, \int_{\{\varphi_1 > \varphi_0\}} (\varphi_1 - \varphi_0)^2 \omega_{\varphi_1}^n \right\}}. \quad \square$$

7 Monge–Ampère Equation, Uniqueness

We assume that M is a compact complex manifold with smooth boundary (which may be empty) with a Kähler form ω. Our goal will be to prove the following two results:

Theorem 18 (Yau [47]). *Assume that M has no boundary. Then for $f \in C^\infty(M)$, $f > 0$ such that $\int_M f\omega^n = V$ there exists, unique up to an additive constant, $\varphi \in C^\infty(M)$ with $\omega + dd^c\varphi > 0$ satisfying the complex Monge–Ampère equation*

$$(\omega + dd^c\varphi)^n = f\omega^n. \tag{29}$$

Theorem 19. *Assume that M has smooth nonempty boundary. Take $f \in C^\infty(M)$, $f > 0$, and let $\psi \in C^\infty(M)$ be such that $\omega + dd^c\psi > 0$ and $(\omega + dd^c\psi)^n \geq f\omega^n$. Then there exists $\varphi \in C^\infty(M)$, $\omega + dd^c\varphi > 0$, satisfying (29) and $\varphi = \psi$ on ∂M.*

Theorem 19 can be rephrased as follows: the Dirichlet problem

$$\begin{cases} \varphi \in C^\infty(M) \\ \omega + dd^c\varphi > 0 \\ (\omega + dd^c\varphi)^n = f\omega^n \\ \varphi = \psi \text{ on } \partial M \end{cases}$$

has a solution provided that it has a smooth subsolution. It is a combination of the results proved in several papers [1, 15, 16, 18, 27, 47].

We will give a proof of Theorem 19 under additional assumption that the boundary of M is *flat*, that is near every boundary point, after a holomorphic change of coordinates, the boundary is of the form $\{\operatorname{Re} z^n = 0\}$. We will use this assumption only for the boundary estimate for second derivatives (see Theorem 27 below), but the result is also true without it (see [27]).

This extra assumption is satisfied in the geodesic equation case, then M is of the form $M' \times \bar{D}$, where M' is a manifold without boundary and D is a bounded domain in $\mathbb{C}$ with smooth boundary. This will immediately give existence of smooth ε-geodesics. (Note that by (20) the geodesic equation is covered here.)

The uniqueness in Theorems 18 and 19 is in fact very simple: if $\varphi, \tilde{\varphi}$ are the solutions then

$$0 = \omega_\varphi^n - \omega_{\tilde{\varphi}}^n = dd^c(\varphi - \tilde{\varphi}) \wedge T,$$

where

$$T = \sum_{p=0}^{n-1} \omega_\varphi^p \wedge \omega_{\tilde{\varphi}}^{n-p-1}.$$

Since $T > 0$, we will get $\varphi - \tilde{\varphi} = const$ in the first case and $\varphi = \tilde{\varphi}$ in the second one.

This argument does not work anymore if we allow the solutions to be degenerate, that is assuming only that $\omega_\varphi \geq 0$, $\omega_{\tilde{\varphi}} \geq 0$. In fact, much more general results hold here. We will allow continuous solutions given by the Bedford Taylor theory [5] (see also [8]) – then ω_φ^n is a measure.

Theorem 20 ([12]). *Assume that M has no boundary. If $\varphi, \tilde{\varphi} \in C(M)$ are such that $\omega_\varphi \geq 0$, $\omega_{\tilde{\varphi}} \geq 0$ and $\omega_\varphi^n = \omega_{\tilde{\varphi}}^n$ then $\varphi - \tilde{\varphi} = const$.*

Proof. Assume $n = 2$, the general case is similar, for details see [12]. Write

$$0 = \omega_\varphi^2 - \omega_{\tilde\varphi}^2 = dd^c\rho \wedge (\omega_\varphi + \omega_{\tilde\varphi}),$$

where $\rho = \varphi - \tilde\varphi$. Therefore

$$0 = -\int_M \rho\, dd^c\rho \wedge (\omega_\varphi + \omega_{\tilde\varphi}) = \int_M d\rho \wedge d^c\rho \wedge (\omega_\varphi + \omega_{\tilde\varphi}),$$

and thus

$$d\rho \wedge d^c\rho \wedge \omega_\varphi = d\rho \wedge d^c\rho \wedge \omega_{\tilde\varphi} = 0. \tag{30}$$

We have to show that

$$d\rho \wedge d^c\rho \wedge \omega = 0. \tag{31}$$

By (30)

$$\int_M d\rho \wedge d^c\rho \wedge \omega = -\int_M d\rho \wedge d^c\rho \wedge dd^c\varphi$$
$$= \int_M d\rho \wedge d^c\varphi \wedge dd^c\rho = \int_M d\rho \wedge d^c\varphi \wedge (\omega_\varphi - \omega_{\tilde\varphi}).$$

By the Schwarz inequality and (30) again

$$\left|\int_M d\rho \wedge d^c\varphi \wedge \omega_\varphi\right|^2 \leq \int_M d\rho \wedge d^c\rho \wedge \omega_\varphi \int_M d\varphi \wedge d^c\varphi \wedge \omega_\varphi = 0.$$

Similarly we show that

$$\int_M d\rho \wedge d^c\varphi \wedge \omega_{\tilde\varphi} = 0$$

and (31) follows. □

Theorem 21 ([15]). *Let M have nonempty boundary. Assume that $\varphi, \tilde\varphi \in C(M)$ are such that $\omega_\varphi \geq 0$, $\omega_{\tilde\varphi} \geq 0$, $\omega_\varphi^n \geq \omega_{\tilde\varphi}^n$ and $\varphi \leq \tilde\varphi$ on ∂M. Then $\varphi \leq \tilde\varphi$ in M.*

Proof. For $\varepsilon > 0$ set $\varphi_\varepsilon := \max\{\varphi - \varepsilon, \tilde\varphi\}$, so that $\varphi_\varepsilon = \tilde\varphi$ near ∂M. Since for continuous plurisubharmonic functions we have

$$(dd^c \max\{u, v\})^n \geq \chi_{\{u \geq v\}}(dd^c u)^n + \chi_{\{u < v\}}(dd^c v)^n$$

(it is a very simple consequence of the continuity of the Monge–Ampère operator, see e.g. Theorem 3.8 in [8]), it follows that $\omega_{\varphi_\varepsilon}^n \geq \omega_{\tilde\varphi}^n$. Therefore, without loss of generality, we may assume that $\varphi \geq \tilde\varphi$ in M, $\varphi = \tilde\varphi$ near ∂M, and we have to show that $\varphi = \tilde\varphi$ in M.

Assume again $n = 2$. Then, since $\rho := \varphi - \tilde\varphi$ vanishes near ∂M, we have

$$0 \leq \int_M \rho(\omega_\varphi^2 - \omega_{\tilde\varphi}^2) = -\int_M d\rho \wedge d^c\rho \wedge (\omega_\varphi + \omega_{\tilde\varphi}).$$

We thus get (30) and the rest of the proof is the same as that of Theorem 20. □

Assuming Theorem 19 and estimates proved in Sects. 8–13, we get Theorem 12. From the comparison principle it follows that ε-geodesics converge uniformly to a weak geodesic which is almost $C^{1,1}$ (that is it satisfies (28)). It is an open problem if it has to be fully $C^{1,1}$ (it was shown in [15] in case the bisectional curvature is nonnegative).

8 Continuity Method

In order to prove existence in Theorems 18 and 19 we fix an integer $k \geq 2$ and $\alpha \in (0, 1)$. Let f_0 denote the r.h.s. of the equation for which we already know the solution: $f_0 = 1$ in the first case and $f_0 = \omega_\psi^n/\omega^n$ in the second one. For $t \in [0, 1]$ set

$$f_t := (1-t)f_0 + tf.$$

By S denote the set of those $t \in [0, 1]$ for which the problem

$$\begin{cases} \varphi_t \in C^{k+2,\alpha}(M) \\ \omega + dd^c\varphi_t > 0 \\ (\omega + dd^c\varphi_t)^n = f_t \omega^n \\ \int_M \varphi_t \omega^n = 0, \end{cases}$$

resp.

$$\begin{cases} \varphi_t \in C^{k+2,\alpha}(M) \\ \omega + dd^c\varphi_t > 0 \\ (\omega + dd^c\varphi_t)^n = f_t \omega^n \\ \varphi_t = \psi \text{ on } \partial M \end{cases}$$

has a solution (by the previous section it has to be unique). We clearly have $0 \in S$ and we have to show that $1 \in S$. For this it will be enough to prove that S is open and closed.

Openness. The Monge–Ampère operator we treat as the mapping

$$\mathcal{M} : \mathcal{A} \ni \varphi \longmapsto \frac{\omega_\varphi^n}{\omega^n} \in \mathcal{B},$$

where

$$\mathcal{A} := \{\varphi \in C^{k+2,\alpha}(M) : \omega_\varphi > 0, \int_M \varphi \omega^n = 0\}$$

$$\mathcal{B} := \{\tilde{f} \in C^{k,\alpha}(M) : \int_M \tilde{f} \omega^n = \int_M \omega^n\},$$

resp.
$$\mathcal{A} := \{\varphi \in C^{k+2,\alpha}(M) : \omega_\varphi > 0, \varphi = \psi \text{ on } \partial M\}$$
$$\mathcal{B} := C^{k,\alpha}(M).$$

Then $\mathcal{A}$ is an open subset of the Banach space

$$\mathcal{E} := \{\eta \in C^{k+2,\alpha}(M) : \int_M \eta \omega^n = 0\},$$

resp. a hyperplane in the Banach space $C^{k+2,\alpha}(M)$ with the tangent space

$$\mathcal{E} := \{\eta \in C^{k+2,\alpha}(M) : \varphi = 0 \text{ on } \partial M\}.$$

On the other hand, $\mathcal{B}$ is a hyperplane of the Banach space $C^{k,\alpha}(M)$ with the tangent space

$$\mathcal{F} := \{\tilde{f} \in C^{k,\alpha}(M) : \int_M \tilde{f} \omega^n = 0\},$$

resp. $\mathcal{B}$ is a Banach space itself and $\mathcal{F} := \mathcal{B}$. We would like to show that for every $\varphi \in \mathcal{A}$ the differential

$$d_\varphi \mathcal{M} : \mathcal{E} \to \mathcal{F}$$

is an isomorphism. But since

$$d_\varphi \mathcal{M}.\eta = \frac{1}{2} \Delta \eta,$$

where the Laplacian is taken w.r.t. ω_φ, it follows from the standard theory of the Laplace equation on Riemannian manifolds. Therefore $\mathcal{M}$ is locally invertible, in particular $\mathcal{M}(\mathcal{A})$ is open in $\mathcal{B}$ and thus S is open in $[0, 1]$.

Closedness. Assume that we knew that

$$||\varphi_t||_{k+2,\alpha} \leq C, \quad t \in S, \tag{32}$$

for some uniform constant C, where $||\cdot||_{k,\alpha} = ||\cdot||_{C^{k,\alpha}(M)}$. Then by the Arzela–Ascoli theorem every sequence in $\{\varphi_t : t \in S\}$ would contain a subsequence whose derivatives of order at most $k+1$ converged uniformly.

The proof of existence of solutions in Theorems 18 and 19 is therefore reduced to (32) for all k big enough. The first step (but historically the latest in the Calabi–Yau case) is the L^∞-estimate, this is done in Sect. 9. The gradient and second derivative estimates are presented in Sects. 10–12. They are all very specific for the complex Monge–Ampère equation and most of them (except for Theorem 25) are applicable also in the degenerate case, that is they do not depend on a lower positive bound for f. Finally, in Sect. 13, we make use of the general Evans–Krylov theory for nonlinear elliptic equations of second order (see e.g. [26], in the boundary case it is due to Caffarelli et al. [16]). This gives a $C^{2,\alpha}$ bound and then higher order estimates follow from the standard Schauder theory of linear elliptic equations of second order with variable coefficients.

9 L^∞-Estimate

If $\partial M \neq \emptyset$ then by the comparison principle, Theorem 21, for any $\varphi \in C(M)$ with $\omega_\varphi \geq 0$, $\omega_\varphi^n \leq \omega_\psi^n$, $\varphi = \psi$ on ∂M, we have

$$\psi \leq \varphi \leq \max_{\partial M} \psi,$$

so we immediately get the L^∞-estimate in the second case. The case $\partial M = \emptyset$ is more difficult and historically turned out to be the main obstacle in proving the Calabi conjecture. Its proof making use of Moser's iteration was in fact the main contribution of Yau [47] (see also [31] for some simplifications).

Theorem 22. *Assume $\partial M = \emptyset$. Take $\varphi \in C(M)$ with $\omega_\varphi \geq 0$, satisfying the Monge–Ampère equation $\omega_\varphi^n = f\omega^n$. Then*

$$\operatorname{osc} \varphi \leq C(M, \omega, \|f\|_\infty).$$

Proof. It will be convenient to assume that $V = \int_M \omega^n = 1$ and that $\max_M \varphi = -1$, so that $\|\varphi\|_p \leq \|\varphi\|_q$ for $p \leq q$ (we use the notation $\|\cdot\|_p = \|\cdot\|_{L^p(M)}$). Write

$$(f - 1)\omega^n = dd^c\varphi \wedge T,$$

where

$$T = \sum_{p=0}^{n-1} \omega_\varphi^p \wedge \omega^{n-p-1}.$$

Note that $T \geq \omega^{n-1}$. Then for $p \geq 2$

$$\int_M (-\varphi)^{p-1}(f - 1)\omega^n = \int_M (-\varphi)^{p-1} dd^c\varphi \wedge T$$

$$= -\int_M d(-\varphi)^{p-1} \wedge d^c\varphi \wedge T$$

$$= (p-1) \int_M (-\varphi)^{p-2} d\varphi \wedge d^c\varphi \wedge T$$

$$\geq (p-1) \int_M (-\varphi)^{p-2} d\varphi \wedge d^c\varphi \wedge \omega^{n-1} \quad (33)$$

$$= \frac{4(p-1)}{p^2} \int_M d(-\varphi)^{p/2} \wedge d^c(-\varphi)^{p/2} \wedge \omega^{n-1}$$

$$\geq \frac{c_n}{p} \|\nabla((-\varphi))^{p/2}\|_2^2.$$

By the Sobolev inequality

$$||\varphi||_{pn/(n-1)}^{p/2} = ||(-\varphi)^{p/2}||_{2n/(n-1)} \leq C(M)\left(||(-\varphi)^{p/2}||_2 + ||\nabla((-\varphi)^{p/2})||_2\right).$$

Combining this with (33) we will get

$$||\varphi||_{pn/(n-1)} \leq (Cp)^{1/p}||\varphi||_p.$$

Setting
$$p_0 := 2, \quad p_{k+1} = np_k/(n-1), \ k = 1, 2, \ldots,$$

we will get
$$||\varphi||_\infty = \lim_{k \to \infty} ||\varphi||_{p_k} \leq \tilde{C}||\varphi||_2$$

and it remains to use the following elementary estimate:

Proposition 23. *Assume that $\partial M = \emptyset$ and let $\varphi \in C(M)$ be such that $\omega_\varphi \geq 0$, $\max_\varphi M = 0$. Then for any $p < \infty$*

$$||\varphi||_p \leq C(M, p).$$

Proof. It will easily follow from local properties of plurisubharmonic functions. For $p = 1$ we can use the following result: if u is a negative subharmonic function in the ball $B(0, 3R)$ in $\mathbb{C}^n$ then

$$||u||_{L^1(B(0,R))} \leq C(n, R) \inf_{B(0,R)} (-u).$$

After covering M with finite number of balls of radius R, a simple procedure starting at the point where $\varphi = 0$ will give us the required estimate for $||\varphi||_1$. The case $p > 1$ is now an immediate consequence of the following fact: if u is a negative plurisubharmonic function in $B(0, 2R)$ then

$$||u||_{L^p(B(0,R))} \leq C(n, p, R) ||u||_{L^1(B(0,2R))}. \qquad \square$$

10 Interior Second Derivative Estimate

It turns out that in case of Theorem 18 one can bypass the gradient estimate. The interior estimate for the second derivative which will be needed in the proofs of both cases was shown independently by Aubin [1] and Yau [47]. We will show the following version from [14]:

Theorem 24. *Assume that $\varphi \in C^4(M)$ satisfies $\omega_\varphi > 0$ and $\omega_\varphi^n = f\omega^n$. Then*

$$\Delta\varphi \leq C, \qquad (34)$$

where C depends only on n, on upper bounds for f, the scalar curvature of M, $\operatorname{osc}\varphi$ and $\sup_{\partial M} \Delta\varphi$ (if $\partial M = \emptyset$ then this is void), and on lower bounds for $f^{1/(n-1)}\Delta(\log f)$ and the bisectional curvature of M.

Proof. By $C_1, C_2, \ldots$ we will denote constants depending only on the required quantities. Set
$$\alpha := \log(\Delta\varphi + 2n) - A\varphi$$
(note that $\Delta\varphi > -2n$), where A under control will be specified later. We may assume that α attains maximum at y in the interior of M, otherwise we are done. Let g be a local potential for ω near y and set $u := g + \varphi$. We choose normal coordinates at y (so that $g_{j\bar{k}} = \delta_{jk}$, $g_{j\bar{k}l} = 0$ at y), so that in addition the matrix $(u_{j\bar{k}})$ is diagonal at y. Then at y

$$\alpha_{p\bar{p}} = \frac{(\Delta u)_{p\bar{p}}}{\Delta u} - \frac{|(\Delta u)_p|^2}{(\Delta u)^2} + A - A u_{p\bar{p}}$$

$$(\Delta u)_p = 2 \sum_j u_{j\bar{j}p}$$

$$(\Delta)_{p\bar{p}} = 2 \sum_j u_{j\bar{j}p\bar{p}} + 2 \sum_j R_{j\bar{j}p\bar{p}} u_{j\bar{j}}.$$

(by (9)). The equation $\omega_\varphi^n = f\omega^n$ now reads

$$\det(u_{p\bar{q}}) = f \det(g_{p\bar{q}}). \tag{35}$$

Differentiating w.r.t. z^j and $\bar{z}^j$ we get

$$u^{p\bar{q}} u_{p\bar{q}j} = (\log f)_j + g^{p\bar{q}} g_{p\bar{q}j} \tag{36}$$

and

$$u^{p\bar{q}} u_{p\bar{q}j\bar{j}} = (\log f)_{j\bar{j}} + u^{p\bar{t}} u^{s\bar{q}} u_{p\bar{q}j} u_{s\bar{t}\bar{j}} + g^{p\bar{q}} g_{p\bar{q}j\bar{j}} - g^{p\bar{t}} g^{s\bar{q}} g_{p\bar{q}j} g_{s\bar{t}\bar{j}}. \tag{37}$$

Therefore at y

$$\sum_p \frac{u_{j\bar{j}p\bar{p}}}{u_{p\bar{p}}} = (\log f)_{j\bar{j}} + \sum_{p,q} \frac{|u_{p\bar{q}j}|^2}{u_{p\bar{p}} u_{q\bar{q}}} - \sum_p R_{j\bar{j}p\bar{p}}$$

and, since $\alpha_{p\bar{p}} \leq 0$,

$$0 \geq \sum_p \frac{\alpha_{p\bar{p}}}{u_{p\bar{p}}} = \frac{1}{\Delta u}\left(\Delta(\log f) + 2\sum_{j,p,q}\frac{|u_{p\bar{q}j}|^2}{u_{p\bar{p}}u_{q\bar{q}}} - S + 2\sum_{j,p}\frac{R_{j\bar{j}p\bar{p}}u_{j\bar{j}}}{u_{p\bar{p}}}\right)$$

$$- \frac{4}{(\Delta u)^2}\sum_p \frac{|\sum_j u_{j\bar{j}p}|^2}{u_{p\bar{p}}} + A\sum_p \frac{1}{u_{p\bar{p}}} - An.$$

By the Schwarz inequality

$$\left|\sum_j u_{j\bar{j}p}\right|^2 \leq \frac{\Delta u}{2}\sum_q \frac{|u_{q\bar{q}p}|^2}{u_{q\bar{q}}}$$

and therefore we can get rid of the terms with third derivatives. We also have

$$\Delta(\log f) \geq -\frac{C_1}{f^{1/(n-1)}},$$

$$2\sum_{j,p}\frac{R_{j\bar{j}p\bar{p}}u_{j\bar{j}}}{u_{p\bar{p}}} \geq -C_2\Delta u \sum_p \frac{1}{u_{p\bar{p}}}$$

(by (12)), and

$$\sum_p \frac{1}{u_{p\bar{p}}} \geq (n-1)\left(\sum_p \frac{u_{p\bar{p}}}{u_{1\bar{1}}\ldots u_{n\bar{n}}}\right)^{1/(n-1)} \geq \left(\frac{\Delta u}{f}\right)^{1/(n-1)}$$

(we may assume $n \geq 2$). Therefore, choosing $A := C_2 + 1$, at y we get

$$-\frac{C_1}{f^{1/(n-1)}\Delta u} - \frac{S}{\Delta u} + \left(\frac{\Delta u}{f}\right)^{1/(n-1)} - C_3 \geq 0.$$

Multiplying by $f^{1/(n-1)}\Delta u$ we will get at y

$$(\Delta u)^{n/(n-1)} - C_4 \Delta u - C_5 \leq 0,$$

and thus
$$\Delta u(y) \leq C_6.$$

Therefore $\alpha \leq \alpha(y) \leq C_7$ and we get (34). $\qquad\square$

An upper bound for $\Delta\varphi$ for functions satisfying $\omega_\varphi \geq 0$ easily gives a bound mixed complex derivatives of φ

$$|\varphi_{j\bar{k}}| \leq C.$$

However, it does not imply the full estimate for the second derivative of φ:

Example. Set $S := \{re^{it} : 0 \leq r \leq 1, \pi/4 \leq t \leq 3\pi/4\}$ and

$$u(z) := \frac{2}{\pi} \int_S \log|z - \zeta| d\lambda(\zeta).$$

Then $u_{z\bar{z}} = \chi_S \in L^\infty(\mathbb{C})$, and thus $u \in W^{2,p}_{loc}(\mathbb{C})$ for every $p < \infty$ (which implies that $u \in C^{1,\alpha}(\mathbb{C})$ for every $\alpha < 1$ by Morrey's embedding theorem). However

$$u_{xx}(0) = \frac{2}{\pi} \int_S \frac{y^2 - x^2}{(x^2 + y^2)^2} d\lambda(z) = 4 \int_0^1 \int_{\pi/4}^{3\pi/4} \frac{\sin^2 t - \cos^2 t}{r^2} dt\, dr = \infty,$$

and $u \notin W^{2,\infty}_{loc}(\mathbb{C}) = C^{1,1}(\mathbb{C})$.

The following estimate will enable to apply the real Evans–Krylov theory (see Sect. 13) directly, without reproving its complex version.

Theorem 25 ([15]). *Assume that $\varphi \in C^4(M)$, $\omega_\varphi > 0$, $\omega_\varphi^n = f\omega^n$. Then*

$$|\nabla^2 \varphi| \leq C, \tag{38}$$

where C depends on n, on upper bounds for $|R|$, $|\nabla R|$, $\|\varphi\|_{C^{0,1}(M)}$, $\Delta\varphi$, $\sup_{\partial M} |\nabla^2 \varphi|$, $\|f\|_{C^{1,1}(M)}$ and a lower (positive) bound for f on M.

Proof. We have to estimate the eigenvalues of the mapping $X \mapsto \nabla_X \nabla\varphi$. Since their sum is under control from below (by $-2n$), it will be enough to get an upper bound. The maximal eigenvalue is given by

$$\beta = \max_{X \in TM \setminus \{0\}} \frac{\langle \nabla_X \nabla\varphi, X \rangle}{|X|^2}.$$

This is a continuous function on M (but not necessarily smooth). Locally we have

$$\nabla_{\partial_j} \nabla\varphi = \partial_j(g^{p\bar{q}}\varphi_p)\partial_{\bar{q}} + \partial_j(g^{p\bar{q}}\varphi_{\bar{q}})\partial_p + g^{p\bar{q}}\varphi_{\bar{q}}\Gamma^s_{jp}\partial_s$$
$$= g^{p\bar{q}}\varphi_{j\bar{q}}\partial_p + (g^{p\bar{q}}\varphi_p)_j \partial_{\bar{q}}.$$

Therefore for a real vector field $X = X^j \partial_j + \bar{X}^k \partial_{\bar{k}}$

$$\langle \nabla_X \nabla\varphi, X \rangle = 2\operatorname{Re} X^j \left(\bar{X}^k \varphi_{j\bar{k}} + X^l g_{l\bar{q}}(g^{p\bar{q}}\varphi_p)_j \right)$$
$$= D_X^2 \varphi + 2\operatorname{Re}\left(X^j X^l g^{p\bar{q}} g_{j\bar{q}l}\varphi_p \right),$$

where D_X denotes Euclidean directional derivative in direction X.

Set
$$\alpha := \beta + \frac{1}{2}|\nabla\varphi|^2.$$

We may assume that α attains maximum y in the interior of M. Near y we choose normal coordinates ($g_{j\bar{k}} = \delta_{jk}$, $g_{j\bar{k}l} = g_{j\bar{k}lm} = 0$ at y) so that in addition the matrix $(\varphi_{j\bar{k}})$ is diagonal at y. Take fixed $X = (X^1, \ldots, X^N) \in \mathbb{C}^N$ such that at y one has $|X|^2 (= 2g_{j\bar{k}} X^j \bar{X}^k) = 1$. Near y define

$$\tilde{\beta} := \frac{\langle \nabla_X \nabla \varphi, X \rangle}{|X|^2}$$

and
$$\tilde{\alpha} := \tilde{\beta} + |\nabla\varphi|^2.$$

Then $\tilde{\beta} \leq \beta$, $\tilde{\beta}(y) = \beta(y)$ and $\tilde{\alpha} \leq \alpha \leq \alpha(y) = \tilde{\alpha}(y)$, so that $\tilde{\alpha}$ (which is defined locally) also has a maximum at y, the same as that of α. The advantage of $\tilde{\alpha}$ is that it is smooth (this argument goes back to [11]). It remains to estimate $\tilde{\beta}(y)$ from above.

The function $u := \varphi + g$ solves (35). Similarly as with (37) we will get at y

$$\sum_p \frac{D_X^2 \varphi_{p\bar{p}}}{u_{p\bar{p}}} \geq D_X^2(\log f) + \sum_p D_X^2 g_{p\bar{p}} - \sum_p \frac{D_X^2 g_{p\bar{p}}}{u_{p\bar{p}}}.$$

Since f is under control from below, we have $D_X^2(\log f) \geq -C_1$ and by Theorem 24

$$\frac{1}{C_2} \leq u_{p\bar{p}} \leq C_3.$$

This, together with the fact that $|R|$ is under control, implies that

$$\sum_p \frac{D_X^2 \varphi_{p\bar{p}}}{u_{p\bar{p}}} \geq -C_4. \tag{39}$$

Using the fact that $|X| = 1$ and $(|X|^2)_p = 0$ at y, combined with (36), at y we will get

$$\tilde{\beta}_{p\bar{p}} = D_X^2 \varphi_{p\bar{p}} + 2\mathrm{Re} \sum_l X^j X^k g_{j\bar{l}k\bar{p}p} \varphi_l$$
$$+ 2\mathrm{Re} \sum_l X^j X^k g_{j\bar{l}k\bar{p}} \varphi_{lp} - X^j \bar{X}^k g_{j\bar{k}p\bar{p}} D_X^2 \varphi \tag{40}$$
$$\geq D_X^2 \varphi_{p\bar{p}} - C_5 - C_6 \tilde{\beta},$$

where we used in addition that $|\nabla R|$ is under control.

Near y we have

$$\frac{1}{2}(|\nabla\varphi|^2)_p = (g^{j\bar{k}})_p \varphi_j \varphi_{\bar{k}} + g^{j\bar{k}} \varphi_{jp}\varphi_{\bar{k}} + g^{j\bar{k}}\varphi_j \varphi_{p\bar{k}}.$$

Therefore at y

$$\frac{1}{2}(|\nabla\varphi|^2)_{p\bar{p}} = \sum_{j,k} R_{j\bar{k}p\bar{p}}\varphi_j\varphi_{\bar{k}} + 2\operatorname{Re}\sum_j \varphi_{jp\bar{p}}\varphi_{\bar{j}} + \sum_j |\varphi_{jp}|^2 + \varphi_{p\bar{p}}^2.$$

Since

$$2\operatorname{Re}\sum_{j,p}\frac{\varphi_{jp\bar{p}}\varphi_{\bar{j}}}{u_{p\bar{p}}} = 2\operatorname{Re}\sum_j (\log f)_j \varphi_{\bar{j}} \geq -C_7$$

and

$$\sum_{j,p}\frac{|\varphi_{jp}|^2}{u_{p\bar{p}}} \geq \frac{1}{C_8}\tilde{\beta}^2 - C_9,$$

from (39), (40) it follows that at y

$$0 \geq \sum_p \frac{\alpha_{p\bar{p}}}{u_{p\bar{p}}} \geq \frac{1}{C_8}\tilde{\beta}^2 - C_{10}\tilde{\beta} - C_{11}. \qquad \square$$

Since the estimate depends on a lower bound for f, Theorem 25 cannot be used in the degenerate case. It is an open problem if one can get rid of this dependence, this would in particular imply full $C^{1,1}$-regularity of weak geodesics obtained by Chen [18]. This was shown only under additional assumption that M has nonnegative bisectional curvature, see [15].

11 Gradient Estimate

If $\partial M = \emptyset$ then Theorem 24 gives an a priori estimate for the Laplacian, and thus also for the gradient. However, if $\partial M \neq \emptyset$ then a direct gradient estimate is necessary because the boundary estimates from Sect. 12 depend on it.

The estimate for $|\nabla\varphi|$ on ∂M follows easily from the comparison principle: if $h \in C^\infty(M)$ is harmonic in the interior of M with $h = \psi$ on ∂M then

$$\psi \leq \varphi \leq h$$

in M. Therefore on ∂M we have

$$|\nabla\varphi| \leq \max\{|\nabla\psi|, |\nabla h|\}.$$

We have the following interior gradient bound from [14] (see also [29, 30]).

Theorem 26. Let $\varphi \in C^3(M)$ be such that $\omega_\varphi > 0$ and $\omega_\varphi^n = f\omega^n$. Then

$$|\nabla\varphi| \leq C, \tag{41}$$

where C depends on n, on upper bounds for osc φ, $\sup_{\partial M} |\nabla\varphi|$, f, $|\nabla(f^{1/n})|$ and on a lower bound for the bisectional curvature of M.

Proof. We may assume that $\inf_M \varphi = 0$ and $C_0 := \sup_M \varphi = \mathrm{osc}\,\varphi$. Set

$$\beta = \frac{1}{2}|\nabla\varphi|^2$$

and

$$\alpha := \log\beta - \gamma \circ \varphi,$$

where $\gamma \in C^\infty([0, C_0])$ with $\gamma' \geq 0$ will be determined later. We may assume that γ attains maximum at y in the interior of M. Near y write $u = \varphi + g$, where g is a local potential for ω. Similarly as before, we may assume that at y we have $g_{j\bar{k}} = \delta_{j\bar{k}}$, $g_{j\bar{k}l} = 0$ and $(u_{j\bar{k}})$ is diagonal.

At y we will get

$$\beta = \sum_j |\varphi_j|^2$$

$$\beta_p = \sum_j \varphi_{jp}\varphi_{\bar{j}} + \varphi_p(u_{p\bar{p}} - 1)$$

$$\beta_{p\bar{p}} = \sum_{j,k} R_{j\bar{k}p\bar{p}}\varphi_j\varphi_{\bar{k}} + 2\mathrm{Re}\sum_j u_{p\bar{p}j}\varphi_{\bar{j}} + \sum_j |\varphi_{jp}|^2 + \varphi_{p\bar{p}}^2.$$

and

$$\alpha_{p\bar{p}} = \frac{\beta_{p\bar{p}}}{\beta} - ((\gamma')^2 + \gamma'')|\varphi_p|^2 - \gamma'\varphi_{p\bar{p}},$$

where for simplicity we denote $\gamma' \circ \varphi$ just by γ' (and similarly for γ''). By (36)

$$\sum_p \frac{u_{p\bar{p}j}}{u_{p\bar{p}}} = (\log f)_j.$$

Since

$$\frac{1}{\beta}\sum_{j,k,p} \frac{R_{j\bar{k}p\bar{p}}\varphi_j\varphi_{\bar{k}}}{u_{p\bar{p}}} \geq -C_1 \sum_p \frac{1}{u_{p\bar{p}}}$$

and (we may assume that $\beta(y) \geq 1$)

$$\frac{2}{\beta}\operatorname{Re}\sum_j (\log f)_j \varphi_{\bar j} \geq -2|\nabla(\log f)| \geq -\frac{C_2}{f^{1/n}} \geq -C_2 \sum_p \frac{1}{u_{p\bar p}},$$

we will obtain at y

$$0 \geq \sum_p \frac{\alpha_{p\bar p}}{u_{p\bar p}} \geq (\gamma' - C_3)\sum_p \frac{1}{u_{p\bar p}} + \frac{1}{\beta}\sum_{j,p}\frac{|\varphi_{jp}|^2}{u_{p\bar p}} - [(\gamma')^2 + \gamma'']\sum_p \frac{|\varphi_p|^2}{u_{p\bar p}} - n\gamma'.$$

We have to estimate the term

$$\frac{1}{\beta}\sum_{j,p}\frac{|\varphi_{jp}|^2}{u_{p\bar p}}$$

from below. For this we will use that fact that $\alpha_p = 0$ at y. Therefore $\beta_p = \gamma'\beta\varphi_p$, that is

$$\sum_j \varphi_{jp}\varphi_{\bar j} = (\gamma'\beta - u_{p\bar p} + 1)\varphi_p.$$

By the Schwarz inequality

$$\left|\sum_j \varphi_{jp}\varphi_{\bar j}\right|^2 \leq \beta \sum_j |\varphi_{jp}|^2,$$

hence

$$\frac{1}{\beta}\sum_{j,p}\frac{|\varphi_{jp}|^2}{u_{p\bar p}} \geq \frac{1}{\beta^2}\sum_p \frac{(\gamma'\beta + 1 - u_{p\bar p})^2|\varphi_p|^2}{u_{p\bar p}} \geq (\gamma')^2 \sum_p \frac{|\varphi_p|^2}{u_{p\bar p}} - 2\gamma' - \frac{2}{\beta}.$$

This gives

$$0 \geq (\gamma' - C_3)\sum_p \frac{1}{u_{p\bar p}} - \gamma''\sum_p \frac{|\varphi_p|^2}{u_{p\bar p}} - (n+2)\gamma' - \frac{2}{\beta}.$$

We now set $\gamma(t) = -t^2/2 + (C_0 + C_3 + 1)t$, so that $\gamma'' = -1$ and $\gamma' \geq 1$ in $[0, C_0]$. We will get

$$\sum_p \frac{1}{u_{p\bar p}} + \sum_p \frac{|\varphi_p|^2}{u_{p\bar p}} \leq C_5.$$

Therefore $u_{p\bar p} \leq C_6$ and $\beta \leq C_7$ at y, and we easily arrive at (41). □

12 Boundary Second Derivative Estimate

In this section we want to show the a priori estimate

$$|\nabla^2 \varphi| \leq C$$

on ∂M. It is due to Caffarelli et al. [16] if the boundary is strongly pseudoconvex and to B. Guan [27] in the general case. We will prove the following local result which is applicable to the case of flat boundary:

Theorem 27. *Write $B_R = B(0, R)$ and $B_R^- = B_R \cap \{x^n \leq 0\}$. Let $u, v \in C^3(B_R^-)$ be such that $(u_{j\bar{k}}) > 0$, $(v_{j\bar{k}}) \geq \lambda(\delta_{jk})$ for some $\lambda > 0$,*

$$\det(u_{j\bar{k}}) = f \leq \det(v_{j\bar{k}}).$$

Assume moreover that $u \geq v$ on B_R^- and $u = v$ on $\{x^n = 0\}$. Then

$$|D^2 u(0)| \leq C,$$

where C depends on n, on upper bounds for $||v||_{C^{2,1}(B_R^-)}$, $||f^{1/n}||_{C^{0,1}(B_R^-)}$, $||u||_{C^{0,1}(B_R^-)}$, and on lower bounds for λ, R.

Proof. If s, t are tangential directions to $\{x^n = 0\}$ then $u_{st}(0) = v_{st}(0)$, so $|u_{st}(0)|$ is under control. The main step in the proof is to estimate the tangential-normal derivative $u_{tx^n}(0)$. Set $r := R/2$ and

$$\tilde{w} := -(u - v) + 2A_1 x^n (r + x^n),$$

where $A_1 > 0$ under control will be determined later. We have $\tilde{w} \leq 0$ in B_r^- and

$$u^{j\bar{k}} \tilde{w}_{j\bar{k}} = -n + u^{j\bar{k}} v_{j\bar{k}} + A_1 u^{n\bar{n}} \geq -n + \lambda \sum u^{j\bar{j}} + A_1 u^{n\bar{n}}.$$

By $\lambda_1 \leq \cdots \leq \lambda_n$ denote the eigenvalues of $(u_{j\bar{k}})$. Then $\sum u^{j\bar{j}} = \sum 1/\lambda_j$ and $u^{n\bar{n}} \geq 1/\lambda_n$. Since $\lambda_1 \ldots \lambda_n = f$, by the inequality between geometric and arithmetic means we will obtain

$$u^{j\bar{k}} \tilde{w}_{j\bar{k}} \geq -n + \frac{\lambda}{2} \sum u^{j\bar{j}} + \frac{\lambda}{2} \sum \frac{1}{\lambda_j} + \frac{A_1}{\lambda_n}$$

$$\geq -n + \frac{\lambda}{2} \sum u^{j\bar{j}} + \frac{n(\lambda/2)^{1-1/n} A_1^{1/n}}{f^{1/n}}$$

$$\geq \frac{\lambda}{2}(1 + \sum u^{j\bar{j}}) \tag{42}$$

for A_1 sufficiently large.

Further define

$$w := \pm(u-v)_t + \frac{1}{2}(u-v)_{y^n}^2 - A_2|z|^2 + A_3\tilde{w},$$

where positive A_2, A_3 under control will be determined later. Since $(u-v)_t = (u-v)_{y^n} = 0$ on $\{x^n = 0\}$, we have $w \le 0$ on $\{x^n = 0\}$. We also have

$$\left| \pm(u-v)_t + \frac{1}{2}(u-v)_{y^n}^2 \right| \le C_1$$

and thus for A_2 sufficiently large $w \le 0$ on $\partial B_r \cap \{x^n \le 0\}$. By (36)

$$u^{j\bar{k}} \left(\pm(u-v)_t + \frac{1}{2}(u-v)_{y^n}^2 \right)_{j\bar{k}} = \pm(\log f)_t \mp u^{j\bar{k}} v_{tj\bar{k}} + (u-v)_{y^n}(\log f)_{y^n}$$
$$+ u^{j\bar{k}}(u-v)_{y^n j}(u-v)_{y^n \bar{k}}$$
$$\ge -C_2(1 + \sum u^{j\bar{j}}),$$

where the last inequality follows from

$$f^{-1/n} \le \frac{1}{n} \sum u^{j\bar{j}}.$$

Therefore, from (42) we get

$$u^{j\bar{k}} w_{j\bar{k}} \ge 0$$

if A_3 is chosen sufficiently large. Now from the maximum principle we obtain $w \le 0$ in B_r^- and thus

$$|(u-v)_{tx^n}(0)| \le A_3\big((u-v)_{x^n}(0) + 2A_1 r\big),$$

so

$$|u_{tx^n}(0)| \le C_3.$$

It remains to estimate the normal-normal derivative $u_{x^n x^n}(0)$. At the origin we can now write

$$f = \det(u_{j\bar{k}}) = u_{n\bar{n}} \det(u_{j\bar{k}})_{j,k \le n-1} + \mathcal{R} = u_{n\bar{n}} \det(v_{j\bar{k}})_{j,k \le n-1} + \mathcal{R},$$

where $|\mathcal{R}|$ is under control. Therefore

$$0 \leq u_{n\bar{n}}(0) \leq \frac{C_4}{\lambda^{n-1}}$$

and the normal-normal estimate follows. □

13 Higher Order Estimates

We will make use of a general (real) theory of nonlinear elliptic equations of second order. They are of the form

$$F(D^2 u, Du, u, x) = 0, \tag{43}$$

where

$$F : \mathbb{R}^{m^2} \times \mathbb{R}^m \times \mathbb{R} \times \Omega \longrightarrow \mathbb{R}$$

(Ω is a domain in $\mathbb{R}^m$) satisfies two basic assumptions:

$$F \text{ is concave in } D^2 u \tag{44}$$

and elliptic, that is

$$\frac{\partial F}{\partial u_{x_s x_t}} \zeta^s \zeta^t \geq \lambda |\zeta|^2, \quad \zeta \in \mathbb{R}^m, \tag{45}$$

for some $\lambda > 0$.

If by $\mathcal{M}_+$ we denote the set of Hermitian positive matrices then, as one can show (see e.g. [13, 24]),

$$(\det A)^{1/n} = \frac{1}{n} \inf\{\text{trace}(AB) : B \in \mathcal{M}_+, \det B = 1\}, \quad A \in \mathcal{M}_+.$$

Moreover, one can also easily prove the following formula for the minimal eigenvalue of $(u_{j\bar{k}}) > 0$

$$\lambda_{min}\left(\frac{\partial \det(u_{j\bar{k}})}{\partial u_{x_s x_t}}\right) = \frac{\det(u_{j\bar{k}})}{4\lambda_{max}(u_{j\bar{k}})},$$

(see e.g. [9]). (Here x_s denote real variables in $\mathbb{C}^n$, $s = 1, \ldots, 2n$.)

By Theorem 24 we can assume that

$$\frac{1}{C}|\zeta|^2 \leq u_{j\bar{k}} \zeta^j \bar{\zeta}^k \leq C|\zeta|^2, \quad \zeta \in \mathbb{C}^n. \tag{46}$$

Therefore, if we define F as

$$F(D^2u, z) := (\det(u_{j\bar{k}}))^{1/n} - f(z)$$

for functions (or rather matrices) satisfying (46) and extend it in a right way to the set of all symmetric real $2n \times 2n$-matrices, then F satisfies (44) and (45).

Theorem 28 ([22, 23, 32, 38, 46]). *Assume that $u \in C^3(\Omega)$ solves (43), where F is C^2 and satisfies (44) and (45). Then for $\Omega' \Subset \Omega$ there exists $\alpha \in (0, 1)$ depending only on upper bounds for $||u||_{C^{1,1}(\Omega)}$, $||F||_{C^{1,1}(\Omega)}$ and a lower bound for λ, and C depending in addition on a lower bound for dist $(\Omega', \partial\Omega)$, such that*

$$||u||_{C^{2,\alpha}(\Omega')} \leq C.$$

Theorem 29 ([16, 32]). *Assume that u, defined in $B_R^+ := B(0, R) \cap \{x^m \geq 0\}$, solves (43) with F satisfying (44) and (45) and $u = \psi$ on $B(0, R) \cap \{x^m = 0\}$. Then there exists $\alpha \in (0, 1)$ and C, depending only on m, λ, R, $||u||_{C^{1,1}}$, $||F||_{C^{1,1}}$ and $||\psi||_{C^{3,1}}$, such that*

$$||u||_{C^{2,\alpha}(B_{R/2}^+)} \leq C.$$

Now the standard Schauder theory applied to (the linearization of) F gives the required a priori estimate (32).

References

1. T. Aubin, Equations du type de Monge-Ampère sur les variétés Kähleriennes compactes. C. R. Acad. Sci. Paris **283**, 119–121 (1976)
2. T. Aubin, Réduction du cas positif de léquation de Monge-Ampère sur les variétés kählériennes compactes à la demonstration dune inégalité. J. Funct. Anal. **57**, 143–153 (1984)
3. S. Bando, T. Mabuchi, Uniqueness of Einstein Kähler metrics modulo connected group actions, in Algebraic Geometry, Sendai, 1985. Advanced Studies in Pure Mathematics, vol. 10 (North-Holland, Amsterdam, 1987), pp. 11–40
4. E. Bedford, M. Kalka, Foliations and complex Monge-Ampère equations. Commun. Pure Appl. Math. **30**, 543–571 (1977)
5. E. Bedford, B.A. Taylor, The Dirichlet problem for a complex Monge-Ampère equation. Invent. Math. **37**, 1–44 (1976)
6. R.J. Berman, S. Boucksom, V. Guedj, A. Zeriahi, A variational approach to complex Monge-Ampère equations [arXiv:0907.4490] (to appear in Publ. math. de)
7. B. Berndtsson, A Brunn-Minkowski type inequality for Fano manifolds and the Bando-Mabuchi uniqueness theorem [arXiv:1103.0923]
8. Z. Błocki, The complex Monge-Ampère operator in hyperconvex domains. Ann. Sc. Norm. Sup. Pisa **23**, 721–747 (1996)
9. Z. Błocki, On the regularity of the complex Monge-Ampère operator, in Complex Geometric Analysis in Pohang, ed. by K.-T. Kim, S.G. Krantz. Contemporary Mathematics, vol. 222 (American Mathematical Society, Providence, 1999), pp. 181–189
10. Z. Błocki, Interior regularity of the complex Monge-Ampère equation in convex domains. Duke Math. J. **105**, 167–181 (2000)
11. Z. Błocki, Regularity of the degenerate Monge-Ampère equation on compact Kähler manifolds. Math. Z. **244**, 153–161 (2003)

12. Z. Błocki, Uniqueness and stability for the Monge-Ampère equation on compact Kähler manifolds. Indiana Univ. Math. J. **52**, 1697–1702 (2003)
13. Z. Błocki, The Calabi-Yau theorem. Course given in Toulouse, January 2005, in *Complex Monge-Ampe`re Equations and Geodesics in the Space of Kähler Metrics*, ed. by V. Guedj. Lecture Notes in Mathematics, vol. 2038 (Springer, Berlin, 2012), pp. 201–227
14. Z. Błocki, A gradient estimate in the Calabi-Yau theorem. Math. Ann. **344**, 317–327 (2009)
15. Z. Błocki, On Geodesics in the Space of Kähler Metrics, in *Advances in Geometric Analysis*, ed. by S. Janeczko, J. Li, D. Phong, Adv. Lect. Math., vol. 21 (International Press, Boston, 2012), pp. 3–20
16. L. Caffarelli, J.J. Kohn, L. Nirenberg, J. Spruck, The Dirichlet problem for non-linear second order elliptic equations II: Complex Monge-Ampère, and uniformly elliptic equations. Commun. Pure Appl. Math. **38**, 209–252 (1985)
17. E. Calabi, The space of Kähler metrics. Proceedings of the International Congress of Mathematicians, Amsterdam, 1954, vol. 2, pp. 206–207
18. X.X. Chen, The space of Kähler metrics. J. Differ. Geom. **56**, 189-234 (2000)
19. X.X. Chen, G. Tian, Geometry of Kähler metrics and foliations by holomorphic discs. Publ. Math. IHES **107**, 1–107 (2008)
20. S.K. Donaldson, Symmetric spaces, Kähler geometry and Hamiltonian dynamics, in *Northern California Symplectic Geometry Seminar*. Am. Math. Soc. Transl. Ser. 2, vol. 196 (American Mathematical Society, Providence, 1999), pp. 13–33
21. S.K. Donaldson, Constant scalar curvature metrics on toric surfaces. GAFA **19**, 83–136 (2009)
22. L.C. Evans, Classical solutions of fully nonlinear, convex, second order elliptic equations. Commun. Pure Appl. Math. **25**, 333–363 (1982)
23. L.C. Evans, Classical solutions of the Hamilton-Jacobi-Bellman equation for uniformly elliptic operators. Trans. Am. Math. Soc. **275**, 245–255 (1983)
24. B. Gaveau, Méthodes de contrôle optimal en analyse complexe I. Résolution d'équations de Monge-Ampère. J. Funct. Anal. **25**, 391–411 (1977)
25. T.W. Gamelin, N. Sibony, Subharmonicity for uniform algebras. J. Funct. Anal. **35**, 64–108 (1980)
26. D. Gilbarg, N.S. Trudinger, in *Elliptic Partial Differential Equations of Second Order*. Classics in Mathematics (Springer, Berlin, 1998)
27. B. Guan, The Dirichlet problem for complex Monge-Ampère equations and regularity of the pluri-complex Green function. Comm. Anal. Geom. **6**, 687–703 (1998)
28. D. Guan, On modified Mabuchi functional and Mabuchi moduli space of Kähler metrics on toric bundles. Math. Res. Lett. **6**, 547–555 (1999)
29. P. Guan, A gradient estimate for complex Monge-Ampère equation. Preprint (2009)
30. A. Hanani, Equations du type de Monge-Ampère sur les variétés hermitiennes compactes. J. Funct. Anal. **137**, 49–75 (1996)
31. J.L. Kazdan, A remark on the proceding paper of Yau. Commun. Pure Appl. Math. **31**, 413–414 (1978)
32. N.V. Krylov, Boundedly inhomogeneous elliptic and parabolic equations. Izv. Akad. Nauk SSSR **46**, 487–523 (1982); English translation: Math. USSR Izv. **20**, 459–492 (1983)
33. L Lempert, L. Vivas, Geodesics in the space of Kähler metrics [arXiv:1105.2188]
34. T. Mabuchi, K-energy maps integrating Futaki invariants. Tohoku Math. J. **38**, 575–593 (1986)
35. T. Mabuchi, Some symplectic geometry on compact Kähler manifolds. I. Osaka J. Math. **24**, 227–252 (1987)
36. Y. Matsushima, Sur la structure du groupe d'homéomorphismes analytiques d'une certaine variété kählérienne. Nagoya Math. J. **11**, 145–150 (1957)
37. D.H. Phong, J. Sturm, Lectures on stability and constant scalar curvature, in *Handbook of Geometric Analysis*, vol. 3, Adv. Lect. Math., vol. 14 (International Press, Boston, 2010), pp. 357–436
38. F. Schulz, Über nichtlineare, konkave elliptische Differentialgleichungen. Math. Z. **191**, 429–448 (1986)

39. S. Semmes, Complex Monge-Ampère and symplectic manifolds. Am. J. Math. **114**, 495–550 (1992)
40. Y.-T. Siu, Lectures on Hermitian-Einstein metrics for stable bundles and Kähler-Einstein metrics, DMV Seminar, 8, Birkhäuser, 1987
41. G. Tian, On Calabi's conjecture for complex surfaces with positive first Chern class. Invent. Math. **101**, 101–172 (1990)
42. G. Tian, The K-energy on hypersurfaces and stability. Comm. Anal. Geom. **2**, 239–265 (1994)
43. G. Tian, Kähler-Einstein metrics with positive scalar curvature. Invent. Math. **137**, 1–37 (1997)
44. G. Tian, *Canonical Metrics in Kähler Geometry*. Lectures in Mathematics (ETH Zurich, Birkhäuser, Zurich, 2000)
45. V. Tosatti, Kähler-Einstein metrics on Fano surfaces. Expo. Math. **30**, 11–31 (2012) [arXiv:1010.1500]
46. N. Trudinger, Fully nonlinear, uniformly elliptic equations under natural structure conditions. Trans. Am. Math. Soc. **278**, 751–769 (1983)
47. S.-T. Yau, On the Ricci curvature of a compact Kähler manifold and the complex Monge-Ampère equation, I. Commun. Pure Appl. Math. **31**, 339–411 (1978)
48. S.-T. Yau, Calabi's conjecture and some new results in algebraic geometry. Proc. Nat. Acad. Sci. USA **74**, 1798–1799 (1977)

Applications of Pluripotential Theory to Algebraic Geometry

Jean-Pierre Demailly

Abstract These lectures are devoted to the study of various contemporary problems of algebraic geometry, using fundamental tools from complex potential theory, namely plurisubharmonic functions, positive currents and Monge-Ampère operators. Since their inception by Oka and Lelong in the mid 1940s, plurisubharmonic functions have been used extensively in many areas of algebraic and analytic geometry, as they are the function theoretic counterpart of pseudoconvexity, the complexified version of convexity. One such application is the theory of L^2 estimates via the Bochner-Kodaira-Hörmander technique, which provides very strong existence theorems for sections of holomorphic vector bundles with positive curvature. One can mention here the foundational work achieved by Bochner, Kodaira, Nakano, Morrey, Kohn, Andreotti-Vesentini, Grauert, Hörmander, Bombieri, Skoda and Ohsawa-Takegoshi in the course of more than four decades. Another development is the theory of holomorphic Morse inequalities (1985), which relate certain curvature integrals with the asymptotic cohomology of large tensor powers of line or vector bundles, and bring a useful complement to the Riemann-Roch formula.

We describe here the main techniques involved in the proof of holomorphic Morse inequalities (Sect. 1) and their link with Monge-Ampère operators and intersection theory. Section 2, especially, gives a fundamental approximation theorem for closed (1, 1)-currents, using a Bergman kernel technique in combination with the Ohsawa-Takegoshi theorem. As an application, we study the geometric properties of positives cones of an algebraic variety (nef and pseudo-effective cone), and derive from there some results about asymptotic cohomology functionals in Sect. 3. The last Sect. 4 provides an application to the study of the Green-Griffiths-Lang conjecture. The latter conjecture asserts that every entire curve drawn on a projective variety of general type should satisfy a global algebraic equation; via a probabilistic

---

J.-P. Demailly (✉)
Université de Grenoble I, Institut Fourier, BP74, 100 rue des Maths, 38402 Saint-Martin d'Hères, France
e-mail: demailly@fourier.ujf-grenoble.fr

curvature estimate, holomorphic Morse inequalities imply that entire curves must at least satisfy a global algebraic differential equation.

1 Holomorphic Morse Inequalities

Holomorphic Morse inequalities provide asymptotic bounds for the cohomology of tensor powers of holomorphic line bundles. They are a very useful complement to the Riemann–Roch formula in many circumstances. They were first introduced in [25], and were largely motivated by Siu's solution [85, 86] of the Grauert–Riemenschneider conjecture, which we reprove here as a special case of a stronger statement. The basic tool is a spectral theorem which describes the eigenvalue distribution of complex Laplace–Beltrami operators. The original proof of [25] was based partly on Siu's techniques and partly on an extension of Witten's analytic proof of standard Morse inequalities [102]. Somewhat later Bismut [8] and Getzler [49] gave new proofs, both relying on an analysis of the heat kernel in the spirit of the Atiyah–Bott–Patodi proof of the Atiyah–Singer index theorem [1]. Although the basic idea is simple, Bismut used deep results arising from probability theory (the Malliavin calculus), while Getzler relied on his supersymmetric symbolic calculus for spin pseudodifferential operators [48].

We present here a slightly more elementary and self-contained proof which was suggested to us by Mohan Ramachandran on the occasion of a visit to Chicago in 1989. The reader is referred to [25, 27] for more details.

1.1 Introduction

1.1.1 Real Morse Inequalities

Let M be a compact C^∞ manifold, $\dim_\mathbb{R} M = m$, and h a Morse function, i.e. a function such that all critical points are non degenerate. The standard (real) Morse inequalities relate the Betti numbers $b_q = \dim H^q_{DR}(M, \mathbb{R})$ and the numbers

$$s_q = \#\text{ critical points of index } q ,$$

where the index of a critical point is the number of negative eigenvalues of the Hessian form $(\partial^2 h/\partial x_i \partial x_j)$. Specifically, the following "strong Morse inequalities" hold:

$$b_q - b_{q-1} + \cdots + (-1)^q b_0 \leq s_q - s_{q-1} + \cdots + (-1)^q s_0 \qquad (1)$$

for each integer $q \geq 0$. As a consequence, one recovers the "weak Morse inequalities" $b_q \leq s_q$ and the expression of the Euler–Poincaré characteristic

$$\chi(M) = b_0 - b_1 + \cdots + (-1)^m b_m = s_0 - s_1 + \cdots + (-1)^m s_m . \qquad (2)$$

Applications of Pluripotential Theory to Algebraic Geometry

These results are purely topological. They are obtained by showing that M can be reconstructed from the structure of the Morse function by attaching cells according to the index of the critical points; real Morse inequalities are then obtained as a consequence of the Mayer–Vietoris exact sequence (see [74]).

1.1.2 Dolbeault Cohomology

Instead of looking at De Rham cohomology, we want to investigate here Dolbeault cohomology, i.e. cohomology of the $\bar{\partial}$-complex. Let X be a compact complex manifold, $n = \dim_\mathbb{C} X$ and E be a holomorphic vector bundle over X with rank $E = r$. Let us recall that there is a canonical $\bar{\partial}$-operator

$$\bar{\partial} : C^\infty(X, \Lambda^{p,q}T_X^* \otimes E) \longrightarrow C^\infty(X, \Lambda^{p,q+1}T_X^* \otimes E) \qquad (3)$$

acting on spaces of (p, q)-forms with values in E. By the Dolbeault isomorphism theorem, there is an isomorphism

$$H_{\bar{\partial}}^{p,q}(X, E) := H_{\bar{\partial}}^q(C^\infty(X, \Lambda^{p,\bullet}T_X^* \otimes E)) \simeq H^q(X, \Omega_X^p \otimes \mathcal{O}(E)) \qquad (4)$$

from the cohomology of the $\bar{\partial}$-complex onto the cohomology of the sheaf of holomorphic p-forms with values in E. In particular, we have

$$H_{\bar{\partial}}^{0,q}(X, E) \simeq H^q(X, \mathcal{O}(E)), \qquad (5)$$

and we will denote as usual $h^q(X, E) = \dim H^q(X, \mathcal{O}(E))$.

1.1.3 Connections and Curvature

Let us consider first a C^∞ complex vector bundle $E \to M$ on a real differential manifold M (without necessarily any holomorphic structure at this point). A *connection* D on E is a linear differential operator

$$D : C^\infty(M, \Lambda^q T_M^* \otimes E) \to C^\infty(M, \Lambda^{q+1} T_M^* \otimes E) \qquad (6)$$

satisfying the Leibniz rule

$$D(f \wedge s) = df \wedge s + (-1)^{\deg f} f \wedge Ds \qquad (7)$$

for all forms $f \in C^\infty(X, \Lambda^p T_M^*), s \in C^\infty(X, \Lambda^q T_M^* \otimes E)$. On an open set $U \subset M$ where E is trivial, $E_{|U} \simeq U \times \mathbb{C}^r$, the Leibniz rule shows that a connection D can be written in a unique way

$$Ds \simeq ds + \Gamma \wedge s \qquad (8)$$

where $\Gamma \in C^\infty(U, \Lambda^1 T_M^* \otimes \mathrm{Hom}(\mathbb{C}^r, \mathbb{C}^r))$ is an arbitrary $r \times r$ matrix of 1-forms and d acts componentwise. It is then easy to check that

$$D^2 s \simeq (d\Gamma + \Gamma \wedge \Gamma) \wedge s \qquad \text{on} \quad U. \tag{9}$$

Therefore $D^2 s = \theta_D \wedge s$ for some global 2-form $\theta_D \in C^\infty(M, \Lambda^2 T_M^* \otimes \mathrm{Hom}(E, E))$, given by $\theta_D \simeq d\Gamma_U + \Gamma_U \wedge \Gamma_U$ on any trivializing open set U with a connection matrix Γ_U.

Definition 1. The (normalized) curvature tensor of D is defined to be $\Theta_D = \frac{i}{2\pi}\theta_D$, in other words

$$\frac{i}{2\pi} D^2 s = \Theta_D \wedge s$$

for any section $s \in C^\infty(M, \Lambda^q T_M^* \otimes E)$.

The main reason for the introduction of the factor $\frac{i}{2\pi}$ is the well known formula for the expression of the Chern classes in the ring of differential forms of even degree: one has

$$\det(\mathrm{Id} + \lambda \Theta_D) = 1 + \lambda \gamma_1(D) + \lambda^2 \gamma_2(D) + \ldots + \lambda^r \gamma_r(D),$$

where $\gamma_j(D)$ is a d-closed differential form of degree $2j$. Moreover, $\gamma_j(D)$ has integral periods, i.e. the De Rham cohomology class $\{\gamma_j(D)\} \in H^{2j}(M, \mathbb{R})$ is the image of an integral class, namely the j-th Chern class $c_j(E) \in H^{2j}(M, \mathbb{Z})$.

1.1.4 Hermitian Connections

Assume now that the fibers of E are endowed with a C^∞ Hermitian metric h, and that the isomorphism $E_{|U} \simeq U \times \mathbb{C}^r$ is given by a C^∞ frame (e_λ). Then we have a canonical sesquilinear pairing

$$C^\infty(M, \Lambda^p T_M^* \otimes E) \times C^\infty(M, \Lambda^q T_M^* \otimes E) \longrightarrow C^\infty(M, \Lambda^{p+q} T_M^*)$$
$$(u, v) \longmapsto \{u, v\}_h$$

given by

$$\{u, v\}_h = \sum_{\lambda, \mu} u_\lambda \wedge \overline{v}_\mu \langle e_\lambda, e_\mu \rangle_h \qquad \text{for} \quad u = \sum u_\lambda \otimes e_\lambda, \ v = \sum v_\mu \otimes e_\mu.$$

The connection D is said to be *Hermitian* (or compatible with the Hermitian metric h) if it satisfies the additional property

$$d\{u, v\}_h = \{Du, v\}_h + (-1)^{\deg u}\{u, Dv\}_h. \tag{10}$$

Applications of Pluripotential Theory to Algebraic Geometry   147

Assuming that (e_λ) is h-orthonormal, one easily checks that D is Hermitian if and only if the associated connection matrix Γ is skew-symmetric, i.e. $\Gamma^* = -\Gamma$. In this case $\theta_D = d\Gamma + \Gamma \wedge \Gamma$ also satisfies $\theta_D^* = -\theta_D$, thus

$$\Theta_D = \frac{i}{2\pi}\theta_D \in C^\infty(M, \Lambda^2 T_M^* \otimes \mathrm{Herm}(E, E)). \tag{11}$$

Special case 2. For a bundle E of rank $r = 1$, the connection matrix Γ of a Hermitian connection D can be more conveniently written $\Gamma = -iA$ where A is a *real* 1-form. Then we have

$$\Theta_D = \frac{i}{2\pi}d\Gamma = \frac{1}{2\pi}dA.$$

Frequently, especially in physics, the real 2-form $B = dA = 2\pi\Theta_D \in C^\infty(M, \Lambda^2 T_M^*)$ is referred to as the *magnetic field*, and the 1-form A as its potential. A phase change $\tilde{s}(x) = s(x)e^{i\alpha(x)}$ in the isomorphism $E_{|U} \simeq U \times \mathbb{C}$ replaces A with the new connection form $\tilde{A} = A + d\alpha$.

1.1.5 Connections on a Hermitian Holomorphic Vector Bundle

If $M = X$ is a complex manifold, every connection D can be split in a unique way as the sum $D = D' + D''$ of a $(1,0)$-connection D' and a $(0,1)$-connection D'':

$$D' : C^\infty(M, \Lambda^{p,q} T_X^* \otimes E) \longrightarrow C^\infty(M, \Lambda^{p+1,q} T_X^* \otimes E),$$
$$D'' : C^\infty(M, \Lambda^{p,q} T_X^* \otimes E) \longrightarrow C^\infty(M, \Lambda^{p,q+1} T_X^* \otimes E).$$

In a local trivialization given by a C^∞ frame, one can write

$$D'u = d'u + \Gamma' \wedge u,$$
$$D''u = d''u + \Gamma'' \wedge u,$$

with $\Gamma = \Gamma' + \Gamma''$ and $d' = \partial$, $d'' = \overline{\partial}$. If (E, h) is a C^∞ Hermitian structure, the connection is Hermitian if and only if $\Gamma' = -(\Gamma'')^*$ in any h-orthonormal frame. Thus there exists a unique Hermitian connection corresponding to a prescribed $(0, 1)$ part D''.

Assume now that the Hermitian bundle (E, h) has a *holomorphic* structure. The unique Hermitian connection D for which $D'' = \overline{\partial}$ is called the *Chern connection* of (E, h). In a local holomorphic frame (e_λ) of $E_{|U}$, the metric h is given by some Hermitian matrix $H = (h_{\lambda\mu})$ where $h_{\lambda\mu} = \langle e_\lambda, e_\mu \rangle_h$. Standard computations yield the expression of the Chern connection:

$$\begin{cases} D's = \partial s + \overline{H}^{-1}\partial\overline{H} \wedge s, \\ D''s = \bar{\partial}s, \\ \theta_D \wedge s = D^2 s = (D'D'' + D''D')s = -\bar{\partial}(\overline{H}^{-1}\partial\overline{H}) \wedge s. \end{cases}$$

Definition 3. The Chern curvature tensor of (E, h) is the curvature tensor of its Chern connection, denoted

$$\theta_{E,h} = D'D'' + D''D' = -\bar{\partial}(\overline{H}^{-1}\partial\overline{H}).$$

In the special case of a rank 1 bundle E, the matrix H is simply a positive function, and it is convenient to introduce its weight φ such that $H = (e^{-\varphi})$ where $\varphi \in C^\infty(U, \mathbb{R})$ depends on the given trivialization $E_{|U} \simeq U \times \mathbb{C}$. We have in this case

$$\Theta_{E,h} = \frac{i}{2\pi}\theta_{E,h} = \frac{i}{2\pi}\partial\bar{\partial}\varphi \quad \text{on} \quad U, \tag{12}$$

and therefore $\Theta_{E,h}$ is a closed real $(1, 1)$-form.

1.1.6 Fundamental Facts of Hodge Theory

Assume here that M is a Riemannian manifold with metric $g = \sum g_{ij}dx_i \otimes dx_j$. Given q-forms u, v on M with values in E, we consider the global L^2 norm and inner product

$$\|u\|^2 = \int_M |u(x)|^2 d\sigma(x), \quad \langle\!\langle u, v \rangle\!\rangle = \int_M \langle u(x), v(x) \rangle \, d\sigma(x), \tag{13}$$

where $|u|$ is the pointwise Hermitian norm and $d\sigma$ the Riemannian volume form. The Laplace Beltrami operator associated with the connection D is

$$\Delta = DD^* + D^*D,$$

acting on any of the spaces $C^\infty(M, \Lambda^q T_M^* \otimes E)$; here

$$D^* : C^\infty(M, \Lambda^q T_M^* \otimes E) \longrightarrow C^\infty(M, \Lambda^{q-1} T_M^* \otimes E) \tag{14}$$

is the (formal) L^2 adjoint of D. The complex Laplace operators $\Delta' = D'D'^* + D'^*D'$ and $\Delta'' = D''D''^* + D''^*D''$ are defined similarly when $M = X$ is a complex manifold. In degree 0 we simply have $\Delta = D^*D$. A well-known calculation shows that the principal symbol of Δ is $\sigma_\Delta(x, \xi) = -|\xi|^2 \text{Id}$ (while $\sigma_{\Delta'}(x, \xi) = \sigma_{\Delta''}(x, \xi) = -\frac{1}{2}|\xi|^2 \text{Id}$). As a consequence Δ, Δ', Δ'' are always *elliptic operators*.

Applications of Pluripotential Theory to Algebraic Geometry

When M is *compact*, the operator Δ acting on any of the spaces of the complex $C^\infty(M, \Lambda^\bullet T_M^* \otimes E)$ has a discrete spectrum

$$\lambda_1 \leq \lambda_2 \leq \cdots \leq \lambda_j \leq \cdots$$

and corresponding eigenfunctions $\psi_j \in C^\infty(M, \Lambda^q T_M^* \otimes E)$, depending of course on q.

Our main goal is to obtain asymptotic formulas for the eigenvalues. For this, we will make an essential use of the *heat operator* $e^{-t\Delta}$. In the above setting, the heat operator is the bounded Hermitian operator associated to the *heat kernel*

$$K_t(x, y) = \sum_{\nu=1}^{+\infty} e^{-\lambda_\nu t} \psi_\nu(x) \otimes \psi_\nu^*(y), \tag{15}$$

i.e.

$$\langle\!\langle u, e^{-t\Delta} v\rangle\!\rangle = \int_{M \times M} \langle u(x), K_t(x, y) \cdot v(y)\rangle\, d\sigma(x)\, d\sigma(y).$$

Standard results of the theory of elliptic operators show that

$$K_t \in C^\infty(]0, +\infty[\times M \times M, \mathrm{Hom}(E, E))$$

and that $K_t(x, y)$ is the solution of the differential equation

$$\frac{\partial}{\partial t} K_t(x, y) = -\Delta_x K_t(x, y), \qquad \lim_{t \to 0_+} K_t(x, y) = \delta_y(x) \quad \text{(Dirac at } y\text{)}, \tag{16}$$

as follows formally from the fact that $\frac{\partial}{\partial t} e^{-t\Delta} = -\Delta e^{-t\Delta}$ and $e^{-0\Delta} = \mathrm{Id}$. The asymptotic distribution of eigenvalues can be recovered from the straightforward formula

$$\sum_{\nu=1}^{+\infty} e^{-\lambda_\nu t} = \int_M \mathrm{tr}_E K_t(x, x) d\sigma(x). \tag{17}$$

In the sequel, we are especially interested in the 0-eigenspace:

Definition 4. The space of Δ-harmonic forms is defined to be

$$\mathcal{H}_\Delta^q(M, E) = \mathrm{Ker}\, \Delta = \{u \in C^\infty(M, \Lambda^q T_M^* \otimes E)\,;\ \Delta u = 0\}.$$

When M is compact, an integration by part shows that

$$\langle\!\langle \Delta u, u\rangle\!\rangle = \|Du\|^2 + \|D^*u\|^2,$$

hence u is Δ-harmonic if and only if $Du = D^*u = 0$. Moreover, as Δ is a self-adjoint operator, standard elliptic theory implies that

$$C^\infty(M, \Lambda^q T_M^* \otimes E) = \mathrm{Ker}\, \Delta \oplus \mathrm{Im}\, \Delta = \mathcal{H}_\Delta^q(M, E) \oplus \mathrm{Im}\, \Delta, \tag{18}$$

and Ker $\Delta = \mathcal{H}_\Delta^q(M, E)$, Im Δ are orthogonal with respect to the L^2 inner product. Clearly Im $\Delta \subset$ Im $D +$ Im D^*, and both images Im D, Im D^* are orthogonal to the space of harmonic forms by what we have just seen. As a consequence, we have

$$\text{Im } \Delta = \text{Im } D + \text{Im } D^*. \tag{19}$$

Hodge isomorphism theorem 5. *Assume that M is compact and that D is an integrable connection,* i.e. $D^2 = 0$ (or $\theta_D = 0$). Then D defines on spaces of sections $C^\infty(M, \Lambda^q T_M^* \otimes E)$ a differential complex which can be seen as a generalization of the De Rham complex. The condition $D^2 = 0$ immediately implies that Im $D \perp$ Im D^* and we conclude from the above discussion that there is an orthogonal direct sum

$$C^\infty(M, \Lambda^q T_M^* \otimes E) = \mathcal{H}_\Delta^q(M, E) \oplus \text{Im } D \oplus \text{Im } D^*. \tag{20}$$

If we put $u = h + Dv + D^*w$ according to this decomposition, then $Du = DD^*w = 0$ if and only if $\|D^*w\| = \langle\!\langle DD^*w, w\rangle\!\rangle = 0$, thus

$$\text{Ker } D = \mathcal{H}_\Delta^q(M, E) \oplus \text{Im } D.$$

This implies the *Hodge isomorphism theorem*

$$H_{\text{DR}}^q(M, E) := \text{Ker } D/\text{Im } D \simeq \mathcal{H}_\Delta^q(M, E). \tag{21}$$

In case $M = X$ is a compact complex manifold, (E, h) a Hermitian holomorphic vector bundle and $D = D' + D''$ the Chern connection, the integrability condition $D''^2 = \overline\partial^2 = 0$ is always satisfied. Thus we get an analogous isomorphism

$$H^q(X, \mathcal{O}(E)) \simeq H_{\overline\partial}^{0,q}(X, E) \simeq \mathcal{H}_{\Delta''}^{0,q}(M, E), \tag{22}_{0,q}$$

and more generally

$$H^q(X, \Omega_X^p \otimes \mathcal{O}(E)) \simeq H_{\overline\partial}^{p,q}(X, E) \simeq \mathcal{H}_{\Delta''}^{p,q}(M, E), \tag{22}_{p,q}$$

where $\mathcal{H}_{\Delta''}^{p,q}(M, E)$ is the space of Δ''-harmonic forms of type (p, q) with values in E.

Corollary 6 (Hodge decomposition theorem). *If (X, ω) is a compact Kähler manifold and (E, h) is a flat Hermitian vector bundle over X (i.e. $D_{E,h}^2 = 0$), then there is an isomorphism*

$$H_{\text{DR}}^k(M, E) \simeq \bigoplus_{p+q=k} H_{\overline\partial}^{p,q}(X, E).$$

In fact, under the condition that ω is Kähler, i.e. $d\omega = 0$, well-known identities of Kähler geometry imply $\Delta' = \Delta'' = \frac{1}{2}\Delta$, and as a consequence

$$\mathcal{H}_\Delta^k(M, E) = \bigoplus_{p+q=k} \mathcal{H}_{\Delta''}^{p,q}(X, E).$$

1.2 Holomorphic Morse Inequalities

1.2.1 Main Statements

Let X be a compact complex n-dimensional manifold, $L \to X$ a holomorphic line bundle and $E \to X$ a holomorphic vector bundle of rank $r = \operatorname{rank} E$. We assume that L is equipped with a smooth Hermitian metric h and denote accordingly $\Theta_{L,h}$ its curvature form; by definition this is a closed real $(1, 1)$-form and its cohomology class $c_1(L)_\mathbb{R} = \{\Theta_{L,h}\} \in H^2_{\mathrm{DR}}(X, \mathbb{R})$ is the first Chern class of L.

q-index sets 7. *We define the q-index sets and $\{\leq q\}$-index sets of (L, h) to be*

$$X(L, h, q) = \left\{ x \in X \,;\, \Theta_{L,h}(x) \text{ has } \begin{array}{l} q \text{ negative eigenvalues} \\ n-q \text{ positive eigenvalues} \end{array} \right\}$$

$$X(L, h, \leq q) = \bigcup_{1 \leq j \leq q} X(L, h, j).$$

Clearly $X(L, h, q)$ and $X(L, h, \leq q)$ are open subsets of X, and we have a partition into "chambers" $X = S \cup \bigcup_{0 \leq q \leq n} X(L, h, q)$ where $S = \{x \in X \,;\, \Theta_{L,h}(x) = 0\}$ is the degeneration set. The following theorem was first proved in [25].

Main Theorem 8. *The cohomology groups of tensor powers $E \otimes L^k$ satisfy the following asymptotic estimates as $k \to +\infty$:*

$(8)_{\mathrm{WM}}$ *Weak Morse inequalities*:

$$h^q(X, E \otimes L^k) \leq r \frac{k^n}{n!} \int_{X(L,h,q)} (-1)^q \Theta_{L,h}^n + o(k^n).$$

$(8)_{\mathrm{SM}}$ *Strong Morse inequalities*:

$$\sum_{0 \leq j \leq q} (-1)^{q-j} h^j(X, E \otimes L^k) \leq r \frac{k^n}{n!} \int_{X(L,h,\leq q)} (-1)^q \Theta_{L,h}^n + o(k^n).$$

$(8)_{\mathrm{RR}}$ *Asymptotic Riemann–Roch formula*:

$$\chi(X, E \otimes L^k) := \sum_{0 \leq j \leq n} (-1)^j h^j(X, E \otimes L^k) = r \frac{k^n}{n!} \int_X \Theta_{L,h}^n + o(k^n).$$

The weak Morse form (8)$_{\text{WM}}$ follows from strong Morse (8)$_{\text{SM}}$ by adding consecutive inequalities for the indices $q-1$ and q, since the signs $(-1)^{q-j}$ and $(-1)^{q-1-j}$ are opposite. Also, (8)$_{\text{RR}}$ is just a weaker formulation of the existence of the Hilbert polynomial, and as such, is a consequence of the Hirzebruch–Riemann–Roch formula; it follows formally from (8)$_{\text{SM}}$ with $q = n$ and $q = n+1$, since $h^{n+1} = 0$ identically and the signs are reversed. Now, by adding (8)$_{\text{SM}}$ for the indices of opposite parity $q+1$ and $q-2$, we find

$$h^{q+1}(X, E \otimes L^k) - h^q(\ldots) + h^{q-1}(\ldots) \leq r \frac{k^n}{n!} \int_{X(L,h,\{q-1,q,q+1\})} (-1)^{q+1} \Theta^n_{L,h} + o(k^n),$$

where $X(L, h, \{q-1, q, q+1\})$ is meant for the union of chambers of indices $q-1$, q, $q+1$. As a consequence, we get lower bounds for the cohomology groups:

$$h^q(X, E \otimes L^k) \geq h^q - h^{q+1} - h^{q-1} \geq r \frac{k^n}{n!} \int_{X(L,h,\{q-1,q,q+1\})} (-1)^q \Theta^n_{L,h} - o(k^n). \quad (23)$$

Another important special case is (8)$_{\text{SM}}$ for $q = 1$, which yields the lower bound

$$h^0(X, E \otimes L^k) \geq h^0 - h^1 \geq r \frac{k^n}{n!} \int_{X(L,h,\leq 1)} \Theta^n_{L,h} - o(k^n). \quad (24)$$

As we will see later in the applications, this lower bound provides a very useful criterion to prove the existence of sections of large tensor powers of a line bundle. □

1.2.2 Heat Kernel and Eigenvalue Distribution

We introduce here a basic heat equation technique, from which all asymptotic eigenvalue estimates can be derived via an explicit formula, known as Mehler's formula.

We start with a compact Riemannian manifold (M, g) with $\dim_{\mathbb{R}} M = m$, and denote by $d\sigma$ its Riemannian volume form. Let (L, h_L) (resp. (E, h_E)) be a Hermitian complex line (resp. vector bundle) on M, equipped with a Hermitian connection D_L (resp. D_E).

We denote by $D_k = D_{E \otimes L^k}$ the associated connection on $E \otimes L^k$, and by $\Delta_k = D_k^* D_k$ the Laplace–Beltrami operator acting on sections of $E \otimes L^k$ (i.e. forms of degree 0). As in Case 2, we introduce the (local) connection form $\Gamma_L = -iA$ of L and the corresponding (global) curvature 2-form $B = dA \in C^\infty(M, \Lambda^2 T_M^*)$, i.e. the "magnetic field" (Γ_E and the corresponding curvature tensor Θ_E of D_E will not play a significant role here). Finally, we assume that an additional section $V \in C^\infty(M, \text{Herm}(E, E))$ is given ("electric field"); for simplicity of notation, we still denote by V the operator $V \otimes \text{Id}_{L^k}$ acting on $E \otimes L^k$.

Applications of Pluripotential Theory to Algebraic Geometry

If $\Omega \subset M$ is a smoothly bounded open subset of M, we consider for u in the Sobolev space $W_0^1(\Omega, E \otimes L^k)$ the quadratic form

$$Q_{k,\Omega}(u) = \int_\Omega \frac{1}{k}|D_k u|^2 - \langle Vu, u\rangle. \qquad (25)$$

Here $W_0^1(\Omega, E \otimes L^k)$ is the closure of the space of smooth sections with compact support in Ω, taken in the Hilbert space $W_{\mathrm{loc}}^1(M, E \otimes L^k)$ of sections that have L_{loc}^2 coefficients as well as their first derivatives. In other words, we consider the densely defined self adjoint operator

$$\square_k = \frac{1}{k} D_k^* D_k - V \qquad (26)$$

acting in the Hilbert space $W_0^1(\Omega, E \otimes L^k)$, i.e. with Dirichlet boundary conditions. Again, $\square_k$ acting on $W_0^1(\Omega, E \otimes L^k)$ has a discrete spectrum whenever Ω is relatively compact (and also sometimes when Ω is unbounded, according to the behavior of B and V at infinity; except otherwise stated, we will assume that we are in this case later on). Then, there is an associated "localized" heat kernel

$$K_{t,k,\Omega}(x,y) = \sum_{\nu=1}^{+\infty} e^{-\lambda_{\nu,k,\Omega} t} \psi_{\nu,k,\Omega}(x) \otimes \psi_{\nu,k,\Omega}^*(y) \qquad (27)$$

where $\psi_{\nu,k,\Omega} \in W_0^1(\Omega, E \otimes L^k)$ are the eigenfunctions and $\lambda_{\nu,k,\Omega}$ their eigenvalues.

We want to study the asymptotic eigenvalue distribution of $\square_k$ as $k \to +\infty$, and more precisely get an asymptotic formula for the corresponding heat kernel $e^{-t\square_k}$. The basic idea is to decompose the proof in three steps:

(α) Convince ourselves that the asymptotic estimates can be "localized", up to lower order error terms.
(β) Show that the local estimates can be obtained by freezing the coefficients of the operators involved at any given point.
(γ) Compute explicitly the heat kernel in the case of connections with constant curvature, assuming moreover that $\Omega \simeq \mathbb{R}^m$ with the flat Euclidean metric.

(α) In order to see that the situation can be localized, we fix a partition of unity (τ_j) relative to an arbitrarily fine finite covering (Ω_j) of $\overline{\Omega}$, such that $\sum \tau_j^2 = 1$ near $\overline{\Omega}$. We consider the continuous injection

$$I_{\Omega,\Omega_j} : W_0^1(\Omega, E \otimes L^k) \to \bigoplus_j W_0^1(\Omega \cap \Omega_j, E \otimes L^k), \qquad u \mapsto (\tau_j u)_j,$$

the inverse of which is $(u_j) \mapsto u = \sum \tau_j u_j$. As $\sum \tau_j d\tau_j = 0$ on Ω, we find

$$\sum_j Q_{k,\Omega_j}(\tau_j u) - Q_{k,\Omega}(u) = \frac{1}{k} \int_\Omega \left(\sum |d\tau_j|^2\right)|u|^2 \leq O\!\left(\frac{1}{k}\right)|u|^2. \qquad (28)$$

By the minimax principle, it follows that the eigenvalues of $\bigoplus Q_{k,\Omega_j | \operatorname{Im} I_{\Omega,\Omega_j}}$ and those of $Q_{k,\Omega}$ differ by at most $O(1/k)$ as $k \to +\infty$. This explains why a localization process is possible, at least as far as the eigenvalue distribution is concerned. For the related heat kernels on small geodesic balls, one can use the following localization principle.

Proposition 9. *Let $\Omega_\rho = B(x^0, \rho)$ be a geodesic ball of (M, g) of radius ρ where $\rho <$ injectivity radius. Then there exist constants C_1 and $\varepsilon_1 > 0$ such that for all $t \in {]}0, \min(k\varepsilon_1, k\rho^2/2m){]}$ and every $x_0 \in M$ we have*

$$\left| K_{t,k,M}(x^0, x^0) - K_{t,k,\Omega_\rho}(x^0, x^0) \right| \leq C_1 \left(\frac{k}{t}\right)^{m/2} \exp\left(-\frac{k\rho^2}{4t} + 2t \sup_{\Omega_\rho} \|V\|\right).$$

A proof of this technical result is given in Thierry Bouche's PhD thesis (cf. [17]). It relies on a use of Kato's inequality (cf. [54]), which amounts to say that we get an upper bound of $K_{t,k,M}$ in the case when the curvature is trivial; one can then use the calculations given below to get the explicit bound, see e.g. (29').

(β) Now, let $x^0 \in M$ be a given point. We choose coordinates $(x_1, \ldots, x_m)$ centered at x^0 such that $(\partial/\partial x_1, \ldots, \partial/\partial x_m)$ is orthonormal at x^0 with respect to the Riemannian metric g. By changing the orthonormal frame of L as in Case 2, we can adjust the connection form $\Gamma_L = -iA$ of L to be given by any local potential $A(x) = \sum_j A_j(x) dx_j$ such that $B = dA$, and we can therefore arrange that $A(x^0) = 0$. Similarly, we can fix a unitary frame of E such that $\Gamma_E(x^0) = 0$. Set $x^0 = 0$ for simplicity. The first term of our Laplace operator $\square_k = \frac{1}{k} D_k^* D_k - V$ is the square of the first order operator

$$k^{-1/2} D_k u(x) = k^{-1/2} \big(du(x) + k \operatorname{Id}_E \otimes \Gamma_L(x) \cdot u(x) + \operatorname{Id}_{L^k} \otimes \Gamma_E(x) \cdot u(x)\big)$$

$$= k^{-1/2} \sum_j \left(\frac{\partial u}{\partial x_j} - ik^{1/2} A_j(x) u(x)\right) dx_j + k^{-1/2} \operatorname{Id}_{L^k} \otimes \Gamma_E(x) \cdot u(x).$$

If we use a rescaling $x = k^{-1/2}\tilde{x}$ and set $\tilde{u}(\tilde{x}) = u(x) = u(k^{-1/2}\tilde{x})$, this operator takes the form

$$\tilde{D}_k \tilde{u}(\tilde{x}) = \sum_j \left(\frac{\partial \tilde{u}}{\partial \tilde{x}_j} - ik^{1/2} A_j(k^{-1/2}\tilde{x}) \tilde{u}(\tilde{x})\right) dx_j + O(k^{-1/2}|\tilde{x}|) \tilde{u}(\tilde{x}) dx.$$

As $A_j(0) = 0$, the term $k^{1/2} A_j(k^{-1/2}\tilde{x})$ converges modulo $O(k^{-1/2}|\tilde{x}|^2)$ terms to the linearized part $\tilde{A}_j(\tilde{x}) = \sum_{i,j} \frac{\partial A_j}{\partial x_i}(0) \tilde{x}_i$. Observe also that the connection form Γ_E of E only contributes for terms of the form $O(k^{-1/2}|\tilde{x}|)$ (and thus will be negligible in the end, together with the quadratic terms of A_j). Our initial operator $\square_k = \frac{1}{k} D_k^* D_k - V$ becomes

$$\tilde{\square}_k = \tilde{D}_k^* \tilde{D}_k - \tilde{V}$$

where $\tilde{V}(\tilde{x}) = V(k^{-1/2}\tilde{x})$ and where the adjoint is computed with respect to the rescaled metric $\tilde{g}(x) = \sum g_{ij}(k^{-1/2}\tilde{x}) \, d\tilde{x}_j d\tilde{x}_j$; here $\tilde{g} \to \sum (d\tilde{x}_j)^2$ as $k \to +\infty$ thanks to the assumption that $g_{ij}(0) = \delta_{ij}$. Modulo lower order terms $O(k^{-1/2}|\tilde{x}|^2)$, $\tilde{D}_k$ is given by a linear connection form

$$\tilde{A}(\tilde{x}) = \sum B_{ij} \tilde{x}_i \, d\tilde{x}_j$$

associated with the constant magnetic field $B(x^0) = \sum_{i,j} B_{ij} dx_i \wedge dx_j$ frozen at $x^0 = 0$. We can moreover choose orthonormal coordinates so that $B(x^0)$ takes the standard form

$$B(x^0) = \sum_{j=1}^{s} B_j \, dx_j \wedge dx_{j+s}$$

where $2s \leq m$ is the rank of the alternate 2-form $B(x^0)$ and B_j the curvature eigenvalues with respect to $g(x^0)$. The corresponding linearized potential is

$$\tilde{A}(\tilde{x}) = \sum_{j=1}^{s} B_j \tilde{x}_j \, d\tilde{x}_{j+s}.$$

The intuition from Physics is that the eigenfunctions represent "waves" of heat propagation of a certain typical wave length λ in the coordinates $\tilde{x}$, and of a corresponding (much shorter) wave length $\lambda k^{-1/2}$ in the original coordinates. At that scale, our space behaves as if the metrics were flat and the curvature constant.

(γ) Let us consider the operators obtained by "freezing" the coefficients at any point x^0, as explained at step (β), although we will not perform the rescaling here. More specifically, we assume that

- L has constant curvature $B = \sum_{j=1}^{s} B_j dx_j \wedge dx_{j+s}$. Then there is a local trivialization in which

$$D_L u = du - iA \wedge u, \qquad A = \sum_{j=1}^{s} B_j x_j dx_{j+s}.$$

- $\Omega \simeq \mathbb{R}^m$ and the metric g is flat: $g = \sum dx_j \otimes dx_j$.
- $E \simeq \Omega \times \mathbb{C}^r$ is a trivial (flat) Hermitian bundle.
- the Hermitian form V is constant. We choose an orthonormal frame of E in which V is diagonal, i.e.

$$\langle Vu, u \rangle = \sum_{1 \leq \lambda \leq r} V_\lambda |u_\lambda|^2.$$

In this ideal situation, the connection D_k on $E \otimes L^k$ can be written $D_k u = du - ikA \wedge u$ and the quadratic form $Q_{k,\Omega}$ is given by

$$Q_{k,\Omega}(u) = \int_{\mathbb{R}^m} \frac{1}{k} \left(\sum_{\substack{1 \leq j \leq s \\ 1 \leq \lambda \leq r}} \left(\left|\frac{\partial u_\lambda}{\partial x_j}\right|^2 + \left|\frac{\partial u_\lambda}{\partial x_{j+s}} - ikB_j x_j u_\lambda\right|^2 \right) + \sum_{\substack{j > 2s \\ 1 \leq \lambda \leq r}} \left|\frac{\partial u_\lambda}{\partial x_j}\right|^2 \right)$$
$$- \sum_{1 \leq \lambda \leq r} V_\lambda |u_\lambda|^2.$$

In this situation, $Q_{k,\Omega}$ is a direct sum of quadratic forms acting on each component u_λ and the computation of $e^{-t\square_k}$ is reduced to the following model cases (29), (30) in dimension 1 or 2:

$$Q(f) = \int_{\mathbb{R}} \left|\frac{df}{dx}\right|^2, \quad \square f = -\frac{d^2 f}{dx^2} \tag{29}$$

As is well known (and although the spectrum is not discrete in that case) the kernel of the "elementary" heat operator $e^{-t\square}$ is given by

$$K_{t,\mathbb{R}}(x,y) = \frac{1}{\sqrt{4\pi t}} e^{-(x-y)^2/4t}, \tag{29'}$$

as follows from solving equation (16). The second model case is:

$$Q(f) = \int_{\mathbb{R}^2} \left|\frac{df}{dx_1}\right|^2 + \left|\frac{df}{dx_2} - iax_1 f\right|^2. \tag{30}$$

A partial Fourier transform $\hat{f}(x_1, \xi_2) = \frac{1}{\sqrt{2\pi}} \int_{\mathbb{R}} f(x_1, x_2) e^{-ix_2 \xi_2} dx_2$ gives

$$Q(f) = \int_{\mathbb{R}^2} \left|\frac{d\hat{f}}{dx_1}(x_1, \xi_2)\right|^2 + a^2 \left(x_1 - \frac{\xi_2}{a}\right)^2 |\hat{f}(x_1, \xi_2)|^2$$

and the change of variables $x_1' = x_1 - \xi_2/a$, $x_2' = \xi_2$ leads (after dropping the second variable x_2') to the so called "harmonic oscillator" energy functional

$$q(g) = \int_{\mathbb{R}} \left|\frac{dg}{dx}\right|^2 + a^2 x^2 |g|^2, \quad \square = -\frac{d^2}{dx^2} + a^2 x^2. \tag{31}$$

The heat kernel of this operator is given by *Mehler's formula*:

$$k_{t,\mathbb{R}}(x,y) = \sqrt{\frac{a}{2\pi \sinh 2at}} \exp\left(-\frac{a}{2}(\coth 2at)(x-y)^2 - a(\tanh at)xy\right), \tag{31'}$$

which actually reduces to (29′) when $a \to 0$. One way of obtaining this relation is to observe that the unitary eigenfunctions of $\Box$ are

$$\left(2^p p! \sqrt{\frac{\pi}{a}}\right)^{-1/2} \Phi_p(\sqrt{a}x), \quad p = 0, 1, 2, \ldots,$$

with associated eigenvalues $(2p + 1)a$, where (Φ_p) is the sequence of functions associated with Hermite polynomials:

$$\Phi_p(x) = e^{x^2/2} \frac{d^p}{dx^p}(e^{-x^2}).$$

In fact, for $a = 1$, easy calculations bearing on derivatives of $e^{x^2/2}$ show that

$$\left(-\frac{d^2}{dx^2} + x^2\right)\Phi_p(x)$$

$$= -e^{x^2/2}\frac{d^{p+2}}{dx^{p+2}}(e^{-x^2}) - 2x\, e^{x^2/2}\frac{d^{p+1}}{dx^{p+1}}(e^{-x^2}) - e^{x^2/2}\frac{d^p}{dx^p}(e^{-x^2}).$$

We can now replace the first term by $e^{x^2/2}\frac{d^{p+1}}{dx^{p+1}}(2x \cdot e^{-x^2})$ and use the Leibniz formula for the differentiation of the product to see that $\Box \Phi_p(x) = (2p+1)\Phi_p(x)$. Therefore

$$k_{t,\mathbb{R}}(x, y) = \sqrt{\frac{a}{\pi}} e^{a(x^2+y^2)/2} \sum_{p=0}^{+\infty} \frac{e^{-(2p+1)at}}{2^p p! a^p} \frac{d^p}{dx^p}(e^{-ax^2}) \frac{d^p}{dy^p}(e^{-ay^2}).$$

The above summation $\Sigma(x, y) = \sum_{p=0}^{+\infty} \ldots$ can be computed via its Fourier transform

$$\hat{\Sigma}(\xi, \eta) = \frac{1}{2a} e^{-at} \sum_{p=0}^{+\infty} \frac{1}{p!}\left(\frac{e^{-2at}}{2a}\right)^p (i\xi)^p (i\eta)^p e^{-\xi^2/4a} e^{-\eta^2/4a}$$

$$= \frac{1}{2a} e^{-at} \exp\left(-\frac{1}{4a}(\xi^2 + \eta^2 + 2e^{-2at}\xi\eta)\right),$$

thus

$$\Sigma(x, y) = \frac{e^{-at}}{\sqrt{1 - e^{-4at}}} \exp\left(-\frac{a}{1 - e^{-4at}}(x^2 + y^2 - 2e^{-2at}xy)\right).$$

and Mehler's formula (31′) follows. Through our change of variables, the heat operator of Q is given by

$$\widehat{K_{t,\mathbb{R}^2} f}(x_1, \xi_2) = \int_{\mathbb{R}} k_{t,\mathbb{R}}\left(x_1 - \frac{\xi_2}{a}, y_1 - \frac{\xi_2}{a}\right) \hat{f}(y_1, \xi_2) dy_1.$$

By an inverse partial Fourier transform left to the reader, we obtain the desired heat kernel expression

$$K_{t,\mathbb{R}^2}(x_1, x_2; y_1, y_2) = \frac{a}{4\pi \sinh at} \exp\left(-\frac{a}{4}(\coth at)\left((x_1 - y_1)^2 + (x_2 - y_2)^2\right)\right)$$

$$\times \exp\left(\frac{i}{2}a(x_1 + y_1)(x_2 - y_2)\right). \tag{30′}$$

The heat kernel associated with a sum of (pairwise commuting) operators $\Box_1, \ldots, \Box_m$ acting on disjoint sets of variables is the product of the corresponding heat kernels $e^{-t\Box_j}$. Let $K^\lambda_{t,k,\Omega}$ be the heat kernel of the component of $Q_{k,\Omega}$ acting on each single entry u_λ. The factor in the heat kernel corresponding to each pair of variables (x_j, x_{j+s}), $1 \leq j \leq s$, is obtained by substituting kB_j to a and t/k to t (the latter rescaling comes from the initial factor $\frac{1}{k}$ in the expression of $Q_{k,\Omega}$). For the other coordinates $j > 2s$ where B has no coefficients, the kernel falls back to the "elementary" heat kernel (29′). Finally, the constant term $-V_\lambda |u_\lambda|^2$ contributes to multiplying the heat kernel by e^{tV_λ}. Therefore we get for the global heat kernel on $\Omega = \mathbb{R}^n$ the explicit formula

$$K^\lambda_{t,k,\mathbb{R}^n}(x, y) = \prod_{j=1}^{s} \frac{kB_j}{4\pi \sinh B_j t} \exp\left(-\frac{kB_j}{4}(\coth B_j t)\left((x_{2j-1} - y_{2j-1})^2 + (x_{2j} - y_{2j})^2\right)\right.$$

$$\left. + \frac{i}{2}kB_j(x_{2j-1} + y_{2j-1})(x_{2j} - y_{2j})\right)$$

$$\times e^{tV_\lambda} \times \frac{1}{(4\pi t/k)^{m-2s/2}} \exp\left(-k \sum_{j>2s}(x_j - y_j)^2/4t\right). \tag{32}$$

On the diagonal of $\mathbb{R}^n \times \mathbb{R}^n$, the global heat kernel $K_{t,k,\mathbb{R}^n}$ is thus given by the rather simple $(\text{Herm}(E) \otimes \text{Id}_{L^k})$-valued tensor depending only on B, V and t/k:

$$K_{t,k,\mathbb{R}^n}(x, x) = \left(\frac{k}{4\pi t}\right)^{m/2} e^{tV} \prod_{j=1}^{s} \frac{B_j t}{\sinh B_j t}. \tag{33}$$

Theorem 10. *Consider the general (variable coefficient) case. For $\delta > 0$ small, the heat kernel of $\Box_k$ over M admits an asymptotic estimate*

$$K_{t,k,M}(x, x) = \left(\frac{k}{4\pi t}\right)^{m/2} e^{tV(x)} \prod_{j=1}^{s} \frac{B_j(x) t}{\sinh B_j(x) t} \left(1 + O(k^{-1/2+\delta})\right)$$

as $k \to +\infty$, where $O(k^{-1/2+\delta})$ is uniform with respect to $x \in M$ and t in a bounded interval $]0, T] \subset]0, +\infty[$ (moreover, for every open set $\Omega \subset M$, a similar estimate is valid for $K_{t,k,\Omega}$ on relatively compact subsets of Ω).

Proof. Notice first that $(t, x) \mapsto \prod_{j=1}^{s} \frac{B_j(x)t}{\sinh B_j(x)t}$ extends as a smooth positive function on $[0, +\infty[\times M$, equal to 1 when $t = 0$: this is in fact the inverse of the square root of the determinant of the positive definite symmetric matrix

$$\frac{\sin(tb(x))}{tb(x)} = \sum_{p=0}^{+\infty} \frac{t^{2p}(-b(x)^2)^p}{(2p+1)!} \geq \mathrm{Id},$$

where $b(x)$ is the antisymmetric endomorphism of T_M associated with the alternate 2-form $B(x)$ and $-b(x)^2 = b(x)^\dagger b(x) \geq 0$.

The only thing one has still to get convinced of is that the kernel of $e^{-t\Box_k} - e^{-t\Box_k^0}$ is $(k/t)^{m/2} O(k^{-1/2+\delta})$ uniformly along the diagonal at any point $(x^0, x^0) \in M \times M$, where $\Box_k^0$ is the operator $\Box_k$ "freezed" at x^0. We can do this in a canonical way by using normal coordinates from the Riemannian exponential mapping

$$\exp_{x^0} : \mathbb{R}^m \simeq T_{M,x^0} \to M,$$

and trivializations of E and L produced by parallel transport along geodesics from x^0 to any point $x \in B(x^0, \rho_0)$, where $\rho_0 = $ injectivity radius of M. In this way, we actually get automatically that $\Gamma_L(x^0) = \Gamma_E(x^0) = 0$. When Supp $u \subset \Omega_\rho := B(x^0, \rho)$, a Taylor expansion yields $D_k u - D_k^0 u = O(|x| + k|x|^2) \cdot u$ and we get the estimates

$$Q_{k,\Omega_\rho}(u) - Q_{k,\Omega_\rho}^0(u) = \int_M \frac{1}{k}(|D_k u|^2 - |D_k^0 u|^2) - \langle (V - V^0)u, u \rangle$$

$$= O\left(\int_M \frac{1}{k}((\rho + k\rho^2)|D_k^0 u||u| + (\rho + k\rho^2)^2 |u|^2) + \rho|u|^2 \right)$$

$$= O\left(\int_M \frac{\varepsilon}{k}|D_k^0 u|^2 + \left(\frac{(\rho + k\rho^2)^2}{k\varepsilon} + \rho \right)|u|^2 \right),$$

$$= O\left(\varepsilon Q_{k,\Omega_\rho}^0(u) + \left(\frac{(\rho + k\rho^2)^2}{k\varepsilon} + \rho + \varepsilon \right)|u|^2 \right)$$

whenever $\varepsilon < 1$, hence there is a constant $C_{\rho,k,\varepsilon} = O\left(\frac{(\rho + k\rho^2)^2}{k\varepsilon} + \rho + \varepsilon \right)$ such that

$$(1-\varepsilon)Q_{k,\Omega_\rho}^0(u) - C_{\rho,k,\varepsilon}|u|^2 \leq Q_{k,\Omega_\rho}(u) \leq (1+\varepsilon)Q_{k,\Omega_\rho}^0(u) + C_{\rho,k,\varepsilon}|u|^2.$$

From this, we conclude that $e^{-t\square_k}$ is squeezed (as a positive bounded self-adjoint operator) between $e^{-C_{\rho,k,\varepsilon}t}e^{-t(1+\varepsilon)\square_k^0}$ and $e^{C_{\rho,k,\varepsilon}t}e^{-t(1-\varepsilon)\square_k^0}$. By definition of the heat kernel we have

$$K_{t,k,\Omega_\rho}(x^0, x^0) = \lim_{\nu \to +\infty} \int_{\Omega_\rho \times \Omega_\rho} K_{t,k,\Omega_\rho}(x, y) u_\nu(x) \overline{u_\nu(y)} \, d\sigma(x) \, d\sigma(y)$$

$$= \lim_{\nu \to +\infty} \langle\!\langle e^{-t\square_k} u_\nu, u_\nu \rangle\!\rangle$$

when $u_\nu \xrightarrow{L^1} \delta_{x^0}$ (Dirac measure), thus

$$e^{-C_{\rho,k,\varepsilon}T} K^0_{(1+\varepsilon)t,k,\Omega_\rho}(x^0, x^0) - K^0_{t,k,\Omega_\rho}(x^0, x^0)$$
$$\leq K_{t,k,\Omega_\rho}(x^0, x^0) - K^0_{t,k,\Omega_\rho}(x^0, x^0)$$
$$\leq e^{C_{\rho,k,\varepsilon}T} K^0_{(1-\varepsilon)t,k,\Omega_\rho}(x^0, x^0) - K^0_{t,k,\Omega_\rho}(x^0, x^0).$$

We take here $\rho = \varepsilon = k^{-1/2+\delta}$, so that $C_{\rho,k,\varepsilon} = O(k^{-1/2+\delta})$. The expected uniform bounds are then obtained by an application of Proposition 9, where the choice $\rho = k^{-1/2+\delta} \gg k^{-1/2}$ ensures that the relative errors

$$K_{t,k,M} - K_{t,k,\Omega_\rho} \quad \text{and} \quad K^0_{t,k,\mathbb{R}^m} - K^0_{t,k,\Omega_\rho}$$

are very small, namely of the order of magnitude $O(\exp(-k^\delta/4T))$. □

As a consequence, we obtain the following estimate for the eigenvalues:

Corollary 11. *The eigenvalues $\lambda_{\nu,k,\Omega}$ of $Q_{k,\Omega}$ satisfy for every $t > 0$ the estimate*

$$\sum_{\nu=1}^{+\infty} e^{-t\lambda_{\nu,k,\Omega}} = (1 + O(k^{-1/2})) \left(\frac{k}{4\pi t}\right)^{m/2} \int_\Omega \mathrm{tr}(e^{tV(x)}) \prod_{j=1}^s \frac{B_j(x)t}{\sinh B_j(x)t} \, d\sigma(x).$$

This result can be also interpreted in terms of the counting function

$$N_{k,\Omega}(\lambda) = \#\{\nu \,;\, \lambda_{\nu,k,\Omega} \leq \lambda\}$$

and of the spectral density measure (a sum of Dirac measures on the real line)

$$\mu_{k,\Omega} = k^{-m/2} \frac{d}{d\lambda} N_{k,\Omega}(\lambda).$$

Notice that the measures $\mu_{k,\Omega}$ are all supported in the fixed interval $[-\nu_0, +\infty[$, where ν_0 is an upper bound for the eigenvalues of $V(x)$, $x \in M$. In these notations, Corollary 11 can be restated:

$$\lim_{k\to+\infty}\int_{-\infty}^{+\infty}e^{-t\lambda}d\mu_{k,\Omega}(\lambda) = \frac{1}{(4\pi t)^{m/2}}\int_\Omega \text{tr}(e^{tV(x)})\prod_{j=1}^s \frac{B_j(x)t}{\sinh B_j(x)t}d\sigma(x).$$

We thus see that the sequence of measures $\mu_{k,\Omega}$ converges weakly to a measure μ_Ω whose Laplace transform is given by the right hand side. Inverting the formula, one obtains:

Corollary 12. *For almost all $\lambda \in \mathbb{R}$*

$$\lim_{k\to+\infty} k^{-m/2}N_{k,\Omega}(\lambda) = \mu_\Omega(]-\infty,\lambda]) = \int_\Omega \sum_{j=1}^r \nu_{B(x)}(V_j(x)+\lambda)d\sigma(x) \quad (34)$$

where $\nu_{B(x)}(\lambda)$ is the function on $M\times\mathbb{R}$ defined by

$$\nu_B(\lambda) = \frac{2^{s-m}\pi^{-m/2}}{\Gamma(\frac{m}{2}-s+1)} B_1\cdots B_s \sum_{(p_1,\ldots,p_s)\in\mathbb{N}^s} \left[\lambda - \sum(2p_j+1)B_j\right]_+^{\frac{m}{2}-s}. \quad (35)$$

Proof. We leave as an exercise to the reader to check that the Laplace transform

$$\int_{-\infty}^{+\infty} e^{-t\lambda}d\nu_B(v+\lambda) = e^{tv}\int_{-\infty}^{+\infty}e^{-t\lambda}d\nu_B(\lambda)$$

is actually equal to

$$\frac{e^{tv}}{(4\pi t)^{m/2}}\prod_{j=1}^s \frac{B_j(x)t}{\sinh B_j(x)t}.\qquad\square$$

1.2.3 Proof of the Holomorphic Morse Inequalities

Let X be a compact complex manifold, L and E holomorphic Hermitian vector bundles of rank 1 and r over X. If X is endowed with a Hermitian metric ω, Hodge theory shows that the Dolbeault cohomology group $H^q(X, E\otimes L^k)$ can be identified with the space of harmonic $(0,q)$-forms with respect to the Laplace–Beltrami operator $\Delta_k'' = \overline{\partial}_k\overline{\partial}_k^* + \overline{\partial}_k^*\overline{\partial}_k$ acting on $E\otimes L^k$. We thus have to estimate the zero-eigenspace of Δ_k''.

In order to apply Corollary 12, we first have to compute Δ_k'' in terms of the Hermitian connection ∇_k on $E\otimes L^k\otimes \Lambda^{0,q}T_X^*$ deduced from the Chern connections of L, E, T_X. What plays now the role of E is the (non holomorphic) bundle $E\otimes \Lambda^{0,q}T_X^*$.

The relation between Δ_k'' and ∇_k is most easily obtained by means of the Bochner–Kodaira–Nakano identity. In order to simplify the exposition, we assume here that the metric ω on X is *Kähler*. For any Hermitian holomorphic line bundle

G on X, the operators Δ' and Δ'' associated with the Chern connection $D = D_G$ are related by the B-K-N identity (cf. [2, 10, 61, 76])

$$\Delta'' = \Delta' + [i\theta_G, \Lambda] \tag{36}$$

where $\theta_G = D_G^2 \in C^\infty(X, \Lambda^{1,1} T_X^* \otimes \operatorname{Hom}(G, G))$ is the curvature tensor and $\Lambda = L^*$ is the adjoint of the Lefschetz operator $Lu = \omega \wedge u$.

The Leibniz rule implies $\theta_{E \otimes L^k} = k\theta_L \otimes \operatorname{Id}_E + \theta_E \otimes \operatorname{Id}_{L^k}$ (omitting the Hermitian metrics for simplicity of notation), thus

$$\Delta_k'' = \Delta_k' + k[i\theta_L, \Lambda] + [i\theta_E, \Lambda].$$

At a given point $z^0 \in X$, we can find a coordinate system $(z_1, \ldots, z_n)$ such that $(\partial/\partial z_j)$ is an orthonormal basis of T_X diagonalizing $i\theta_L(z^0)$, in such a way that

$$\omega(z^0) = \frac{i}{2} \sum_{1 \le j \le n} dz_j \wedge d\bar{z}_j, \qquad i\theta_L(z^0) = \frac{i}{2} \sum_{1 \le j \le n} \alpha_j dz_j \wedge d\bar{z}_j$$

where $\alpha_1, \ldots, \alpha_n$ are the curvature eigenvalues of $i\theta_L(z^0)$. A standard formula gives the expression of the curvature term $[i\theta_L, \Lambda]u$ for any (p, q)-form u. In fact, for $u = \sum u_{I,J,\lambda} dz_I \wedge d\bar{z}_J \otimes e_\lambda$, we have

$$\langle [i\theta_L, \Lambda]u, u \rangle = \sum_{I,J,\lambda} (\alpha_J - \alpha_{\complement I}) |u_{I,J,\lambda}|^2$$

where $\alpha_J = \sum_{j \in J} \alpha_j$. In the case of a $(0, q)$-form $u = \sum u_{J,\lambda} d\bar{z}_J \otimes e_\lambda$ we simply have $\Delta_k' u = D_k'^* D_k' u = \nabla_k'^* \nabla_k' u$ and

$$\Delta_k'' = \nabla_k'^* \nabla_k' - kV' + [i\theta_E, \Lambda], \tag{37'}$$

$$\langle V'u, u \rangle = \sum_{J,\lambda} \alpha_{\complement J} |u_{J,\lambda}|^2 \quad (\text{here } I = \emptyset).$$

This is not yet what was needed, since only the $(1, 0)$ part ∇_k' appears. To get the $(0, 1)$ component, we consider u as a (n, q) form with values in $E \otimes L^k \otimes \Lambda^n T_X$. We then get $\Delta_k' u = D_k' D_k'^* u$ where

$$D_k'^* u = -\sum \partial u_{I,J,\lambda}/\partial \bar{z}_j \, dz_1 \wedge \cdots \widehat{dz_j} \cdots \wedge dz_n \wedge d\bar{z}_J \otimes e_\lambda$$

in normal coordinates. Thus $\Delta_k' u = \nabla_k''^* \nabla_k'' u$ and

$$\Delta_k'' = \nabla_k''^* \nabla_k'' + kV'' + [i\theta_{E \otimes \Lambda^n T_X}, \Lambda], \tag{37''}$$

$$\langle V''u, u \rangle = \sum_{J,\lambda} \alpha_J |u_{J,\lambda}|^2 \quad (\text{here } I = \{1, \ldots, n\}).$$

If the metric ω is non Kähler, we get additional torsion terms, but these terms are independent of k. A combination of (37') and (37'') yields

$$\frac{2}{k}\Delta_k'' = \frac{1}{k}\nabla_k^*\nabla_k - V + \frac{1}{k}W \tag{38}$$

where W is a Hermitian form independent of k and

$$\langle Vu, u\rangle = \sum_{J,\lambda}(\alpha_{{\complement}J} - \alpha_J)|u_{J,\lambda}|^2.$$

Now apply Theorem 10 and observe that W does not give any significant contribution to the heat kernel as $k \to +\infty$. We write here $z_j = x_j + iy_j$ and the "magnetic field"

$$B = i\theta_L = \sum_{1 \le j \le n} \alpha_j dx_j \wedge dy_j.$$

The curvature eigenvalues are given by $B_j = |\alpha_j|$. We denote $s = s(x)$ the rank of $B(x)$ and order the eigenvalues so that

$$|\alpha_1| \ge \cdots \ge |\alpha_s| > 0 = \alpha_{s+1} = \cdots = \alpha_n.$$

The eigenvalues of V acting on $E \otimes \Lambda^n T_X^*$ are the coefficients $\alpha_{{\complement}J} - \alpha_J$, counted with multiplicity r. Therefore

Theorem 13. *The heat kernel associated with $e^{-\frac{2t}{k}\Delta_k''}$ in bidegree $(0, q)$ satisfies*

$$K_t^k(x,x) \sim k^n \frac{r \sum_{|J|=q} e^{t(\alpha_{{\complement}J}(x) - \alpha_J(x))}}{(4\pi)^n t^{n-s}} \prod_{j=1}^s \frac{|\alpha_j(x)|}{\sinh|\alpha_j(x)|t}$$

as $k \to +\infty$. In particular, if $\lambda_1^{k,q} \le \lambda_2^{k,q} \le \cdots$ are the eigenvalues of $\frac{1}{k}\Delta_k''$ in bidegree $(0, q)$, we have

$$\sum_{\nu=1}^{+\infty} e^{-2t\lambda_\nu^{k,q}} \sim rk^n \sum_{|J|=q} \int_X \frac{e^{t(\alpha_{{\complement}J}(x) - \alpha_J(x))}}{(4\pi)^n t^{n-s}} \prod_{j=1}^s \frac{|\alpha_j(x)|}{\sinh|\alpha_j(x)|t}$$

for every $t > 0$.

At this point, the main idea is to use the eigenspaces to construct a finite dimensional subcomplex of the Dolbeault complex possessing the same cohomology groups. This was already the basic idea in Witten's analytic proof of the standard Morse inequalities [102]. We denote by

$$\mathcal{H}_\lambda^{k,q}, \quad \text{resp.} \quad \mathcal{H}_{\le \lambda}^{k,q}$$

the λ-eigenspace of $\frac{1}{k}\Delta''_k$ acting on $C^\infty(X, \Lambda^{0,q}T^*_X \otimes E \otimes L^k)$, resp. the direct sum of eigenspaces corresponding to all eigenvalues $\leq \lambda$. As $\bar\partial_k$ and Δ''_k commute, we see that $\bar\partial(\mathcal{H}^{k,q}_\lambda) \subset \mathcal{H}^{k,q+1}_\lambda$, thus $\mathcal{H}^{k,\bullet}_\lambda$ and $\mathcal{H}^{k,\bullet}_{\leq\lambda}$ are finite dimensional subcomplexes of the Dolbeault complex

$$\bar\partial \,:\, C^\infty(X, \Lambda^{0,\bullet}T^*_X E \otimes L^k).$$

Since $\bar\partial_k\bar\partial^*_k + \bar\partial^*_k\bar\partial_k = \Delta''_k = k\lambda\,\mathrm{Id}$ on $\mathcal{H}^{k,\bullet}_\lambda$, we see that $\mathcal{H}^{k,\bullet}_\lambda$ has trivial cohomology for $\lambda \neq 0$. Since $\mathcal{H}^{k,\bullet}_0$ is the space of harmonic forms, we see that $\mathcal{H}^{k,\bullet}_{\leq\lambda}$ has the same cohomology as the Dolbeault complex for $\lambda > 0$. We will call this complex the Witten $\bar\partial$-complex. We need an elementary lemma of linear algebra.

Lemma 14. Set $h^q_k = \dim H^q(X, E \otimes L^k)$. Then for every $t > 0$

$$h^q_k - h^{q-1}_k + \cdots + (-1)^q h^0_k \leq \sum_{\ell=0}^{q}(-1)^{q-\ell}\sum_{j=1}^{+\infty} e^{-t\lambda^{k,\ell}_j}.$$

Proof. The left hand side is the contribution of the 0 eigenvalues in the right hand side. All we have to check is that the contribution of the other eigenvalues is ≥ 0. The contribution of the eigenvalues such that $\lambda^{k,\ell}_j = \lambda > 0$ is

$$e^{-t\lambda}\sum_{\ell=0}^{q}(-1)^{q-\ell}\dim\mathcal{H}^{k,\ell}_\lambda.$$

As $\mathcal{H}^{k,\bullet}_\lambda$ is exact, one easily sees that the last sum is equal to the dimension of $\bar\partial\mathcal{H}^{k,q}_\lambda \subset \mathcal{H}^{k,q+1}_\lambda$, hence ≥ 0. □

Combining Theorem 13 with Lemma 14, we get

$$h^q_k - h^{q-1}_k + \cdots + (-1)^q h^0_k$$

$$\leq o(k^n) + rk^n \sum_{\ell=0}^{q}(-1)^{q-\ell}\sum_{|J|=\ell}\int_X \frac{\prod_{j\leq s}|\alpha_j|\cdot e^{t(\alpha_{\complement J}-\alpha_J-\sum|\alpha_j|)}}{2^{2n-s}\pi^n t^{n-s}\prod_{j\leq s}(1-e^{-2t|\alpha_j|})}.$$

This inequality is valid for any $t > 0$, so we can let t tend to $+\infty$. It is clear that $\alpha_{\complement J} - \alpha_J - \sum|\alpha_j|$ is always ≤ 0, thus the integrand tends to 0 at every point where $s < n$. When $s = n$, we have $\alpha_{\complement J}(x) - \alpha_J x) - \sum|\alpha_j(x)| = 0$ if and only if $\alpha_j(x) > 0$ for every $j \in \complement J$ and $\alpha_j(x) < 0$ for every $j \in J$. This implies $x \in X(L, h, \ell)$; in this case there is only one multi-index J satisfying the above conditions and the limit is

$$(2\pi)^{-n}|\alpha_1\cdots\alpha_n| = (2\pi)^{-n}|(i\theta_{L,h})^n| = |\Theta^n_{L,h}|,$$

as $\Theta_{L,h} = \frac{i}{2\pi}\theta_{L,h}$ by definition. By the monotone convergence theorem, our sum of integrals converges to

$$\sum_{\ell=0}^{q}(-1)^{q-\ell}\int_{X(L,h,\ell)}(2\pi)^{-n}|\alpha_1\cdots\alpha_n|d\sigma = \frac{1}{n!}\int_{X(L,h,\leq q)}(-1)^q\Theta_{L,h}^n.$$

The Main Theorem 8 follows. □

1.3 Applications to Algebraic Geometry

1.3.1 Solution of the Grauert–Riemenschneider Conjecture

Let L be a holomorphic line bundle over a compact connected complex manifold X of dimension n and $V_k = H^0(X, L^k)$. Denote by $Z(V_k)$ the set of common zeroes of all sections in V_k, and fix a basis $(\sigma_0, \ldots, \sigma_N)$ of V_k. There is a canonical holomorphic map

$$\Phi_{kL} : X \smallsetminus Z(V_k) \longrightarrow \mathbb{P}(V_k), \qquad x \mapsto [\sigma_0(x) : \ldots : \sigma_N(x)] \tag{39}$$

sending a point $x \in X \smallsetminus Z(V_k)$ to the hyperplane $H \subset V_k$ of sections $\sigma = \sum \lambda_j \sigma_j \in V_k$ such that $\sigma(x) = \sum \lambda_j \sigma_j(x) = 0$; it is therefore given by $x \mapsto [\sigma_0(x) : \ldots : \sigma_N(x)]$ in projective coordinates on $\mathbb{P}(V_k) \simeq \mathbb{P}^N$. The pull-back $\Phi_{kL}^*\mathcal{O}(d)$ can be identified with the restriction of L^{kd} to $X \smallsetminus Z(V_k)$; indeed, to any homogeneous polynomial $P(w_0, \ldots, w_N) \in H^0(\mathbb{P}^N, \mathcal{O}(d))$ of degree d, one can associate a section

$$s = P(\sigma_0, \ldots, \sigma_N) \in H^0(X, L^{kd}). \tag{40}$$

When L possesses a smooth Hermitian metric h with $\Theta_{L,h} > 0$, one can construct many sections of high tensor powers L^k (e.g. by Hörmander's L^2 estimates [56], [4] for $\bar{\partial}$). For $k \geq k_0$ large enough, the "base locus" $Z(V_k)$ is empty, the sections in V_k separate any two points of X and generate all 1-jets at any point. Then Φ_{kL} gives an embedding of X in some projective space $\mathbb{P}^N$, for $N = N(k)$ and $k \geq k_0$. In this way, the theory of L^2 estimates implies the *Kodaira embedding theorem*: a compact complex manifold X is projective algebraic if and only if X possesses a Hermitian line bundle (L, h) with C^∞ positive curvature.

The Grauert–Riemenschneider conjecture [50] is an attempt to characterize the more general class of Moishezon varieties in terms of semi-positive line bundles. Let us first recall a few definitions. The algebraic dimension $a(X)$ is the transcendence degree of the field $\mathcal{M}(X)$ of meromorphic functions on X. A well-known theorem of Siegel [83] asserts that $0 \leq a(X) \leq n$ (see Corollary 17 below). A compact manifold or variety X is said to be *Moishezon* if $a(X) = n$.

By definition, the *Kodaira dimension* $\kappa(L)$ is the supremum of the dimension of the images $Y_k = \Phi_{kL}(X \smallsetminus Z(V_k)) \subset \mathbb{P}(V_k^*)$ for all integers $k > 0$ [one defines $\kappa(L) = -\infty$ when $V_k = 0$ for all k, in which case we always have $Y_k = \emptyset$]. Since the field of meromorphic functions on X obtained by restriction of rational functions of $\mathbb{P}(V_k^*)$ to Y_k has transcendence degree at least equal to $\dim Y_k$, we infer that

$$-\infty \leq \kappa(L) = \sup \dim Y_k \leq a(X) \leq n. \tag{41}$$

Definition 15. The line bundle $L \to X$ is said to be *big* if $\kappa(L)$ is maximal, i.e. $\kappa(L) = n = \dim X$.

The following standard lemma is needed (cf. [80, 83]).

Lemma 16 (Serre–Siegel). *For every line bundle $L \to X$, there exist constants $C \geq c > 0$ and $k_0 \in \mathbb{N}^*$ such that*

$$\dim H^0(X, L^k) \leq C\, k^{\kappa(L)} \qquad \text{for all } k \geq 1,$$
$$\dim H^0(X, L^k) \geq c\, k^{\kappa(L)} \qquad \text{for all } k \geq 1 \text{ multiple of } k_0.$$

Proof. The lower bound is obtained by taking k_0 such that $p := \dim Y_{k_0} = \kappa(L)$. Then, by the rank theorem, there exists a point $x_0 \in X \smallsetminus Z(V_{k_0})$ and a basis $(\sigma_0, \ldots, \sigma_N)$ of $H^0(X, L^{k_0})$ such that $\sigma_0(x_0) \neq 0$ and

$$\big(d(\sigma_1/\sigma_0) \wedge \ldots \wedge d(\sigma_p/\sigma_0)\big)(x_0) \neq 0.$$

Then by taking $s = P(\sigma_0, \ldots, \sigma_p, 0, \ldots, 0)$ in (40), we obtain an injection of the space of homogeneous polynomials of degree d in $p+1$ variables into $H^0(X, L^{k_0 d})$, whence

$$h^0(X, L^{k_0 d}) \geq \binom{d+p}{p} \geq d^p/p!.$$

The proof of the upper bound proceeds as follows: select a Hermitian metric h, on L and a finite family of coordinate balls $B_j = B(z_j, r_j)$ such that $B'_j = B(z_j, r_j/2)$ cover X, and $L_{|B_j}$ is trivial for each j. By moving a little bit the points z_j, we may assume that Φ_{kL} has maximal rank at all points z_j for all k (the bad set is at most a countable union of analytic sets, so it is nowhere dense). If L^k has many sections, one can solve a linear system in many unknowns to get a section s vanishing at a high order m at all centers z_j. Then the Schwarz lemma gives

$$\|s\|_{h,\infty} = \sup_j \|s\|_{h,B'_j} \leq 2^{-m} C(h)^k \sup_j \|s\|_{h,B_j} \leq 2^{-m} C(h)^k \|s\|_{h,\infty}$$

where $C(h)$ is a bound for the oscillation of the metric h on B_j, which we may assume to be finite after possibly shrinking B_j. Thus $m \leq k \log C(h)/\log 2$ if $s \neq 0$. Since the sections of L^k are constant along the fibers of Φ_{kL}, only $m^{\dim Y_k} \#\{z_j\}$ equations transversally to the fibers are needed to make s vanish at order m. Therefore we can choose $m \approx (h^0(X, L^k)/\#\{z_j\})^{1/\dim Y_k}$ and still get a non zero section, so that

$$h^0(X, L^k) \approx \#\{z_j\} \cdot m^{\dim Y_k} \leq C k^{\kappa(L)}.$$
□

Corollary 17 (Siegel). *For every compact complex manifold X*

$$a(X) := \operatorname{tr deg}_{\mathbb{C}} \mathcal{M}(X) \leq n.$$

Proof. Fix s algebraically independent elements $f_1, \ldots, f_s \in \mathcal{M}(X)$ and let D be the sup of the pole divisors of the f_j's. To every polynomial $P(f_1, \ldots, f_s)$ of degree $\leq k$ corresponds injectively a section $\sigma_P = P(f_1, \ldots, f_s) \in H^0(X, \mathcal{O}(kD))$. A dimension count implies

$$\frac{k^s}{s!} \leq \binom{k+s}{s} \leq C k^{\kappa(\mathcal{O}(D))} \leq C k^n$$

by Lemma 16. Therefore $s \leq n$. □

Now, the Grauert–Riemenschneider conjecture [50] can be stated as follows.

Grauert-Riemenschneider conjecture 18. *A compact complex variety Y is Moishezon if and only if there is a proper non singular modification $X \to Y$ and a Hermitian line bundle (L, h) over X such that the curvature form $\Theta_{L,h}$ is > 0 on a dense open subset of X.*

Proof. When Y is Moishezon, it is well known that there exists a projective algebraic modification X; therefore we can even take L to be ample and then there exists h such that $\Theta_{L,h} > 0$ everywhere on X.

The converse statement was proved by Siu in [85, 86], assuming only $\Theta_{L,h} \geq 0$ everywhere and $\Theta_{L,h} > 0$ in at least one point. Morse inequalities provide in fact a much stronger criterion, requiring only the positivity of some curvature integral:

Theorem 19. *If a Hermitian line bundle (L, h) on X satisfies the integral condition*

$$\int_{X(L,h,\leq 1)} (\Theta_{L,h})^n > 0,$$

then $\kappa(L) = n$, in particular X is Moishezon.

In fact, the lower bound (24) applied with $E = \mathcal{O}_X$ implies immediately that $h^0(X, L^k) \geq c k^n$, hence $\kappa(L) = n$. Now, if X is a modification of Y, we have $\mathcal{M}(Y) \simeq \mathcal{M}(X)$, so $a(X) = a(Y)$, and Y has to be Moishezon. □

1.3.2 Cohomology Estimates for nef Line Bundles

On a projective algebraic manifold X, a line bundle L is said to be *nef* if $L \cdot C \geq 0$ for every algebraic curve $C \subset X$. If ω is a given Kähler or Hermitian $(1, 1)$-form on X, it can be shown (cf. [26]) that L is nef if and only if for every $\varepsilon > 0$ there exists a smooth Hermitian metric h_ε such that $\Theta_{L,h_\varepsilon} \geq -\varepsilon\omega$ on X; in fact, the latter property clearly implies

$$L \cdot C = \int_C \Theta_{L,h_\varepsilon} \geq -\varepsilon \int_C \omega \implies L \cdot C \geq 0$$

for every curve C. Conversely, if $L \cdot C \geq 0$ for every curve C, the well-known Kleiman criterion (cf. [53]) implies that $kL + A$ is ample for every ample divisor A. Hence there exists a smooth Hermitian metric h_k on L such that

$$\Theta_{kL+A} = k\Theta_{L,h_k} + \Theta_{A,h_A} > 0 \implies \Theta_{L,h_k} \geq -\frac{1}{k}\omega, \text{ where } \omega = \Theta_{A,h_A} > 0.$$

Therefore, one can introduce the following *definition of nefness* on an arbitrary compact complex manifold.

Definition 20. Let X be a compact complex manifold and ω a given smooth positive $(1, 1)$-form on X. A line bundle $L \to X$ is said to be *nef* if for every $\varepsilon > 0$ there exists a smooth Hermitian metric h_ε on L such that $\Theta_{L,h_\varepsilon} \geq -\varepsilon\omega$ everywhere on X.

A consequence of holomorphic Morse inequalities 21. *If X is compact Kähler and L is nef, for every holomorphic vector bundle E on X one has*

$$h^q(X, \mathcal{O}(E) \otimes \mathcal{O}(kL)) = o(k^n) \quad \text{for all } q \geq 1.$$

Proof. Let ω be a Kähler metric. The nefness of L implies that there exists a smooth Hermitian metric h_ε on L such that $\Theta_{L,h_\varepsilon} \geq -\varepsilon\omega$. On $X(L, h_\varepsilon, 1)$ we have exactly 1 negative eigenvalue λ_1 which is belongs to $[-\varepsilon, 0[$ and the other ones λ_j ($j \geq 2$) are positive. The product $\lambda_1 \cdots \lambda_n$ satisfies $|\lambda_1 \cdots \lambda_n| \leq \varepsilon \prod_{j \geq 2}(\lambda_j + \varepsilon)$, hence

$$\frac{1}{n!}\left|\Theta^n_{L,h_\varepsilon}\right| \leq \frac{1}{(n-1)!}\varepsilon\omega \wedge (\Theta_{L,h_\varepsilon} + \varepsilon\omega)^{n-1} \quad \text{on } X(L, h_\varepsilon, 1).$$

By integrating, we find

$$\int_{X(L,h_\varepsilon,1)} \Theta^n_{L,h_\varepsilon} \leq n\varepsilon \int_X \omega \wedge (c_1(L) + \varepsilon\omega)^{n-1}$$

and the result follows. $\square$

Note 22. When X is non Kähler, D. Popovici [79] has announced bounds for the Monge–Ampère masses of Θ_{L,h_ε} which still imply the result, but the proof is much

Applications of Pluripotential Theory to Algebraic Geometry

harder in that case. On the other hand, when X is projective algebraic, an elementary hyperplane section argument and an induction on dimension easily implies the stronger upper bounds

$$h^q(X, \mathcal{O}(E) \otimes \mathcal{O}(kL)) = O(k^{n-q}) \qquad \text{for all } q \geq 0. \qquad (42)$$

Hint. By Serre duality, it is enough to show that

$$h^q(X, \mathcal{O}(F) \otimes \mathcal{O}(-kL)) = O(k^q) \qquad \text{for every } q \geq 0$$

and every holomorphic vector bundle F. Choose a very ample line bundle A so big that $F' = F^* \otimes \mathcal{O}(A)$ is Nakano positive, and apply the Nakano vanishing theorem and Serre duality to see that $H^q(X, \mathcal{O}(F) \otimes \mathcal{O}(-A) \otimes \mathcal{O}(-kL)) = 0$ for all k and $q \geq 1$. Use the exact sequence $0 \to \mathcal{O}_X(-A) \to \mathcal{O}_X \to \mathcal{O}_A \to 0$, take the tensor product with $\mathcal{O}(F) \otimes \mathcal{O}(-kL)$ and apply induction. □

It is unknown whether the accurate bound (42) holds true on a general compact complex manifold, even when X is assumed to be Kähler.

1.3.3 Distortion Inequalities for Asymptotic Fubini–Study Metrics

Another application of the heat kernel estimates is a generalization of G. Kempf's distortion inequalities [58, 60] to all projective algebraic manifolds. In this generality, the result was obtained by Th. Bouche [17], and in less generality (but with somewhat stronger estimates) by G. Tian [98].

Let L be a positive Hermitian line bundle over a projective manifold X, equipped with a Hermitian metric ω. Then $V_k = H^0(X, L^k)$ has a natural Hermitian metric given by the global L^2 norm of sections. For $k \geq k_0$ large enough, Φ_{kL} is an embedding and L^k can be identified to the pull-back $\Phi_k^* \mathcal{O}(1)$. We want to compare the original metric $|\bullet|$ of L and the metric $|\bullet|_{\text{FS}}$ induced by the Fubini–Study metric of $\mathcal{O}(1)$.

Let $(s_1, \ldots, s_N)$ be an orthonormal basis of $H^0(X, L^k)$. It is not difficult to check that

$$|\xi|_{\text{FS}}^2 = \frac{|\xi|^2}{|s_1(x)|^2 + \cdots + |s_N(x)|^2} \quad \text{for } \xi \in L_x^k,$$

thus all that we need is to get an estimate of $\sum |s_j(x)|^2$. However, this sum is the contribution of the 0 eigenvalue in the heat kernel

$$K_t^k(x, x) = \sum_{j=1}^{+\infty} e^{-2t\lambda_j^k} |\psi_j(x)|^2$$

associated to $\frac{2}{k}\square_k''$ in bidegree $(0,0)$. We observe that non zero eigenvalues λ_j^k are also eigenvalues in bidegree $(0,1)$, since $\overline{\partial}$ is injective on the corresponding eigenspaces. The associated eigenfunctions are $\overline{\partial}\psi_j/\sqrt{k\lambda_j^k}$, for

$$\|\overline{\partial}\psi_j\|^2 = \langle\!\langle \Delta_k''\psi_j, \psi_j\rangle\!\rangle = k\lambda_j^k.$$

Thus the summation

$$\sum_{j=1}^{+\infty} e^{-2t\lambda_j^k} |\overline{\partial}\psi_j(x)|^2$$

is bounded by the heat kernel in bidegree $(0,1)$, which is itself bounded by $k^n e^{-ct}$ with $c > 0$ (note that $\alpha_{\complement J} - \alpha_J - \sum |\alpha_j| < 0$ on X for $|J| = 1$). Taking $t = k^\varepsilon$ with ε small, one can check that all estimates remain uniformly valid and that the contribution of the non zero eigenfunctions in $K_t^k(x,x)$ becomes negligible in C^0 norm. Then Theorem 13 shows that

$$\sum |s_j(x)|^2 \sim K_t^k(x,x) \sim k^n (2\pi)^{-n} |\alpha_1(x) \cdots \alpha_n(x)|$$

as $t = k^\varepsilon \to +\infty$. For $\xi \in L_x^k$ we get therefore the C^0 uniform estimate

$$\frac{|\xi|^2}{|\xi|_{\mathrm{FS}}^2} \sim \left(\frac{k}{2\pi}\right)^n |\alpha_1(x) \cdots \alpha_n(x)| \qquad \text{as} \quad k \to +\infty. \tag{43}$$

As a consequence, the Fubini–Study metric on L induced by Φ_{kL} converges uniformly to the original metric. G. Tian [98] proved that this last convergence statement holds in norm C^4. It is now known that there is in fact an asymptotic expansion in $1/k$, and therefore C^∞ convergence; this holds true even in the almost complex setting, see [15, 82].

1.3.4 Algebraic Counterparts of the Holomorphic Morse Inequalities

One difficulty in the application of the analytic form of the inequalities is that the curvature integral is in general quite uneasy to compute, since it is neither a topological nor an algebraic invariant. However, the Morse inequalities can be reformulated in a more algebraic setting in which only algebraic invariants are involved. We give here two such reformulations—after they were found via analysis in [30], F. Angelini [5] gave a purely algebraic proof (see also [88, 100] for related ideas).

Theorem 23. *Let $L = F - G$ be a holomorphic line bundle over a compact Kähler manifold X, where F and G are numerically effective line bundles. Then for every $q = 0, 1, \ldots, n = \dim X$, there is an asymptotic strong Morse inequality*

$$\sum_{0\leq j\leq q}(-1)^{q-j}h^j(X,kL) \leq \frac{k^n}{n!}\sum_{0\leq j\leq q}(-1)^{q-j}\binom{n}{j}F^{n-j}\cdot G^j + o(k^n).$$

Proof. By adding ε times a Kähler metric ω to the curvature forms of F and G, $\varepsilon > 0$ one can write $\Theta_L = \tilde\Theta_{F,\varepsilon} - \tilde\Theta_{G,\varepsilon}$ where $\tilde\Theta_{F,\varepsilon} = \frac{i}{2\pi}\Theta_F + \varepsilon\omega$ and $\tilde\Theta_{G,\varepsilon} = \frac{i}{2\pi}\Theta_G + \varepsilon\omega$ are positive definite. Let $\lambda_1 \geq \cdots \geq \lambda_n > 0$ be the eigenvalues of $\tilde\Theta_{G,\varepsilon}$ with respect to $\tilde\Theta_{F,\varepsilon}$. Then the eigenvalues of $\frac{i}{2\pi}\Theta_L$ with respect to $\tilde\Theta_{F,\varepsilon}$ are the real numbers $1 - \lambda_j$ and the set $X(L,h,\leq q)$ is the set $\{\lambda_{q+1} < 1\}$ of points $x \in X$ such that $\lambda_{q+1}(x) < 1$. The strong Morse inequalities yield

$$\sum_{0\leq j\leq q}(-1)^{q-j}h^j(X,kL) \leq \frac{k^n}{n!}\int_{\{\lambda_{q+1}<1\}}(-1)^q\prod_{1\leq j\leq n}(1-\lambda_j)\tilde\Theta_{F,\varepsilon}^n + o(k^n).$$

On the other hand we have

$$\binom{n}{j}\tilde\Theta_{F,\varepsilon}^{n-j}\wedge \tilde\Theta_{G,\varepsilon}^j = \sigma_n^j(\lambda)\tilde\Theta_{F,\varepsilon}^n,$$

where $\sigma_n^j(\lambda)$ is the j-th elementary symmetric function in $\lambda_1,\ldots,\lambda_n$, hence

$$\sum_{0\leq j\leq q}(-1)^{q-j}\binom{n}{j}F^{n-j}\cdot G^j = \lim_{\varepsilon\to 0}\int_X\sum_{0\leq j\leq q}(-1)^{q-j}\sigma_n^j(\lambda)\tilde\Theta_{F,\varepsilon}^n.$$

Thus, to prove the lemma, we only have to check that

$$\sum_{0\leq j\leq n}(-1)^{q-j}\sigma_n^j(\lambda) - 1\!\!1_{\{\lambda_{q+1}<1\}}(-1)^q\prod_{1\leq j\leq n}(1-\lambda_j) \geq 0$$

for all $\lambda_1 \geq \cdots \geq \lambda_n \geq 0$, where $1\!\!1_{\{\ldots\}}$ denotes the characteristic function of a set. This is easily done by induction on n (just split apart the parameter λ_n and write $\sigma_n^j(\lambda) = \sigma_{n-1}^j(\lambda) + \sigma_{n-1}^{j-1}(\lambda)\lambda_n$). □

In the case $q = 1$, we get an especially interesting lower bound (this bound has been observed and used by S. Trapani [100] in a similar context).

Consequence 24. $h^0(X,kL) - h^1(X,kL) \geq \frac{k^n}{n!}(F^n - nF^{n-1}\cdot G) - o(k^n)$. *Therefore some multiple kL has a section as soon as $F^n - nF^{n-1}\cdot G > 0$.*

Remark 25. The weaker inequality

$$h^0(X,kL) \geq \frac{k^n}{n!}(F^n - nF^{n-1}\cdot G) - o(k^n)$$

is easy to prove if X is projective algebraic. Indeed, by adding a small ample $\mathbb{Q}$-divisor to F and G, we may assume that F, G are ample. Let $m_0 G$ be very ample and let k' be the smallest integer $\geq k/m_0$. Then $h^0(X, kL) \geq h^0(X, kF - k'm_0 G)$. We select k' smooth members G_j, $1 \leq j \leq k'$ in the linear system $|m_0 G|$ and use the exact sequence

$$0 \to H^0(X, kF - \sum G_j) \to H^0(X, kF) \to \bigoplus H^0(G_j, kF_{|G_j}).$$

Kodaira's vanishing theorem yields $H^q(X, kF) = 0$ and $H^q(G_j, kF_{|G_j}) = 0$ for $q \geq 1$ and $k \geq k_0$. By the exact sequence combined with Riemann–Roch, we get

$$h^0(X, kL) \geq h^0(X, kF - \sum G_j)$$

$$\geq \frac{k^n}{n!} F^n - O(k^{n-1}) - \sum \left(\frac{k^{n-1}}{(n-1)!} F^{n-1} \cdot G_j - O(k^{n-2}) \right)$$

$$\geq \frac{k^n}{n!} \left(F^n - n \frac{k' m_0}{k} F^{n-1} \cdot G \right) - O(k^{n-1})$$

$$\geq \frac{k^n}{n!} \left(F^n - n F^{n-1} \cdot G \right) - O(k^{n-1}).$$

(This simple proof is due to F. Catanese.) □

Corollary 26. *Suppose that F and G are nef and that F is big. Some multiple of $mF - G$ has a section as soon as*

$$m > n \frac{F^{n-1} \cdot G}{F^n}.$$

In the last condition, the factor n is sharp: this is easily seen by taking $X = \mathbb{P}_1^n$ and $F = \mathcal{O}(a, \ldots, a)$ and $G = \mathcal{O}(b_1, \ldots, b_n)$ over $\mathbb{P}_1^n$; the condition of the corollary is then $m > \sum b_j/a$, whereas $k(mF - G)$ has a section if and only if $m \geq \sup b_j/a$; this shows that we cannot replace n by $n(1 - \varepsilon)$.

1.4 Morse Inequalities on q-Convex Varieties

Thierry Bouche [16] has obtained an extension of holomorphic Morse inequalities to the case of strongly q-convex manifolds. We explain here the main ideas involved.

A complex (non compact) manifold X of dimension n is strongly q-convex in the sense of Andreotti and Grauert [3] if there exists a C^∞ exhaustion function ψ on X such that $i \partial \bar{\partial} \psi$ has at least $n - q + 1$ positive eigenvalues outside a compact subset of X. In this case, the Andreotti–Grauert theorem shows that all cohomology groups $H^m(X, \mathscr{F})$ with values in a coherent analytic sheaf are finite dimensional for $m \geq q$.

Theorem 27. *Let L, E be holomorphic vector bundles over X with rank $L = 1$, rank $E = r$. Assume that X is strongly q-convex and that L has a Hermitian metric h for which $\Theta_{L,h}$ has at least $n - p + 1$ nonnegative eigenvalues outside a compact subset $K \subset X$. Then for all $m \geq p + q - 1$ the following strong Morse inequalities hold:*

$$\sum_{\ell=m}^{n}(-1)^{\ell-m}\dim H^{\ell}(X, E \otimes L^{k}) \leq r\frac{k^{n}}{n!}\int_{X(L,h,\geq m)}(-1)^{m}\Theta_{L,h}^{n} + o(k^{n}).$$

Proof. For every $c \in \mathbb{R}$, we consider the sublevel sets

$$X_{c} = \{x \in X \; ; \; \psi(x) < c\}.$$

Select c_0 such that $i\partial\bar{\partial}\psi$ has $n - q + 1$ positive eigenvalues on $X \smallsetminus X_c$. One can choose a Hermitian metric ω_0 on X in such a way that the eigenvalues $\gamma_1^0 \leq \cdots \leq \gamma_n^0$ of $i\partial\bar{\partial}\psi$ with respect to ω_0 satisfy

$$-\frac{1}{n} \leq \gamma_1^0 \leq \cdots \leq \gamma_{q-1}^0 \leq 1 \quad \text{and} \quad \gamma_q^0 = \cdots = \gamma_n^0 = 1 \text{ on } X \smallsetminus X_{c_0}; \quad (44)$$

this can be achieved by taking ω_0 equal to $i\partial\bar{\partial}\psi$ on a C^∞ subbundle of T_X of rank $n - q + 1$ on which $i\partial\bar{\partial}\psi$ is positive, and ω_0 very large on the orthogonal complement. We set $\omega = e^\rho \omega_0$ where ρ is a function increasing so fast at infinity that ω will be complete.

More important, we multiply the metric of L by a weight $e^{-\chi \circ \psi}$ where χ is a convex increasing function. The resulting Hermitian line bundle is denoted (L_χ, h_χ). For any $(0, m)$ form u with values in $E \otimes L^k$, viewed as an (n, m) form with values in $E \otimes L^k \otimes \Lambda^n T_X$, the Bochner–Kodaira–Nakano formula implies an inequality

$$\langle\!\langle \Delta_k'' u, u \rangle\!\rangle \geq \int_X k \langle [i\Theta_{L_\chi, h_\chi}), \Lambda]u, u \rangle + \langle Wu, u \rangle$$

where W depends only on the curvature of $E \otimes \Lambda^n T_X$ and the torsion of ω. By the formulas of Sect. 1.2.3, we have

$$\langle [i\Theta_{L_\chi, h_\chi}), \Lambda]u, u \rangle \geq (\alpha_1 + \cdots + \alpha_m)|u|^2$$

where $\alpha_1 \leq \cdots \leq \alpha_n$ are the eigenvalues of

$$i\Theta_{L_\chi, h_\chi} = i\Theta_{L,h} + i\partial\bar{\partial}(\chi \circ \psi) \geq i\Theta_{L,h} + (\chi' \circ \psi)i\partial\bar{\partial}\psi.$$

If β is the lowest eigenvalue of $i\Theta_{L,h}$ with respect to ω, we find

$$\alpha_j \geq \beta + (\chi' \circ \psi)\gamma_j^0/e^\rho,$$

$$\alpha_1 + \cdots + \alpha_m \geq m\beta + (\chi' \circ \psi)(\gamma_1^0 + \cdots + \gamma_m^0)/e^\rho,$$

and by (44) we get for all $m \geq q$:

$$\alpha_1 + \cdots + \alpha_m \geq m\beta + \frac{1}{n} e^{-\rho} \chi' \circ \psi \text{ on } X \smallsetminus X_{c_0}.$$

It follows that one can choose χ increasing very fast in such a way that the Bochner inequality becomes

$$\langle \Delta_k'' u, u \rangle \geq k \int_{X \smallsetminus X_{c_0}} A(x) |u(x)|^2 - C_1 \int_X |u(x)|^2 \qquad (45)$$

where $A \geq 1$ is a function tending to $+\infty$ at infinity on X and $C_1 \geq 0$. Now, Rellich's lemma easily shows that Δ_k'' has a compact resolvent. Hence the spectrum of Δ_k'' is discrete and its eigenspaces are finite dimensional. Standard arguments also show the following:

Lemma 28. *When χ increases sufficiently fast at infinity, the space $\mathcal{H}^m(X, L_\chi^k \otimes E)$ of L^2-harmonic forms of bidegree $(0, m)$ for Δ_k'' is isomorphic to the cohomology group $H^m(X, E \otimes L^k)$ for all $k \in \mathbb{N}$ and $m \geq q$.*

For a domain $\Omega \subset\subset X$, we consider the quadratic form

$$Q_\Omega^{k,m}(u) = \frac{1}{k} \int_\Omega |\overline{\partial}_k u|^2 + |\overline{\partial}_k^* u|^2$$

with Dirichlet boundary conditions on $\partial\Omega$. We denote by $\mathcal{H}_{\leq \lambda, \Omega}^{k,m}$ the direct sum of all eigenspaces of $Q_\Omega^{k,m}$ corresponding to eigenvalues $\leq \lambda$ (i.e. $\leq k\lambda$ for Δ_k'').

Lemma 29. *For every $\lambda \geq 0$ and $\varepsilon > 0$, there exists a domain $\Omega \subset\subset X$ and an integer k_0 such that*

$$\dim \mathcal{H}_{\leq \lambda, \Omega}^{k,m} \leq \dim \mathcal{H}_{\leq \lambda, X}^{k,m} \leq \dim \mathcal{H}_{\leq \lambda+\varepsilon, \Omega}^{k,m} \text{ for } k \geq k_0.$$

Proof. The left hand inequality is a straightforward consequence of the minimax principle, because the domain of the global quadratic form $Q_\Omega^{k,m}$ is contained in the domain of $Q_X^{k,m}$.

For the other inequality, let $u \in \mathcal{H}_{\leq \lambda, X}^{k,m}$. Then (45) gives

$$k \int_{X \smallsetminus X_{c_0}} A |u|^2 - C_1 \int_{X_{c_0}} |u|^2 \leq k\lambda \int_X |u|^2.$$

Choose $c_2 > c_1 > c_0$ so that $A(x) \geq a$ on $X \smallsetminus X_{c_1}$ and a cut-off function φ with compact support in X_{c_2} such that $0 \leq \varphi \leq 1$ and $\varphi = 1$ on X_{c_1}. Then we find

$$\int_{X \smallsetminus X_{c_1}} |u|^2 \leq \frac{C_1 + k\lambda}{ka} \int_X |u|^2.$$

Applications of Pluripotential Theory to Algebraic Geometry 175

For a large enough, we get $\int_{X \smallsetminus X_{c_1}} |u|^2 \leq \varepsilon \|u\|^2$. Set $\Omega = X_{c_2}$. Then

$$Q_\Omega^{k,m}(\varphi u) = \frac{1}{k} \int_\Omega |\bar\partial\varphi \wedge u + \varphi \bar\partial_k u|^2 + |\varphi \bar\partial_k^* u - \partial\varphi \lrcorner u|^2$$

$$\leq (1+\varepsilon) Q_X^{k,m}(u) + \frac{C_2}{k}\left(1 + \frac{1}{\varepsilon}\right)\|u\|^2$$

$$\leq (1+\varepsilon)(\lambda + \frac{C_2}{k\varepsilon})\|u\|^2.$$

As $\|\varphi u\|^2 \geq \int_{X_{c_1}} |u|^2 \geq (1-\varepsilon)\|u\|^2$, we infer

$$Q_\Omega^{k,m}(\varphi u) \leq \frac{1+\varepsilon}{1-\varepsilon}\left(\lambda + \frac{C_2}{k\varepsilon}\right)\|\varphi u\|^2.$$

If ε is replaced by a suitable smaller number and k taken large enough, we obtain $Q_\Omega^{k,m}(v) \leq (\lambda + \varepsilon)\|v\|^2$ for all $v \in \varphi \mathscr{H}_{\leq \lambda, X}^{k,m}$. Then the right hand inequality in Lemma 29 follows by the minimax principle. □

Now, Corollary 12 easily computes the counting function $N_\Omega^{k,m}$ for the eigenvalues:

$$\lim_{\lambda \to 0_+} \lim_{k \to +\infty} k^{-n} N_\Omega^{k,m}(\lambda) = \frac{r}{n!} \int_{X(L_\chi, h_\chi, m)} (-1)^m \left(\frac{i}{2\pi} \theta_{L_\chi, h_\chi}\right)^n.$$

Applying this to the Witten complex $\mathscr{H}_{\leq \lambda, X}^{k, \bullet}$, we easily infer the inequality of Theorem 27, except that $c(L)$ is replaced by $c(L_\chi)$. However, up to now, the inequality is valid for all $m \geq q$. Take the convex function χ equal to 0 on $]-\infty, c_0]$. Then

$$\Theta_{L_\chi, h_\chi} = \frac{i}{2\pi} \theta_{L_\chi, h_\chi} = \Theta_{L,h} + \frac{i}{2\pi} \partial\bar\partial(\chi \circ \psi)$$

coincides with $\Theta_{L,h}$ on X_{c_0} and has at most $(p-1)+(q-1)$ negative eigenvalues on $X \smallsetminus X_{c_0}$. Hence $X(L_\chi, h_\chi, m) = X(L, h, m)$ for $m \geq p+q-1$ and $\Theta_{L_\chi, h_\chi} = \Theta_{L,h}$ on these sets. Theorem 27 is proved. □

As a corollary, one obtains a general a priori estimate for the Monge–Ampère operator $(i\partial\bar\partial)^n$ on q-convex manifolds.

Corollary: calculus inequalities 30. *Let X be a strongly q-convex manifold and φ a C^∞ function on X, weakly p-convex outside a compact subset of X. For $\ell = 0, 1, \ldots, n$, let G_ℓ be the open set of points where $i\partial\bar\partial\varphi$ is non degenerate and admits ℓ negative eigenvalues. Then for all $m \geq p = q - 1$*

$$\sum_{\ell=m}^n \int_{G_\ell} (i\partial\bar\partial\varphi)^m \text{ has the sign of } (-1)^m.$$

This result has been first obtained by Y.T. Siu [87] for q-convex domains in a Stein manifold. At that time, the q-convex case of the inequalities was not yet available and Siu had to rely on a rather sophisticated approximation argument of Stein manifolds by algebraic varieties; the proof could then be reduced to the compact case.

The general statement given above is in fact a direct consequence of Theorem 27: take for L the trivial bundle $L = \mathcal{O}_X$ equipped with the metric defined by the weight $e^{-\varphi}$ and $E = \mathcal{O}_X$. Since $H^m(X, L^k) = H^m(X, \mathcal{O}_X)$ is independent of k and finite dimensional, Theorem 27 implies

$$k^n \sum_{\ell=m}^{n} \int_{G_\ell} (-1)^m (i \partial \overline{\partial} \varphi)^n \geq \text{constant} - o(k^n)$$

for all $k \geq k_0$ and $m \geq p + q - 1$, whence the result. □

2 Approximation of Currents and Intersection Theory

2.1 Introduction

Many concepts described in this section (e.g. pseudo-effectivity) are quite general and make sense on an arbitrary compact complex manifold X—no projective or Kähler assumption is needed. In this general context, it is better to work with $\partial\overline{\partial}$-cohomology classes instead of De Rham cohomology classes: we define the *Bott–Chern cohomology* of X to be

$$H^{p,q}_{\mathrm{BC}}(X, \mathbb{C}) = \{d\text{-closed } (p,q)\text{-forms}\} / \{\partial\overline{\partial}\text{-exact } (p,q)\text{-forms}\}. \qquad (46)$$

It is easily shown that these cohomology groups are finite dimensional and can be computed either with spaces of smooth forms or with currents; in fact, they can be computed by certain complexes of sheaves of forms or currents that both provide fine resolutions of the same sheaves of holomorphic or anti-holomorphic forms. Our statement therefore follows formally from general results of sheaf theory. Also, finiteness can be obtained by the usual Cartan–Serre proof based on Montel's theorem for Čech cohomology. In both cases, the quotient topology of $H^{p,q}_{\mathrm{BC}}(X, \mathbb{C})$ induced by the Fréchet topology of smooth forms or by the weak topology of currents is Hausdorff. Clearly, $H^{\bullet}_{\mathrm{BC}}(X, \mathbb{C})$ is a bigraded algebra, and it is trivial by definition that there are always canonical morphisms

$$H^{p,q}_{\mathrm{BC}}(X, \mathbb{C}) \to H^{p,q}_{\overline{\partial}}(X, \mathbb{C}), \qquad \bigoplus_{p+q=k} H^{p,q}_{\mathrm{BC}}(X, \mathbb{C}) \to H^{k}_{\mathrm{DR}}(X, \mathbb{C}). \qquad (47)$$

By Hodge decomposition and by the well-known $\partial\bar{\partial}$-lemma of Kähler geometry, these morphisms are isomorphisms when X is Kähler; especially, we get a canonical algebra isomorphism

$$H^\bullet_{\mathrm{DR}}(X,\mathbb{C}) \simeq \bigoplus_{p,q} H^{p,q}_{\bar\partial}(X,\mathbb{C}) \qquad \text{if } X \text{ is Kähler}. \tag{48}$$

We will see in paragraph 5 (Remark 63) that this is true more generally if X is in the Fujiki class $\mathscr{C}$, i.e., the class of manifolds bimeromorphic to Kähler manifolds.

2.2 Pseudo-Effective Line Bundles and Singular Hermitian Metrics

Let L be a holomorphic line bundle on a compact complex manifold X. It is important for many applications to allow singular Hermitian metrics.

Definition 31. A singular Hermitian metric h on L is a Hermitian metric such that, for any trivialisation $L_{|U} \simeq U \times \mathbb{C}$, the metric is given by $h = e^{-\varphi}$, $\varphi \in L^1_{\mathrm{loc}}(U)$.

The curvature tensor

$$\Theta_{L,h} = \frac{i}{2\pi}\partial\bar{\partial}\varphi = -\frac{i}{2\pi}\partial\bar{\partial}\log h \tag{49}$$

can then be computed in the sense of distributions, and defines in this way a (global) closed $(1,1)$-current on X. It defines a (real) cohomology class $\{\Theta_{L,h}\} \in H^{1,1}_{\mathrm{BC}}(X,\mathbb{C})$ which is mapped to the first Chern class $c_1(L)$ by the canonical morphisms (2). We will therefore still denote this Bott–Chern class by $c_1(L)$. The positive case is of special interest.

Definition 32. We say that L pseudo-effective if $c_1(L) \in H^{1,1}_{\mathrm{BC}}(X,\mathbb{C})$ is the cohomology class of some closed positive current T, i.e. if L can be equipped with a singular Hermitian metric h with $T = \Theta_{L,h} \geq 0$ as a current, in other words, if the weight functions φ can be chosen to be plurisubharmonic on each trivialization open set U.

The locus where h has singularities turns out to be extremely important. One way is to introduce multiplier ideal sheaves following A. Nadel [75]. The main idea actually goes back to the fundamental works of Bombieri [12] and H. Skoda [94].

Definition 33. Let φ be a psh (plurisubharmonic) function on an open subset $\Omega \subset X$. To φ we associate the ideal subsheaf $\mathscr{I}(\varphi) \subset \mathscr{O}_\Omega$ of germs of holomorphic functions $f \in \mathscr{O}_{\Omega,x}$ such that $|f|^2 e^{-\varphi}$ is integrable with respect to the Lebesgue measure in some local coordinates near x.

The zero variety $V(\mathscr{I}(\varphi))$ is thus the set of points in a neighborhood of which $e^{-\varphi}$ is non integrable. The following result implies that this is always an analytic set.

Proposition 34 ([75]). *For any psh function φ on $\Omega \subset X$, the sheaf $\mathscr{I}(\varphi)$ is a coherent sheaf of ideals over Ω. Moreover, if Ω is a bounded Stein open set, the sheaf $\mathscr{I}(\varphi)$ is generated by any Hilbert basis of the L^2 space $\mathscr{H}^2(\Omega, \varphi)$ of holomorphic functions f on Ω such that $\int_\Omega |f|^2 e^{-\varphi} d\lambda < +\infty$.*

Proof. Since the result is local, we may assume that Ω is a bounded pseudoconvex open set in $\mathbb{C}^n$. By the strong Noetherian property of coherent sheaves, the family of sheaves generated by finite subsets of $\mathscr{H}^2(\Omega, \varphi)$ has a maximal element on each compact subset of Ω, hence $\mathscr{H}^2(\Omega, \varphi)$ generates a coherent ideal sheaf $\mathscr{J} \subset \mathscr{O}_\Omega$. It is clear that $\mathscr{J} \subset \mathscr{I}(\varphi)$; in order to prove the equality, we need only check that $\mathscr{J}_x + \mathscr{I}(\varphi)_x \cap \mathfrak{m}_{\Omega,x}^{s+1} = \mathscr{I}(\varphi)_x$ for every integer s, in view of the Krull lemma. Let $f \in \mathscr{I}(\varphi)_x$ be defined in a neighborhood V of x and let θ be a cut-off function with support in V such that $\theta = 1$ in a neighborhood of x. We solve the equation $\bar\partial u = g := \bar\partial(\theta f)$ by means of Hörmander's L^2 estimates [4, 56], applied with the strictly psh weight

$$\tilde\varphi(z) = \varphi(z) + (n+s)\log|z-x|^2 + |z|^2.$$

We get a solution u such that $\int_\Omega |u|^2 e^{-\varphi} |z-x|^{-2(n+s)} d\lambda < \infty$, thus $F = \theta f - u$ is holomorphic, $F \in \mathscr{H}^2(\Omega, \varphi)$ and $f_x - F_x = u_x \in \mathscr{I}(\varphi)_x \cap \mathfrak{m}_{\Omega,x}^{s+1}$. This proves the coherence. Now, $\mathscr{J}$ is generated by any Hilbert basis of $\mathscr{H}^2(\Omega, \varphi)$, because it is well-known that the space of sections of any coherent sheaf is a Fréchet space, therefore closed under local L^2 convergence. □

Another important way of measuring singularities is via Lelong numbers—a natural generalization of the concept of multiplicity to psh functions. Recall that the Lelong number of a function $\varphi \in \mathrm{Psh}(\Omega)$ at a point x_0 is defined to be

$$\nu(\varphi, x_0) = \liminf_{z \to x_0} \frac{\varphi(z)}{\log|z - x_0|} = \lim_{r \to 0_+} \frac{\sup_{B(x_0, r)} \varphi}{\log r}. \tag{50}$$

In particular, if $\varphi = \log|f|$ with $f \in \mathscr{O}(\Omega)$, then $\nu(\varphi, x_0)$ is equal to the vanishing order

$$\mathrm{ord}_{x_0}(f) = \sup\{k \in \mathbb{N}; D^\alpha f(x_0) = 0, \forall |\alpha| < k\}.$$

The link with multiplier ideal sheaves is provided by the following standard result due to Skoda [94].

Lemma 35. *Let φ be a psh function on an open set Ω and let $x \in \Omega$.*

(a) *If $\nu(\varphi, x) < 2$, then $e^{-\varphi}$ is Lebesgue integrable on a neighborhood of x, in particular $\mathscr{I}(\varphi)_x = \mathscr{O}_{\Omega,x}$.*
(b) *More generally, if $\nu(\varphi, x) \geq 2(n + s)$ for some integer $s \geq 0$, then*

$$e^{-\varphi} \geq c|z - x|^{-2n-2s}, \qquad c > 0$$

in a neighborhood of x, and $\mathscr{I}(\varphi)_x \subset \mathfrak{m}_{\Omega,x}^{s+1}$, where $\mathfrak{m}_{\Omega,x}$ is the maximal ideal of $\mathcal{O}_{\Omega,x}$. In particular $e^{-\varphi}$ is non integrable at x if $\nu(\varphi, x) \geq 2n$.

(c) *The zero variety $V(\mathscr{I}(\varphi))$ of $\mathscr{I}(\varphi)$ satisfies*

$$V_{2n}(\varphi) \subset V(\mathscr{I}(\varphi)) \subset E_2(\varphi)$$

where $E_c(\varphi) = \{x \in X \,;\, \nu(\varphi, x) \geq c\}$ is the c-upperlevel set of Lelong numbers of φ.

The only non trivial part is Lemma 35(a); the proof relies on the Bochner–Martinelli representation formula for $T = \frac{i}{\pi}\partial\bar{\partial}\varphi$ (see [94]). One should observe that Lemma 35(a) (resp. (b)) is optimal, as one can see by taking $\varphi(z) = \lambda \log|z_1|$, resp. $\varphi(z) = \lambda \log|z|$, on $\Omega = \mathbb{C}^n$.

2.3 Hermitian Metrics with Minimal Singularities and Analytic Zariski Decomposition

We show here by a general "abstract" method that a pseudo-effective line bundle always has a Hermitian metric $h_{\min}$ with minimal singularities among those with nonnegative curvature $\Theta_{L,h} \geq 0$ in the sense of currents. The following definition was introduced in [44].

Definition 36. Let L be a pseudo-effective line bundle on a compact complex manifold X. Consider two Hermitian metrics h_1, h_2 on L with curvature $\Theta_{L,h_j} \geq 0$ in the sense of currents.

(a) We will write $h_1 \preccurlyeq h_2$, and say that h_1 is less singular than h_2, if there exists a constant $C > 0$ such that $h_1 \leq Ch_2$.
(b) We will write $h_1 \sim h_2$, and say that h_1, h_2 are equivalent with respect to singularities, if there exists a constant $C > 0$ such that $C^{-1}h_2 \leq h_1 \leq Ch_2$.

Of course $h_1 \preccurlyeq h_2$ if and only if the associated weights in suitable trivializations locally satisfy $\varphi_2 \leq \varphi_1 + C$. This implies in particular $\nu(\varphi_1, x) \leq \nu(\varphi_2, x)$ at each point. The above definition is motivated by the following observation.

Theorem 37. *For every pseudo-effective line bundle L over a compact complex manifold X, there exists up to equivalence of singularities a unique class of Hermitian metrics h with minimal singularities such that $\Theta_{L,h} \geq 0$.*

Proof. The proof is almost trivial. We fix once for all a smooth metric h_∞ (whose curvature is of random sign and signature), and we write singular metrics of L under the form $h = h_\infty e^{-\psi}$. The condition $\Theta_{L,h} \geq 0$ is equivalent to $\frac{i}{2\pi}\partial\bar{\partial}\psi \geq -u$ where $u = \Theta_{L,h_\infty}$. This condition implies that ψ is plurisubharmonic up to the addition of the weight φ_∞ of h_∞, and therefore locally bounded from above. Since we are concerned with metrics only up to equivalence of singularities, it is always possible to adjust ψ by a constant in such a way that $\sup_X \psi = 0$. We now set

$$h_{\min} = h_\infty e^{-\psi_{\min}}, \qquad \psi_{\min}(x) = \sup_\psi \psi(x)$$

where the supremum is extended to all functions ψ such that $\sup_X \psi = 0$ and $\frac{i}{2\pi}\partial\bar\partial\psi \geq -u$. By standard results on plurisubharmonic functions (see Lelong [69]), $\psi_{\min}$ still satisfies $\frac{i}{2\pi}\partial\bar\partial\psi_{\min} \geq -u$ (i.e. the weight $\varphi_\infty + \psi_{\min}$ of $h_{\min}$ is plurisubharmonic), and $h_{\min}$ is obviously the metric with minimal singularities that we were looking for. [In principle one should take the upper semicontinuous regularization $\psi_{\min}^*$ of $\psi_{\min}$ to really get a plurisubharmonic weight, but since $\psi_{\min}^*$ also participates to the upper envelope, we obtain here $\psi_{\min} = \psi_{\min}^*$ automatically]. □

Remark 38. In general, the supremum $\psi = \sup_{j \in I} \psi_j$ of a locally dominated family of plurisubharmonic functions ψ_j is not plurisubharmonic strictly speaking, but its "upper semi-continuous regularization" $\psi^*(z) = \limsup_{\zeta \to z} \psi(\zeta)$ is plurisubharmonic and coincides almost everywhere with ψ, with $\psi^* \geq \psi$. However, in the context of (41), ψ^* still satisfies $\psi^* \leq 0$ and $\frac{i}{2\pi}\partial\bar\partial\psi \geq -u$, hence ψ^* participates to the upper envelope. As a consequence, we have $\psi^* \leq \psi$ and thus $\psi = \psi^*$ is indeed plurisubharmonic. Under a strict positivity assumption, namely if L is a big line bundle (i.e. the curvature can be taken to be strictly positive in the sense of currents, see Definition 42(d) and Theorem 43(b), then $h_{\min}$ can be shown to possess some regularity properties. The reader may consult [7] for a rather general (but certainly non trivial) proof that $\psi_{\min}$ possesses locally bounded second derivatives $\partial^2 \psi_{\min}/\partial z_j \partial\bar z_k$ outside an analytic set $Z \subset X$; in other words, $\Theta_{L,h_{\min}}$ has locally bounded coefficients on $X \smallsetminus Z$. □

Definition 39. Let L be a pseudo-effective line bundle. If h is a singular Hermitian metric such that $\Theta_{L,h} \geq 0$ and

$$H^0(X, mL \otimes \mathscr{I}(h^{\otimes m})) \simeq H^0(X, mL) \qquad \text{for all } m \geq 0,$$

we say that h is an *analytic Zariski decomposition* of L.

In other words, we require that h has singularities so mild that the vanishing conditions prescribed by the multiplier ideal sheaves $\mathscr{I}(h^{\otimes m})$ do not kill any sections of L and its multiples.

Exercise 40. A special case is when there is an isomorphism $pL = A + E$ where A and E are effective divisors such that $H^0(X, mpL) = H^0(X, mA)$ for all m and $\mathscr{O}(A)$ is generated by sections. Then A possesses a smooth Hermitian metric h_A, and this metric defines a singular Hermitian metric h on L with poles $\frac{1}{p}E$ and curvature $\frac{1}{p}\Theta_{A,h_A} + \frac{1}{p}[E]$. Show that this metric h is an analytic Zariski decomposition.
Note: When X projective and there is a decomposition $pL = A + E$ with A nef (see Definition 20), E effective and $H^0(X, mpL) = H^0(X, mA)$ for all m, one says

that the $\mathbb{Q}$-divisor equality $L = \frac{1}{p}A + \frac{1}{p}E$ is an *algebraic Zariski decomposition* of L. It can be shown that Zariski decompositions exist in dimension 2, but in higher dimension they do not exist in general. ☐

Theorem 41. *The metric $h_{\min}$ with minimal singularities provides an analytic Zariski decomposition.*

It follows that an analytic Zariski decomposition always exists (while algebraic decompositions do not exist in general, especially in dimension 3 and more).

Proof. Let $\sigma \in H^0(X, mL)$ be any section. Then we get a singular metric h on L by putting $|\xi|_h = |\xi/\sigma(x)^{1/m}|$ for $\xi \in L_x$, and it is clear that $|\sigma|_{h^m} = 1$ for this metric. Hence $\sigma \in H^0(X, mL \otimes \mathcal{I}(h^{\otimes m}))$, and a fortiori $\sigma \in H^0(X, mL \otimes \mathcal{I}(h_{\min}^{\otimes m}))$ since $h_{\min}$ is less singular than h. ☐

2.4 Description of Positive Cones (Kähler and Projective Cases)

Let us recall that an integral cohomology class in $H^2(X, \mathbb{Z})$ is the first Chern class of a holomorphic (or algebraic) line bundle if and only if it lies in the *Neron–Severi* group

$$\mathrm{NS}(X) = \mathrm{Ker}\left(H^2(X, \mathbb{Z}) \to H^2(X, \mathcal{O}_X)\right) \tag{51}$$

(this fact is just an elementary consequence of the exponential exact sequence $0 \to \mathbb{Z} \to \mathcal{O} \to \mathcal{O}^* \to 0$). If X is compact Kähler, as we will suppose from now on in this section, this is the same as saying that the class is of type $(1, 1)$ with respect to Hodge decomposition.

Let us consider the real vector space $\mathrm{NS}_{\mathbb{R}}(X) = \mathrm{NS}(X) \otimes_{\mathbb{Z}} \mathbb{R}$, which can be viewed as a subspace of the space $H^{1,1}(X, \mathbb{R})$ of real $(1, 1)$ cohomology classes. Its dimension is by definition the Picard number

$$\rho(X) = \mathrm{rank}_{\mathbb{Z}} \mathrm{NS}(X) = \dim_{\mathbb{R}} \mathrm{NS}_{\mathbb{R}}(X). \tag{52}$$

We thus have $0 \leq \rho(X) \leq h^{1,1}(X)$, and the example of complex tori shows that all intermediate values can occur when $n = \dim X \geq 2$.

The positivity concepts for line bundles considered in Sects. 1.3.2 and 2.2 possess in fact natural generalizations to $(1, 1)$ classes which are not necessarily integral or rational—and this works at least in the category of compact Kähler manifolds (in fact, by using Bott–Chern cohomology, one could even extend these concepts to arbitrary compact complex manifolds).

Definition 42. Let (X, ω) be a compact Kähler manifold.

(a) The Kähler cone is the set $\mathcal{K} \subset H^{1,1}(X, \mathbb{R})$ of cohomology classes $\{\omega\}$ of Kähler forms. This is an open convex cone.

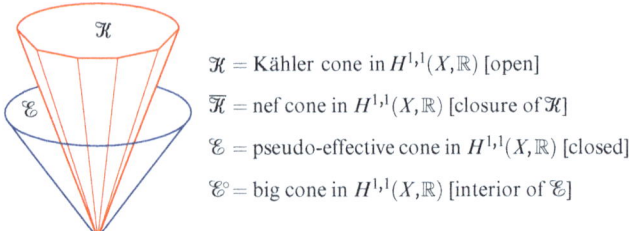

Fig. 1 Positive cones in a compact Kähler manifold

(b) The closure $\overline{\mathcal{K}}$ of the Kähler cone consists of classes $\{\alpha\} \in H^{1,1}(X,\mathbb{R})$ such that for every $\varepsilon > 0$ the sum $\{\alpha + \varepsilon\omega\}$ is Kähler, or equivalently, for every $\varepsilon > 0$, there exists a smooth function φ_ε on X such that $\alpha + i\partial\overline{\partial}\varphi_\varepsilon \geq -\varepsilon\omega$. We say that $\overline{\mathcal{K}}$ is the cone of *nef* (1, 1)-classes.
(c) The pseudo-effective cone is the set $\mathcal{E} \subset H^{1,1}(X,\mathbb{R})$ of cohomology classes $\{T\}$ of closed positive currents of type (1, 1). This is a closed convex cone.
(d) The interior $\mathcal{E}°$ of $\mathcal{E}$ consists of classes which still contain a closed positive current after one subtracts $\varepsilon\{\omega\}$ for $\varepsilon > 0$ small, in other words, they are classes of closed (1, 1)-currents T such that $T \geq \varepsilon\omega$. Such a current will be called a Kähler current, and we say that $\{T\} \in H^{1,1}(X,\mathbb{R})$ is a big (1, 1)-class.

The openness of $\mathcal{K}$ is clear by definition, and the closedness of $\mathcal{E}$ is a consequence of the fact that bounded sets of currents are weakly compact (as follows from the similar weak compactness property for bounded sets of positive measures). It is then clear that $\overline{\mathcal{K}} \subset \mathcal{E}$.

In spite of the fact that cohomology groups can be defined either in terms of forms or currents, it turns out that the cones $\overline{\mathcal{K}}$ and $\mathcal{E}$ are in general different. To see this, it is enough to observe that a Kähler class $\{\alpha\}$ satisfies $\int_Y \alpha^p > 0$ for every p-dimensional analytic set. On the other hand, if X is the surface obtained by blowing-up $\mathbb{P}^2$ in one point, then the exceptional divisor $E \simeq \mathbb{P}^1$ has a cohomology class $\{\alpha\}$ such that $\int_E \alpha = E^2 = -1$, hence $\{\alpha\} \notin \overline{\mathcal{K}}$, although $\{\alpha\} = \{[E]\} \in \mathcal{E}$.

In case X is projective, all Chern classes $c_1(L)$ of line bundles lie by definition in $NS(X)$, and likewise, all classes of real divisors $D = \sum c_j D_j$, $c_j \in \mathbb{R}$, lie in $NS_{\mathbb{R}}(X)$. In order to deal with such *algebraic classes*, we therefore introduce the intersections

$$\mathcal{K}_{NS} = \mathcal{K} \cap NS_{\mathbb{R}}(X), \qquad \mathcal{E}_{NS} = \mathcal{E} \cap NS_{\mathbb{R}}(X),$$

and refer to classes of $H^{1,1}(X,\mathbb{R})$ not contained in $NS_{\mathbb{R}}(X)$ as *transcendental classes*.

A very important fact is that all four cones $\mathcal{K}_{NS}$, $\mathcal{E}_{NS}$, $\overline{\mathcal{K}}_{NS}$, $\mathcal{E}°_{NS}$ have simple algebraic interpretations.

Fig. 2 Positive algebraic classes in a projective manifold

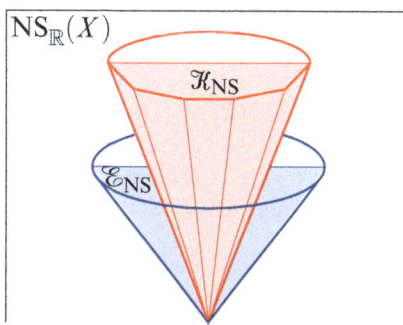

Theorem 43. *Let X be a projective manifold. Then*

(a) *$\mathcal{K}_{NS}$ is equal to the open cone $\mathrm{Amp}(X)$ generated by classes of* ample *(or very ample) divisors A (recall that a divisor A is said to be very ample if the linear system $H^0(X, \mathcal{O}(A))$ provides an embedding of X in projective space).*
(b) *The interior $\mathcal{E}_{NS}^\circ$ is the cone $\mathrm{Big}(X)$ generated by classes of* big *divisors, namely divisors D such that $h^0(X, \mathcal{O}(kD)) \geq c\, k^{\dim X}$ for k large.*
(c) *$\mathcal{E}_{NS}$ is the closure $\overline{\mathrm{Eff}(X)}$ of the cone generated by classes of* effective *divisors, i.e. divisors $D = \sum c_j D_j$, $c_j \in \mathbb{R}_+$.*
(d) *The closed cone $\overline{\mathcal{K}}_{NS}$ consists of the closure $\overline{\mathrm{Nef}(X)}$ of the cone generated by* nef *divisors D (or* nef *line bundles L), namely effective integral divisors D such that $D \cdot C \geq 0$ for every curve C, also equal to $\overline{\mathrm{Amp}(X)}$.*

In other words, the terminology "nef", "big", "pseudo-effective" used for classes of the full transcendental cones appear to be a natural extrapolation of the algebraic case.

Proof. First notice that since all of our cones $\mathscr{C}$ have non empty interior in $\mathrm{NS}_\mathbb{R}(X)$ (which is a rational vector space in terms of a basis of elements in $H^2(X, \mathbb{Q})$), the rational points $\mathscr{C}_\mathbb{Q} := \mathscr{C} \cap \mathrm{NS}_\mathbb{Q}(X)$, $\mathrm{NS}_\mathbb{Q}(X) = \mathrm{NS}(X) \otimes_\mathbb{Z} \mathbb{Q}$, are dense in each of them. (a) is therefore just Kodaira's embedding theorem when we look at rational points, and properties (b) and (d) are obtained easily by passing to the closure of the open cones. We will now give details of the proof only for (b) which is possibly slightly more involved.

By looking at points of $\mathcal{E}_\mathbb{Q}^\circ = \mathcal{E}^\circ \cap \mathrm{NS}_\mathbb{Q}(X)$ and multiplying by a denominator, it is enough to check that a line bundle L such that $c_1(L) \in \mathcal{E}^\circ$ is big. However, this means that L possesses a singular Hermitian metric h_L such that $\Theta_{L,h_L} \geq \varepsilon\omega$ for some Kähler metric ω. For some integer $p_0 > 0$, we can then produce a singular Hermitian metric with positive curvature and with a given logarithmic pole $h_L^{p_0} e^{-\theta(z) \log |z - x_0|^2}$ in a neighborhood of every point $x_0 \in X$ (here θ is a smooth cut-off function supported on a neighborhood of x_0). Then Hörmander's L^2 existence theorem [4, 56] can be used to produce sections of L^k which generate all jets of order $(k/p_0) - n$ at points x_0, so that L is big.

Conversely, if L is big and A is a (smooth) very ample divisor, the exact sequence
$0 \to \mathscr{O}_X(kL - A) \to \mathscr{O}_X(kL) \to \mathscr{O}_A(kL_{\restriction A}) \to 0$ and the estimates

$$h^0(X, \mathscr{O}_X(kL)) \geq ck^n, \quad h^0(A, \mathscr{O}_A(kL_{\restriction A})) = O(k^{n-1})$$

imply that $\mathscr{O}_X(kL - A)$ has a section for k large, thus $kL - A \equiv E$ for some effective divisor E. This means that there exists a singular metric h_L on L such that

$$\Theta_{L,h_L} = \frac{1}{k}\left(\Theta_{A,h_A} + [E]\right) \geq \frac{1}{k}\omega$$

where $\omega = \Theta_{A,h_A}$, hence $c_1(L) \in \mathscr{E}^\circ$. $\square$

Corollary 44. *If L is nef, then L is big (i.e. $\kappa(L) = n$) if and only if $L^n > 0$. Moreover, if L is nef and big, then for every $\delta > 0$, L has a singular metric $h = e^{-\varphi}$ such that $\max_{x \in X} \nu(\varphi, x) \leq \delta$ and $i\Theta_{L,h} \geq \varepsilon\omega$ for some $\varepsilon > 0$. The metric h can be chosen to be smooth on the complement of a fixed divisor E, with logarithmic poles along E.*

Proof. By holomorphic Morse inequalities 21 and the Riemann–Roch formula, we have

$$h^0(X, kL) = \chi(X, kL) + o(k^n) = k^n L^n/n! + o(k^n),$$

whence the first statement. By the proof of Theorem 43(b), there exists a singular metric h_1 on L such that

$$\frac{i}{2\pi}\Theta_{L,h_1} = \frac{1}{k}\left(\frac{i}{2\pi}\Theta_{A,h_A} + [E]\right) \geq \frac{1}{k}\omega, \qquad \omega = \frac{i}{2\pi}\Theta_{A,h_A}.$$

Now, for every $\varepsilon > 0$, there is a smooth metric h_ε on L such that $\frac{i}{2\pi}\Theta_{L,h_\varepsilon} \geq -\varepsilon\omega$. The convex combination of metrics $h'_\varepsilon = h_1^{k\varepsilon} h_\varepsilon^{1-k\varepsilon}$ is a singular metric with poles along E which satisfies

$$\frac{i}{2\pi}\Theta_{L,h'_\varepsilon} \geq \varepsilon(\omega + [E]) - (1 - k\varepsilon)\varepsilon\omega \geq k\varepsilon^2\omega.$$

Its Lelong numbers are $\varepsilon\nu(E, x)$ and they can be made smaller than δ by choosing $\varepsilon > 0$ small. $\square$

We still need a few elementary facts about the numerical dimension of nef line bundles.

Definition 45. Let L be a nef line bundle on a compact Kähler manifold (X, ω). One defines the numerical dimension of L to be

$$\mathrm{nd}(L) = \max\{k = 0, \ldots, n \,;\, c_1(L)^k \neq 0 \text{ in } H^{2k}(X, \mathbb{R})\}.$$

Notice that if L is nef, each power $c_1(L)^k$ can be represented by a closed positive current $\Theta_k \in c_1(L)^k$ obtained as a weak limit of powers of smooth positive forms

$$\Theta_k = \lim_{m \to +\infty} \left(\alpha + \frac{1}{m}\omega + \partial\bar{\partial}\varphi_m\right)^k, \qquad \alpha \in c_1(L).$$

Such a weak limit exists since $\int_X \left(\alpha + \frac{1}{m}\omega + \partial\bar{\partial}\varphi_m\right)^k \wedge \omega^{n-k}$ is uniformly bounded as $m \to +\infty$. Then we see that

$$\int_X c_1(L)^k \wedge \omega^{n-k} = \int_X \Theta_k \wedge \omega^{n-k} > 0 \iff \Theta_k \neq 0 \iff c_1(L)^k \neq 0.$$

By Corollary 44, we have $\kappa(L) = n$ if and only if $\mathrm{nd}(L) = n$. In general, we merely have an inequality.

Proposition 46. *If L is a nef line bundle on a compact Kähler manifold (X, ω), then $\kappa(L) \leq \mathrm{nd}(L)$.*

Proof. We consider arbitrary irreducible analytic subsets $Z \subset X$ and prove by induction on $p = \dim Z$ that $\kappa(L_{|Z}) \leq \mathrm{nd}(L_{|Z})$ where $\mathrm{nd}(L_{|Z})$ is the supremum of all integers k such that $c_1(L_{|Z})^k \neq 0$, i.e. $\int_X [Z] \wedge c_1(L)^k \wedge \omega^{p-k} > 0$. This will prove our statement when $Z = X$, $p = n$. The statement is trivial if $p = 0$, so we suppose now that $p > 0$. We can also assume that $r = \kappa(L_{|Z}) > 0$, otherwise there is nothing to prove. This implies that a sufficient large multiple $m_0 L$ has at least two independent sections σ_0, σ_1 on Z. Consider the linear system $|a_0\sigma_0 + a_1\sigma_1|$, $a = [a_0 : a_1] \in \mathbb{P}^1_\mathbb{C}$, and take $Y = Y_a \subset Z$ to be an irreducible component of the divisor of $\sigma_a := a_0\sigma_0 + a_1\sigma_1$ which is not a fixed component when a varies. For m sufficiently divisible, $\Phi_{mL_{|Z}}$ has rank r at a generic (smooth) point of Z, hence the rank of $(\Phi_{mL_{|Z}})_{|Y}$ is $\geq r' := \min(r, p-1)$ if $a \in \mathbb{P}^1_\mathbb{C}$ is itself generic. A fortiori rank$(\Phi_{mL_{|Y}}) \geq r'$ (we may even have sections on Y which do not extend to Z). By the induction hypothesis we find

$$\int_X [Y] \wedge c_1(L)^{r'} \wedge \omega^{p-1-r'} > 0.$$

Now, we use the fact that $[Z] \wedge c_1(m_0 L) - [Y]$ can be represented by an effective cycle (the sum of all components $\neq Y$ in the divisor of our generic section σ_a). This implies

$$\int_X [Z] \wedge c_1(L)^{r'+1} \wedge \omega^{p-1-r'} \geq \frac{1}{m_0} \int_X [Y] \wedge c_1(L)^{r'} \wedge \omega^{p-1-r'} > 0.$$

If $r = p$, we have $r' = p-1$, hence $r'+1 = r$ and we are done. If $r < p$, we have $r' = r$ and then we use the obvious inequality $\alpha \leq C_0 \omega$ for some representative $\alpha \in c_1(L)$ and some $C_0 > 0$ to conclude that

$$\int_X [Z] \wedge c_1(L)^r \wedge \omega^{p-r} \geq \frac{1}{C_0} \int_X [Z] \wedge c_1(L)^{r+1} \wedge \omega^{p-1-r} > 0. \qquad \square$$

Remark 47. It may happen that $\kappa(L) <$ nd(L): take e.g.

$$L \to X = X_1 \times X_2$$

equal to the total tensor product of an ample line bundle L_1 on a projective manifold X_1 and of a unitary flat line bundle L_2 on an elliptic curve X_2 given by a representation $\pi_1(X_2) \to U(1)$ such that no multiple kL_2 with $k \neq 0$ is trivial. Then $H^0(X, kL) = H^0(X_1, kL_1) \otimes H^0(X_2, kL_2) = 0$ for $k > 0$, and thus $\kappa(L) = -\infty$. However $c_1(L) = \mathrm{pr}_1^* c_1(L_1)$ has numerical dimension equal to $\dim X_1$. The same example shows that the Kodaira dimension may increase by restriction to a subvariety (if $Y = X_1 \times \{\text{point}\}$, then $\kappa(L_{\restriction Y}) = \dim Y$).

2.5 Approximation of Plurisubharmonic Functions via Bergman Kernels

We prove here, as an application of the Ohsawa–Takegoshi L^2 extension theorem [78], that every psh function on a pseudoconvex open set $\Omega \subset \mathbb{C}^n$ can be approximated very accurately by functions of the form $c \log |f|$, where $c > 0$ and f is a holomorphic function. The main idea is taken from [28]. For other applications to algebraic geometry, see [29] and Demailly–Kollár [38]. We first recall the statement of the generalized L^2 extension theorem; its proof relies on a subtle enhancement of the Bochner–Kodaira technique, and we refer to the literature for details.

Theorem 48 (Ohsawa–Takegoshi [78], Manivel [70]). *Let X be a complex n-dimensional manifold possessing a smooth plurisubharmonic exhaustion function ("weakly pseudoconvex" or "weakly 1-convex" manifold), and a Kähler metric ω. Let L (resp. E) be a Hermitian holomorphic line bundle (resp. a Hermitian holomorphic vector bundle of rank r over X), and s a global holomorphic section of E. Assume that s is generically transverse to the zero section, and let*

$$Y = \{x \in X \;;\; s(x) = 0, \Lambda^r ds(x) \neq 0\}, \qquad p = \dim Y = n - r.$$

Finally, let φ be an arbitrary plurisubharmonic function on X. Assume that the $(1,1)$-form $\Theta_L + r\frac{i}{2\pi}\partial\overline{\partial}(\log|s|^2 + \varphi)$ is semi-positive and that there is a continuous function $\alpha \geq 1$ such that the following two inequalities hold everywhere on X:

(a) $\Theta_L + r \dfrac{i}{2\pi} \partial\overline{\partial}(\log|s|^2 + \varphi) \geq \alpha^{-1} \dfrac{\{\Theta_E s, s\}}{|s|^2}$,

(b) $|s| \leq e^{-\alpha}$.

Then for every holomorphic section f_Y of the line bundle $\Lambda^n T_X^* \otimes L$ over Y such that $\int_Y |f_Y|^2 e^{-\varphi} |\Lambda^r(ds)|^{-2} dV_\omega < +\infty$, there exists a holomorphic extension f_X of f_Y over X such that

$$\int_X \frac{|f_X|^2 e^{-\varphi}}{|s|^{2r}(-\log|s|)^2} dV_{X,\omega} \leq C_r \int_Y \frac{|f_Y|^2 e^{-\varphi}}{|\Lambda^r(ds)|^2} dV_{Y,\omega},$$

where C_r is a numerical constant depending only on r.

Theorem 49. *Let φ be a plurisubharmonic function on a bounded pseudoconvex open set $\Omega \subset \mathbb{C}^n$. For every $m > 0$, let $\mathcal{H}_\Omega(m\varphi)$ be the Hilbert space of holomorphic functions f on Ω such that $\int_\Omega |f|^2 e^{-2m\varphi} d\lambda < +\infty$ and let $\varphi_m = \frac{1}{2m} \log \sum |\sigma_\ell|^2$ where (σ_ℓ) is an orthonormal basis of $\mathcal{H}_\Omega(m\varphi)$. Then there are constants $C_1, C_2 > 0$ independent of m such that*

(a) $\varphi(z) - \dfrac{C_1}{m} \leq \varphi_m(z) \leq \sup\limits_{|\zeta - z| < r} \varphi(\zeta) + \dfrac{1}{m} \log \dfrac{C_2}{r^n}$ *for every $z \in \Omega$ and $r < d(z, \partial\Omega)$. In particular, φ_m converges to φ pointwise and in L^1_{loc} topology on Ω when $m \to +\infty$ and*

(b) $\nu(\varphi, z) - \dfrac{n}{m} \leq \nu(\varphi_m, z) \leq \nu(\varphi, z)$ *for every $z \in \Omega$.*

Proof. (a) Note that $\sum |\sigma_\ell(z)|^2$ is the square of the norm of the evaluation linear form $\text{ev}_z : f \mapsto f(z)$ on $\mathcal{H}_\Omega(m\varphi)$, since $\sigma_\ell(z) = \text{ev}_z(\sigma_\ell)$ is the ℓ-th coordinate of ev_z in the orthonormal basis (σ_ℓ). In other words, we have

$$\sum |\sigma_\ell(z)|^2 = \sup_{f \in B(1)} |f(z)|^2$$

where $B(1)$ is the unit ball of $\mathcal{H}_\Omega(m\varphi)$ (The sum is called the *Bergman kernel* associated with $\mathcal{H}_\Omega(m\varphi)$). As φ is locally bounded from above, the L^2 topology is actually stronger than the topology of uniform convergence on compact subsets of Ω. It follows that the series $\sum |\sigma_\ell|^2$ converges uniformly on Ω and that its sum is real analytic. Moreover, by what we just explained, we have

$$\varphi_m(z) = \sup_{f \in B(1)} \frac{1}{m} \log |f(z)|.$$

For $z_0 \in \Omega$ and $r < d(z_0, \partial\Omega)$, the mean value inequality applied to the psh function $|f|^2$ implies

$$|f(z_0)|^2 \leq \frac{1}{\pi^n r^{2n}/n!} \int_{|z-z_0|<r} |f(z)|^2 d\lambda(z)$$

$$\leq \frac{1}{\pi^n r^{2n}/n!} \exp\left(2m \sup_{|z-z_0|<r} \varphi(z)\right) \int_\Omega |f|^2 e^{-2m\varphi} d\lambda.$$

If we take the supremum over all $f \in B(1)$ we get

$$\varphi_m(z_0) \leq \sup_{|z-z_0|<r} \varphi(z) + \frac{1}{2m} \log \frac{1}{\pi^n r^{2n}/n!}.$$

and the second inequality in (a) is proved—as we see, this is an easy consequence of the mean value inequality. Conversely, the Ohsawa–Takegoshi L^2 extension Theorem 48 applied to the 0-dimensional subvariety $\{z_0\} \subset \Omega$ and to the trivial bundles $L = \Omega \times \mathbb{C}$ and $E = \Omega \times \mathbb{C}^n$, with the section $s(z) = z - z_0$ of E, shows that for any $a \in \mathbb{C}$ there is a holomorphic function f on Ω such that $f(z_0) = a$ and

$$\int_\Omega |f|^2 e^{-2m\varphi} d\lambda \leq C_3 |a|^2 e^{-2m\varphi(z_0)},$$

where C_3 only depends on n and diam Ω. We fix a such that the right hand side is 1. Then $\|f\| \leq 1$ and so we get

$$\varphi_m(z_0) \geq \frac{1}{m} \log |f(z_0)| = \frac{1}{m} \log |a| = \varphi(z) - \frac{\log C_3}{2m}.$$

The inequalities given in (a) are thus proved. Taking $r = 1/m$, we find that

$$\lim_{m \to +\infty} \sup_{|\zeta - z| < 1/m} \varphi(\zeta) = \varphi(z)$$

by the upper semicontinuity of φ, and so $\lim \varphi_m(z) = \varphi(z)$, since $\lim \frac{1}{m} \log(C_2 m^n) = 0$.

(b) The above estimates imply

$$\sup_{|z-z_0|<r} \varphi(z) - \frac{C_1}{m} \leq \sup_{|z-z_0|<r} \varphi_m(z) \leq \sup_{|z-z_0|<2r} \varphi(z) + \frac{1}{m} \log \frac{C_2}{r^n}.$$

After dividing by $\log r < 0$ when $r \to 0$, we infer

$$\frac{\sup_{|z-z_0|<2r} \varphi(z) + \frac{1}{m} \log \frac{C_2}{r^n}}{\log r} \leq \frac{\sup_{|z-z_0|<r} \varphi_m(z)}{\log r} \leq \frac{\sup_{|z-z_0|<r} \varphi(z) - \frac{C_1}{m}}{\log r},$$

and from this and definition (50), it follows immediately that

$$\nu(\varphi, x) - \frac{n}{m} \leq \nu(\varphi_m, z) \leq \nu(\varphi, z). \qquad \square$$

Theorem 49 implies in a straightforward manner the deep result of [84] on the analyticity of the Lelong number upperlevel sets.

Corollary 50 ([84]). *Let φ be a plurisubharmonic function on a complex manifold X. Then, for every $c > 0$, the Lelong number upperlevel set*

$$E_c(\varphi) = \{z \in X \; ; \; \nu(\varphi, z) \geq c\}$$

is an analytic subset of X.

Proof. Since analyticity is a local property, it is enough to consider the case of a psh function φ on a pseudoconvex open set $\Omega \subset \mathbb{C}^n$. The inequalities obtained in Theorem 49(b) imply that

$$E_c(\varphi) = \bigcap_{m \geq m_0} E_{c-n/m}(\varphi_m).$$

Now, it is clear that $E_c(\varphi_m)$ is the analytic set defined by the equations $\sigma_\ell^{(\alpha)}(z) = 0$ for all multi-indices α such that $|\alpha| < mc$. Thus $E_c(\varphi)$ is analytic as a (countable) intersection of analytic sets. □

Remark 51. It can be easily shown that the Lelong numbers of any closed positive (p, p)-current coincide (at least locally) with the Lelong numbers of a suitable plurisubharmonic potential φ (see [94]). Hence Siu's theorem also holds true for the Lelong number upperlevel sets $E_c(T)$ of any closed positive (p, p)-current T.

Theorem 49 motivates the following definition.

Definition 52. A plurisubharmonic function φ on a complex manifold X is said to have analytic singularities if it can be written locally near every point $x_0 \in X$ as

$$\varphi(z) = c \log \sum_{1 \leq j \leq N} |g_j(z)|^2 + O(1), \qquad \text{i.e. up to equivalence of singularities,}$$

with a family of holomorphic functions (g_j) defined near x_0 and $c > 0$. Also, a closed positive $(1, 1)$ current T is said to have analytic singularities if its plurisubharmonic potential has analytic singularities. We also refer to this situation by saying that φ or T have logarithmic poles. When X is algebraic, we say that the singularities are algebraic if $c \in \mathbb{Q}_+$ and the (g_j) are sections of some algebraic line bundle $\mathcal{O}(D)$, $x_0 \notin \operatorname{Supp} D$.

Notice that by Noetherianity, a convergent series $\log \sum_{j \in \mathbb{N}} |g_j|^2$ can be replaced by a finite sum up to equivalence of singularities, thus Theorem 49 always produces plurisubharmonic functions φ_m with analytic singularities.

2.6 Global Approximation of Closed (1,1)-Currents on a Compact Complex Manifold

We take here X to be an arbitrary compact complex manifold (no Kähler assumption is needed). Now, let T be a closed $(1, 1)$-current on X. We assume that T is *quasi-positive*, i.e. that there exists a $(1, 1)$-form γ with continuous coefficients such that $T \geq \gamma$; the case of positive currents ($\gamma = 0$) is of course the most important.

Lemma 53. *There exists a smooth closed $(1,1)$-form α representing the same $\partial\bar\partial$-cohomology class as T and a quasi-psh function φ on X such that $T = \alpha + \frac{i}{\pi}\partial\bar\partial\varphi$. (We say that a function φ is quasi-psh if its complex Hessian is bounded below by a $(1,1)$-form with locally bounded coefficients, that is, if $i\partial\bar\partial\varphi$ is quasi-positive).*

Proof. Select an open covering (U_j) of X by coordinate balls such that $T = \frac{i}{\pi}\partial\bar\partial\varphi_j$ over U_j, and construct a global function $\varphi = \sum \theta_j \varphi_j$ by means of a partition of unity $\{\theta_j\}$ subordinate to U_j. Now, we observe that $\varphi - \varphi_k$ is smooth on U_k because all differences $\varphi_j - \varphi_k$ are smooth in the intersections $U_j \cap U_k$, and we have the equality $\varphi - \varphi_k = \sum \theta_j(\varphi_j - \varphi_k)$. Therefore $\alpha := T - \frac{i}{\pi}\partial\bar\partial\varphi$ is smooth. □

By replacing T with $T - \alpha$ and γ with $\gamma - \alpha$, we can assume without loss of generality that $\{T\} = 0$, i.e. that $T = \frac{i}{\pi}\partial\bar\partial\varphi$ with a quasi-psh function φ on X such that $\frac{i}{\pi}\partial\bar\partial\varphi \geq \gamma$.

Our goal is to approximate T in the weak topology by currents $T_m = \frac{i}{\pi}\partial\bar\partial\varphi_m$ such their potentials φ_m have analytic singularities in the sense of Definition 52, more precisely, defined on a neighborhood V_{x_0} of any point $x_0 \in X$ in the form $\varphi_m(z) = c_m \log \sum_j |\sigma_{j,m}|^2 + O(1)$, where $c_m > 0$ and the $\sigma_{j,m}$ are holomorphic functions on V_{x_0}.

We select a finite covering (W_ν) of X with open coordinate charts, and shrink them a little to be on the safe side. Given $\delta > 0$, we take in each W_ν a maximal family of points with (coordinate) distance to the boundary $\geq 3\delta$ and mutual distance $\geq \delta/2$. In this way, we get for $\delta > 0$ small a finite covering of X by open balls U'_j of radius δ (actually every point is even at distance $\leq \delta/2$ of one of the centers, otherwise the family of points would not be maximal), such that the concentric ball U_j of radius 2δ is relatively compact in the corresponding chart W_ν. Let $\tau_j : U_j \longrightarrow B(a_j, 2\delta)$ be the isomorphism given by the coordinates of W_ν; by taking $\delta > 0$ small enough, we can assume that the coordinates of U_j extend to $U_j \cup U_k$ whenever $U_j \cap U_k \neq \emptyset$. Let $\varepsilon(\delta)$ be a modulus of continuity for γ on the sets U_j, such that $\lim_{\delta \to 0} \varepsilon(\delta) = 0$ and $\gamma_x - \gamma_{x'} \leq \frac{1}{2}\varepsilon(\delta) \omega_x$ for all $x, x' \in U_j$. We denote by γ_j the $(1,1)$-form with constant coefficients on $B(a_j, 2\delta)$ such that $\tau_j^* \gamma_j$ coincides with $\gamma - \varepsilon(\delta) \omega$ at $\tau_j^{-1}(a_j)$. Then we have

$$0 \leq \gamma - \tau_j^* \gamma_j \leq 2\varepsilon(\delta) \omega \quad \text{on } U_j \tag{53}$$

for $\delta > 0$ small. We set $\varphi_j = \varphi \circ \tau_j^{-1}$ on $B(a_j, 2\delta)$ and let q_j be the homogeneous quadratic function in $z - a_j$ such that $\frac{i}{\pi}\partial\bar\partial q_j = \gamma_j$ on $B(a_j, 2\delta)$. Then $\varphi_j - q_j$ is plurisubharmonic on $B(a_j, 2\delta)$ since

$$\frac{i}{\pi}\partial\bar\partial((\varphi_j - q_j) \circ \tau_j) = T - \tau_j^* \gamma_j \geq \gamma - \tau_j^* \gamma_j \geq 0. \tag{54}$$

We let $U'_j \subset\subset U''_j \subset\subset U_j$ be the concentric balls of radii δ, 1.5δ, 2δ respectively. On each open set U_j the function $\psi_j := \varphi - q_j \circ \tau_j = (\varphi_j - q_j) \circ \tau_j$ is plurisubharmonic, so Theorem 49 applied with $\Omega = U_j \simeq B(a_j, 2\delta)$ produces functions

Applications of Pluripotential Theory to Algebraic Geometry 191

$$\psi_{j,m} = \frac{1}{2m} \log \sum_\ell |\sigma_{j,\ell}|^2, \quad (\sigma_{j,\ell}) = \text{basis of } \mathcal{H}_{U_j}(m\psi_j). \tag{55}$$

The functions $\psi_{j,m} + q_j \circ \tau_j$ on U_j then have to be glued together by a partition of unity technique. For this, we rely on the following "discrepancy" lemma, estimating the variation of the approximating functions on overlapping balls.

Lemma 54. *There is a constant C independent of m and δ such that the quasi-psh functions $w_{j,m} = 2m(\psi_{j,m} + q_j \circ \tau_j)$, i.e.*

$$w_{j,m}(x) = 2m\, q_j \circ \tau_j(x) + \log \sum_\ell |\sigma_{j,\ell}(x)|^2, \quad x \in U_j'',$$

satisfy

$$|w_{j,m} - w_{k,m}| \leq C\left(\log \delta^{-1} + m\varepsilon(\delta)\delta^2\right) \quad \text{on } U_j'' \cap U_k''.$$

Proof. The details will be left as an exercise to the reader. The main idea is the following: for any holomorphic function $f_j \in \mathcal{H}_{U_j}(m\psi_j)$, a $\overline{\partial}$ equation $\overline{\partial} u = \overline{\partial}(\theta f_j)$ can be solved on U_k, where θ is a cut-off function with support in $U_j'' \cap U_k''$, on a ball of radius $< \delta/4$, equal to 1 on the ball of radius $\delta/8$ centered at a given point $x_0 \in U_j'' \cap U_k''$, with $|\overline{\partial}\theta| = O(\delta^{-1})$. We apply the L^2 estimate with respect to the weight $(n+1)\log|x - x_0|^2 + 2m\psi_k$, where the first term is picked up so as to force the solution u to vanish at x_0, in such a way that $F_k = u - \theta f_j$ is holomorphic and $F_k(x_0) = f_j(x_0)$. The discrepancy between the weights on U_j'' and U_k'' is given by

$$\psi_j - \psi_k = -(q_j \circ \tau_j - q_k \circ \tau_k).$$

By re-centering the quadratic functions at $\tau_j(x_0)$, resp. $\tau_k(x_0)$, we can write

$$q_j \circ \tau_j - q_k \circ \tau_k = \operatorname{Re} G_{jk} + R_{jk}$$

where G_{jk} is holomorphic on $U_j \cup U_k$ [equal to a difference of linear forms in the coordinates of $B(a_j, 2\delta)$ and $B(a_k, 2\delta)$], $G_{jk}(x_0) = q_j \circ \tau_j(x_0) - q_k \circ \tau_k(x_0)$ and $R_{jk} = O(\varepsilon(\delta)\delta^2)$ is a remainder term coming from the change of coordinates and the slight discrepancy between $\partial\overline{\partial}(q_j \circ \tau_j)$ and $\partial\overline{\partial}(q_k \circ \tau_k)$ at the common point x_0, with $R_{jk}(x_0) = 0$. In this way, we get

$$|e^{mG_{jk}}|^2 e^{-m\psi_k} = e^{-m\psi_j - 2mR_{jk}},$$

so that we have a uniform control of the L^2 norm of the solution $f_k = e^{mG_{jk}} F_k = e^{mG_{jk}}(u - \theta f_j)$ of the form

$$\int_{U_k} |f_k|^2 e^{-2m\psi_k} \leq C\delta^{-2n-4} e^{mO(\varepsilon(\delta)\delta^2)} \int_{U_j} |f_j|^2 e^{-2m\psi_j}.$$

The required estimate follows, using the equality

$$e^{2m\psi_{j,m}(x)} = \sum_\ell |\sigma_{j,\ell}(x)|^2 = \sup_{f\in\mathcal{H}_{U_j}(m\psi_j),\, \|f\|\le 1} |f(x)|^2 \quad \text{on } U_j,$$

and the analogous equality on U_k. □

Now, the actual gluing of our quasi-psh functions is performed using the following elementary partition of unity calculation.

Lemma 55. *Let $U'_j \Subset U''_j$ be locally finite open coverings of a complex manifold X by relatively compact open sets, and let θ_j be smooth nonnegative functions with support in U''_j, such that $\theta_j \le 1$ on U''_j and $\theta_j = 1$ on U'_j. Let $A_j \ge 0$ be such that*

$$i(\theta_j \partial\bar\partial \theta_j - \partial\theta_j \wedge \bar\partial\theta_j) \ge -A_j \omega \quad \text{on } U''_j \smallsetminus U'_j$$

for some positive $(1,1)$-form ω. Finally, let w_j be quasi-psh functions on U_j with the property that $i\partial\bar\partial w_j \ge \gamma$ for some real $(1,1)$-form γ on M, and let C_j be constants such that

$$w_j(x) \le C_j + \sup_{k\ne j,\, U'_k \ni x} w_k(x) \quad \text{on } U''_j \smallsetminus U'_j.$$

Then the function $w = \log\left(\sum \theta_j^2 e^{w_j}\right)$ is quasi-psh and satisfies

$$i\partial\bar\partial w \ge \gamma - 2\left(\sum_j \mathbb{1}_{U''_j \smallsetminus U'_j} A_j e^{C_j}\right)\omega.$$

Proof. If we set $\alpha_j = \theta_j \partial w_j + 2\partial\theta_j$, a straightforward computation shows that

$$\partial w = \frac{\sum(\theta_j^2 \partial w_j + 2\theta_j \partial\theta_j)e^{w_j}}{\sum \theta_j^2 e^{w_j}} = \frac{\sum \theta_j e^{w_j} \alpha_j}{\sum \theta_j^2 e^{w_j}},$$

$$\partial\bar\partial w = \frac{\sum(\alpha_j \wedge \bar\alpha_j + \theta_j^2 \partial\bar\partial w_j + 2\theta_j \partial\bar\partial\theta_j - 2\partial\theta_j \wedge \bar\partial\theta_j)e^{w_j}}{\sum \theta_j^2 e^{w_j}} - \frac{\sum_{j,k} \theta_j e^{w_j} \theta_k e^{w_k} \alpha_j \wedge \bar\alpha_k}{\left(\sum \theta_j^2 e^{w_j}\right)^2}$$

$$= \frac{\sum_{j<k}|\theta_j \alpha_k - \theta_k \alpha_j|^2 e^{w_j} e^{w_k}}{\left(\sum \theta_j^2 e^{w_j}\right)^2} + \frac{\sum \theta_j^2 e^{w_j} \partial\bar\partial w_j}{\sum \theta_j^2 e^{w_j}} + \frac{\sum(2\theta_j \partial\bar\partial\theta_j - 2\partial\theta_j \wedge \bar\partial\theta_j)e^{w_j}}{\sum \theta_j^2 e^{w_j}}$$

by using the Legendre identity. The first term in the last line is nonnegative and the second one is $\ge \gamma$. In the third term, if x is in the support of $\theta_j \partial\bar\partial\theta_j - \partial\theta_j \wedge \bar\partial\theta_j$, then $x \in U''_j \smallsetminus U'_j$ and so $w_j(x) \le C_j + w_k(x)$ for some $k \ne j$ with $U'_k \ni x$ and $\theta_k(x) = 1$. This gives

$$i \frac{\sum(2\theta_j \partial\bar\partial\theta_j - 2\partial\theta_j \wedge \bar\partial\theta_j)e^{w_j}}{\sum \theta_j^2 e^{w_j}} \ge -2\sum_j \mathbb{1}_{U''_j \smallsetminus U'_j} e^{C_j} A_j \omega.$$

The expected lower bound follows. □

Applications of Pluripotential Theory to Algebraic Geometry

We apply Lemma 55 to functions $\tilde{w}_{j,m}$ which are just slight modifications of the functions $w_{j,m} = 2m(\psi_{j,m} + q_j \circ \tau_j)$ occurring in Lemma 54:

$$\tilde{w}_{j,m}(x) = w_{j,m}(x) + 2m\left(\frac{C_1}{m} + C_3\varepsilon(\delta)(\delta^2/2 - |\tau_j(x)|^2)\right)$$

$$= 2m\left(\psi_{j,m}(x) + q_j \circ \tau_j(x) + \frac{C_1}{m} + C_3\varepsilon(\delta)(\delta^2/2 - |\tau_j(x)|^2)\right)$$

where $x \mapsto z = \tau_j(x)$ is a local coordinate identifying U_j to $B(0, 2\delta)$, C_1 is the constant occurring in Lemma 54 and C_3 is a sufficiently large constant. It is easy to see that we can take $A_j = C_4\delta^{-2}$ in Lemma 55. We have

$$\tilde{w}_{j,m} \geq w_{j,m} + 2C_1 + m\frac{C_3}{2}\varepsilon(\delta)\delta^2 \quad \text{on } B(x_j, \delta/2) \subset U'_j,$$

since $|\tau_j(x)| \leq \delta/2$ on $B(x_j, \delta/2)$, while

$$\tilde{w}_{j,m} \leq w_{j,m} + 2C_1 - mC_3\varepsilon(\delta)\delta^2 \quad \text{on } U''_j \smallsetminus U'_j.$$

For $m \geq m_0(\delta) = (\log \delta^{-1}/(\varepsilon(\delta)\delta^2)$, Lemma 54 implies $|w_{j,m} - w_{k,m}| \leq C_5 m\varepsilon(\delta)\delta^2$ on $U''_j \cap U''_k$. Hence, for C_3 large enough, we get

$$\tilde{w}_{j,m}(x) \leq \sup_{k \neq j,\, B(x_k, \delta/2) \ni x} w_{k,m}(x) \leq \sup_{k \neq j,\, U'_k \ni x} w_{k,m}(x) \quad \text{on } U''_j \smallsetminus U'_j,$$

and we can take $C_j = 0$ in the hypotheses of Lemma 55. The associated function $w = \log\left(\sum \theta_j^2 e^{\tilde{w}_{j,m}}\right)$ is given by

$$w = \log \sum_j \theta_j^2 \exp\left(2m\left(\psi_{j,m} + q_j \circ \tau_j + \frac{C_1}{m} + C_3\varepsilon(\delta)(\delta^2/2 - |\tau_j|^2)\right)\right).$$

If we define $\varphi_m = \frac{1}{2m}w$, we get

$$\varphi_m(x) := \frac{1}{2m}w(x) \geq \psi_{j,m}(x) + q_j \circ \tau_j(x) + \frac{C_1}{m} + \frac{C_3}{4}\varepsilon(\delta)\delta^2 > \varphi(x)$$

in view of Lemma 54, by picking an index j such that $x \in B(x_j, \delta/2)$. In the opposite direction, the maximum number N of overlapping balls U_j does not depend on δ, and we thus get

$$w \leq \log N + 2m\left(\max_j \{\psi_{j,m}(x) + q_j \circ \tau_j(x)\} + \frac{C_1}{m} + \frac{C_3}{2}\varepsilon(\delta)\delta^2\right).$$

By definition of ψ_j we have $\sup_{|\zeta-x|<r} \psi_j(\zeta) \leq \sup_{|\zeta-x|<r} \varphi(\zeta) - q_j \circ \tau_j(x) + C_5 r$ thanks to the uniform Lipschitz continuity of $q_j \circ \tau_j$, thus by Lemma 54 again we find

$$\varphi_m(x) \leq \frac{\log N}{2m} + \sup_{|\zeta-x|<r} \varphi(\zeta) + \frac{C_1}{m} + \frac{1}{m} \log \frac{C_2}{r^n} + \frac{C_3}{2} \varepsilon(\delta) \delta^2 + C_5 r.$$

By taking for instance $r = 1/m$ and $\delta = \delta_m \to 0$, we see that φ_m converges to φ. On the other hand (53) implies $\frac{i}{\pi} \partial\bar\partial q_j \circ \tau_j(x) = \tau_j^* \gamma_j \geq \gamma - 2\varepsilon(\delta)\omega$, thus

$$\frac{i}{\pi} \partial\bar\partial \tilde w_{j,m} \geq 2m(\gamma - C_6 \varepsilon(\delta)\omega).$$

Lemma 55 then produces the lower bound

$$\frac{i}{\pi} \partial\bar\partial w \geq 2m(\gamma - C_6 \varepsilon(\delta)\omega) - C_7 \delta^{-2}\omega,$$

whence

$$\frac{i}{\pi} \partial\bar\partial \varphi_m \geq \gamma - C_8 \varepsilon(\delta)\omega$$

for $m \geq m_0(\delta) = (\log \delta^{-1})/(\varepsilon(\delta)\delta^2)$. We can fix $\delta = \delta_m$ to be the smallest value of $\delta > 0$ such that $m_0(\delta) \leq m$, then $\delta_m \to 0$ and we have obtained a sequence of quasi-psh functions φ_m satisfying the following properties.

Theorem 56. *Let φ be a quasi-psh function on a compact complex manifold X such that $\frac{i}{\pi}\partial\bar\partial\varphi \geq \gamma$ for some continuous $(1,1)$-form γ. Then there is a sequence of quasi-psh functions φ_m such that φ_m has the same singularities as a logarithm of a sum of squares of holomorphic functions and a decreasing sequence $\varepsilon_m > 0$ converging to 0 such that*

(a) $\varphi(x) < \varphi_m(x) \leq \sup_{|\zeta-x|<r} \varphi(\zeta) + C\left(\frac{|\log r|}{m} + r + \varepsilon_m\right)$
with respect to coordinate open sets covering X. In particular, φ_m converges to φ pointwise and in $L^1(X)$ and

(b) $\nu(\varphi,x) - \frac{n}{m} \leq \nu(\varphi_m, x) \leq \nu(\varphi, x)$ *for every $x \in X$;*

(c) $\frac{i}{\pi}\partial\bar\partial\varphi_m \geq \gamma - \varepsilon_m \omega.$

In particular, we can apply this to an arbitrary positive or quasi-positive closed $(1,1)$-current $T = \alpha + \frac{i}{\pi}\partial\bar\partial\varphi$.

Corollary 57. *Let T be a quasi-positive closed $(1,1)$-current on a compact complex manifold X such that $T \geq \gamma$ for some continuous $(1,1)$-form γ. Then there is a sequence of currents T_m whose local potentials have the same singularities* as $1/m$ times a logarithm of a sum of squares of holomorphic functions and a decreasing sequence $\varepsilon_m > 0$ converging to 0 such that

(a) T_m converges weakly to T,
(b) $\nu(T, x) - \dfrac{n}{m} \leq \nu(T_m, x) \leq \nu(T, x)$ for every $x \in X$,
(c) $T_m \geq \gamma - \varepsilon_m \omega$.

We say that our currents T_m are approximations of T with logarithmic poles.

By using blow-ups of X, the structure of the currents T_m can be better understood. In fact, consider the coherent ideals $\mathcal{J}_m$ generated locally by the holomorphic functions $(\sigma_{j,m}^{(k)})$ on U_k in the local approximations

$$\varphi_{k,m} = \frac{1}{2m} \log \sum_j |\sigma_{j,m}^{(k)}|^2 + O(1)$$

of the potential φ of T on U_k. These ideals are in fact globally defined, because the local ideals $\mathcal{J}_m^{(k)} = (\sigma_{j,m}^{(k)})$ are integrally closed, and they coincide on the intersections $U_k \cap U_\ell$ as they have the same order of vanishing by the proof of Lemma 54. By Hironaka [55], we can find a composition of blow-ups with smooth centers $\mu_m : \tilde{X}_m \to X$ such that $\mu_m^* \mathcal{J}_m$ is an invertible ideal sheaf associated with a normal crossing divisor E_m. Now, we can write

$$\mu_m^* \varphi_{k,m} = \varphi_{k,m} \circ \mu_m = \frac{1}{m} \log |s_{E_m}| + \tilde{\varphi}_{k,m}$$

where s_{E_m} is the canonical section of $\mathcal{O}(-E_m)$ and $\tilde{\varphi}_{k,m}$ is a smooth potential. This implies

$$\mu_m^* T_m = \frac{1}{m}[E_m] + \beta_m \tag{56}$$

where $[E_m]$ is the current of integration over E_m and β_m is a smooth closed $(1, 1)$-form which satisfies the lower bound $\beta_m \geq \mu_m^*(\gamma - \varepsilon_m \omega)$. (Recall that the pull-back of a closed $(1, 1)$-current by a holomorphic map f is always well-defined, by taking a local plurisubharmonic potential φ such that $T = i\partial\bar{\partial}\varphi$ and writing $f^*T = i\partial\bar{\partial}(\varphi \circ f)$). In the remainder of this section, we derive from this a rather important geometric consequence, first appeared in [42]). We need two related definitions.

Definition 58. A Kähler current on a compact complex space X is a closed positive current T of bidegree $(1, 1)$ which satisfies $T \geq \varepsilon \omega$ for some $\varepsilon > 0$ and some smooth positive Hermitian form ω on X.

Definition 59. A compact complex manifold is said to be in the Fujiki class $\mathscr{C}$ if it is bimeromorphic to a Kähler manifold (or equivalently, using Hironaka's desingularization theorem, if it admits a proper Kähler modification).

Theorem 60. *A compact complex manifold X is bimeromorphic to a Kähler manifold (i.e. $X \in \mathscr{C}$) if and only if it admits a Kähler current.*

Proof. If X is bimeromorphic to a Kähler manifold Y, Hironaka's desingularization theorem implies that there exists a blow-up $\tilde{Y}$ of Y (obtained by a sequence of blow-ups with smooth centers) such that the bimeromorphic map from Y to X can be resolved into a modification $\mu : \tilde{Y} \to X$. Then $\tilde{Y}$ is Kähler and the push-forward $T = \mu_* \tilde{\omega}$ of a Kähler form $\tilde{\omega}$ on $\tilde{Y}$ provides a Kähler current on X. In fact, if ω is a smooth Hermitian form on X, there is a constant C such that $\mu^* \omega \leq C \tilde{\omega}$ (by compactness of $\tilde{Y}$), hence

$$T = \mu_* \tilde{\omega} \geq \mu_*(C^{-1}\mu^*\omega) = C^{-1}\omega.$$

Conversely, assume that X admits a Kähler current $T \geq \varepsilon \omega$. By Corollary 57(c), there exists a Kähler current $\tilde{T} = T_m \geq \frac{\varepsilon}{2}\omega$ (with $m \gg 1$ so large that $\varepsilon_m \leq \varepsilon/2$) in the same $\partial\bar{\partial}$-cohomology class as T, possessing logarithmic poles. Observation (56) implies the existence of a composition of blow-ups $\mu : \tilde{X} \to X$ such that

$$\mu^* \tilde{T} = [\tilde{E}] + \tilde{\beta} \qquad \text{on } \tilde{X},$$

where $\tilde{E}$ is a $\mathbb{Q}$-divisor with normal crossings and $\tilde{\beta}$ a smooth closed $(1, 1)$-form such that $\tilde{\beta} \geq \frac{\varepsilon}{2}\mu^*\omega$. In particular $\tilde{\beta}$ is positive outside the exceptional locus of μ. This is not enough yet to produce a Kähler form on $\tilde{X}$, but we are not very far. Suppose that $\tilde{X}$ is obtained as a tower of blow-ups

$$\tilde{X} = X_N \to X_{N-1} \to \cdots \to X_1 \to X_0 = X,$$

where X_{j+1} is the blow-up of X_j along a smooth center $Y_j \subset X_j$. Denote by $S_{j+1} \subset X_{j+1}$ the exceptional divisor, and let $\mu_j : X_{j+1} \to X_j$ be the blow-up map. Now, we use the following simple

Lemma 61. *For every Kähler current T_j on X_j, there exists $\varepsilon_{j+1} > 0$ and a smooth form u_{j+1} in the $\partial\bar{\partial}$-cohomology class of $[S_{j+1}]$ such that*

$$T_{j+1} = \mu_j^* T_j - \varepsilon_{j+1} u_{j+1}$$

is a Kähler current on X_{j+1}.

Proof. The line bundle $\mathcal{O}(-S_{j+1})|S_{j+1}$ is equal to $\mathcal{O}_{P(N_j)}(1)$ where N_j is the normal bundle to Y_j in X_j. Pick an arbitrary smooth Hermitian metric on N_j, use this metric to get an induced Fubini–Study metric on $\mathcal{O}_{P(N_j)}(1)$, and finally extend this metric as a smooth Hermitian metric on the line bundle $\mathcal{O}(-S_{j+1})$. Such a metric has positive curvature along tangent vectors of X_{j+1} which are tangent to the fibers of $S_{j+1} = P(N_j) \to Y_j$. Assume furthermore that $T_j \geq \delta_j \omega_j$ for some Hermitian form ω_j on X_j and a suitable $0 < \delta_j \ll 1$. Then

$$\mu_j^* T_j - \varepsilon_{j+1} u_{j+1} \geq \delta_j \mu_j^* \omega_j - \varepsilon_{j+1} u_{j+1}$$

Applications of Pluripotential Theory to Algebraic Geometry 197

where $\mu_j^*\omega_j$ is semi-positive on X_{j+1}, positive definite on $X_{j+1} \smallsetminus S_{j+1}$, and also positive definite on tangent vectors of $T_{X_{j+1}|S_{j+1}}$ which are not tangent to the fibers of $S_{j+1} \to Y_j$. The statement is then easily proved by taking $\varepsilon_{j+1} \ll \delta_j$ and by using an elementary compactness argument on the unit sphere bundle of $T_{X_{j+1}}$ associated with any given Hermitian metric. □

End of proof of Theorem 60. If $\tilde{u}_j$ is the pull-back of u_j to the final blow-up $\tilde{X}$, we conclude inductively that $\mu^*\tilde{T} - \sum \varepsilon_j \tilde{u}_j$ is a Kähler current. Therefore the smooth form

$$\tilde{\omega} := \tilde{\beta} - \sum \varepsilon_j \tilde{u}_j = \mu^*\tilde{T} - \sum \varepsilon_j \tilde{u}_j - [\tilde{E}]$$

is Kähler and we see that $\tilde{X}$ is a Kähler manifold. □

Remark 62. A special case of Theorem 60 is the following characterization of Moishezon varieties (i.e. manifolds which are bimeromorphic to projective algebraic varieties or, equivalently, whose algebraic dimension is equal to their complex dimension): *A compact complex manifold X is Moishezon if and only if X possesses a Kähler current T such that the De Rham cohomology class $\{T\}$ is rational, i.e. $\{T\} \in H^2(X, \mathbb{Q})$.* In fact, in the above proof, we get an integral current T if we take the push forward $T = \mu_*\tilde{\omega}$ of an integral ample class $\{\tilde{\omega}\}$ on Y, where $\mu : Y \to X$ is a projective model of Y. Conversely, if $\{T\}$ is rational, we can take the ε_j's to be rational in Lemma 61. This produces at the end a Kähler metric $\tilde{\omega}$ with rational De Rham cohomology class on $\tilde{X}$. Therefore $\tilde{X}$ is projective by the Kodaira embedding theorem. This result was already observed in [59] (see also [13, 14] and Sect. 3.7 for a more general perspective based on a singular holomorphic Morse inequalities).

Remark 63. Hodge decomposition also holds true for manifolds $X \in \mathscr{C}$. In fact let $\mu : \tilde{X} \to X$ be a modification such that $\tilde{X}$ is Kähler. Then there are natural morphisms

$$\mu^* : H^{p,q}_{\bar{\partial}}(X, \mathbb{C}) \to H^{p,q}_{\bar{\partial}}(\tilde{X}, \mathbb{C}), \qquad \mu_* : H^{p,q}_{\bar{\partial}}(\tilde{X}, \mathbb{C}) \to H^{p,q}_{\bar{\partial}}(X, \mathbb{C})$$

induced respectively by the pull-back of smooth forms (resp. the direct image of currents). Clearly, $\mu_* \circ \mu^* = \mathrm{Id}$, therefore μ^* is injective and μ_* surjective, and similar results hold true for Bott–Chern cohomology or De Rham cohomology. It follows easily from this that the $\partial\bar{\partial}$-lemma still holds true for $X \in \mathscr{C}$, and that there are isomorphisms

$$H^{p,q}_{\mathrm{BC}}(X, \mathbb{C}) \to H^{p,q}_{\bar{\partial}}(X, \mathbb{C}), \qquad \bigoplus_{p+q=k} H^{p,q}_{\mathrm{BC}}(X, \mathbb{C}) \to H^k_{\mathrm{DR}}(X, \mathbb{C}).$$

2.7 Zariski Decomposition and Mobile Intersections

Let X be compact Kähler and let $\alpha \in \mathscr{E}^\circ$ be in the *interior* of the pseudo–effective cone. In analogy with the algebraic context such a class α is called "big", and it can then be represented by a *Kähler current* T, i.e. a closed positive $(1, 1)$-current T such that $T \geq \delta \omega$ for some smooth Hermitian metric ω and a constant $\delta \ll 1$. We first need a variant of the approximation theorem proved in Paragraph 5.

Regularization theorem for currents 64. *Let X be a compact complex manifold equipped with a Hermitian metric ω. Let $T = \alpha + i\partial\bar{\partial}\varphi$ be a closed $(1, 1)$-current on X, where α is smooth and φ is a quasi-plurisubharmonic function. Assume that $T \geq \gamma$ for some real $(1, 1)$-form γ on X with real coefficients. Then there exists a sequence $T_m = \alpha + i\partial\bar{\partial}\varphi_m$ of closed $(1, 1)$-currents such that*

(a) *φ_m (and thus T_m) is smooth on the complement $X \smallsetminus Z_m$ of an analytic set Z_m, and the Z_m's form an increasing sequence*

$$Z_0 \subset Z_1 \subset \cdots \subset Z_m \subset \cdots \subset X.$$

(b) *There is a uniform estimate $T_m \geq \gamma - \delta_m \omega$ with $\lim \downarrow \delta_m = 0$ as m tends to $+\infty$.*

(c) *The sequence (φ_m) is non increasing, and we have $\lim \downarrow \varphi_m = \varphi$. As a consequence, T_m converges weakly to T as m tends to $+\infty$.*

(d) *Near Z_m, the potential φ_m has logarithmic poles, namely, for every $x_0 \in Z_m$, there is a neighborhood U of x_0 such that $\varphi_m(z) = \lambda_m \log \sum_\ell |g_{m,\ell}|^2 + O(1)$ for suitable holomorphic functions $(g_{m,\ell})$ on U and $\lambda_m > 0$. Moreover, there is a (global) proper modification $\mu_m : \tilde{X}_m \to X$ of X, obtained as a sequence of blow-ups with smooth centers, such that $\varphi_m \circ \mu_m$ can be written locally on $\tilde{X}_m$ as*

$$\varphi_m \circ \mu_m(w) = \lambda_m \Big(\sum n_\ell \log |\tilde{g}_\ell|^2 + f(w) \Big)$$

where $(\tilde{g}_\ell = 0)$ are local generators of suitable (global) divisors E_ℓ on $\tilde{X}_m$ such that $\sum E_\ell$ has normal crossings, n_ℓ are positive integers, and the f's are smooth functions on $\tilde{X}_m$.

Sketch of proof. We essentially repeat the proofs of Theorems 49 and 56 with additional considerations. One fact that does not follow readily from these proofs is the monotonicity of the sequence φ_m (which we will not really need anyway— it can be obtained by applying Theorem 49 with 2^m instead of m, and by using the Ohsawa–Takegoshi L^2 extension theorem 48 for potentials $2^m \varphi(x) + 2^m \varphi(y)$ on the diagonal of $X \times X$, so that the restriction is $2^{m+1}\varphi(x)$ on the diagonal; we refer e.g. to [44] for details). The map μ_m is obtained by blowing-up the (global) ideals $\mathscr{J}_m$ defined by the holomorphic functions $(g_{j,m})$ in the local approximations $\varphi_m \sim \frac{1}{2m} \log \sum_j |g_{j,m}|^2$. By Hironaka [55], we can achieve that $\mu_m^* \mathscr{J}_m$ is an invertible ideal sheaf associated with a normal crossing divisor. □

Corollary 65. *If T is a Kähler current, then one can write $T = \lim T_m$ for a sequence of Kähler currents T_m which have logarithmic poles with coefficients in $\frac{1}{m}\mathbb{Z}$, i.e. there are modifications $\mu_m : X_m \to X$ such that*

$$\mu_m^* T_m = [E_m] + \beta_m$$

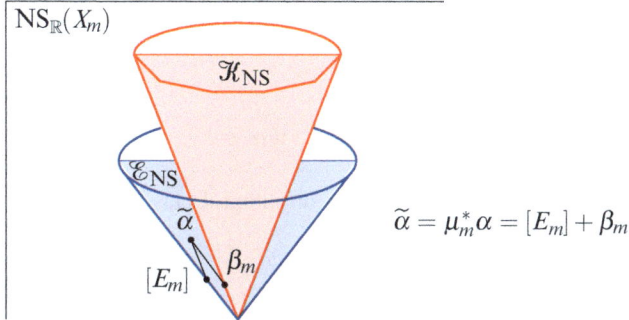

Fig. 3 Approximate Zariski decomposition

where E_m is an effective $\mathbb{Q}$-divisor on X_m with coefficients in $\frac{1}{m}\mathbb{Z}$ (the "fixed part") and β_m is a closed semi-positive form (the "mobile part").

Proof. We apply Theorem 64 with $\gamma = \varepsilon\omega$ and m so large that $\delta_m \leq \varepsilon/2$. Then T_m has analytic singularities and $T_m \geq \frac{\varepsilon}{2}\omega$, so we get a composition of blow-ups $\mu_m : X_m \to X$ such
$$\mu_m^* T_m = [E_m] + \beta_m,$$
where E_m is an effective $\mathbb{Q}$-divisor and $\beta_m \geq \frac{\varepsilon}{2}\mu_m^*\omega$. In particular, β_m is strictly positive outside the exceptional divisors, by playing with the multiplicities of the components of the exceptional divisors in E_m, we could even achieve that β_m is a Kähler class on X_m. Notice also that by construction, μ_m is obtained by blowing-up the multiplier ideal sheaves $\mathscr{I}(mT) = \mathscr{I}(m\varphi)$ associated to a potential φ of T. □

The more familiar algebraic analogue would be to take $\alpha = c_1(L)$ with a big line bundle L and to blow-up the base locus of $|mL|$, $m \gg 1$, to get a $\mathbb{Q}$-divisor decomposition

$$\mu_m^* L \sim E_m + D_m, \qquad E_m \text{ effective}, \quad D_m \text{ base point free.} \tag{57}$$

(One says that D_m is base point free if $H^0(X, \mathscr{O}(D_m))$ is generated by sections, in other words if D_m is entirely "mobile" in the linear system $|D_m|$). Such a blow-up is usually referred to as a "log resolution" of the linear system $|mL|$, and we say that $E_m + D_m$ is an approximate Zariski decomposition of L. We will also use this terminology for Kähler currents with logarithmic poles.

Definition 66. We define the *volume*, or mobile self-intersection of a class $\alpha \in H^{1,1}(X, \mathbb{R})$ to be

$$\mathrm{Vol}(\alpha) = \sup_{T \in \alpha} \int_{X \smallsetminus \mathrm{Sing}(T)} T^n = \sup_{T \in \alpha} \int_{\widetilde{X}} \beta^n > 0,$$

where the supremum is taken over all Kähler currents $T \in \alpha$ with logarithmic poles, and $\mu^* T = [E] + \beta$ with respect to some modification $\mu : \tilde{X} \to X$. Correspondingly, we set

$$\mathrm{Vol}(\alpha) = 0 \quad \text{if } \alpha \notin \mathcal{E}^\circ.$$

In the special case where $\alpha = c_1(L)$ is an integral class, we have the following interpretation of the volume.

Theorem 67. *If L is a big line bundle and $\mu_m^* L \sim E_m + D_m$ is a log resolution of $|mL|$, we have*

$$\mathrm{Vol}(c_1(L)) = \lim_{m \to +\infty} D_m^n = \lim_{m \to +\infty} \frac{n!}{m^n} h^0(X, mL),$$

Sketch of proof. Given a Kähler current $T \in c_1(L)$ with logarithmic pole, we can always take a blow-up $\mu : \tilde{X} \to X$ so that $\mu^* T = [E] + \beta$ where E is an effective $\mathbb{R}$-divisor and $\beta \geq 0$. By using a perturbation technique as in Lemma 61, we can always assume that E is a $\mathbb{Q}$-divisor and that β is Kähler. Then $\{\beta\} = \mu^* c_1(L) - \{[E]\}$ is a rational class and therefore β is the first Chern class $c_1(A)$ of an ample $\mathbb{Q}$-divisor on $\tilde{X}$. When m is a multiple of a suitable denominator m_0 and $m = qm_0 + r$, $0 \leq r < m_0$, we get by the elementary Riemann–Roch formula

$$h^0(X, mL) \geq h^0(\tilde{X}, m\mu^* L - m_0[m/m_0]E) = h^0(\tilde{X}, m_0[m/m_0]A + r\mu^* L)$$

$$\sim \frac{m^n}{n!} \int_{\tilde{X}} \beta^n,$$

hence $\liminf \frac{n!}{m^n} h^0(X, mL) \geq \mathrm{Vol}(c_1(L))$ by taking the supremum over all such currents T. In the other direction, the inequality $\limsup \frac{n!}{m^n} h^0(X, mL) \leq \mathrm{Vol}(c_1(L))$ is obtained by subtracting a small rational multiple εA of an ample line bundle A. One shows that multiples of $L - \varepsilon A$ roughly have the same number of sections as those of L by an exact sequence argument similar to what was done in the proof of Theorem 43(b). By a result of Fujita [47] (cf. also [37]), the volume of the base point free part $D_{m,\varepsilon}$ in a log resolution of $|m(L - \varepsilon A)|$ approximates $\limsup \frac{n!}{m^n} h^0(X, m(L - \varepsilon A))$, so we get $\mu_{m,\varepsilon}^* L = E_{m,\varepsilon} + (D_{m,\varepsilon} + \varepsilon A)$ where $D_{m,\varepsilon} + A$ is ample. The positive $(1,1)$-current $T_{m,\varepsilon} = (\mu_{m,\varepsilon})_* \Theta_{D_{m,\varepsilon} + \varepsilon A}$ is a Kähler current with logarithmic poles and its volume approaches $\limsup \frac{n!}{m^n} h^0(X, mL)$ when $\varepsilon \ll 1$ and m is large. $\square$

In these terms, we get the following statement.

Proposition 68. *Let L be a big line bundle on the projective manifold X. Let $\varepsilon > 0$. Then there exists a modification $\mu : X_\varepsilon \to X$ and a decomposition $\mu^*(L) = E + \beta$ with E an effective $\mathbb{Q}$-divisor and β a big and nef $\mathbb{Q}$-divisor such that*

$$\mathrm{Vol}(L) - \varepsilon \leq \mathrm{Vol}(\beta) \leq \mathrm{Vol}(L).$$

Applications of Pluripotential Theory to Algebraic Geometry

It is very useful to observe that the supremum in Definition 66 is actually achieved by a collection of currents whose singularities satisfy a filtering property. Namely, if $T_1 = \alpha + i\partial\bar\partial\varphi_1$ and $T_2 = \alpha + i\partial\bar\partial\varphi_2$ are two Kähler currents with logarithmic poles in the class of α, then

$$T = \alpha + i\partial\bar\partial\varphi, \qquad \varphi = \max(\varphi_1, \varphi_2) \tag{58}$$

is again a Kähler current with weaker singularities than T_1 and T_2. One could define as well

$$T = \alpha + i\partial\bar\partial\varphi, \qquad \varphi = \frac{1}{2m}\log(e^{2m\varphi_1} + e^{2m\varphi_2}), \tag{58'}$$

where $m = \mathrm{lcm}(m_1, m_2)$ is the lowest common multiple of the denominators occurring in T_1, T_2. Now, take a simultaneous log-resolution $\mu_m : X_m \to X$ for which the singularities of T_1 and T_2 are resolved as $\mathbb{Q}$-divisors E_1 and E_2. Then clearly the associated divisor in the decomposition $\mu_m^* T = [E] + \beta$ is given by $E = \min(E_1, E_2)$. By doing so, the volume $\int_{X_m} \beta^n$ gets increased, as we shall see in the proof of Theorem 69 below.

Theorem 69 (Boucksom [18]). *Let X be a compact Kähler manifold. We denote here by $H^{k,k}_{\geq 0}(X)$ the cone of cohomology classes of type (k,k) which have non-negative intersection with all closed semi-positive smooth forms of bidegree $(n-k, n-k)$.*

(a) *For each integer $k = 1, 2, \ldots, n$, there exists a canonical "mobile intersection product"*

$$\mathcal{E} \times \cdots \times \mathcal{E} \to H^{k,k}_{\geq 0}(X), \quad (\alpha_1, \ldots, \alpha_k) \mapsto \langle \alpha_1 \cdot \alpha_2 \cdots \alpha_{k-1} \cdot \alpha_k \rangle$$

such that $\mathrm{Vol}(\alpha) = \langle \alpha^n \rangle$ whenever α is a big class.

(b) *The product is increasing, homogeneous of degree 1 and superadditive in each argument, i.e.*

$$\langle \alpha_1 \cdots (\alpha'_j + \alpha''_j) \cdots \alpha_k \rangle \geq \langle \alpha_1 \cdots \alpha'_j \cdots \alpha_k \rangle + \langle \alpha_1 \cdots \alpha''_j \cdots \alpha_k \rangle.$$

It coincides with the ordinary intersection product when the $\alpha_j \in \mathcal{K}$ are nef classes.

(c) *The mobile intersection product satisfies the Hovanskii–Teissier inequalities [57, 96, 97]*

$$\langle \alpha_1 \cdot \alpha_2 \cdots \alpha_n \rangle \geq (\langle \alpha_1^n \rangle)^{1/n} \cdots (\langle \alpha_n^n \rangle)^{1/n} \qquad (\text{with } \langle \alpha_j^n \rangle = \mathrm{Vol}(\alpha_j)).$$

(d) *For $k = 1$, the above "product" reduces to a (non linear) projection operator*

$$\mathcal{E} \to \mathcal{E}_1, \qquad \alpha \to \langle \alpha \rangle$$

onto a certain convex subcone $\mathcal{E}_1$ of $\mathcal{E}$ such that $\overline{\mathcal{K}} \subset \mathcal{E}_1 \subset \mathcal{E}$. Moreover, there is a "divisorial Zariski decomposition"

$$\alpha = \{N(\alpha)\} + \langle \alpha \rangle$$

where $N(\alpha)$ is a uniquely defined effective divisor which is called the "negative divisorial part" of α. The map $\alpha \mapsto N(\alpha)$ is homogeneous and subadditive, and $N(\alpha) = 0$ if and only if $\alpha \in \mathcal{E}_1$.

(e) *The components of $N(\alpha)$ always consist of divisors whose cohomology classes are linearly independent, especially $N(\alpha)$ has at most $\rho = \operatorname{rank}_{\mathbb{Z}} \operatorname{NS}(X)$ components.*

Proof. We essentially repeat the arguments developed in [18], with some simplifications arising from the fact that X is supposed to be Kähler from the beginning.

(a) First assume that all classes α_j are big, i.e. $\alpha_j \in \mathcal{E}^\circ$. Fix a smooth closed $(n-k, n-k)$ *semi-positive* form u on X. We select Kähler currents $T_j \in \alpha_j$ with logarithmic poles, and a simultaneous log-resolution $\mu : \tilde{X} \to X$ such that

$$\mu^* T_j = [E_j] + \beta_j.$$

We consider the direct image current $\mu_*(\beta_1 \wedge \cdots \wedge \beta_k)$ (which is a closed positive current of bidegree (k,k) on X) and the corresponding integrals

$$\int_{\tilde{X}} \beta_1 \wedge \cdots \wedge \beta_k \wedge \mu^* u \geq 0.$$

If we change the representative T_j with another current T'_j, we may always take a simultaneous log-resolution such that $\mu^* T'_j = [E'_j] + \beta'_j$, and by using (58') we can always assume that $E'_j \leq E_j$. Then $D_j = E_j - E'_j$ is an effective divisor and we find $[E_j] + \beta_j \equiv [E'_j] + \beta'_j$, hence $\beta'_j \equiv \beta_j + [D_j]$. A substitution in the integral implies

$$\int_{\tilde{X}} \beta'_1 \wedge \beta_2 \wedge \cdots \wedge \beta_k \wedge \mu^* u$$

$$= \int_{\tilde{X}} \beta_1 \wedge \beta_2 \wedge \cdots \wedge \beta_k \wedge \mu^* u + \int_{\tilde{X}} [D_1] \wedge \beta_2 \wedge \cdots \wedge \beta_k \wedge \mu^* u$$

$$\geq \int_{\tilde{X}} \beta_1 \wedge \beta_2 \wedge \cdots \wedge \beta_k \wedge \mu^* u.$$

Similarly, we can replace successively all forms β_j by the β'_j, and by doing so, we find

$$\int_{\tilde{X}} \beta'_1 \wedge \beta'_2 \wedge \cdots \wedge \beta'_k \wedge \mu^* u \geq \int_{\tilde{X}} \beta_1 \wedge \beta_2 \wedge \cdots \wedge \beta_k \wedge \mu^* u.$$

We claim that the closed positive currents $\mu_*(\beta_1 \wedge \cdots \wedge \beta_k)$ are uniformly bounded in mass. In fact, if ω is a Kähler metric in X, there exists a constant $C_j \geq 0$ such that $C_j\{\omega\} - \alpha_j$ is a Kähler class. Hence $C_j\omega - T_j \equiv \gamma_j$ for some Kähler form γ_j on X. By pulling back with μ, we find $C_j\mu^*\omega - ([E_j] + \beta_j) \equiv \mu^*\gamma_j$, hence

$$\beta_j \equiv C_j\mu^*\omega - ([E_j] + \mu^*\gamma_j).$$

By performing again a substitution in the integrals, we find

$$\int_{\tilde{X}} \beta_1 \wedge \cdots \wedge \beta_k \wedge \mu^*u \leq C_1 \cdots C_k \int_{\tilde{X}} \mu^*\omega^k \wedge \mu^*u = C_1 \cdots C_k \int_X \omega^k \wedge u$$

and this is true especially for $u = \omega^{n-k}$. We can now arrange that for each of the integrals associated with a countable dense family of forms u, the supremum is achieved by a sequence of currents $(\mu_m)_*(\beta_{1,m} \wedge \cdots \wedge \beta_{k,m})$ obtained as direct images by a suitable sequence of modifications $\mu_m : \tilde{X}_m \to X$. By extracting a subsequence, we can achieve that this sequence is weakly convergent and we set

$$\langle \alpha_1 \cdot \alpha_2 \cdots \alpha_k \rangle = \lim_{m \to +\infty} \uparrow \{(\mu_m)_*(\beta_{1,m} \wedge \beta_{2,m} \wedge \cdots \wedge \beta_{k,m})\}$$

(the monotonicity is not in terms of the currents themselves, but in terms of the integrals obtained when we evaluate against a smooth closed semi-positive form u). By evaluating against a basis of positive classes $\{u\} \in H^{n-k,n-k}(X)$, we infer by Serre duality that the class of $\langle \alpha_1 \cdot \alpha_2 \cdots \alpha_k \rangle$ is uniquely defined (although, in general, the representing current is not unique).
(b) It is indeed clear from the definition that the mobile intersection product is homogeneous, increasing and superadditive in each argument, at least when the α_j's are in $\mathscr{E}^\circ$. However, we can extend the product to the closed cone $\mathscr{E}$ by monotonicity, by setting

$$\langle \alpha_1 \cdot \alpha_2 \cdots \alpha_k \rangle = \lim_{\delta \downarrow 0} \downarrow \langle (\alpha_1 + \delta\omega) \cdot (\alpha_2 + \delta\omega) \cdots (\alpha_k + \delta\omega) \rangle$$

for arbitrary classes $\alpha_j \in \mathscr{E}$ (again, monotonicity occurs only where we evaluate against closed semi-positive forms u). By weak compactness, the mobile intersection product can always be represented by a closed positive current of bidegree (k,k).
(c) The Hovanskii–Teissier inequalities are a direct consequence of the fact that they hold true for nef classes, so we just have to apply them to the classes $\beta_{j,m}$ on $\tilde{X}_m$ and pass to the limit.
(d) When $k = 1$ and $\alpha \in \mathscr{E}^0$, we have

$$\alpha = \lim_{m \to +\infty} \{(\mu_m)_* T_m\} = \lim_{m \to +\infty} (\mu_m)_*[E_m] + \{(\mu_m)_*\beta_m\}$$

and $\langle \alpha \rangle = \lim_{m \to +\infty} \{(\mu_m)_* \beta_m\}$ by definition. However, the images $F_m = (\mu_m)_* F_m$ are effective $\mathbb{Q}$-divisors in X, and the filtering property implies that F_m is a decreasing sequence. It must therefore converge to a (uniquely defined) limit $F = \lim F_m := N(\alpha)$ which is an effective $\mathbb{R}$-divisor, and we get the asserted decomposition in the limit.

Since $N(\alpha) = \alpha - \langle \alpha \rangle$ we easily see that $N(\alpha)$ is subadditive and that $N(\alpha) = 0$ if α is the class of a smooth semi-positive form. When α is no longer a big class, we define

$$\langle \alpha \rangle = \lim_{\delta \downarrow 0} \downarrow \langle \alpha + \delta \omega \rangle, \qquad N(\alpha) = \lim_{\delta \downarrow 0} \uparrow N(\alpha + \delta \omega)$$

(the subadditivity of N implies $N(\alpha + (\delta + \varepsilon)\omega) \leq N(\alpha + \delta \omega)$). The divisorial Zariski decomposition follows except maybe for the fact that $N(\alpha)$ might be a convergent countable sum of divisors. However, this will be ruled out when (e) is proved. As $N(\bullet)$ is subadditive and homogeneous, the set $\mathscr{E}_1 = \{\alpha \in \mathscr{E} \,;\, N(\alpha) = 0\}$ is a closed convex cone, and we find that $\alpha \mapsto \langle \alpha \rangle$ is a projection of $\mathscr{E}$ onto $\mathscr{E}_1$ (according to [18], $\mathscr{E}_1$ consists of those pseudo-effective classes which are "nef in codimension 1").

(e) Let $\alpha \in \mathscr{E}^\circ$, and assume that $N(\alpha)$ contains linearly dependent components F_j. Then already all currents $T \in \alpha$ should be such that $\mu^* T = [E] + \beta$ where $F = \mu_* E$ contains those linearly dependent components. Write $F = \sum \lambda_j F_j$, $\lambda_j > 0$ and assume that

$$\sum_{j \in J} c_j F_j \equiv 0$$

for a certain non trivial linear combination. Then some of the coefficients c_j must be negative (and some other positive). Then E is numerically equivalent to

$$E' \equiv E + t\mu^*\left(\sum \lambda_j F_j\right),$$

and by choosing $t > 0$ appropriate, we obtain an effective divisor E' which has a zero coefficient on one of the components $\mu^* F_{j_0}$. By replacing E with $\min(E, E')$ via (58'), we eliminate the component $\mu^* F_{j_0}$. This is a contradiction since $N(\alpha)$ was supposed to contain F_{j_0}. □

Definition 70. For a class $\alpha \in H^{1,1}(X, \mathbb{R})$, we define the numerical dimension $\mathrm{nd}(\alpha)$ to be $\mathrm{nd}(\alpha) = -\infty$ if α is not pseudo-effective, and

$$\mathrm{nd}(\alpha) = \max\{p \in \mathbb{N} \,;\, \langle \alpha^p \rangle \neq 0\}, \qquad \mathrm{nd}(\alpha) \in \{0, 1, \ldots, n\}$$

if α is pseudo-effective.

By the results of [42], a class is big ($\alpha \in \mathscr{E}^\circ$) if and only if $\mathrm{nd}(\alpha) = n$. Classes of numerical dimension 0 can be described much more precisely, again following Boucksom [18].

Theorem 71. *Let X be a compact Kähler manifold. Then the subset $\mathscr{D}_0$ of irreducible divisors D in X such that $\mathrm{nd}(D) = 0$ is countable, and these divisors are rigid as well as their multiples. If $\alpha \in \mathscr{E}$ is a pseudo-effective class of numerical dimension 0, then α is numerically equivalent to an effective $\mathbb{R}$-divisor $D = \sum_{j \in J} \lambda_j D_j$, for some finite subset $(D_j)_{j \in J} \subset \mathscr{D}_0$ such that the cohomology classes $\{D_j\}$ are linearly independent and some $\lambda_j > 0$. If such a linear combination is of numerical dimension 0, then so is any other linear combination of the same divisors.*

Proof. It is immediate from the definition that a pseudo-effective class is of numerical dimension 0 if and only if $\langle \alpha \rangle = 0$, in other words if $\alpha = N(\alpha)$. Thus $\alpha \equiv \sum \lambda_j D_j$ as described in Theorem 71, and since $\lambda_j \langle D_j \rangle \leq \langle \alpha \rangle$, the divisors D_j must themselves have numerical dimension 0. There is at most one such divisor D in any given cohomology class in $NS(X) \cap \mathscr{E} \subset H^2(X, \mathbb{Z})$, otherwise two such divisors $D \equiv D'$ would yield a blow-up $\mu : \tilde{X} \to X$ resolving the intersection, and by taking $\min(\mu^* D, \mu^* D')$ via (58'), we would find $\mu^* D \equiv E + \beta$, $\beta \neq 0$, so that $\{D\}$ would not be of numerical dimension 0. This implies that there are at most countably many divisors of numerical dimension 0, and that these divisors are rigid as well as their multiples. □

Remark 72. If L is an arbitrary holomorphic line bundle, we define its numerical dimension to be $\mathrm{nd}(L) = \mathrm{nd}(c_1(L))$. Using the canonical maps $\Phi_{|mL|}$ and pulling-back the Fubini–Study metric it is immediate to see that $\mathrm{nd}(L) \geq \kappa(L)$.

The above general concept of numerical dimension leads to a very natural formulation of the abundance conjecture for Kähler varieties.

Generalized Abundance Conjecture 73. *Let X be an arbitrary compact Kähler manifold X.*

(a) *The Kodaira dimension of X should be equal to its numerical dimension: $\kappa(K_X) = \mathrm{nd}(K_X)$.*
(b) *More generally, let Δ be a $\mathbb{Q}$-divisor which is klt (Kawamata log terminal, i.e. such that $c_X(\Delta) > 1$). Then $\kappa(K_X + \Delta) = \mathrm{nd}(K_X + \Delta)$.*

Remark 74. It is obvious that abundance holds in the case $\mathrm{nd}(K_X) = -\infty$ (if L is not pseudo-effective, no multiple of L can have sections), or in the case $\mathrm{nd}(K_X) = n$ which implies K_X big (the latter property follows e.g. from the solution of the Grauert–Riemenschneider conjecture in the form proven in [25], see also [42]).

In the remaining cases, the most tractable situation is the case when $\mathrm{nd}(K_X) = 0$. In fact Theorem 71 then gives $K_X \equiv \sum \lambda_j D_j$ for some effective divisor with numerically independent components, $\mathrm{nd}(D_j) = 0$. It follows that the λ_j are rational and therefore

$$K_X \sim \sum \lambda_j D_j + F \qquad \text{where } \lambda_j \in \mathbb{Q}^+, \mathrm{nd}(D_j) = 0 \text{ and } F \in \mathrm{Pic}^0(X). \quad (*)$$

If we assume additionally that $q(X) = h^{0,1}(X)$ is zero, then mK_X is linearly equivalent to an integral divisor for some multiple m, and it follows immediately that $\kappa(X) = 0$. The case of a general projective manifold with $\mathrm{nd}(K_X) = 0$ and positive irregularity $q(X) > 0$ has been solved by Campana–Peternell [23], Proposition 46. It would be interesting to understand the Kähler case as well.

2.8 The Orthogonality Estimate

The goal of this section is to show that, in an appropriate sense, approximate Zariski decompositions are almost orthogonal.

Theorem 75. *Let X be a projective manifold, and let $\alpha = \{T\} \in \mathcal{E}^\circ_{\mathrm{NS}}$ be a big class represented by a Kähler current T. Consider an approximate Zariski decomposition*

$$\mu_m^* T_m = [E_m] + [D_m]$$

Then

$$(D_m^{n-1} \cdot E_m)^2 \leq 20\,(C\omega)^n \big(\mathrm{Vol}(\alpha) - D_m^n\big)$$

where $\omega = c_1(H)$ is a Kähler form and $C \geq 0$ is a constant such that $\pm \alpha$ is dominated by $C\omega$ (i.e., $C\omega \pm \alpha$ is nef). In other words, E_m and D_m become "more and more orthogonal" as D_m^n approaches the volume.

Proof. For every $t \in [0, 1]$, we have

$$\mathrm{Vol}(\alpha) = \mathrm{Vol}(E_m + D_m) \geq \mathrm{Vol}(tE_m + D_m).$$

Now, by our choice of C, we can write E_m as a difference of two nef divisors

$$E_m = \mu^* \alpha - D_m = \mu_m^*(\alpha + C\omega) - (D_m + C\mu_m^*\omega). \qquad \square$$

Lemma 76. *For all nef $\mathbb{R}$-divisors A, B we have*

$$\mathrm{Vol}(A - B) \geq A^n - nA^{n-1} \cdot B$$

as soon as the right hand side is positive.

Proof. In case A and B are integral divisors, this is a consequence of holomorphic Morse inequalities (cf. Consequence 24). If A and B are $\mathbb{Q}$-divisors, we conclude by the homogeneity of the volume. The general case of $\mathbb{R}$-divisors follows by approximation (actually, as it is defined to be a supremum, the volume function can easily be shown to be lower semi-continuous, but it is in fact even continuous, cf. [[18], 3.1.26]). $\qquad \square$

Remark 77. We hope that Lemma 76 also holds true on an arbitrary Kähler manifold for arbitrary nef (non necessarily integral) classes. This would follow from

Conjecture 95 generalizing holomorphic Morse inequalities to non integral classes, exactly by the same proof as Theorem 23.

Lemma 78. *Let $\beta_1, \ldots, \beta_n$ and $\beta'_1, \ldots, \beta'_n$ be nef classes on a compact Kähler manifold $\tilde{X}$ such that each difference $\beta'_j - \beta_j$ is pseudo-effective. Then the n-th intersection products satisfy*

$$\beta_1 \cdots \beta_n \leq \beta'_1 \cdots \beta'_n.$$

Proof. We can proceed step by step and replace just one β_j by $\beta'j \equiv \beta_j + T_j$ where T_j is a closed positive $(1, 1)$-current and the other classes $\beta'_k = \beta_k, k \neq j$ are limits of Kähler forms. The inequality is then obvious. □

End of proof of Theorem 75. In order to exploit the lower bound of the volume, we write

$$tE_m + D_m = A - B, \qquad A = D_m + t\mu_m^*(\alpha + C\omega), \qquad B = t(D_m + C\mu_m^*\omega).$$

By our choice of the constant C, both A and B are nef. Lemma 76 and the binomial formula imply

$$\mathrm{Vol}(tE_m + D_m) \geq A^n - nA^{n-1} \cdot B$$

$$= D_m^n + nt\, D_m^{n-1} \cdot \mu_m^*(\alpha + C\omega) + \sum_{k=2}^{n} t^k \binom{n}{k} D_m^{n-k} \cdot \mu_m^*(\alpha + C\omega)^k$$

$$- nt\, D_m^{n-1} \cdot (D_m + C\mu_m^*\omega)$$

$$- nt^2 \sum_{k=1}^{n-1} t^{k-1} \binom{n-1}{k} D_m^{n-1-k} \cdot \mu_m^*(\alpha + C\omega)^k \cdot (D_m + C\mu_m^*\omega).$$

Now, we use the obvious inequalities

$$D_m \leq \mu_m^*(C\omega), \qquad \mu_m^*(\alpha + C\omega) \leq 2\mu_m^*(C\omega), \qquad D_m + C\mu_m^*\omega \leq 2\mu_m^*(C\omega)$$

in which all members are nef (and where the inequality $\leq$ means that the difference of classes is pseudo-effective). We use Lemma 78 to bound the last summation in the estimate of the volume, and in this way we get

$$\mathrm{Vol}(tE_m + D_m) \geq D_m^n + ntD_m^{n-1} \cdot E_m - nt^2 \sum_{k=1}^{n-1} 2^{k+1} t^{k-1} \binom{n-1}{k} (C\omega)^n.$$

We will always take t smaller than $1/10n$ so that the last summation is bounded by $4(n-1)(1 + 1/5n)^{n-2} < 4ne^{1/5} < 5n$. This implies

$$\mathrm{Vol}(tE_m + D_m) \geq D_m^n + nt\, D_m^{n-1} \cdot E_m - 5n^2t^2(C\omega)^n.$$

Now, the choice $t = \frac{1}{10n}(D_m^{n-1} \cdot E_m)((C\omega)^n)^{-1}$ gives by substituting

$$\frac{1}{20}\frac{(D_m^{n-1} \cdot E_m)^2}{(C\omega)^n} \leq \mathrm{Vol}(E_m + D_m) - D_m^n \leq \mathrm{Vol}(\alpha) - D_m^n$$

(and we have indeed $t \leq \frac{1}{10n}$ by Lemma 78), whence Theorem 75. Of course, the constant 20 is certainly not optimal. □

Corollary 79. *If $\alpha \in \mathscr{E}_{\mathrm{NS}}$, then the divisorial Zariski decomposition $\alpha = N(\alpha) + \langle \alpha \rangle$ is such that*

$$\langle \alpha^{n-1} \rangle \cdot N(\alpha) = 0.$$

Proof. By replacing α with $\alpha + \delta c_1(H)$, one sees that it is sufficient to consider the case where α is big. Then the orthogonality estimate implies

$$(\mu_m)_*(D_m^{n-1}) \cdot (\mu_m)_* E_m = D_m^{n-1} \cdot (\mu_m)^*(\mu_m)_* E_m$$
$$\leq D_m^{n-1} \cdot E_m \leq C(\mathrm{Vol}(\alpha) - D_m^n)^{1/2}.$$

Since $\langle \alpha^{n-1} \rangle = \lim(\mu_m)_*(D_m^{n-1})$, $N(\alpha) = \lim(\mu_m)_* E_m$ and $\lim D_m^n = \mathrm{Vol}(\alpha)$, we get the desired conclusion in the limit. □

2.9 Dual of the Pseudo-Effective Cone

We consider here the Serre duality pairing

$$H^{1,1}(X,\mathbb{R}) \times H^{n-1,n-1}(X,\mathbb{R}) \longrightarrow \mathbb{R}, \qquad (\alpha,\beta) \longmapsto \alpha \cdot \beta = \int_X \alpha \wedge \beta. \quad (59)$$

When restricted to real vector subspaces generated by integral classes, it defines a perfect pairing

$$\mathrm{NS}_{\mathbb{R}} \times \mathrm{NS}_{\mathbb{R}}^{n-1,n-1}(X) \longrightarrow \mathbb{R} \qquad (60)$$

where $\mathrm{NS}_{\mathbb{R}} \subset H^{1,1}(X,\mathbb{R})$ and $\mathrm{NS}_{\mathbb{R}}^{n-1,n-1}(X) \subset H^{n-1,n-1}(X,\mathbb{R})$. Next, we introduce the concept of mobile curves.

Definition 80. *Let X be a smooth projective variety.*

(a) *One defines $\mathrm{NE}(X) \subset \mathrm{NS}_{\mathbb{R}}^{n-1,n-1}(X)$ to be the convex cone generated by cohomology classes of all effective curves in $H^{n-1,n-1}(X,\mathbb{R})$.*
(b) *We say that C is a mobile curve if $C = C_{t_0}$ is a member of an analytic family $\{C_t\}_{t \in S}$ such that $\bigcup_{t \in S} C_t = X$ and, as such, is a reduced irreducible 1-cycle. We define the mobile cone $\mathrm{ME}(X)$, to be the convex cone generated by all mobile curves.*

(c) If X is projective, we say that an effective 1-cycle C is a strongly mobile if we have
$$C = \mu_*(\tilde{A}_1 \cap \cdots \cap \tilde{A}_{n-1})$$
for suitable very ample divisors $\tilde{A}_j$ on $\tilde{X}$, where $\mu : \tilde{X} \to X$ is a modification. We let $\mathrm{ME}^s(X)$ be the convex cone generated by all strongly mobile effective 1-cycles (notice that by taking $\tilde{A}_j$ general enough these classes can be represented by reduced irreducible curves; also, by Hironaka, one could just restrict oneself to compositions of blow-ups with smooth centers).

Clearly, we have
$$\mathrm{ME}^s(X) \subset \mathrm{ME}(X) \subset \mathrm{NE}(X) \subset \mathrm{NS}_{\mathbb{R}}^{n-1,n-1}(X). \tag{61}$$

Another simple observation is:

Proposition 81. *One has $\alpha \cdot C \geq 0$ whenever $\{\alpha\} \in \mathcal{E}$ and $\{C\} \in \mathrm{ME}(X)$. In other words $\mathcal{E}_{\mathrm{NS}} = \mathcal{E} \cap \mathrm{NS}_{\mathbb{R}}(X)$ is contained in the dual cone $(\mathrm{ME}(X))^{\vee}$.*

Proof. If the class $\{\alpha\}$ is represented by a closed positive current T and $C = C_{t_0}$ belongs to a covering family $(C_t)_{t \in S}$, it is easy to see that $T_{|C_t}$ is locally well defined and nonnegative as soon as C_t is not contained in the set of poles of a local potential φ of T. However, this occurs only when t belongs to a pluripolar set $P \subset S$, hence for $t \in S \smallsetminus P$ we have
$$\alpha \cdot C = \int_{C_t} T_{|C_t} \geq 0. \qquad \square$$

The following statement was first proved in [19].

Theorem 82. *If X is projective, the cones $\mathcal{E}_{\mathrm{NS}} = \overline{\mathrm{Eff}(X)}$ and $\overline{\mathrm{ME}^s(X)}$ are dual with respect to Serre duality, and we have $\overline{\mathrm{ME}^s(X)} = \overline{\mathrm{ME}(X)}$.*

In other words, a line bundle L is pseudo-effective if (and only if) $L \cdot C \geq 0$ for all *mobile curves*, i.e., $L \cdot C \geq 0$ for every very generic curve C (not contained in a countable union of algebraic subvarieties). In fact, by definition of $\mathrm{ME}^s(X)$, it is enough to consider only those curves C which are images of generic complete intersection of very ample divisors on some variety $\tilde{X}$, under a modification $\mu : \tilde{X} \to X$. By a standard blowing-up argument, it also follows that a line bundle L on a normal Moishezon variety is pseudo-effective if and only if $L \cdot C \geq 0$ for every mobile curve C.

Proof. By Proposition 81 we have $\mathcal{E}_{\mathrm{NS}} \subset (\mathrm{ME}(X))^{\vee}$ and (61) implies $(\mathrm{ME}(X))^{\vee} \subset (\mathrm{ME}^s(X))^{\vee}$, therefore
$$\mathcal{E}_{\mathrm{NS}} \subset (\mathrm{ME}^s(X))^{\vee}. \tag{62}$$
If we show that $\mathcal{E}_{\mathrm{NS}} = (\mathrm{ME}^s(X))^{\vee}$, we get at the same time $(\mathrm{ME}^s(X))^{\vee} = (\mathrm{ME}(X))^{\vee}$, and therefore by biduality (Hahn–Banach theorem) we will infer

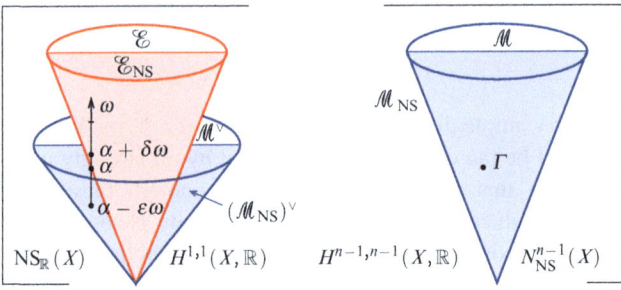

Fig. 4 Duality theorem for positive cones

$\overline{ME^s(X)} = \overline{ME(X)}$. Now, if the inclusion were strict in (62), there would be an element $\alpha \in \partial \mathscr{E}_{NS}$ on the boundary of $\mathscr{E}_{NS}$ which is in the interior of $ME^s(X)^\vee$. Let $\omega = c_1(H)$ be an ample class. Since $\alpha \in \partial \mathscr{E}_{NS}$, the class $\alpha + \delta \omega$ is big for every $\delta > 0$, and since $\alpha \in ((ME^s(X))^\vee)^\circ$ we still have $\alpha - \varepsilon \omega \in (ME^s(X))^\vee$ for $\varepsilon > 0$ small. Therefore

$$\alpha \cdot \Gamma \geq \varepsilon \omega \cdot \Gamma \tag{63}$$

for every strongly mobile curve Γ, and therefore for every $\Gamma \in \overline{ME^s(X)}$. We are going to contradict (63). Since $\alpha + \delta \omega$ is big, we have an approximate Zariski decomposition

$$\mu_\delta^*(\alpha + \delta \omega) = E_\delta + D_\delta.$$

We pick $\Gamma = (\mu_\delta)_*(D_\delta^{n-1}) \in \overline{ME^s(X)}$. By the Hovanskii–Teissier concavity inequality

$$\omega \cdot \Gamma \geq (\omega^n)^{1/n} (D_\delta^n)^{(n-1)/n}.$$

On the other hand

$$\begin{aligned}
\alpha \cdot \Gamma &= \alpha \cdot (\mu_\delta)_*(D_\delta^{n-1}) \\
&= \mu_\delta^* \alpha \cdot D_\delta^{n-1} \leq \mu_\delta^*(\alpha + \delta \omega) \cdot D_\delta^{n-1} \\
&= (E_\delta + D_\delta) \cdot D_\delta^{n-1} = D_\delta^n + D_\delta^{n-1} \cdot E_\delta.
\end{aligned}$$

By the orthogonality estimate, we find

$$\frac{\alpha \cdot \Gamma}{\omega \cdot \Gamma} \leq \frac{D_\delta^n + \bigl(20(C\omega)^n(\mathrm{Vol}(\alpha + \delta \omega) - D_\delta^n)\bigr)^{1/2}}{(\omega^n)^{1/n}(D_\delta^n)^{(n-1)/n}}$$

$$\leq C'(D_\delta^n)^{1/n} + C'' \frac{(\mathrm{Vol}(\alpha + \delta \omega) - D_\delta^n)^{1/2}}{(D_\delta^n)^{(n-1)/n}}.$$

However, since $\alpha \in \partial \mathscr{E}_{NS}$, the class α cannot be big so

$$\lim_{\delta \to 0} D_\delta^n = \mathrm{Vol}(\alpha) = 0.$$

We can also take D_δ to approximate $\text{Vol}(\alpha + \delta\omega)$ in such a way that $(\text{Vol}(\alpha+\delta\omega) - D_\delta^n)^{1/2}$ tends to 0 much faster than D_δ^n. Notice that $D_\delta^n \geq \delta^n \omega^n$, so in fact it is enough to take
$$\text{Vol}(\alpha + \delta\omega) - D_\delta^n \leq \delta^{2n},$$
which gives $(\alpha \cdot \Gamma)/(\omega \cdot \Gamma) \leq (C' + C'')\delta$. This contradicts (63) for δ small. □

3 Asymptotic Cohomology Functionals and Monge–Ampère Operators

The goal of this section is to show that there are strong relations between certain Monge–Ampère integrals appearing in holomorphic Morse inequalities, and asymptotic cohomology estimates for tensor powers of holomorphic line bundles. Especially, we prove that these relations hold without restriction for projective surfaces, and in the special case of the volume, i.e. of asymptotic 0-cohomology, for all projective manifolds. These results can be seen as a partial converse to the Andreotti–Grauert vanishing theorem.

3.1 Introduction and Main Definitions

Throughout this section, X denotes a compact complex manifold, $n = \dim_{\mathbb{C}} X$ its complex dimension and $L \to X$ a holomorphic line bundle. In order to estimate the growth of cohomology groups, it is interesting to consider appropriate "asymptotic cohomology functions". Following partly notation and concepts introduced by A. Küronya [46, 63], we introduce

Definition 83. Let X be a compact complex manifold and let $L \to X$ be a holomorphic line bundle.

(a) The q-th asymptotic cohomology functional is defined as
$$\hat{h}^q(X, L) := \limsup_{k \to +\infty} \frac{n!}{k^n} h^q(X, L^{\otimes k}).$$

(b) The q-th asymptotic holomorphic Morse sum of L is
$$\hat{h}^{\leq q}(X, L) := \limsup_{k \to +\infty} \frac{n!}{k^n} \sum_{0 \leq j \leq q} (-1)^{q-j} h^j(X, L^{\otimes k}).$$

When the lim sup's are limits, we have the obvious relation
$$\hat{h}^{\leq q}(X, L) = \sum_{0 \leq j \leq q} (-1)^{q-j} \hat{h}^j(X, L).$$

Clearly, Definition 83 can also be given for a $\mathbb{Q}$-line bundle L or a $\mathbb{Q}$-divisor D, and in the case $q = 0$ one gets by Theorem 67 what is called the volume of L (see also [18, 37, 67]):

$$\text{Vol}(X, L) := \hat{h}^0(X, L) = \limsup_{k \to +\infty} \frac{n!}{k^n} h^0(X, L^{\otimes k}). \tag{64}$$

3.2 Extension of the Functionals to Real Cohomology Classes

We are going to show that the $\hat{h}^q$ functional induces a continuous map

$$\text{DNS}_{\mathbb{R}}(X) \ni \alpha \mapsto \hat{h}^q_{\text{DNS}}(X, \alpha), \tag{65}$$

which is defined on the "divisorial Néron–Severi space" $\text{DNS}_{\mathbb{R}}(X) \subset H^{1,1}_{\text{BC}}(X, \mathbb{R})$, i.e. the vector space spanned by real linear combinations of classes of divisors in the real Bott–Chern cohomology group of bidegree $(1, 1)$. Here $H^{p,q}_{\text{BC}}(X, \mathbb{C})$ is defined as the quotient of d-closed (p, q)-forms by $\partial\bar{\partial}$-exact (p, q)-forms, and there is a natural conjugation $H^{p,q}_{\text{BC}}(X, \mathbb{C}) \to H^{q,p}_{\text{BC}}(X, \mathbb{C})$ which allows us to speak of real classes when $q = p$. Notice that $H^{p,q}_{\text{BC}}(X, \mathbb{C})$ coincides with the usual Dolbeault cohomology group $H^{p,q}(X, \mathbb{C})$ when X is Kähler, and that $\text{DNS}_{\mathbb{R}}(X)$ coincides with the usual Néron–Severi space

$$\text{NS}_{\mathbb{R}}(X) = \mathbb{R} \otimes_{\mathbb{Q}} \left(H^2(X, \mathbb{Q}) \cap H^{1,1}(X, \mathbb{C}) \right) \tag{66}$$

when X is projective (the inclusion can be strict in general, e.g. on complex 2-tori which only have indefinite integral $(1, 1)$-classes, cf. [BL04]).

For $\alpha \in \text{NS}_{\mathbb{R}}(X)$ (resp. $\alpha \in \text{DNS}_{\mathbb{R}}(X)$), we set

$$\hat{h}^q_{\text{NS}}(X, \alpha) \left(\text{resp. } \hat{h}^q_{\text{DNS}}(X, \alpha) \right) = \limsup_{k \to +\infty, \frac{1}{k} c_1(L) \to \alpha} \frac{n!}{k^n} h^q(X, L)$$

$$= \inf_{\varepsilon > 0, k_0 > 0} \sup_{k \geq k_0, \|\frac{1}{k} c_1(L) - \alpha\| \leq \varepsilon} \frac{n!}{k^n} h^q(X, L). \tag{67}$$

when the pair (k, L) runs over $\mathbb{N}^* \times \text{Pic}(X)$, resp. over $\mathbb{N}^* \times \text{Pic}_D(X)$ where $\text{Pic}_D(X) \subset \text{Pic}(X)$ is the subgroup generated by "divisorial line bundles", i.e. line bundles of the form $\mathcal{O}_X(D)$. Similar definitions can be given for the Morse sum functionals $\hat{h}^{\leq q}_{\text{NS}}(X, \alpha)$ and $\hat{h}^{\leq q}_{\text{DNS}}(X, \alpha)$. Clearly $\hat{h}^{\leq q}_{\text{DNS}}(X, \alpha) \leq \hat{h}^{\leq q}_{\text{NS}}(X, \alpha)$ on $\text{DNS}_{\mathbb{R}}(X)$, but we do not know at this point whether this is always an equality. From the very definition, $\hat{h}^q_{\text{NS}}$, $\hat{h}^{\leq q}_{\text{NS}}$ (and likewise $\hat{h}^q_{\text{DNS}}$, $\hat{h}^{\leq q}_{\text{DNS}}$) are upper semi-continuous functions which are positively homogeneous of degree n, namely

Applications of Pluripotential Theory to Algebraic Geometry

$$\hat{h}_{\text{NS}}^q(X, \lambda\alpha) = \lambda^n \hat{h}_{\text{NS}}^q(X, \alpha) \tag{68}$$

for all $\alpha \in \text{NS}_{\mathbb{R}}(X)$ and all $\lambda \geq 0$. Notice that $\hat{h}_{\text{NS}}^q(X, \alpha)$ and $\hat{h}_{\text{NS}}^{\leq q}(X, \alpha)$ are always finite thanks to holomorphic Morse inequalities (see below).

Proposition 84.

(a) For $L \in \text{Pic}_D(X)$, one has $\hat{h}^q(X, L) = \hat{h}^q(X, c_1(L))$, $\hat{h}^{\leq q}(X, L) = \hat{h}_{\text{DNS}}^{\leq q}(X, c_1(L))$, in particular asymptotic cohomology depends only on the numerical class of L.
(b) The map $\alpha \mapsto \hat{h}_{\text{DNS}}^q(X, \alpha)$ is (locally) Lipschitz continuous on $\text{DNS}_{\mathbb{R}}(X)$.
(c) When $q = 0$, $\hat{h}_{\text{DNS}}^0(X, \alpha)$ and $\hat{h}_{\text{NS}}^0(X, \alpha)$ coincide on $\text{DNS}_{\mathbb{R}}(X)$ and the limsups are limits.

The proof is derived from arguments quite similar to those already developed in [63] (see also [34] for the non projective situation). If $D = \sum p_j D_j$ is an integral divisor, we define its norm to be $\|D\| = \sum |p_j| \text{Vol}_\omega(D_j)$, where the volume of an irreducible divisor is computed by means of a given Hermitian metric ω on X; in other words, this is precisely the mass of the current of integration $[D]$ with respect to ω. Clearly, since X is compact, we get equivalent norms for all choices of Hermitian metrics ω on X. We can also use ω to fix a normalized metric on $H_{\text{BC}}^{1,1}(X, \mathbb{R})$. Elementary properties of potential theory show that $\|c_1(\mathcal{O}(D))\| \leq C \|D\|$ for some constant $C > 0$ (but the converse inequality is of course wrong in most cases). Proposition 84 is a simple consequence of the more precise cohomology estimates (1.9) which will be obtained below. The special case $q = 0$ is easier, in fact, one can get non zero values for $\hat{h}^0(X, L)$ only when L is big, i.e. when X is Moishezon (so that we are always reduced to the divisorial situation); the fact that limsups are limits was proved in Theorem 67. We postpone the proof to Sect. 19, which will provide stronger results based on approximate Zariski decomposition.

Lemma 85. *Let X be a compact complex n-fold. Then for every coherent sheaf $\mathcal{F}$ on X, there is a constant $C_{\mathcal{F}} > 0$ such that for every holomorphic line bundle L on X we have*

$$h^q(X, \mathcal{F} \otimes \mathcal{O}_X(L)) \leq C_{\mathcal{F}}(\|c_1(L)\| + 1)^p$$

where $p = \dim \text{Supp}\, \mathcal{F}$.

Proof. We prove the result by induction on p; it is indeed clear for $p = 0$ since we then have cohomology only in degree 0 and the dimension of $H^0(X, \mathcal{F} \otimes \mathcal{O}_X(L))$ does not depend on L when $\mathcal{F}$ has finite support. Let us consider the support Y of $\mathcal{F}$ and a resolution of singularity $\mu : \hat{Y} \to Y$ of the corresponding (reduced) analytic space. Then $\mathcal{F}$ is an $\mathcal{O}_Y$-module for some non necessarily reduced complex structure $\mathcal{O}_Y = \mathcal{O}_X/\mathcal{J}$ on Y. We can look at the reduced structure $\mathcal{O}_{Y,\text{red}} = \mathcal{O}_X/\mathcal{I}$, $\mathcal{I} = \sqrt{\mathcal{J}}$, and filter $\mathcal{F}$ by $\mathcal{I}^k \mathcal{F}$, $k \geq 0$. Since $\mathcal{I}^k \mathcal{F}/\mathcal{I}^{k+1}\mathcal{F}$ is a coherent $\mathcal{O}_{Y,\text{red}}$-module, we can easily reduce the situation to the case where Y is reduced and $\mathcal{F}$ is an $\mathcal{O}_Y$-module. In that case the cohomology

$$H^q(X, \mathscr{F} \otimes \mathscr{O}_X(L)) = H^q(Y, \mathscr{F} \otimes \mathscr{O}_Y(L_{|Y}))$$

just lives on the reduced space Y.

Now, we have an injective sheaf morphism $\mathscr{F} \to \mu_*\mu^*\mathscr{F}$ whose cokernel $\mathscr{G}$ has support in dimension $< p$. By induction on p, we conclude from the exact sequence that

$$\left|h^q(X, \mathscr{F} \otimes \mathscr{O}_X(L)) - h^q(X, \mu_*\mu^*\mathscr{F} \otimes \mathscr{O}_X(L))\right| \leq C_1(\|c_1(L)\| + 1)^{p-1}.$$

The functorial morphisms

$$\mu^* : H^q(Y, \mathscr{F} \otimes \mathscr{O}_Y(L_{|Y})) \to H^q(\hat{Y}, \mu^*\mathscr{F} \otimes \mathscr{O}_{\hat{Y}}(\mu^*L)_{|Y}),$$

$$\mu_* : H^q(\hat{Y}, \mu^*\mathscr{F} \otimes \mathscr{O}_{\hat{Y}}(\mu^*L)_{|Y}) \to H^q(Y, \mu_*\mu^*\mathscr{F} \otimes \mathscr{O}_Y(L_{|Y}))$$

yield a composition

$$\mu_* \circ \mu^* : H^q(Y, \mathscr{F} \otimes \mathscr{O}_Y(L_{|Y})) \to H^q(Y, \mu_*\mu^*\mathscr{F} \otimes \mathscr{O}_Y(L_{|Y}))$$

induced by the natural injection $\mathscr{F} \to \mu_*\mu^*\mathscr{F}$. This implies

$$h^q(Y, \mathscr{F} \otimes \mathscr{O}_Y(L_{|Y})) \leq h^q(\hat{Y}, \mu^*\mathscr{F} \otimes \mathscr{O}_{\hat{Y}}(\mu^*L_{|Y})) + C_1(\|c_1(L)\| + 1)^{p-1}.$$

By taking a suitable modification $\mu' : Y' \to Y$ of the desingularization $\hat{Y}$, we can assume that $(\mu')^*\mathscr{F}$ is locally free modulo torsion. Then we are reduced to the case where $\mathscr{F}' = (\mu')^*\mathscr{F}$ is a locally free sheaf on a smooth manifold Y', and $L' = (\mu')^*L_{|Y}$. In this case, we apply Morse inequalities to conclude that $h^q(Y', \mathscr{F}' \otimes \mathscr{O}_{Y'}(L')) \leq C_2(\|c_1(L')\| + 1)^p$. Since $\|c_1(L')\| \leq C_3\|c_1(L)\|$ by pulling-back, the statement follows easily. □

Corollary 86. *For every irreducible divisor D on X, there exists a constant C_D such that*

$$h^q(D, \mathscr{O}_D(L_{|D})) \leq C_D(\|c_1(L)\| + 1)^{n-1}$$

Proof. It is enough to apply Lemma 85 with $\mathscr{F} = (i_D)_*\mathscr{O}_D$ where $i_D : D \to X$ is the injection. □

Remark 87. It is very likely that one can get an "elementary" proof of Lemma 85 without invoking resolutions of singularities, e.g. by combining the Cartan–Serre finiteness argument along with the standard Serre–Siegel proof based ultimately on the Schwarz lemma. In this context, one would invoke L^2 estimates to get explicit bounds for the homotopy operators between Čech complexes relative to two coverings $\mathscr{U} = (B(x_j, r_j))$, $\mathscr{U}' = (B(x_j, r_j/2))$ of X by concentric balls. By exercising enough care in the estimates, it is likely that one could reach an explicit dependence $C_D \leq C'\|D\|$ for the constant C_D of Corollary 86. The proof would of course become much more technical than the rather naive brute force approach we have used.

Theorem 88. *Let X be a compact complex manifold. Fix a finitely generated subgroup Γ of the group of $\mathbb{Z}$-divisors on X. Then there are constants C, C' depending only on X, its Hermitian metric ω and the subgroup Γ, satisfying the following properties.*

(a) *Let L and $L' = L \otimes \mathcal{O}(D)$ be holomorphic line bundles on X, where $D \in \Gamma$ is an integral divisor. Then*

$$\left| h^q(X, L') - h^q(X, L) \right| \leq C(\|c_1(L)\| + \|D\|)^{n-1} \|D\|.$$

(b) *On the subspace $\mathrm{DNS}_{\mathbb{R}}(X)$, the asymptotic q-cohomology function $\hat{h}^q_{\mathrm{DNS}}$ satisfies a global estimate*

$$\left| \hat{h}^q_{\mathrm{DNS}}(X, \beta) - \hat{h}^q_{\mathrm{DNS}}(X, \alpha) \right| \leq C'(\|\alpha\| + \|\beta\|)^{n-1} \|\beta - \alpha\|.$$

In particular (without any further assumption on X), $\hat{h}^q_{\mathrm{DNS}}$ is locally Lipschitz continuous on $\mathrm{DNS}_{\mathbb{R}}(X)$.

Proof. (a) We want to compare the cohomology of L and $L' = L \otimes \mathcal{O}(D)$ on X. For this we write $D = D_+ - D_-$, and compare the cohomology of the pairs L and $L_1 = L \otimes \mathcal{O}(-D_-)$ one hand, and of L' and $L_1 = L' \otimes \mathcal{O}(-D_+)$ on the other hand. Since $\|c_1(\mathcal{O}(D))\| \leq C\|D\|$ by elementary potential theory, we see that is enough to consider the case of a negative divisor, i.e. $L' = L \otimes \mathcal{O}(-D)$, $D \geq 0$. If D is an irreducible divisor, we use the exact sequence

$$0 \to L \otimes \mathcal{O}(-D) \to L \to \mathcal{O}_D \otimes L_{|D} \to 0$$

and conclude by Corollary 86 that

$$\left| h^q(X, L \otimes \mathcal{O}(-D)) - h^q(X, L) \right| \leq h^q(D, \mathcal{O}_D \otimes L_{|D}) + h^{q-1}(D, \mathcal{O}_D \otimes L_{|D})$$

$$\leq 2C_D(\|c_1(L)\| + 1)^{n-1}.$$

For $D = \sum p_j D_j \geq 0$, we easily get by induction

$$\left| h^q(X, L \otimes \mathcal{O}(-D)) - h^q(X, L) \right| \leq 2 \sum_j p_j C_{D_j} \left(\|c_1(L)\| + \sum_k p_k \|\nabla_k\| + 1 \right)^{n-1}.$$

If we knew that $C_D \leq C'\|D\|$ as expected in Remark 1.6, then the argument would be complete without any restriction on D. The trouble disappears if we fix D in a finitely generated subgroup Γ of divisors, because only finitely many irreducible components appear in that case, and so we have to deal with only finitely many constants C_{D_j}. Property 9(a) is proved.

(b) Fix once for all a finite set of divisors $(\Delta_j)_{1 \leq j \leq t}$ providing a basis of $\mathrm{DNS}_{\mathbb{R}}(X) \subset H_{\mathrm{BC}}^{1,1}(X, \mathbb{R})$. Take two elements α and β in $\mathrm{DNS}_{\mathbb{R}}(X)$, and fix $\varepsilon > 0$. Then $\beta - \alpha$ can be ε-approximated by a $\mathbb{Q}$-divisor $\sum \lambda_j D_j$, $\lambda_j \in \mathbb{Q}$, and we can find a pair (k, L) with k arbitrarily large such that $\frac{1}{k} c_1(L)$ is ε-close to α and $n!/k^n h^q(X, L)$ approaches $\hat{h}_{\mathrm{DNS}}^q(X, \alpha)$ by ε. Then $\frac{1}{k} L + \sum \lambda_j \Delta_j$ approaches β as closely as we want. When approximating $\beta - \alpha$, we can arrange that $k \lambda_j$ is an integer by taking k large enough. Then β is approximated by $\frac{1}{k} c_1(L')$ with $L' = L \otimes \mathcal{O}(\sum k \lambda_j \Delta_j)$. Property (a) implies

$$h^q(X, L') - h^q(X, L) \geq -C \Big(\|c_1(L)\| + \Big\| \sum k \lambda_j \Delta_j \Big\| \Big)^{n-1} \Big\| \sum k \lambda_j \Delta_j \Big\|$$
$$\geq -C k^n \big(\|\alpha\| + \varepsilon + \|\beta - \alpha\| + \varepsilon \big)^{n-1} \big(\|\beta - \alpha\| + \varepsilon \big).$$

We multiply the previous inequality by $n!/k^n$ and get in this way

$$\frac{n!}{k^n} h^q(X, L') \geq \hat{h}_{\mathrm{DNS}}^q(X, \alpha) - \varepsilon - C' \big(\|\alpha\| + \|\beta\| + \varepsilon \big)^{n-1} \big(\|\beta - \alpha\| + \varepsilon \big).$$

By taking the limsup and letting $\varepsilon \to 0$, we finally obtain

$$\hat{h}_{\mathrm{DNS}}^q(X, \beta) - \hat{h}_{\mathrm{DNS}}^q(X, \alpha) \geq -C' \big(\|\alpha\| + \|\beta\| \big)^{n-1} \|\beta - \alpha\|.$$

Property 9(b) follows by exchanging the roles of α and β. □

3.3 *Transcendental Asymptotic Cohomology Functions*

Our ambition is to extend the function $\hat{h}_{\mathrm{NS}}^q$ in a natural way to the full cohomology group $H_{\mathrm{BC}}^{1,1}(X, \mathbb{R})$. The main trouble, already when X is projective algebraic, is that the Picard number $\rho(X) = \dim_{\mathbb{R}} \mathrm{NS}_{\mathbb{R}}(X)$ may be much smaller than $\dim_{\mathbb{R}} H_{\mathrm{BC}}^{1,1}(X, \mathbb{R})$, namely, there can be rather few integral classes of type $(1, 1)$ on X. It is well known for instance that $\rho(X) = 0$ for a generic complex torus of dimension $n \geq 2$, while $\dim_{\mathbb{R}} H_{\mathrm{BC}}^{1,1}(X, \mathbb{R}) = n^2$. However, if we look at the natural morphism

$$H_{\mathrm{BC}}^{1,1}(X, \mathbb{R}) \to H_{\mathrm{DR}}^2(X, \mathbb{R}) \simeq H^2(X, \mathbb{R})$$

to de Rham cohomology, then $H^2(X, \mathbb{Q})$ is dense in $H^2(X, \mathbb{R})$. Therefore, given a class $\alpha \in H_{\mathrm{BC}}^{1,1}(X, \mathbb{R})$ and a smooth d-closed $(1, 1)$-form u in α, we can find an infinite sequence $\frac{1}{k} L_k$ ($k \in S \subset \mathbb{N}$) of topological $\mathbb{Q}$-line bundles, equipped with Hermitian metrics h_k and compatible connections ∇_k such that the curvature forms $\frac{1}{k} \Theta_{\nabla_k}$ converge to u. By using Kronecker's approximation with respect to the integral lattice $H^2(X, \mathbb{Z})/\mathrm{torsion} \subset H^2(X, \mathbb{R})$, we can even achieve a fast diophantine approximation

$$\|\Theta_{\nabla_k} - ku\| \leq C k^{-1/b_2} \tag{69}$$

for a suitable infinite subset $k \in S \subset \mathbb{N}$ of multipliers. Then in particular

$$\|\Theta_{\nabla_k}^{0,2}\| = \|\Theta_{\nabla_k}^{0,2} - k\, u^{0,2}\| \leq C k^{-1/b_2}, \tag{70}$$

and we see that (L_k, h_k, ∇_k) is a C^∞ Hermitian line bundle which is extremely close to being holomorphic, since $(\nabla_k^{0,1})^2 = \Theta_{\nabla_k}^{0,2}$ is very small. We fix a Hermitian metric ω on X and introduce the complex Laplace–Beltrami operator

$$\overline{\square}_{k,q} = (\nabla_k^{0,1})(\nabla_k^{0,1})^* + (\nabla_k^{0,1})^*(\nabla_k^{0,1}) \quad \text{acting on } L^2(X, \Lambda^{0,q} T_X^* \otimes L_k).$$

We look at its eigenspaces with respect to the L^2 metric induced by ω on X and h_k on L_k. In the holomorphic case, Hodge theory tells us that the 0-eigenspace is isomorphic to $H^q(X, \mathcal{O}(L_k))$, but in the "almost holomorphic case" the 0-eigenvalues deviate from 0, essentially by a shift of the order of magnitude of $\|\Theta_{\nabla_k}^{0,2}\| \sim k^{-1/b_2}$ (see also the PhD thesis of L. Laeng [64, Chap. 4], for more details). It is thus natural to introduce in this case

Definition 89. Let X be a compact complex manifold and $\alpha \in H^{1,1}_{\mathrm{BC}}(X, \mathbb{R})$ an arbitrary Bott–Chern $(1, 1)$-class. We define the "transcendental" asymptotic q-cohomology functions to be

(a) $\hat{h}_{\mathrm{tr}}^q(X, \alpha) = \inf\limits_{u \in \alpha} \limsup\limits_{\varepsilon \to 0, k \to +\infty, L_k, h_k, \nabla_k, \frac{1}{k}\Theta_{\nabla_k} \to u} \dfrac{n!}{k^n} N(\overline{\square}_{k,q}, \leq k\varepsilon)$

(b) $\hat{h}_{\mathrm{tr}}^{\leq q}(X, \alpha) = \inf\limits_{u \in \alpha} \limsup\limits_{\varepsilon \to 0, k \to +\infty, L_k, h_k, \nabla_k, \frac{1}{k}\Theta_{\nabla_k} \to u} \dfrac{n!}{k^n} \sum\limits_{0 \leq j \leq q} (-1)^{q-j} N(\overline{\square}_{k,j}, \leq k\varepsilon)$

where the lim sup runs over all 5-tuples $(\varepsilon, k, L_k, h_k, \nabla_k)$, and where $N(\overline{\square}_{k,q}, k\varepsilon)$ denotes the sum of dimensions of all eigenspaces of eigenvalues at most equal to $k\varepsilon$ for the Laplace–Beltrami operator $\overline{\square}_{k,q}$ on $L^2(X, \Lambda^{0,q} T_X^* \otimes L_k)$ associated with (L_k, h_k, ∇_k) and the base Hermitian metric ω.

The word "transcendental" refers here to the fact that we deal with classes α of type $(1, 1)$ which are not algebraic or even analytic. Of course, in the definition, we could have restricted the limsup to families satisfying a better approximation property $\|\frac{1}{k}\Theta_{\nabla_k} - u\| \leq C k^{-1-1/b_2}$ for some large constant C (this would lead a priori to a smaller limsup, but there is enough stability in the parameter dependence of the spectrum for making such a change irrelevant). The minimax principle easily shows that Definition 36 does not depend on ω, as the eigenvalues are at most multiplied or divided by constants under a change of base metric. When $\alpha \in \mathrm{NS}_\mathbb{R}(X)$, by restricting our families $\{(\varepsilon, k, L_k, h_k, \nabla_k)\}$ to the case of holomorphic line bundles only, we get the obvious inequalities

$$\hat{h}_{\mathrm{NS}}^q(X, \alpha) \leq \hat{h}_{\mathrm{tr}}^q(X, \alpha), \quad \forall \alpha \in \mathrm{NS}_\mathbb{R}(X), \tag{71}$$

$$\hat{h}_{\mathrm{NS}}^{\leq q}(X, \alpha) \leq \hat{h}_{\mathrm{tr}}^{\leq q}(X, \alpha), \quad \forall \alpha \in \mathrm{NS}_\mathbb{R}(X). \tag{72}$$

It is natural to raise the question whether these inequalities are always equalities. Hopefully, the calculation of the quantities $\lim_{k\to+\infty} \frac{n!}{k^n} N(\overline{\square}_{k,q}, \leq k\varepsilon)$ is a problem of spectral theory which is completely understood thanks to Sect. 1 (see also [25, 91]). In fact, by Corollary I (1.13), the above limit can be evaluated explicitly for any value of $\varepsilon \in \mathbb{R}$, except possibly for a countable number of values of ε for which jumps occur; one only has to take care that the non-integrability of $\overline{\partial}$ due to the diophantine approximation does not contribute asymptotically to the eigenvalue distribution, a fact which follows immediately from (40) (cf. [64]).

Theorem 90. *With the above notations and assumptions, let us introduce at each point x in X the "spectral density function", defined as a finite sum*

$$\nu_u(\lambda) = \frac{n!\,(4\pi)^{s-n}}{(n-s)!}|u_1|\ldots|u_s| \sum_{(p_1,\ldots,p_s)\in\mathbb{N}^s} \left(\lambda - \sum_{j=1}^{s}(2p_j+1)|u_j|\right)_+^{n-s}$$

where $s = s(x)$ is the rank of the real $(1,1)$-form u at x, and u_j, $1 \leq j \leq s$, its non zero eigenvalues with respect to the base Hermitian metric ω, and $u_{s+1} = \ldots = u_n = 0$. For each multi-index $J \subset \{1,2,\ldots,n\}$, let us set $u_J = \sum_{j\in J} u_j$. Then the asymptotic spectrum of $\overline{\square}_{k,q}$ admits the estimate

$$\lim_{k\to+\infty} \frac{n!}{k^n} N(\overline{\square}_{k,q}, \leq k\lambda) = \int_X \sum_{|J|=q} \nu_u(\lambda + u_{\complement J} - u_J)\, dV_\omega$$

except possibly for a countable number of values of λ which are discontinuities of the right hand integral as an increasing integral of λ.

Corollary 91. *We have (as a limit rather than just a $\limsup$) the spectral estimate*

$$\lim_{\varepsilon\to 0, k\to+\infty, L_k, h_k, \nabla_k, \frac{1}{k}\Theta_{\nabla_k}\to u} \frac{n!}{k^n} N(\overline{\square}_{k,q}, \leq k\varepsilon) = \int_{X(u,q)} (-1)^q u^n.$$

Coming back to the transcendental asymptotic cohomology functions, we get the following fundamental result, which gives in some sense an explicit formula for $\hat{h}^q_{\mathrm{tr}}(X,\alpha)$ and $\hat{h}^{\leq q}_{\mathrm{tr}}(X,\alpha)$ in terms of Monge–Ampère operators.

Theorem 92. *The $\limsup$'s defining $\hat{h}^q_{\mathrm{tr}}(X,\alpha)$ and $\hat{h}^{\leq q}_{\mathrm{tr}}(X,\alpha)$ are limits, and we have*

(a) $\hat{h}^q_{\mathrm{tr}}(X,\alpha) = \inf_{u\in\alpha} \int_{X(u,q)} (-1)^q u^n$ *(u smooth).*

(b) $\hat{h}^{\leq q}_{\mathrm{tr}}(X,\alpha) = \inf_{u\in\alpha} \int_{X(u,\leq q)} (-1)^q u^n$ *(u smooth).*

Now, if $L \to X$ is a holomorphic line bundle, we have by definition

$$\hat{h}^{\leq q}(X,L) \leq \hat{h}^{\leq q}_{\mathrm{DNS}}(X,c_1(L)) \leq \hat{h}^{\leq q}_{\mathrm{NS}}(X,c_1(L)) \leq \inf_{u\in c_1(L)} \int_{X(u,\leq q)} (-1)^q u^n \quad (73)$$

(u smooth), where the last inequality is a consequence of holomorphic Morse inequalities. We hope for the following conjecture which would imply that we always have equalities.

Conjecture 93. For every holomorphic line bundle $L \to X$ on a compact complex manifold X, we have

(a) $\hat{h}^q(X, L) = \inf_{u \in \alpha} \int_{X(u,q)} (-1)^q u^n$, u smooth,

(b) $\hat{h}^{\leq q}(X, L) = \inf_{u \in \alpha} \int_{X(u,\leq q)} (-1)^q u^n$, u smooth.

Since the right hand side is easily seen to depend continuously on $\alpha \in H^{1,1}_{BC}(X, \mathbb{C})$, one would get:

Corollary of the conjecture 94. *If Conjecture 93 holds true, then*

(a) $\quad \hat{h}^q_{NS}(X, \alpha) = \hat{h}^q_{tr}(X, \alpha) \quad$ and $\quad$ (b) $\quad \hat{h}^{\leq q}_{NS}(X, \alpha) = \hat{h}^{\leq q}_{tr}(X, \alpha)$

for all classes $\alpha \in \mathrm{NS}_{\mathbb{R}}(X)$.

In general, equalities 93(a, b) seem rather hard to prove. In some sense, they would stand as an asymptotic converse of the Andreotti–Grauert theorem [3]: under a suitable q-convexity assumption, the latter asserts the vanishing of related cohomology groups in degree q; here, conversely, assuming a known growth of these groups in degree q, we expect to be able to say something about the q-index sets of suitable Hermitian metrics on the line bundles under consideration. The only cases where we have a positive answer to Question 2.8 are when X is projective and $q = 0$ or $\dim X \leq 2$ (see Theorems 97 and 98 below). In the general setting of compact complex manifolds, we also hope for the following "transcendental" case of holomorphic Morse inequalities.

Conjecture 95. Let X be a compact complex n-fold and α an arbitrary cohomology class in $H^{1,1}_{BC}(X, \mathbb{R})$. Then the volume, defined as the supremum

$$\mathrm{Vol}(\alpha) := \sup_{0 < T \in \alpha} \int_{X \smallsetminus \mathrm{Sing}(T)} T^n, \tag{74}$$

extended to all Kähler currents $T \in \alpha$ with analytic singularities (see Definition II (4.4)), satisfies

$$\mathrm{Vol}(\alpha) \geq \sup_{u \in \alpha} \int_{X(u,0) \cup X(u,1)} u^n \tag{75}$$

where u runs over all smooth closed $(1, 1)$ forms. In particular, if the right hand side is positive, then α contains a Kähler current.

By the holomorphic Morse inequalities, Conjecture 95 holds true in case α is an integral class. Our hope is that the general case can be attained by the diophantine approximation technique described earlier; there are however major hurdles, see [64] for a few hints on these issues.

3.4 Invariance by Modification

We end this section by the observation that the asymptotic cohomology functions are invariant by modification, namely that for every modification $\mu : \tilde{X} \to X$ and every line bundle L we have e.g.

$$\hat{h}^q(X, L) = \hat{h}^q(\tilde{X}, \mu^* L). \tag{76}$$

In fact the Leray spectral sequence provides an E_2 term

$$E_2^{p,q} = H^p(X, R^q\mu_* \mathcal{O}_{\tilde{X}}(\mu^* L^{\otimes k})) = H^p(X, \mathcal{O}_X(L^{\otimes k}) \otimes R^q\mu_* \mathcal{O}_{\tilde{X}}).$$

Since $R^q\mu_* \mathcal{O}_{\tilde{X}}$ is equal to $\mathcal{O}_X$ for $q = 0$ and is supported on a proper analytic subset of X for $q \geq 1$, one infers that $h^p(X, \mathcal{O}_X(L^{\otimes k} \otimes R^q\mu_* \mathcal{O}_{\tilde{X}})) = O(k^{n-1})$ for all $q \geq 1$. The spectral sequence implies that

$$h^q(X, L^{\otimes k}) - \hat{h}^q(\tilde{X}, \mu^* L^{\otimes k}) = O(k^{n-1}).$$

We claim that the Morse integral infimums are also invariant by modification.

Proposition 96. *Let (X, ω) be a compact Kähler manifold, $\alpha \in H^{1,1}(X, \mathbb{R})$ a real cohomology class and $\mu : \tilde{X} \to X$ a modification. Then*

(a) $\displaystyle\inf_{u \in \alpha} \int_{X(u,q)} (-1)^q u^n = \inf_{v \in \mu^* \alpha} \int_{X(v,q)} (-1)^q v^n,$

(b) $\displaystyle\inf_{u \in \alpha} \int_{X(u,\leq q)} (-1)^q u^n = \inf_{v \in \mu^* \alpha} \int_{X(v,\leq q)} (-1)^q v^n.$

Proof. Given $u \in \alpha$ on X, we obtain Morse integrals with the same values by taking $v = \mu^* u$ on $\tilde{X}$, hence the infimum on $\tilde{X}$ is smaller or equal to what is on X. Conversely, we have to show that given a smooth representative $v \in \mu^* \alpha$ on $\tilde{X}$, one can find a smooth representative $u \in X$ such that the Morse integrals do not differ much. We can always assume that $\tilde{X}$ itself is Kähler, since by Hironaka [55] any modification $\tilde{X}$ is dominated by a composition of blow-ups of X. Let us fix some $u_0 \in \alpha$ and write

$$v = \mu^* u_0 + dd^c \varphi, \qquad d^c = \frac{i}{4\pi}(\bar{\partial} - \partial), \qquad dd^c = \frac{i}{2\pi}\partial\bar{\partial},$$

where φ is a smooth function on $\tilde{X}$. We adjust φ by a constant in such a way that $\varphi \geq 1$ on $\tilde{X}$. There exists an analytic set $S \subset X$ such that $\mu : \tilde{X} \smallsetminus \mu^{-1}(S) \to X \smallsetminus S$ is a biholomorphism, and a quasi-psh function ψ_S which is smooth on $X \smallsetminus S$ and has $-\infty$ logarithmic poles on S (see e.g. [24]). We define

$$\tilde{u} = \mu^* u_0 + dd^c \max{}_{\varepsilon_0}(\varphi + \delta \psi_S \circ \mu, \, 0) = v + dd^c \max{}_{\varepsilon_0}(\delta \psi_S \circ \mu, \, -\varphi) \tag{77}$$

Applications of Pluripotential Theory to Algebraic Geometry   221

where $\max_{\varepsilon_0}$, $0 < \varepsilon_0 < 1$, is a regularized max function and $\delta > 0$ is very small. By construction $\tilde{u}$ coincides with $\mu^* u_0$ in a neighborhood of $\mu^{-1}(S)$ and therefore $\tilde{u}$ descends to a smooth closed $(1,1)$-form u on X which coincides with u_0 near S, so that $\tilde{u} = \mu^* u$. Clearly $\tilde{u}$ converges uniformly to v on every compact subset of $\tilde{X} \smallsetminus \mu^{-1}(S)$ as $\delta \to 0$, so we only have to show that the Morse integrals are small (uniformly in δ) when restricted to a suitable small neighborhood of the exceptional set $E = \mu^{-1}(S)$. Take a sufficiently large Kähler metric $\tilde{\omega}$ on $\tilde{X}$ such that

$$-\frac{1}{2}\tilde{\omega} \leq v \leq \frac{1}{2}\tilde{\omega}, \quad -\frac{1}{2}\tilde{\omega} \leq dd^c \varphi \leq \frac{1}{2}\tilde{\omega}, \quad -\tilde{\omega} \leq dd^c \psi_S \circ \mu.$$

Then $\tilde{u} \geq -\tilde{\omega}$ and $\tilde{u} \leq \tilde{\omega} + \delta\, dd^c \psi_S \circ \mu$ everywhere on $\tilde{X}$. As a consequence

$$|\tilde{u}^n| \leq \big(\tilde{\omega} + \delta(\tilde{\omega} + dd^c \psi_S \circ \mu)\big)^n$$
$$\leq \tilde{\omega}^n + n\delta(\tilde{\omega} + dd^c \psi_S \circ \mu) \wedge \big(\tilde{\omega} + \delta(\tilde{\omega} + dd^c \psi_S \circ \mu)\big)^{n-1}$$

thanks to the inequality $(a+b)^n \leq a^n + nb(a+b)^{n-1}$. For any neighborhood V of $\mu^{-1}(S)$ this implies

$$\int_V |\tilde{u}^n| \leq \int_V \tilde{\omega}^n + n\delta(1+\delta)^{n-1} \int_{\tilde{X}} \tilde{\omega}^n$$

by Stokes formula. We thus see that the integrals are small if V and δ are small. The reader may be concerned that Monge–Ampère integrals were used with an unbounded potential ψ_S, but in fact, for any given δ, all the above formulas and estimates are still valid when we replace ψ_S by $\max_{\varepsilon_0}(\psi_S, -(M+2)/\delta)$ with $M = \max_{\tilde{X}} \varphi$, especially formula (77) shows that the form $\tilde{u}$ is unchanged. Therefore our calculations can be handled by using merely smooth potentials. $\square$

3.5 Proof of the Infimum Formula for the Volume

We prove here

Theorem 97. *Let $L \to X$ be a holomorphic line bundle on a projective algebraic manifold X. Then*

$$\mathrm{Vol}(X, L) = \inf_{u \in c_1(L)} \int_{X(u,0)} u^n.$$

It is enough to show the inequality

$$\inf_{u \in c_1(L)} \int_{X(u,0)} u^n \leq \mathrm{Vol}(X, L) \tag{78}$$

and for this, we have to construct metrics approximating the volume. Let us first assume that L is a big line bundle, i.e. that $\mathrm{Vol}(X, L) > 0$. We have seen in Definition 66 and Theorem 67 (cf. also [18]) that $\mathrm{Vol}(X, L)$ is obtained as the supremum of $\int_{X \smallsetminus \mathrm{Sing}(T)} T^n$ for Kähler currents $T = -\frac{i}{2\pi} \partial \bar{\partial} h$ with analytic singularities in $c_1(L)$; this means that locally $h = e^{-\varphi}$ where φ is a strictly plurisubharmonic function which has the same singularities as $c \log \sum |g_j|^2$ where $c > 0$ and the g_j are holomorphic functions. By [28], there exists a blow-up $\mu : \tilde{X} \to X$ such that $\mu^* T = [E] + \beta$ where E is a normal crossing divisor on $\tilde{X}$ and $\beta \geq 0$ smooth. Moreover, by [19] we have the orthogonality estimate

$$[E] \cdot \beta^{n-1} = \int_E \beta^{n-1} \leq C \left(\mathrm{Vol}(X, L) - \beta^n \right)^{1/2}, \tag{79}$$

while

$$\beta^n = \int_{\tilde{X}} \beta^n = \int_{X \smallsetminus \mathrm{Sing}(T)} T^n \quad \text{approaches } \mathrm{Vol}(X, L). \tag{80}$$

In other words, E and β become "more and more orthogonal" as β^n approaches the volume (these properties are summarized by saying that $\mu^* T = [E] + \beta$ defines an approximate Zariski decomposition of $c_1(L)$, cf. also [47]). By subtracting to β a small linear combination of the exceptional divisors and increasing accordingly the coefficients of E, we can achieve that the cohomology class $\{\beta\}$ contains a positive definite form β' on $\tilde{X}$ (i.e. the fundamental form of a Kähler metric); we refer e.g. to [42, proof of Lemma 29] for details. This means that we can replace T by a cohomologous current such that the corresponding form β is actually a Kähler metric, and we will assume for simplicity of notation that this situation occurs right away for T. Under this assumption, there exists a smooth closed $(1, 1)$-form v belonging to the Bott–Chern cohomology class of $[E]$, such that we have identically $(v - \delta \beta) \wedge \beta^{n-1} = 0$ where

$$\delta = \frac{[E] \cdot \beta^{n-1}}{\beta^n} \leq C' (\mathrm{Vol}(X, L) - \beta^n)^{1/2} \tag{81}$$

for some constant $C' > 0$. In fact, given an arbitrary smooth representative $v_0 \in \{[E]\}$, the existence of $v = v_0 + i \partial \bar{\partial} \psi$ amounts to solving a Laplace equation $\Delta \psi = f$ with respect to the Kähler metric β, and the choice of δ ensures that we have $\int_X f \beta^n = 0$ and hence that the equation is solvable. Then $\tilde{u} := v + \beta$ is a smooth closed $(1, 1)$-form in the cohomology class $\mu^* c_1(L)$, and its eigenvalues with respect to β are of the form $1 + \lambda_j$ where λ_j are the eigenvalues of v. The Laplace equation is equivalent to the identity $\sum \lambda_j = n\delta$. Therefore

$$\sum_{1 \leq j \leq n} \lambda_j \leq C'' (\mathrm{Vol}(X, L) - \beta^n)^{1/2}. \tag{82}$$

The inequality between arithmetic means and geometric means implies

$$\prod_{1\leq j\leq n}(1+\lambda_j) \leq \left(1+\frac{1}{n}\sum_{1\leq j\leq n}\lambda_j\right)^n \leq 1+C_3(\text{Vol}(X,L)-\beta^n)^{1/2}$$

whenever all factors $(1+\lambda_j)$ are nonnegative. By 2.2(i) we get

$$\inf_{u\in c_1(L)}\int_{X(u,0)} u^n \leq \int_{\tilde{X}(\tilde{u},0)} \tilde{u}^n$$

$$\leq \int_{\tilde{X}} \beta^n\left(1+C_3(\text{Vol}(X,L)-\beta^n)^{1/2}\right)$$

$$\leq \text{Vol}(X,L)+C_4(\text{Vol}(X,L)-\beta^n)^{1/2}.$$

As β^n approaches $\text{Vol}(X,L)$, this implies inequality (4.1).

We still have to treat the case when L is not big, i.e. $\text{Vol}(X,L)=0$. Let A be an ample line bundle and let $t_0 \geq 0$ be the infimum of real numbers such that $L+tA$ is a big $\mathbb{Q}$-line bundle for t rational, $t > t_0$. The continuity of the volume function implies that $0 < \text{Vol}(X, L+tA) \leq \varepsilon$ for $t > t_0$ sufficiently close to t_0. By what we have just proved, there exists a smooth form $u_t \in c_1(L+tA)$ such that $\int_{X(u_t,0)} u_t^n \leq 2\varepsilon$. Take a Kähler metric $\omega \in c_1(A)$ and define $u = u_t - t\omega$. Then clearly

$$\int_{X(u,0)} u^n \leq \int_{X(u_t,0)} u_t^n \leq 2\varepsilon,$$

hence

$$\inf_{u\in c_1(L)}\int_{X(u,0)} u^n = 0.$$

Inequality (4.1) is now proved in all cases. □

3.6 Estimate of the First Cohomology Group on a Projective Surface

Our goal here is to show the following result.

Theorem 98. *Let $L \to X$ be a holomorphic line bundle on a complex projective surface. Then both weak and strong inequalities (23)(i) and (23)(ii) are equalities for $q = 0, 1, 2$, and the lim sup's involved in $\hat{h}^q(X,L)$ and $\hat{h}^{\leq q}(X,L)$ are limits.*

We start with a projective non singular variety X of arbitrary dimension n, and will later restrict ourselves to the case when X is a surface. The proof again consists of using (approximate) Zariski decomposition, but now we try to compute more explicitly the resulting curvature forms and Morse integrals; this will turn out to be much easier on surfaces.

Assume first that L is a *big* line bundle on X. As in Sect. 3, we can find an approximate Zariski decomposition, i.e. a blow-up $\mu : \tilde{X} \to X$ and a current $T \in c_1(L)$ such $\mu^* T = [E] + \beta$, where E an effective divisor and β a Kähler metric on $\tilde{X}$ such that

$$\text{Vol}(X, L) - \eta < \beta^n < \text{Vol}(X, L), \qquad \eta \ll 1. \tag{83}$$

(On a projective surface, one could even get exact Zariski decomposition, but we want to remain general as long as possible). By blowing-up further, we may assume that E is a normal crossing divisor. We select a Hermitian metric h on $\mathscr{O}(E)$ and take

$$u_\varepsilon = \frac{i}{2\pi} \partial\bar{\partial} \log(|\sigma_E|_h^2 + \varepsilon^2) + \Theta_{\mathscr{O}(E),h} + \beta \in \mu^* c_1(L) \tag{84}$$

where $\sigma_E \in H^0(\tilde{X}, \mathscr{O}(E))$ is the canonical section and $\Theta_{\mathscr{O}(E),h}$ the Chern curvature form. Clearly, by the Lelong–Poincaré equation, u_ε converges to $[E] + \beta$ in the weak topology as $\varepsilon \to 0$. Straightforward calculations yield

$$u_\varepsilon = \frac{i}{2\pi} \frac{\varepsilon^2 D_h^{1,0} \sigma_E \wedge \overline{D_h^{1,0} \sigma_E}}{(\varepsilon^2 + |\sigma_E|^2)^2} + \frac{\varepsilon^2}{\varepsilon^2 + |\sigma_E|^2} \Theta_{E,h} + \beta.$$

The first term converges to $[E]$ in the weak topology, while the second, which is close to $\Theta_{E,h}$ near E, converges pointwise everywhere to 0 on $\tilde{X} \smallsetminus E$. A simple asymptotic analysis shows that

$$\left(\frac{i}{2\pi} \frac{\varepsilon^2 D_h^{1,0} \sigma_E \wedge \overline{D_h^{1,0} \sigma_E}}{(\varepsilon^2 + |\sigma_E|^2)^2} + \frac{\varepsilon^2}{\varepsilon^2 + |\sigma_E|^2} \Theta_{E,h} \right)^p \to [E] \wedge \Theta_{E,h}^{p-1}$$

in the weak topology for $p \geq 1$, hence

$$\lim_{\varepsilon \to 0} u_\varepsilon^n = \beta^n + \sum_{p=1}^{n} \binom{n}{p} [E] \wedge \Theta_{E,h}^{p-1} \wedge \beta^{n-p}. \tag{85}$$

In arbitrary dimension, the signature of u_ε is hard to evaluate, and it is also non trivial to decide the sign of the limiting measure $\lim u_\varepsilon^n$. However, when $n = 2$, we get the simpler formula

$$\lim_{\varepsilon \to 0} u_\varepsilon^2 = \beta^2 + 2[E] \wedge \beta + [E] \wedge \Theta_{E,h}.$$

In this case, E can be assumed to be an exceptional divisor (otherwise some part of it would be nef and could be removed from the poles of T). Hence the matrix $(E_j \cdot E_k)$ is negative definite and we can find a smooth Hermitian metric h on $\mathscr{O}(E)$ such that $(\Theta_{E,h})_{|E} < 0$, i.e. $\Theta_{E,h}$ has one negative eigenvalue everywhere along E.

Lemma 99. *One can adjust the metric h of $\mathscr{O}(E)$ in such a way that $\Theta_{E,h}$ is negative definite on a neighborhood of the support $|E|$ of the exceptional divisor, and $\Theta_{E,h} + \beta$ has signature $(1,1)$ there. (We do not care about the signature far away from $|E|$).*

Proof. At a given point $x_0 \in X$, let us fix coordinates and a positive quadratic form q on $\mathbb{C}^2$. If we put $\psi_\varepsilon(z) = \varepsilon\chi(z)\log(1+\varepsilon^{-1}q(z))$ with a suitable cut-off function χ, then the Hessian form of ψ_ε is equal to q at x_0 and decays rapidly to $O(\varepsilon\log\varepsilon)|dz|^2$ away from x_0. In this way, after multiplying h with $e^{\pm\psi_\varepsilon(z)}$, we can replace the curvature $\Theta_{E,h}(x_0)$ with $\Theta_{E,h}(x_0) \pm q$ without substantially modifying the form away from x_0. This allows to adjust $\Theta_{E,h}$ to be equal to (say) $-\frac{1}{4}\beta(x_0)$ at any singular point $x_0 \in E_j \cap E_k$ in the support of $|E|$, while keeping $\Theta_{E,h}$ negative definite along E. In order to adjust the curvature at smooth points $x \in |E|$, we replace the metric h with $h'(z) = h(z)\exp(-c(z)|\sigma_E(z)|^2)$. Then the curvature form $\Theta_{E,h}$ is replaced by $\Theta_{E,h'}(x) = \Theta_{E_h}(x) + c(x)|d\sigma_E|^2$ at $x \in |E|$ (notice that $d\sigma_E(x) = 0$ if $x \in \mathrm{Sing}|E|$), and we can always select a real function c so that $\Theta_{E,h'}$ is negative definite with one negative eigenvalue between $-1/2$ and 0 at any point of $|E|$. Then $\Theta_{E,h'} + \beta$ has signature $(1,1)$ near $|E|$. □

With this choice of the metric, we see that for $\varepsilon > 0$ small, the sum

$$\frac{\varepsilon^2}{\varepsilon^2 + |\sigma_E|^2}\Theta_{E,h} + \beta$$

is of signature $(2,0)$ or $(1,1)$ (or degenerate of signature $(1,0)$), the non positive definite points being concentrated in a neighborhood of E. In particular the index set $X(u_\varepsilon, 2)$ is empty, and also

$$u_\varepsilon \le \frac{i}{2\pi} \frac{\varepsilon^2 D_h^{1,0}\sigma_E \wedge \overline{D_h^{1,0}\sigma_E}}{(\varepsilon^2 + |\sigma_E|^2)^2} + \beta$$

on a neighborhood V of $|E|$, while u_ε converges uniformly to β on $\widetilde{X} \smallsetminus V$. This implies that

$$\beta^2 \le \liminf_{\varepsilon \to 0} \int_{X(u_\varepsilon, 0)} u_\varepsilon^2 \le \limsup_{\varepsilon \to 0} \int_{X(u_\varepsilon, 0)} u_\varepsilon^2 \le \beta^2 + 2\beta \cdot E.$$

Since $\int_{\widetilde{X}} u_\varepsilon^2 = L^2 = \beta^2 + 2\beta \cdot E + E^2$ we conclude by taking the difference that

$$-E^2 - 2\beta \cdot E \le \liminf_{\varepsilon \to 0} \int_{X(u_\varepsilon, 1)} -u_\varepsilon^2 \le \limsup_{\varepsilon \to 0} \int_{X(u_\varepsilon, 1)} -u_\varepsilon^2 \le -E^2.$$

Let us recall that $\beta \cdot E \le C(\mathrm{Vol}(X,L) - \beta^2)^{1/2} = 0(\eta^{1/2})$ is small by (84) and the orthogonality estimate. The asymptotic cohomology is given here by $\hat{h}^2(X,L) = 0$ since $h^2(X, L^{\otimes k}) = H^0(X, K_X \otimes L^{\otimes -k}) = 0$ for $k \ge k_0$, and we have by Riemann–Roch

$$\hat{h}^1(X,L) = \hat{h}^0(X,L) - L^2 = \mathrm{Vol}(X,L) - L^2 = -E^2 - \beta \cdot E + O(\eta).$$

Here we use the fact that $\frac{n!}{k^n}h^0(X, L^{\otimes k})$ converges to the volume when L is big. All this shows that equality occurs in the Morse inequalities (67) when we pass to the infimum. By taking limits in the Neron–Severi space $\mathrm{NS}_\mathbb{R}(X) \subset H^{1,1}(X, \mathbb{R})$, we further see that equality occurs as soon as L is pseudo-effective, and the same is true if $-L$ is pseudo-effective by Serre duality. It remains to treat the case when neither L nor $-L$ are pseudo-effective. Then $\hat{h}^0(X, L) = \hat{h}^2(X, L) = 0$, and asymptotic cohomology appears only in degree 1, with $\hat{h}^1(X, L) = -L^2$ by Riemann–Roch. Fix an ample line bundle A and let $t_0 > 0$ be the infimum of real numbers such that $L + tA$ is big for t rational, $t > t_0$, resp. let $t_0' > 0$ be the infimum of real numbers t' such that $-L + t'A$ is big for $t' > t_0'$. Then for $t > t_0$ and $t' > t_0'$, we can find a modification $\mu : \tilde{X} \to X$ and currents $T \in c_1(L + tA)$, $T' \in c_1(-L + t'A)$ such that

$$\mu^* T = [E] + \beta, \qquad \mu^* T' = [F] + \gamma$$

where β, γ are Kähler forms and E, F normal crossing divisors. By taking a suitable linear combination $t'(L + tA) - t(-L + t'A)$ the ample divisor A disappears, and we get

$$\frac{1}{t+t'}\Big(t'[E] + t'\beta - t[F] - t\gamma\Big) \in \mu^* c_1(L).$$

After replacing E, F, β, γ by suitable multiples, we obtain an equality

$$[E] - [F] + \beta - \gamma \in \mu^* c_1(L).$$

We may further assume by subtracting that the divisors E, F have no common components. The construction shows that $\beta^2 \leq \mathrm{Vol}(X, L + tA)$ can be taken arbitrarily small (as well of course as γ^2), and the orthogonality estimate implies that we can assume $\beta \cdot E$ and $\gamma \cdot F$ to be arbitrarily small. Let us introduce metrics h_E on $\mathcal{O}(E)$ and h_F on $\mathcal{O}(F)$ as in Lemma 99, and consider the forms

$$u_\varepsilon = + \frac{i}{2\pi} \frac{\varepsilon^2 D^{1,0}_{h_E}\sigma_E \wedge \overline{D^{1,0}_{h_E}\sigma_E}}{(\varepsilon^2 + |\sigma_E|^2)^2} + \frac{\varepsilon^2}{\varepsilon^2 + |\sigma_E|^2}\Theta_{E,h_E} + \beta$$

$$- \frac{i}{2\pi} \frac{\varepsilon^2 D^{1,0}_{h_F}\sigma_F \wedge \overline{D^{1,0}_{h_F}\sigma_F}}{(\varepsilon^2 + |\sigma_F|^2)^2} - \frac{\varepsilon^2}{\varepsilon^2 + |\sigma_F|^2}\Theta_{F,h_F} - \gamma \in \mu^* c_1(L).$$

Observe that u_ε converges uniformly to $\beta - \gamma$ outside of every neighborhood of $|E| \cup |F|$. Assume that $\Theta_{E,h_E} < 0$ on $V_E = \{|\sigma_E| < \varepsilon_0\}$ and $\Theta_{F,h_F} < 0$ on $V_F = \{|\sigma_F| < \varepsilon_0\}$. On $V_E \cup V_F$ we have

$$u_\varepsilon \leq \frac{i}{2\pi} \frac{\varepsilon^2 D^{1,0}_{h_E}\sigma_E \wedge \overline{D^{1,0}_{h_E}\sigma_E}}{(\varepsilon^2 + |\sigma_E|^2)^2} - \frac{\varepsilon^2}{\varepsilon^2 + |\sigma_F|^2}\Theta_{F,h_F} + \beta + \frac{\varepsilon^2}{\varepsilon_0^2}\Theta^+_{E,h_E}$$

where Θ^+_{E,h_E} is the positive part of Θ_{E,h_E} with respect to β. One sees immediately that this term is negligible. The first term is the only one which is not uniformly bounded, and actually it converges weakly to the current $[E]$. By squaring, we find

$$\limsup_{\varepsilon \to 0} \int_{X(u_\varepsilon,0)} u_\varepsilon^2 \leq \int_{X(\beta-\gamma,0)} (\beta-\gamma)^2 + 2\beta \cdot E.$$

Notice that the term $-\frac{\varepsilon^2}{\varepsilon^2+|\sigma_F|^2}\Theta_{F,h_F}$ does not contribute to the limit as it converges boundedly almost everywhere to 0, the exceptions being points of $|F|$, but this set is of measure zero with respect to the current $[E]$. Clearly we have $\int_{X(\beta-\gamma,0)}(\beta-\gamma)^2 \leq \beta^2$ and therefore

$$\limsup_{\varepsilon \to 0} \int_{X(u_\varepsilon,0)} u_\varepsilon^2 \leq \beta^2 + 2\beta \cdot E.$$

Similarly, by looking at $-u_\varepsilon$, we find

$$\limsup_{\varepsilon \to 0} \int_{X(u_\varepsilon,2)} u_\varepsilon^2 \leq \gamma^2 + 2\gamma \cdot F.$$

These lim sup's are small and we conclude that the essential part of the mass is concentrated on the 1-index set, as desired. □

Remark 100. It is interesting to put these results in perspective with the algebraic version Theorem 23 of holomorphic Morse inequalities. When X is projective, the algebraic Morse inequalities used in combination with the birational invariance of the Morse integrals imply the inequalities

(a) $\displaystyle\inf_{u \in c_1(L)} \int_{X(u,q)} (-1)^q u^n \leq \inf_{\mu^*(L) \simeq \mathcal{O}(F-G)} \binom{n}{q} F^{n-q} \cdot G^q$,

(b) $\displaystyle\inf_{u \in c_1(L)} \int_{X(u,\leq q)} (-1)^q u^n \leq \inf_{\mu^*(L) \simeq \mathcal{O}(F-G)} \sum_{0 \leq j \leq q} (-1)^{q-j} \binom{n}{j} F^{n-j} \cdot G^j$,

where the infimums on the right hand side are taken over all modifications $\mu : \tilde{X} \to X$ and all decompositions $\mu^*L = \mathcal{O}(F-G)$ of μ^*L as a difference of two nef $\mathbb{Q}$-divisors F, G on $\tilde{X}$. Again, a natural question is to know whether these infimums derived from algebraic intersection numbers are equal to the asymptotic cohomology functionals $\hat{h}^q(X,L)$ and $\hat{h}^{\leq q}(X,L)$. A positive answer would of course automatically yield a positive answer to the equality cases in 2.9(a) and (b). However, the Zariski decompositions involved in our proofs of equality for $q = 0$ or $n \leq 2$ produce certain effective exceptional divisors which are not nef. It is unclear how to write those effective divisors as a difference of nef divisors. This fact raises a lot of doubts upon the sufficiency of taking merely differences of nef divisors in the infimums (a) and (b), and it is likely that one needs a more subtle formula. □

3.7 Singular Holomorphic Morse Inequalities

The goal of this short section is to extend holomorphic Morse inequalities to the case of singular Hermitian metrics, following Bonavero's PhD thesis [13] (cf. also [14]). We always assume that our Hermitian metrics h are given by quasi-psh weights φ. By Theorem 88, one can always approximate the weight by an arbitrary close quasi-psh weight φ with analytic singularities, modulo smooth functions.

Theorem 101. *Let (L,h) be a holomorphic line bundle on a compact complex n-fold X, and let E be an arbitrary holomorphic vector bundle of rank r. Assume that locally $h = e^{-\varphi}$ has analytic singularities, and that φ is quasi-psh of the form*

$$h = c \log \sum |g_j|^2 \mod C^\infty, \qquad c > 0,$$

in such a way that for a suitable modification $\mu : \widetilde{X} \to X$ one has $\mu^* \Theta_{L,h} = [D] + \beta$ where D is an effective divisor and β a smooth form on $\widetilde{X}$. Let $S = \mu(\operatorname{Supp} D)$ be the singular set of h. Then we have the following asymptotic estimates for the cohomology twisted by the appropriate multiplier ideal sheaves:

(a) $h^q(X, E \otimes L^k \otimes \mathscr{I}(h^k)) \leq r \dfrac{k^n}{n!} \displaystyle\int_{X(L,h,q) \smallsetminus S} (-1)^q \Theta_{L,h}^n + o(k^n)$.

(b) $\displaystyle\sum_{0 \leq j \leq q} (-1)^{q-j} h^j(X, E \otimes L^k \otimes \mathscr{I}(h^k)) \leq r \dfrac{k^n}{n!} \int_{X(L,h,\leq q) \smallsetminus S} (-1)^q \Theta_{L,h}^n + o(k^n)$.

Proof. For this, we observe that the Morse integrals are given by

$$\int_{\widetilde{X}(\beta,q)} (-1)^q \beta^n,$$

thanks to a change of variable $z = \mu(x)$. In fact, by our assumption $\Theta_{L,h}$ is smooth on $X \smallsetminus S$, and its pull-back $\mu^* \Theta_{L,h}$ coincides with the smooth form β on the complement $\widetilde{X} \smallsetminus \operatorname{Supp} D$ (and $\operatorname{Supp} D$ is a negligible set with respect to the integration of the smooth (n,n) form β^n on $\widetilde{X}$.) Now, a straightforward L^2 argument in the change of variable (cf. [33]) yields the direct image formula

$$K_X \otimes \mathscr{I}(h^k) = \mu_*\big(K_{\widetilde{X}} \otimes \mathscr{I}(\mu^* h^k)\big). \tag{86}$$

Let us introduce the relative canonical sheaf $K_{\widetilde{X}/X} = K_{\widetilde{X}} \otimes \mu^* K_X^{-1} = \mathcal{O}(\operatorname{div}(\operatorname{Jac} \mu))$ and let us put

$$\widetilde{L} = \mu^* L, \quad \widetilde{h} = \mu^* h, \quad \widetilde{E} = \mu^* E \otimes K_{\widetilde{X}/X}.$$

Then $\tilde{h}$ has divisorial singularities and therefore $\mathscr{I}(\tilde{h}^k) = \mathscr{O}(-\lfloor kD \rfloor)$ where $\lfloor \ldots \rfloor$ means the integral part of a divisor. The projection formula for direct images yields

$$\mu_*\big(\tilde{E} \otimes \tilde{L}^k \otimes \mathscr{I}(\tilde{h}^k)\big) = E \otimes L^k \otimes \mathscr{I}(h^k),$$

$$R^q \mu_*\big(\tilde{E} \otimes \tilde{L}^k \otimes \mathscr{I}(\tilde{h}^k)\big) = E \otimes L^k \otimes K_X^{-1} \otimes R^q \mu_*\big(K_{\tilde{X}} \otimes \mathscr{I}(\tilde{h}^k)\big).$$

However, for $k \geq k_0$ large enough, the multiplicities of $\lfloor kD \rfloor$ are all > 0 for each of the components of D, hence $\mathscr{I}(h^k) = \mathscr{O}(-\lfloor kD \rfloor)$ is relatively ample with respect to the morphism $\mu : \tilde{X} \to X$. From this, e.g. by an application of Hörmander's L^2 estimates (see [13] for more details), we conclude that $R^q \mu_*\big(K_{\tilde{X}} \otimes \mathscr{I}(\tilde{h}^k)\big) = 0$ for $k \geq k_0$. The Leray spectral sequence then implies

$$H^q\big(X, E \otimes L^k \otimes \mathscr{I}(h^k)\big) \simeq H^q\big(\tilde{X}, \tilde{E} \otimes \tilde{L}^k \otimes \mathscr{I}(\tilde{h}^k)\big). \tag{87}$$

This reduces the proof to the case of divisorial singularities. Let us next assume that D is a $\mathbb{Q}$-divisor. Let a be a denominator for D, and put $k = a\ell + b$, $0 \leq b \leq a-1$. Then

$$\tilde{E} \otimes \tilde{L}^k \otimes \mathscr{I}(\tilde{h}^k) = \tilde{E} \otimes \tilde{L}^{a\ell+b} \otimes \mathscr{O}(-a\ell D - \lfloor bD \rfloor) = F_b \otimes G^\ell$$

where

$$F_b = \tilde{E} \otimes \tilde{L}^b \otimes \mathscr{O}(-\lfloor bD \rfloor), \qquad G = \tilde{L}^a \otimes \mathscr{O}(-aD).$$

By construction, we get a smooth Hermitian metric h_G on G such that $\Theta_{G, h_G} = a\beta$. In this case, the proof is reduced to the standard case of holomorphic Morse inequalities, applied to the smooth Hermitian line bundle (G, h_G) on $\tilde{X}$ and the finite family of rank r vector bundles F_b, $0 \leq b \leq a-1$. The result is true even when D is a real divisor. In fact, we can then perturb the coefficients of D by small ε's to get a rational divisor D_ε, and we then have to change the smooth part of $\Theta_{\tilde{L}, \tilde{h}}$ to $\beta_\varepsilon = \beta + O(\varepsilon)$ (again smooth); actually $\beta_\varepsilon - \beta$ can be taken to be a linear combination by coefficients $O(\varepsilon)$ of given smooth forms representing the Chern classes $c_1(\mathscr{O}(D_j))$ of the components of D. The Morse integrals are then perturbed by $O(\varepsilon)$. On the other hand, Theorem 88 shows that the cohomology groups in the right hand side of (87) are perturbed by εk^n. The result follows as $\varepsilon \to 0$, thanks to the already settled rational case. □

4 Morse Inequalities and the Green–Griffiths–Lang Conjecture

The goal of this section is to study the existence and properties of entire curves $f : \mathbb{C} \to X$ drawn in a complex irreducible n-dimensional variety X, and more specifically to show that they must satisfy certain global algebraic or differential equations as soon as X is projective of general type. By means of holomorphic Morse inequalities and a probabilistic analysis of the cohomology of jet spaces, we are able to prove a significant step of a generalized version of the Green–Griffiths–Lang conjecture on the algebraic degeneracy of entire curves.

4.1 Introduction

Let X be a complex n-dimensional manifold; most of the time we will assume that X is compact and even projective algebraic. By an "entire curve" we always mean a non constant holomorphic map defined on the whole complex line $\mathbb{C}$, and we say that it is algebraically degenerate if its image is contained in a proper algebraic subvariety of the ambient variety. If $\mu : \tilde{X} \to X$ is a modification and $f : \mathbb{C} \to X$ is an entire curve whose image $f(\mathbb{C})$ is not contained in the image $\mu(E)$ of the exceptional locus, then f admits a unique lifting $\tilde{f} : \mathbb{C} \to \tilde{X}$. For this reason, the study of the algebraic degeneration of f is a birationally invariant problem, and singularities do not play an essential role at this stage. We will therefore assume that X is non singular, possibly after performing a suitable composition of blow-ups. We are interested more generally in the situation where the tangent bundle T_X is equipped with a *linear subspace* $V \subset T_X$, that is, an irreducible complex analytic subset of the total space of T_X such that (0.1) all fibers $V_x := V \cap T_{X,x}$ are vector subspaces of $T_{X,x}$. Then the problem is to study entire curves $f : \mathbb{C} \to X$ which are tangent to V, i.e. such that $f_* T_\mathbb{C} \subset V$. We will refer to a pair (X, V) as being a *directed variety* (or *directed manifold*). A morphism of directed varieties $\Phi : (X, V) \to (Y, W)$ is a holomorphic map $\Phi : X \to Y$ such that $\Phi_* V \subset W$; by the irreducibility, it is enough to check this condition over the dense open subset $X \smallsetminus \mathrm{Sing}(V)$ where V is actually a subbundle. Here $\mathrm{Sing}(V)$ denotes the indeterminacy set of the associated meromorphic map $\alpha : X \dashrightarrow G_r(T_X)$ to the Grassmannian bundle of r-planes in T_X, $r = \mathrm{rank}\, V$; we thus have $V_{|X \smallsetminus \mathrm{Sing}(V)} = \alpha^* S$ where $S \to G_r(T_X)$ is the tautological subbundle of $G_r(T_X)$. In that way, we get a category, and we will be mostly interested in the subcategory whose objects (X, V) are projective algebraic manifolds equipped with algebraic linear subspaces. Notice that an entire curve $f : \mathbb{C} \to X$ tangent to V is just a morphism $f : (\mathbb{C}, T_\mathbb{C}) \to (X, V)$.

The case where $V = T_{X/S}$ is the relative tangent space of some fibration $X \to S$ is of special interest, and so is the case of a foliated variety (this is the situation where the sheaf of sections $\mathcal{O}(V)$ satisfies the Frobenius integrability condition $[\mathcal{O}(V), \mathcal{O}(V)] \subset \mathcal{O}(V)$); however, it is very useful to allow as well non integrable linear subspaces V. We refer to $V = T_X$ as being the *absolute case*. Our main target is the following deep conjecture concerning the algebraic degeneracy of entire curves, which generalizes similar statements made in [51] (see also [65, 66]).

Generalized Green-Griffiths-Lang conjecture 102. *Let (X, V) be a projective directed manifold such that the canonical sheaf K_V is big (in the absolute case*

$V = T_X$, this means that X is a variety of general type, and in the relative case we will say that (X, V) is of general type). Then there should exist an algebraic subvariety $Y \subsetneq X$ such that every non constant entire curve $f : \mathbb{C} \to X$ tangent to V is contained in Y.

The precise meaning of K_V and of its bigness will be explained below—our definition *does not coincide* with other frequently used definitions and is in our view better suited to the study of entire curves of (X, V). One says that (X, V) is Brody-hyperbolic when there are no entire curves tangent to V. According to (generalized versions of) conjectures of Kobayashi [Kob70, Kob76] the hyperbolicity of (X, V) should imply that K_V is big, and even possibly ample, in a suitable sense. It would then follow from Conjecture 102 that (X, V) is hyperbolic if and only if for every irreducible variety $Y \subset X$, the linear subspace $V_{\tilde{Y}} = \overline{T_{\tilde{Y} \smallsetminus E} \cap \mu_*^{-1} V} \subset T_{\tilde{Y}}$ has a big canonical sheaf whenever $\mu : \tilde{Y} \to Y$ is a desingularization and E is the exceptional locus.

The most striking fact known at this date on the Green–Griffiths–Lang conjecture is a recent result of Diverio et al. [41] in the absolute case, confirming the statement when $X \subset \mathbb{P}_{\mathbb{C}}^{n+1}$ is a generic non singular hypersurface of large degree d, with a (non optimal) sufficient lower bound $d \geq 2^{n^5}$. Their proof is based in an essential way on a strategy developed by Siu [90, 91], combined with techniques of [31]. Notice that if the Green–Griffiths–Lang conjecture holds true, a much stronger and probably optimal result would be true, namely all smooth hypersurfaces of degree $d \geq n + 3$ would satisfy the expected algebraic degeneracy statement. Moreover, by results of Clemens [21] and Voisin [101], a (very) generic hypersurface of degree $d \geq 2n + 1$ would in fact be hyperbolic for every $n \geq 2$. Such a generic hyperbolicity statement has been obtained unconditionally by McQuillan [71, 72] when $n = 2$ and $d \geq 35$, and by Demailly-El Goul [36] when $n = 2$ and $d \geq 21$. Recently Diverio–Trapani [45] proved the same result when $n = 3$ and $d \geq 593$. By definition, proving the algebraic degeneracy means finding a non zero polynomial P on X such that all entire curves $f : \mathbb{C} \to X$ satisfy $P(f) = 0$. All known methods of proof are based on establishing first the existence of certain algebraic differential equations $P(f ; f', f'', \ldots, f^{(k)}) = 0$ of some order k, and then trying to find enough such equations so that they cut out a proper algebraic locus $Y \subsetneq X$.

Let $J_k V$ be the space of k-jets of curves $f : (\mathbb{C}, 0) \to X$ tangent to V. One defines the sheaf $\mathcal{O}(E_{k,m}^{GG} V^*)$ of jet differentials of order k and degree m to be the sheaf of holomorphic functions $P(z; \xi_1, \ldots \xi_k)$ on $J_k V$ which are homogeneous polynomials of degree m on the fibers of $J^k V \to X$ with respect to local coordinate derivatives $\xi_j = f^{(j)}(0)$ (see below in case V has singularities). The degree m considered here is the weighted degree with respect to the natural $\mathbb{C}^*$ action on $J^k V$ defined by $\lambda \cdot f(t) := f(\lambda t)$, i.e. by reparametrizing the curve with a homothetic change of variable. Since $(\lambda \cdot f)^{(j)}(t) = \lambda^j f^{(j)}(\lambda t)$, the weighted action is given in coordinates by

$$\lambda \cdot (\xi_1, \xi_2, \ldots, \xi_k) = (\lambda \xi_1, \lambda^2 \xi_2, \ldots, \lambda^k \xi_k). \tag{88}$$

One of the major tool of the theory is the following result due to Green–Griffiths [51] (see also [9, 31, 32, 89, 92, 93]).

Fundamental vanishing theorem 103. *Let (X, V) be a directed projective variety and $f : (\mathbb{C}, T_\mathbb{C}) \to (X, V)$ an entire curve tangent to V. Then for every global section $P \in H^0(X, E_{k,m}^{GG} V^* \otimes \mathcal{O}(-A))$ where A is an ample divisor of X, one has $P(f; f', f'', \ldots, f^{(k)}) = 0$.*

Let us give the proof of vanishing Theorem 103 in a special case. We interpret here $E_{k,m}^{GG} V^* \otimes \mathcal{O}(-A)$ as the bundle of differential operators whose coefficients vanish along A. By a well-known theorem of Brody [20], for every entire curve $f : (\mathbb{C}, T_\mathbb{C}) \to (X, V)$, one can extract a convergent "renormalized" sequence $g = \lim f \circ h_\nu$ where h_ν are suitable homographic functions, in such a way that g is an entire curve with bounded derivative $\sup_{t \in \mathbb{C}} \|g'(t)\|_\omega < +\infty$ (with respect to any given Hermitian metric ω on X); the image $g(\mathbb{C})$ is then contained in the cluster set $\overline{f(\mathbb{C})}$, but it is possible that $\overline{g(\mathbb{C})} \subsetneq \overline{f(\mathbb{C})}$. Then Cauchy inequalities imply that all derivatives $g^{(j)}$ are bounded, and therefore, by compactness of X, $u = P(g; g', g'', \ldots, g^{(k)})$ is a bounded holomorphic function on $\mathbb{C}$. However, after raising P to a power, we may assume that A is very ample, and after moving $A \in |A|$, that Supp A intersects $g(\mathbb{C})$. Then u vanishes somewhere, hence $u \equiv 0$ by Liouville's theorem. The proof for the general case is more subtle and makes use of Nevanlinna's second main theorem (see the above references).

It is expected that the global sections of $H^0(X, E_{k,m}^{GG} V^* \otimes \mathcal{O}(-A))$ are precisely those which ultimately define the algebraic locus $Y \subsetneq X$ where the curve f should lie. The problem is then reduced to the question of showing that there are many non zero sections of $H^0(X, E_{k,m}^{GG} V^* \otimes \mathcal{O}(-A))$, and further, understanding what is their joint base locus. The first part of this program is the main result of this section.

Theorem 104. *Let (X, V) be a directed projective variety such that K_V is big and let A be an ample divisor. Then for $k \gg 1$ and $\delta \in \mathbb{Q}_+$ small enough, $\delta \leq c(\log k)/k$, the number of sections $h^0(X, E_{k,m}^{GG} V^* \otimes \mathcal{O}(-m\delta A))$ has maximal growth, i.e. is larger that $c_k m^{n+kr-1}$ for some $m \geq m_k$, where $c, c_k > 0$, $n = \dim X$ and $r = \operatorname{rank} V$. In particular, entire curves $f : (\mathbb{C}, T_\mathbb{C}) \to (X, V)$ satisfy (many) algebraic differential equations.*

The statement is very elementary to check when $r = \operatorname{rank} V = 1$, and therefore when $n = \dim X = 1$. In higher dimensions $n \geq 2$, only very partial results were known at this point, concerning merely the absolute case $V = T_X$. In dimension 2, Theorem 104 is a consequence of the Riemann–Roch calculation of Green–Griffiths [51], combined with a vanishing theorem due to Bogomolov [11]—the latter actually only applies to the top cohomology group H^n, and things become much more delicate when estimates of intermediate cohomology groups are needed. In higher dimensions, Diverio [39, 40] proved the existence of sections of $H^0(X, E_{k,m}^{GG} V^* \otimes \mathcal{O}(-1))$ whenever X is a hypersurface of $\mathbb{P}_\mathbb{C}^{n+1}$ of high degree $d \geq d_n$, assuming $k \geq n$ and $m \geq m_n$. More recently, Merker [73] was able to treat the case of arbitrary hypersurfaces of general type, i.e. $d \geq n+3$, assuming this time k to be very large. The latter result is obtained through explicit algebraic calculations of the spaces of sections, and the proof is computationally very intensive. Bérczi [6] also obtained related results with a different approach based on residue formulas, assuming $d \geq 2^{7n \log n}$.

All these approaches are algebraic in nature, and use only the algebraic version of holomorphic Morse inequalities (Sect. 1.3.4). Here, however, our techniques are based on more elaborate curvature estimates in the spirit of Cowen–Griffiths [22]. They require the stronger analytic form of holomorphic Morse inequalities (see Sects. 1 and 3.7)—and we do not know how to translate our method in an algebraic setting. Notice that holomorphic Morse inequalities are essentially insensitive to singularities, as we can pass to non singular models and blow-up X as much as we want: if $\mu : \tilde{X} \to X$ is a modification then $\mu_* \mathscr{O}_{\tilde{X}} = \mathscr{O}_X$ and $R^q \mu_* \mathscr{O}_{\tilde{X}}$ is supported on a codimension 1 analytic subset (even codimension 2 if X is smooth). As already observed in Sect. 3.4, it follows from the Leray spectral sequence that the cohomology estimates for L on X or for $\tilde{L} = \mu^* L$ on $\tilde{X}$ differ by negligible terms, i.e.

$$h^q(\tilde{X}, \tilde{L}^{\otimes m}) - h^q(X, L^{\otimes m}) = O(m^{n-1}).$$

Finally, singular holomorphic Morse inequalities (see Sect. 3.7) allow us to work with singular Hermitian metrics h; this is the reason why we will only require to have big line bundles rather than ample line bundles. In the case of linear subspaces $V \subset T_X$, we introduce singular Hermitian metrics as follows.

Definition 105. A singular Hermitian metric on a linear subspace $V \subset T_X$ is a metric h on the fibers of V such that the function $\log h : \xi \mapsto \log |\xi|_h^2$ is locally integrable on the total space of V.

Such a metric can also be viewed as a singular Hermitian metric on the tautological line bundle $\mathscr{O}_{P(V)}(-1)$ on the projectivized bundle $P(V) = V \smallsetminus \{0\}/\mathbb{C}^*$, and therefore its dual metric h^* defines a curvature current $\Theta_{\mathscr{O}_{P(V)}(1),h^*}$ of type $(1,1)$ on $P(V) \subset P(T_X)$, such that

$$p^* \Theta_{\mathscr{O}_{P(V)}(1),h^*} = \frac{i}{2\pi} \partial\bar{\partial} \log h, \quad \text{where } p : V \smallsetminus \{0\} \to P(V).$$

If $\log h$ is quasi-plurisubharmonic (or quasi-psh, which means psh modulo addition of a smooth function) on V, then $\log h$ is indeed locally integrable, and we have moreover

$$\Theta_{\mathscr{O}_{P(V)}(1),h^*} \geq -C\omega \tag{89}$$

for some smooth positive $(1,1)$-form on $P(V)$ and some constant $C > 0$; conversely, if (89) holds, then $\log h$ is quasi-psh.

Definition 106. We will say that a singular Hermitian metric h on V is *admissible* if h can be written as $h = e^\varphi h_{0|V}$ where h_0 is a smooth positive definite Hermitian on T_X and φ is a quasi-psh weight with analytic singularities on X, as in Definition 105. Then h can be seen as a singular Hermitian metric on $\mathscr{O}_{P(V)}(1)$, with the property that it induces a smooth positive definite metric on a Zariski open set $X' \subset X \smallsetminus$

Sing(V); we will denote by Sing(h) ⊃ Sing(V) the complement of the largest such Zariski open set X'.

If h is an admissible metric, we define $\mathcal{O}_h(V^*)$ to be the sheaf of germs of holomorphic sections of $V^*_{|X \smallsetminus \mathrm{Sing}(h)}$ which are h^*-bounded near Sing(h); by the assumption on the analytic singularities, this is a coherent sheaf (as the direct image of some coherent sheaf on $P(V)$), and actually, since $h^* = e^{-\varphi} h_0^*$, it is a subsheaf of the sheaf $\mathcal{O}(V^*) := \mathcal{O}_{h_0}(V^*)$ associated with a smooth positive definite metric h_0 on T_X. If r is the generic rank of V and m a positive integer, we define similarly $K^m_{V,h}$ to be sheaf of germs of holomorphic sections of $(\det V^*_{|X'})^{\otimes m} = (\Lambda^r V^*_{|X'})^{\otimes m}$ which are $\det h^*$-bounded, and $K^m_V := K^m_{V,h_0}$.

If V is defined by $\alpha : X \dashrightarrow G_r(T_X)$, there always exists a modification $\mu : \tilde{X} \to X$ such that the composition $\alpha \circ \mu : \tilde{X} \to G_r(\mu^* T_X)$ becomes holomorphic, and then $\mu^* V_{|\mu^{-1}(X \smallsetminus \mathrm{Sing}(V))}$ extends as a locally trivial subbundle of $\mu^* T_X$ which we will simply denote by $\mu^* V$. If h is an admissible metric on V, then $\mu^* V$ can be equipped with the metric $\mu^* h = e^{\varphi \circ \mu} \mu^* h_0$ where $\mu^* h_0$ is smooth and positive definite. We may assume that $\varphi \circ \mu$ has divisorial singularities (otherwise just perform further blow-ups of $\tilde{X}$ to achieve this). We then see that there is an integer m_0 such that for all multiples $m = pm_0$ the pull-back $\mu^* K^m_{V,h}$ is an invertible sheaf on $\tilde{X}$, and $\det h^*$ induces a smooth non singular metric on it (when $h = h_0$, we can even take $m_0 = 1$). By definition we always have $K^m_{V,h} = \mu_*(\mu^* K^m_{V,h})$ for any $m \geq 0$. In the sequel, however, we think of $K_{V,h}$ not really as a coherent sheaf, but rather as the "virtual" $\mathbb{Q}$-line bundle $\mu_*(\mu^* K^{m_0}_{V,h})^{1/m_0}$, and we say that $K_{V,h}$ is big if $h^0(X, K^m_{V,h}) \geq cm^n$ for $m \geq m_1$, with $c > 0$, i.e. if the invertible sheaf $\mu^* K^{m_0}_{V,h}$ is big in the usual sense.

At this point, it is important to observe that "our" canonical sheaf K_V differs from the sheaf $\mathscr{K}_V := i_* \mathcal{O}(K_V)$ associated with the injection $i : X \smallsetminus \mathrm{Sing}(V) \hookrightarrow X$, which is usually referred to as being the "canonical sheaf", at least when V is the space of tangents to a foliation. In fact, $\mathscr{K}_V$ is always an invertible sheaf and there is an obvious inclusion $K_V \subset \mathscr{K}_V$. More precisely, the image of $\mathcal{O}(\Lambda^r T_X^*) \to \mathscr{K}_V$ is equal to $\mathscr{K}_V \otimes_{\mathcal{O}_X} \mathscr{J}$ for a certain coherent ideal $\mathscr{J} \subset \mathcal{O}_X$, and the condition to have h_0-bounded sections on $X \smallsetminus \mathrm{Sing}(V)$ precisely means that our sections are bounded by $\mathrm{Const} \sum |g_j|$ in terms of the generators (g_j) of $\mathscr{K}_V \otimes_{\mathcal{O}_X} \mathscr{J}$, i.e. $K_V = \mathscr{K}_V \otimes_{\mathcal{O}_X} \overline{\mathscr{J}}$ where $\overline{\mathscr{J}}$ is the integral closure of $\mathscr{J}$. More generally,

$$K^m_{V,h} = \mathscr{K}^m_V \otimes_{\mathcal{O}_X} \overline{\mathscr{J}}^{m/m_0}_{h,m_0}$$

where $\overline{\mathscr{J}}^{m/m_0}_{h,m_0} \subset \mathcal{O}_X$ is the (m/m_0)-integral closure of a certain ideal sheaf $\mathscr{J}_{h,m_0} \subset \mathcal{O}_X$, which can itself be assumed to be integrally closed; in our previous discussion, μ is chosen so that $\mu^* \mathscr{J}_{h,m_0}$ is invertible on $\tilde{X}$.

The discrepancy already occurs e.g. with the rank 1 linear space $V \subset T_{\mathbb{P}^n_\mathbb{C}}$ consisting at each point $z \neq 0$ of the tangent to the line $(0z)$ (so that necessarily $V_0 = T_{\mathbb{P}^n_\mathbb{C},0}$). As a sheaf (and not as a linear space), $i_* \mathcal{O}(V)$ is the invertible sheaf generated by the vector field $\xi = \sum z_j \partial/\partial z_j$ on the affine open set $\mathbb{C}^n \subset \mathbb{P}^n_\mathbb{C}$, and therefore $\mathscr{K}_V := i_* \mathcal{O}(V^*)$ is generated over $\mathbb{C}^n$ by the unique 1-form u such

Applications of Pluripotential Theory to Algebraic Geometry   235

that $u(\xi) = 1$. Since ξ vanishes at 0, the generator u is *unbounded* with respect to a smooth metric h_0 on $T_{\mathbb{P}^n_{\mathbb{C}}}$, and it is easily seen that K_V is the non invertible sheaf $K_V = \mathscr{K}_V \otimes \mathfrak{m}_{\mathbb{P}^n_{\mathbb{C}},0}$. We can make it invertible by considering the blow-up $\mu : \tilde{X} \to X$ of $X = \mathbb{P}^n_{\mathbb{C}}$ at 0, so that $\mu^* K_V$ is isomorphic to $\mu^* \mathscr{K}_V \otimes \mathscr{O}_{\tilde{X}}(-E)$ where E is the exceptional divisor. The integral curves C of V are of course lines through 0, and when a standard parametrization is used, their derivatives do not vanish at 0, while the sections of $i_* \mathscr{O}(V)$ do—another sign that $i_* \mathscr{O}(V)$ and $i_* \mathscr{O}(V^*)$ are the *wrong objects* to consider. Another standard example is obtained by taking a generic pencil of elliptic curves $\lambda P(z) + \mu Q(z) = 0$ of degree 3 in $\mathbb{P}^2_{\mathbb{C}}$, and the linear space V consisting of the tangents to the fibers of the rational map $\mathbb{P}^2_{\mathbb{C}} \dashrightarrow \mathbb{P}^1_{\mathbb{C}}$ defined by $z \mapsto Q(z)/P(z)$. Then V is given by

$$0 \longrightarrow i_* \mathscr{O}(V) \longrightarrow \mathscr{O}(T_{\mathbb{P}^2_{\mathbb{C}}}) \xrightarrow{PdQ - QdP} \mathscr{O}_{\mathbb{P}^2_{\mathbb{C}}}(6) \otimes \mathscr{J}_S \longrightarrow 0$$

where $S = \mathrm{Sing}(V)$ consists of the nine points $\{P(z) = 0\} \cap \{Q(z) = 0\}$, and $\mathscr{J}_S$ is the corresponding ideal sheaf of S. Since $\det \mathscr{O}(T_{\mathbb{P}^2}) = \mathscr{O}(3)$, we see that $\mathscr{K}_V = \mathscr{O}(3)$ is ample, which seems to contradict (2) since all leaves are elliptic curves. There is however no such contradiction, because $K_V = \mathscr{K}_V \otimes \mathscr{J}_S$ is not big in our sense (it has degree 0 on all members of the elliptic pencil). A similar example is obtained with a generic pencil of conics, in which case $\mathscr{K}_V = \mathscr{O}(1)$ and card $S = 4$.

For a given admissible Hermitian structure (V, h), we define similarly the sheaf $E^{\mathrm{GG}}_{k,m} V_h^*$ to be the sheaf of polynomials defined over $X \smallsetminus \mathrm{Sing}(h)$ which are "h-bounded". This means that when they are viewed as polynomials $P(z; \xi_1, \ldots, \xi_k)$ in terms of $\xi_j = (\nabla^{1,0}_{h_0})^j f(0)$ where $\nabla^{1,0}_{h_0}$ is the $(1,0)$-component of the induced Chern connection on (V, h_0), there is a uniform bound

$$\left| P(z; \xi_1, \ldots, \xi_k) \right| \leq C \left(\sum \|\xi_j\|_h^{1/j} \right)^m \tag{90}$$

near points of $X \smallsetminus X'$ (see Sect. 2 for more details on this). Again, by a direct image argument, one sees that $E^{\mathrm{GG}}_{k,m} V_h^*$ is always a coherent sheaf. The sheaf $E^{\mathrm{GG}}_{k,m} V^*$ is defined to be $E^{\mathrm{GG}}_{k,m} V_h^*$ when $h = h_0$ (it is actually independent of the choice of h_0, as follows from arguments similar to those given in Sect. 2). Notice that this is exactly what is needed to extend the proof of the vanishing Theorem 103 to the case of a singular linear space V; the value distribution theory argument can only work when the functions $P(f; f', \ldots, f^{(k)})(t)$ do not exhibit poles, and this is guaranteed here by the boundedness assumption.

Our strategy can be described as follows. We consider the Green–Griffiths bundle of k-jets $X^{\mathrm{GG}}_k = J^k V \smallsetminus \{0\}/\mathbb{C}^*$, which by (88) consists of a fibration in *weighted projective spaces*, and its associated tautological sheaf

$$L = \mathscr{O}_{X^{\mathrm{GG}}_k}(1),$$

viewed rather as a virtual $\mathbb{Q}$-line bundle $\mathcal{O}_{X_k^{GG}}(m_0)^{1/m_0}$ with $m_0 = \text{lcm}(1, 2, \ldots, k)$. Then, if $\pi_k : X_k^{GG} \to X$ is the natural projection, we have

$$E_{k,m}^{GG} = (\pi_k)_* \mathcal{O}_{X_k^{GG}}(m) \quad \text{and} \quad R^q(\pi_k)_* \mathcal{O}_{X_k^{GG}}(m) = 0 \text{ for } q \geq 1.$$

Hence, by the Leray spectral sequence we get for every invertible sheaf F on X the isomorphism

$$H^q(X, E_{k,m}^{GG} V^* \otimes F) \simeq H^q(X_k^{GG}, \mathcal{O}_{X_k^{GG}}(m) \otimes \pi_k^* F). \tag{91}$$

The latter group can be evaluated thanks to holomorphic Morse inequalities. In fact we can associate with any admissible metric h on V a metric (or rather a natural family) of metrics on $L = \mathcal{O}_{X_k^{GG}}(1)$. The space X_k^{GG} always possesses quotient singularities if $k \geq 2$ (and even some more if V is singular), but we do not really care since Morse inequalities still work in this setting. As we will see, it is then possible to get nice asymptotic formulas as $k \to +\infty$. They appear to be of a *probabilistic nature* if we take the components of the k-jet (i.e. the successive derivatives $\xi_j = f^{(j)}(0), 1 \leq j \leq k$) as random variables. This probabilistic behaviour was somehow already visible in the Riemann–Roch calculation of [51]. In this way, assuming K_V big, we produce a lot of sections $\sigma_j = H^0(X_k^{GG}, \mathcal{O}_{X_k^{GG}}(m) \otimes \pi_k^* F)$, corresponding to certain divisors $Z_j \subset X_k^{GG}$. The hard problem which is left in order to complete a proof of the generalized Green–Griffiths–Lang conjecture is to compute the base locus $Z = \bigcap Z_j$ and to show that $Y = \pi_k(Z) \subset X$ must be a proper algebraic variety. Unfortunately we cannot address this problem at present.

4.2 Hermitian Geometry of Weighted Projective Spaces

The goal of this section is to introduce natural Kähler metrics on weighted projective spaces, and to evaluate the corresponding volume forms. Here we put $d^c = \frac{i}{4\pi}(\overline{\partial} - \partial)$ so that $dd^c = \frac{i}{2\pi} \partial \overline{\partial}$. The normalization of the d^c operator is chosen such that we have precisely $(dd^c \log |z|^2)^n = \delta_0$ for the Monge–Ampère operator in $\mathbb{C}^n$; also, for every holomorphic or meromorphic section σ of a Hermitian line bundle (L, h) the Lelong–Poincaré can be formulated

$$dd^c \log |\sigma|_h^2 = [Z_\sigma] - \Theta_{L,h}, \tag{92}$$

where $\Theta_{L,h} = \frac{i}{2\pi} D_{L,h}^2$ is the $(1, 1)$-curvature form of L and Z_σ the zero divisor of σ. The closed $(1, 1)$-form $\Theta_{L,h}$ is a representative of the first Chern class $c_1(L)$. Given a k-tuple of "weights" $a = (a_1, \ldots, a_k)$, i.e. of integers $a_s > 0$ with $\gcd(a_1, \ldots, a_k) = 1$, we introduce the weighted projective space $P(a_1, \ldots, a_k)$ to be the quotient of $\mathbb{C}^k \smallsetminus \{0\}$ by the corresponding weighted $\mathbb{C}^*$ action:

Applications of Pluripotential Theory to Algebraic Geometry 237

$$P(a_1,\ldots,a_k) = \mathbb{C}^k \smallsetminus \{0\}/\mathbb{C}^*, \qquad \lambda \cdot z = (\lambda^{a_1} z_1, \ldots, \lambda^{a_k} z_k). \tag{93}$$

As is well known, this defines a toric $(k-1)$-dimensional algebraic variety with quotient singularities. On this variety, we introduce the possibly singular (but almost everywhere smooth and non degenerate) Kähler form $\omega_{a,p}$ defined by

$$\pi_a^* \omega_{a,p} = dd^c \varphi_{a,p}, \qquad \varphi_{a,p}(z) = \frac{1}{p} \log \sum_{1 \le s \le k} |z_s|^{2p/a_s}, \tag{94}$$

where $\pi_a : \mathbb{C}^k \smallsetminus \{0\} \to P(a_1,\ldots,a_k)$ is the canonical projection and $p > 0$ is a positive constant. It is clear that $\varphi_{p,a}$ is real analytic on $\mathbb{C}^k \smallsetminus \{0\}$ if p is an integer and a common multiple of all weights a_s. It is at least C^2 if p is real and $p \ge \max(a_s)$, which will be more than sufficient for our purposes (but everything would still work for any $p > 0$). The resulting metric is in any case smooth and positive definite outside of the coordinate hyperplanes $z_s = 0$, and these hyperplanes will not matter here since they are of capacity zero with respect to all currents $(dd^c \varphi_{a,p})^\ell$. In order to evaluate the volume $\int_{P(a_1,\ldots,a_k)} \omega_{a,p}^{k-1}$, one can observe that

$$\int_{P(a_1,\ldots,a_k)} \omega_{a,p}^{k-1} = \int_{z \in \mathbb{C}^k, \varphi_{a,p}(z)=0} \pi_a^* \omega_{a,p}^{k-1} \wedge d^c \varphi_{a,p}$$

$$= \int_{z \in \mathbb{C}^k, \varphi_{a,p}(z)=0} (dd^c \varphi_{a,p})^{k-1} \wedge d^c \varphi_{a,p}$$

$$= \frac{1}{p^k} \int_{z \in \mathbb{C}^k, \varphi_{a,p}(z)<0} (dd^c e^{p\varphi_{a,p}})^k. \tag{95}$$

The first equality comes from the fact that $\{\varphi_{a,p}(z) = 0\}$ is a circle bundle over $P(a_1,\ldots,a_k)$, together with the identities $\varphi_{a,p}(\lambda \cdot z) = \varphi_{a,p}(z) + \log|\lambda|^2$ and $\int_{|\lambda|=1} d^c \log|\lambda|^2 = 1$. The third equality can be seen by Stokes formula applied to the $(2k-1)$-form

$$(dd^c e^{p\varphi_{a,p}})^{k-1} \wedge d^c e^{p\varphi_{a,p}} = e^{p\varphi_{a,p}} (dd^c \varphi_{a,p})^{k-1} \wedge d^c \varphi_{a,p}$$

on the pseudoconvex open set $\{z \in \mathbb{C}^k \,;\, \varphi_{a,p}(z) < 0\}$. Now, we find

$$(dd^c e^{p\varphi_{a,p}})^k = \left(dd^c \sum_{1 \le s \le k} |z_s|^{2p/a_s}\right)^k = \prod_{1 \le s \le k} \left(\frac{p}{a_s} |z_s|^{\frac{p}{a_s}-1}\right)(dd^c|z|^2)^k, \tag{96}$$

$$\int_{z \in \mathbb{C}^k, \varphi_{a,p}(z)<0} (dd^c e^{p\varphi_{a,p}})^k = \prod_{1 \le s \le k} \frac{p}{a_s} = \frac{p^k}{a_1 \ldots a_k}. \tag{97}$$

In fact, (96) and (97) are clear when $p = a_1 = \ldots = a_k = 1$ (this is just the standard calculation of the volume of the unit ball in $\mathbb{C}^k$); the general case follows by substituting formally $z_s \mapsto z_s^{p/a_s}$, and using rotational invariance together with the observation that the arguments of the complex numbers z_s^{p/a_s} now run in the interval $[0, 2\pi p/a_s[$ instead of $[0, 2\pi[$ (say). As a consequence of (95) and (97), we obtain the well known value

$$\int_{P(a_1,\ldots,a_k)} \omega_{a,p}^{k-1} = \frac{1}{a_1 \ldots a_k}, \tag{98}$$

for the volume. Notice that this is independent of p (as it is obvious by Stokes theorem, since the cohomology class of $\omega_{a,p}$ does not depend on p). When p tends to $+\infty$, we have $\varphi_{a,p}(z) \mapsto \varphi_{a,\infty}(z) = \log\max_{1 \le s \le k} |z_s|^{2/a_s}$ and the volume form $\omega_{a,p}^{k-1}$ converges to a rotationally invariant measure supported by the image of the polycircle $\prod\{|z_s| = 1\}$ in $P(a_1, \ldots, a_k)$. This is so because not all $|z_s|^{2/a_s}$ are equal outside of the image of the polycircle, thus $\varphi_{a,\infty}(z)$ locally depends only on $k-1$ complex variables, and so $\omega_{a,\infty}^{k-1} = 0$ there by log homogeneity.

Our later calculations will require a slightly more general setting. Instead of looking at $\mathbb{C}^k$, we consider the weighted $\mathbb{C}^*$ action defined by

$$\mathbb{C}^{|r|} = \mathbb{C}^{r_1} \times \ldots \times \mathbb{C}^{r_k}, \quad \lambda \cdot z = (\lambda^{a_1} z_1, \ldots, \lambda^{a_k} z_k). \tag{99}$$

Here $z_s \in \mathbb{C}^{r_s}$ for some k-tuple $r = (r_1, \ldots, r_k)$ and $|r| = r_1 + \ldots + r_k$. This gives rise to a weighted projective space

$$P(a_1^{[r_1]}, \ldots, a_k^{[r_k]}) = P(a_1, \ldots, a_1, \ldots, a_k, \ldots, a_k),$$

$$\pi_{a,r} : \mathbb{C}^{r_1} \times \ldots \times \mathbb{C}^{r_k} \smallsetminus \{0\} \longrightarrow P(a_1^{[r_1]}, \ldots, a_k^{[r_k]}) \tag{100}$$

obtained by repeating r_s times each weight a_s. On this space, we introduce the degenerate Kähler metric $\omega_{a,r,p}$ such that

$$\pi_{a,r}^* \omega_{a,r,p} = dd^c \varphi_{a,r,p}, \quad \varphi_{a,r,p}(z) = \frac{1}{p} \log \sum_{1 \le s \le k} |z_s|^{2p/a_s} \tag{101}$$

where $|z_s|$ stands now for the standard Hermitian norm $(\sum_{1 \le j \le r_s} |z_{s,j}|^2)^{1/2}$ on $\mathbb{C}^{r_s}$. This metric is cohomologous to the corresponding "polydisc-like" metric $\omega_{a,p}$ already defined, and therefore Stokes theorem implies

$$\int_{P(a_1^{[r_1]},\ldots,a_k^{[r_k]})} \omega_{a,r,p}^{|r|-1} = \frac{1}{a_1^{r_1} \ldots a_k^{r_k}}. \tag{102}$$

Since $(dd^c \log |z_s|^2)^{r_s} = 0$ on $\mathbb{C}^{r_s} \smallsetminus \{0\}$ by homogeneity, we conclude as before that the weak limit $\lim_{p \to +\infty} \omega_{a,r,p}^{|r|-1} = \omega_{a,r,\infty}^{|r|-1}$ associated with

$$\varphi_{a,r,\infty}(z) = \log \max_{1 \le s \le k} |z_s|^{2/a_s} \tag{103}$$

Applications of Pluripotential Theory to Algebraic Geometry   239

is a measure supported by the image of the product of unit spheres $\prod S^{2r_s-1}$ in $P(a_1^{[r_1]},\ldots,a_k^{[r_k]})$, which is invariant under the action of $U(r_1) \times \ldots \times U(r_k)$ on $\mathbb{C}^{r_1} \times \ldots \times \mathbb{C}^{r_k}$, and thus coincides with the Hermitian area measure up to a constant determined by condition (30). In fact, outside of the product of spheres, $\varphi_{a,r,\infty}$ locally depends only on at most $k-1$ factors and thus, for dimension reasons, the top power $(dd^c\varphi_{a,r,\infty})^{|r|-1}$ must be zero there. In the next section, the following change of variable formula will be needed. For simplicity of exposition we restrict ourselves to continuous functions, but a standard density argument would easily extend the formula to all functions that are Lebesgue integrable with respect to the volume form $\omega_{a,r,p}^{|r|-1}$.

Proposition 107. *Let $f(z)$ be a bounded function on $P(a_1^{[r_1]},\ldots,a_k^{[r_k]})$ which is continuous outside of the hyperplane sections $z_s = 0$. We also view f as a $\mathbb{C}^*$-invariant continuous function on $\prod(\mathbb{C}^{r_s} \smallsetminus \{0\})$. Then*

$$\int_{P(a_1^{[r_1]},\ldots,a_k^{[r_k]})} f(z) \, \omega_{a,r,p}^{|r|-1}$$

$$= \frac{(|r|-1)!}{\prod_s a_s^{r_s}} \int_{(x,u) \in \Delta_{k-1} \times \prod S^{2r_s-1}} f(x_1^{a_1/2p} u_1, \ldots, x_k^{a_k/2p} u_k) \prod_{1 \le s \le k} \frac{x_s^{r_s-1}}{(r_s-1)!} \, dx \, d\mu(u)$$

where Δ_{k-1} is the $(k-1)$-simplex $\{x_s \ge 0, \sum x_s = 1\}$, $dx = dx_1 \wedge \ldots \wedge dx_{k-1}$ its standard measure, and where $d\mu(u) = d\mu_1(u_1) \ldots d\mu_k(u_k)$ is the rotation invariant probability measure on the product $\prod_s S^{2r_s-1}$ of unit spheres in $\mathbb{C}^{r_1} \times \ldots \times \mathbb{C}^{r_k}$. As a consequence

$$\lim_{p \to +\infty} \int_{P(a_1^{[r_1]},\ldots,a_k^{[r_k]})} f(z) \, \omega_{a,r,p}^{|r|-1} = \frac{1}{\prod_s a_s^{r_s}} \int_{\prod S^{2r_s-1}} f(u) \, d\mu(u).$$

Proof. The area formula of the disc $\int_{|\lambda|<1} dd^c |\lambda|^2 = 1$ and a consideration of the unit disc bundle over $P(a_1^{[r_1]},\ldots,a_k^{[r_k]})$ imply that

$$I_p := \int_{P(a_1^{[r_1]},\ldots,a_k^{[r_k]})} f(z) \, \omega_{a,r,p}^{|r|-1} = \int_{z \in \mathbb{C}^{|r|}, \varphi_{a,r,p}(z) < 0} f(z) \, (dd^c \varphi_{a,r,p})^{|r|-1} \wedge dd^c e^{\varphi_{a,r,p}}.$$

Now, a straightforward calculation on $\mathbb{C}^{|r|}$ gives

$$(dd^c e^{p\varphi_{a,r,p}})^{|r|} = \left(dd^c \sum_{1 \le s \le k} |z_s|^{2p/a_s} \right)^{|r|}$$

$$= \prod_{1 \le s \le k} \left(\frac{p}{a_s} \right)^{r_s+1} |z_s|^{2r_s(p/a_s-1)} (dd^c |z|^2)^{|r|}.$$

On the other hand, we have $(dd^c|z|^2)^{|r|} = \frac{|r|!}{r_1!...r_k!} \prod_{1 \leq s \leq k}(dd^c|z_s|^2)^{r_s}$ and

$$(dd^c e^{p\varphi_{a,r,p}})^{|r|} = \left(p\, e^{p\varphi_{a,r,p}}(dd^c\varphi_{a,r,p} + p\, d\varphi_{a,r,p} \wedge d^c\varphi_{a,r,p})\right)^{|r|}$$
$$= |r|p^{|r|+1}e^{|r|p\varphi_{a,r,p}}(dd^c\varphi_{a,r,p})^{|r|-1} \wedge d\varphi_{a,r,p} \wedge d^c\varphi_{a,r,p}$$
$$= |r|p^{|r|+1}e^{(|r|p-1)\varphi_{a,r,p}}(dd^c\varphi_{a,r,p})^{|r|-1} \wedge dd^c e^{\varphi_{a,r,p}},$$

thanks to the homogeneity relation $(dd^c\varphi_{a,r,p})^{|r|} = 0$. Putting everything together, we find

$$I_p = \int_{z \in \mathbb{C}^{|r|}, \varphi_{a,r,p}(z)<0} \frac{(|r|-1)!\, p^{k-1} f(z)}{(\sum_s |z_s|^{2p/a_s})^{|r|-1/p}} \prod_s \frac{(dd^c|z_s|^2)^{r_s}}{r_s!\, a_s^{r_s+1}|z_s|^{2r_s(1-p/a_s)}}.$$

A standard calculation in polar coordinates with $z_s = \rho_s u_s$, $u_s \in S^{2r_s-1}$, yields

$$\frac{(dd^c|z_s|^2)^{r_s}}{|z_s|^{2r_s}} = 2r_s \frac{d\rho_s}{\rho_s} d\mu_s(u_s)$$

where μ_s is the $U(r_s)$-invariant probability measure on S^{2r_s-1}. Therefore

$$I_p = \int_{\varphi_{a,r,p}(z)<0} \frac{(|r|-1)!\, p^{k-1} f(\rho_1 u_1, \ldots, \rho_k u_k)}{(\sum_{1 \leq s \leq k} \rho_s^{2p/a_s})^{|r|-1/p}} \prod_s \frac{2\rho_s^{2pr_s/a_s}\frac{d\rho_s}{\rho_s} d\mu_s(u_s)}{(r_s-1)!\, a_s^{r_s+1}}$$

$$= \int_{u_s \in S^{2r_s-1}, \sum t_s < 1} \frac{(|r|-1)!\, p^{-1} f(t_1^{a_1/2p} u_1, \ldots, t_k^{a_k/2p} u_k)}{(\sum_{1 \leq s \leq k} t_s)^{|r|-1/p}} \prod_s \frac{t_s^{r_s-1} dt_s\, d\mu_s(u_s)}{(r_s-1)!\, a_s^{r_s}}$$

by putting $t_s = |z_s|^{2p/a_s} = \rho_s^{2p/a_s}$, i.e. $\rho_s = t_s^{a_s/2p}$, $t_s \in\,]0, 1]$. We use still another change of variable $t_s = tx_s$ with $t = \sum_{1 \leq s \leq k} t_s$ and $x_s \in\,]0, 1]$, $\sum_{1 \leq s \leq k} x_s = 1$. Then

$$dt_1 \wedge \ldots \wedge dt_k = t^{k-1}\, dx\, dt \qquad \text{where } dx = dx_1 \wedge \ldots \wedge dx_{k-1}.$$

The $\mathbb{C}^*$ invariance of f shows that

$$I_p = \int_{\substack{u_s \in S^{2r_s-1} \\ \sum x_s = 1, t \in]0,1]}} (|r|-1)!\, f(x_1^{a_s/2p} u_1, \ldots, x_k^{a_k/2p} u_k) \prod_{1 \leq s \leq k} \frac{x_s^{r_s-1} d\mu_s(u_s)}{(r_s-1)!\, a_s^{r_s}} \frac{dx\, dt}{p\, t^{1-1/p}}$$

$$= \int_{\substack{u_s \in S^{2r_s-1} \\ \sum x_s = 1}} (|r|-1)!\, f(x_1^{a_s/2p} u_1, \ldots, x_k^{a_k/2p} u_k) \prod_{1 \leq s \leq k} \frac{x_s^{r_s-1} d\mu_s(u_s)}{(r_s-1)!\, a_s^{r_s}}\, dx.$$

This is equivalent to the formula given in Proposition 107. We have $x_s^{2a_s/p} \to 1$ as $p \to +\infty$, and by Lebesgue's bounded convergence theorem and Fubini's formula, we get

$$\lim_{p\to+\infty} I_p = \frac{(|r|-1)!}{\prod_s a_s^{r_s}} \int_{(x,u)\in\Delta_{k-1}\times\prod S^{2r_s-1}} f(u) \prod_{1\le s\le k} \frac{x_s^{r_s-1}}{(r_s-1)!} dx\, d\mu(u).$$

It can be checked by elementary integrations by parts and induction on k, $r_1, \ldots, r_k$ that

$$\int_{x\in\Delta_{k-1}} \prod_{1\le s\le k} x_s^{r_s-1} dx_1 \ldots dx_{k-1} = \frac{1}{(|r|-1)!} \prod_{1\le s\le k} (r_s-1)!. \tag{104}$$

This implies that $(|r|-1)! \prod_{1\le s\le k} \frac{x_s^{r_s-1}}{(r_s-1)!} dx$ is a probability measure on Δ_{k-1} and that

$$\lim_{p\to+\infty} I_p = \frac{1}{\prod_s a_s^{r_s}} \int_{u\in\prod S^{2r_s-1}} f(u)\, d\mu(u).$$

Even without an explicit check, (33) also follows from the fact that we must have equality for $f(z) \equiv 1$ in the latter equality, if we take into account the volume formula (30). □

4.3 Probabilistic Estimate of the Curvature of k-Jet Bundles

Let (X, V) be a compact complex directed non singular variety. To avoid any technical difficulty at this point, we first assume that V is a holomorphic vector subbundle of T_X, equipped with a smooth Hermitian metric h.

According to the notation already specified in the introduction, we denote by $J^k V$ the bundle of k-jets of holomorphic curves $f : (\mathbb{C}, 0) \to X$ tangent to V at each point. Let us set $n = \dim_\mathbb{C} X$ and $r = \mathrm{rank}_\mathbb{C} V$. Then $J^k V \to X$ is an algebraic fiber bundle with typical fiber $\mathbb{C}^{rk}$ (see below). It has a canonical $\mathbb{C}^*$-action defined by $\lambda \cdot f : (\mathbb{C}, 0) \to X$, $(\lambda \cdot f)(t) = f(\lambda t)$. Fix a point x_0 in X and a local holomorphic coordinate system $(z_1, \ldots, z_n)$ centered at x_0 such that V_{x_0} is the vector subspace $\langle \partial/\partial z_1, \ldots, \partial/\partial z_r \rangle$ at x_0. Then, in a neighborhood U of x_0, V admits a holomorphic frame of the form

$$\frac{\partial}{\partial z_\beta} + \sum_{r+1\le\alpha\le n} a_{\alpha\beta}(z)\frac{\partial}{\partial z_\alpha}, \quad 1\le\beta\le r,\ a_{\alpha\beta}(0)=0. \tag{105}$$

Let $f(t) = (f_1(t), \ldots, f_n(t))$ be a k-jet of curve tangent to V starting from a point $f(0) = x \in U$. Such a curve is entirely determined by its initial point and by the projection $\tilde{f}(t) := (f_1(t), \ldots, f_r(t))$ to the first r-components, since the condition $f'(t) \in V_{f(t)}$ implies that the other components must satisfy the ordinary differential equation

$$f'_\alpha(t) = \sum_{1 \le \beta \le r} a_{\alpha\beta}(f(t)) f'_\beta(t).$$

This implies that the k-jet of f is entirely determined by the initial point x and the Taylor expansion

$$\tilde{f}(t) - \tilde{x} = \xi_1 t + \xi_2 t^2 + \ldots + \xi_k t^k + O(t^{k+1}) \tag{106}$$

where $\xi_s = (\xi_{s\alpha})_{1 \le \alpha \le r} \in \mathbb{C}^r$. The $\mathbb{C}^*$ action $(\lambda, f) \mapsto \lambda \cdot f$ is then expressed in coordinates by the weighted action

$$\lambda \cdot (\xi_1, \xi_2, \ldots, \xi_k) = (\lambda \xi_1, \lambda^2 \xi_2, \ldots, \lambda^k \xi_k) \tag{107}$$

associated with the weight $a = (1^{[r]}, 2^{[r]}, \ldots, k^{[r]})$. The quotient projectivized k-jet bundle

$$X_k^{\mathrm{GG}} := (J^k V \smallsetminus \{0\})/\mathbb{C}^* \tag{108}$$

considered by Green and Griffiths [51] is therefore in a natural way a $P(1^{[r]}, 2^{[r]}, \ldots, k^{[r]})$ weighted projective bundle over X. As such, it possesses a canonical sheaf $\mathscr{O}_{X_k^{\mathrm{GG}}}(1)$ such that $\mathscr{O}_{X_k^{\mathrm{GG}}}(m)$ is invertible when m is a multiple of $\mathrm{lcm}(1, 2, \ldots, k)$. Under the natural projection $\pi_k : X_k^{\mathrm{GG}} \to X$, the direct image $(\pi_k)_* \mathscr{O}_{X_k^{\mathrm{GG}}}(m)$ coincides with the sheaf of sections of the bundle $E_{k,m}^{\mathrm{GG}} V^*$ of jet differentials of order k and degree m, namely polynomials

$$P(z; \xi_1, \ldots, \xi_k) = \sum_{\alpha_\ell \in \mathbb{N}^r, 1 \le \ell \le k} a_{\alpha_1 \ldots \alpha_k}(z) \xi_1^{\alpha_1} \ldots \xi_k^{\alpha_k} \tag{109}$$

of weighted degree $|\alpha_1| + 2|\alpha_2| + \ldots + k|\alpha_k| = m$ on $J^k V$ with holomorphic coefficients. The jet differentials operate on germs of curves as differential operators

$$P(f)(t) = \sum a_{\alpha_1 \ldots \alpha_k}(f(t)) \, f'(t)^{\alpha_1} \ldots f^{(k)}(t)^{\alpha_k}. \tag{110}$$

In the sequel, we do not make any further use of coordinate frames as (105), because they need not be related in any way to the Hermitian metric h of V. Instead, we choose a local holomorphic coordinate frame $(e_\alpha(z))_{1 \le \alpha \le r}$ of V on a neighborhood U of x_0, such that

$$\langle e_\alpha(z), e_\beta(z) \rangle = \delta_{\alpha\beta} + \sum_{1 \le i,j \le n, 1 \le \alpha,\beta \le r} c_{ij\alpha\beta} z_i \bar{z}_j + O(|z|^3) \tag{111}$$

for suitable complex coefficients $(c_{ij\alpha\beta})$. It is a standard fact that such a normalized coordinate system always exists, and that the Chern curvature tensor $\frac{i}{2\pi} D_{V,h}^2$ of (V, h) at x_0 is then given by

$$\Theta_{V,h}(x_0) = -\frac{i}{2\pi} \sum_{i,j,\alpha,\beta} c_{ij\alpha\beta} \, dz_i \wedge d\bar{z}_j \otimes e_\alpha^* \otimes e_\beta. \tag{112}$$

Also, instead of defining the vectors $\xi_s \in \mathbb{C}^r$ as in (40), we consider a local holomorphic connection ∇ on $V_{|U}$ (e.g. the one which turns (e_α) into a parallel frame), and take $\xi_k = \nabla^k f(0) \in V_x$ defined inductively by $\nabla^1 f = f'$ and $\nabla^s f = \nabla_{f'}(\nabla^{s-1} f)$. This is just another way of parameterizing the fibers of $J^k V$ over U by the vector bundle $V_{|U}^k$. Notice that this is highly dependent on ∇ (the bundle $J^k V$ actually does not carry a vector bundle or even affine bundle structure); however, the expression of the weighted action (107) is unchanged in this new setting. Now, we fix a finite open covering $(U_\alpha)_{\alpha \in I}$ of X by open coordinate charts such that $V_{|U_\alpha}$ is trivial, along with holomorphic connections ∇_α on $V_{|U_\alpha}$. Let θ_α be a partition of unity of X subordinate to the covering (U_α). Let us fix $p > 0$ and small parameters $1 = \varepsilon_1 \gg \varepsilon_2 \gg \ldots \gg \varepsilon_k > 0$. Then we define a global weighted exhaustion on $J^k V$ by putting for any k-jet $f \in J_x^k V$

$$\Psi_{h,p,\varepsilon}(f) := \left(\sum_{\alpha \in I} \theta_\alpha(x) \sum_{1 \leq s \leq k} \varepsilon_s^{2p} \|\nabla_\alpha^s f(0)\|_{h(x)}^{2p/s} \right)^{1/p} \tag{113}$$

where $\| \ \|_{h(x)}$ is the Hermitian metric h of V evaluated on the fiber V_x, $x = f(0)$. The function $\Psi_{h,p,\varepsilon}$ satisfies the fundamental homogeneity property

$$\Psi_{h,p,\varepsilon}(\lambda \cdot f) = \Psi_{h,p,\varepsilon}(f) |\lambda|^2 \tag{114}$$

with respect to the $\mathbb{C}^*$ action on $J^k V$, in other words, it induces a Hermitian metric on the dual L^* of the tautological $\mathbb{Q}$-line bundle $L_k = \mathcal{O}_{X_k^{GG}}(1)$ over X_k^{GG}. The curvature of L_k is given by

$$\pi_k^* \Theta_{L_k, \Psi_{h,p,\varepsilon}^*} = dd^c \log \Psi_{h,p,\varepsilon} \tag{115}$$

where $\pi_k : J^k V \smallsetminus \{0\} \to X_k^{GG}$ is the canonical projection. Our next goal is to compute precisely the curvature and to apply holomorphic Morse inequalities to $L \to X_k^{GG}$ with the above metric. It might look a priori like an untractable problem, since the definition of $\Psi_{h,p,\varepsilon}$ is a rather unnatural one. However, the "miracle" is that the asymptotic behavior of $\Psi_{h,p,\varepsilon}$ as $\varepsilon_s/\varepsilon_{s-1} \to 0$ is in some sense uniquely defined and very natural. It will lead to a computable asymptotic formula, which is moreover simple enough to produce useful results.

Lemma 108. *On each coordinate chart U equipped with a holomorphic connection ∇ of $V_{|U}$, let us define the components of a k-jet $f \in J^k V$ by $\xi_s = \nabla^s f(0)$, and consider the rescaling transformation*

$$\rho_{\nabla,\varepsilon}(\xi_1, \xi_2, \ldots, \xi_k) = (\varepsilon^1 \xi_1, \varepsilon^2 \xi_2, \ldots, \varepsilon^k \xi_k) \quad \text{on } J_x^k V, \ x \in U$$

(it commutes with the $\mathbb{C}^*$-action but is otherwise unrelated and not canonically defined over X as it depends on the choice of ∇). Then, if p is a multiple of

lcm$(1, 2, \ldots, k)$ and $\varepsilon_s/\varepsilon_{s-1} \to 0$ for all $s = 2, \ldots, k$, the rescaled function $\Psi_{h,p,\varepsilon} \circ \rho_{\nabla,\varepsilon}^{-1}(\xi_1, \ldots, \xi_k)$ converges towards

$$\left(\sum_{1 \le s \le k} \|\xi_s\|_h^{2p/s} \right)^{1/p}$$

on every compact subset of $J^k V_{|U} \smallsetminus \{0\}$, uniformly in C^∞ topology.

Proof. Let $U \subset X$ be an open set on which $V_{|U}$ is trivial and equipped with some holomorphic connection ∇. Let us pick another holomorphic connection $\tilde{\nabla} = \nabla + \Gamma$ where $\Gamma \in H^0(U, \Omega_X^1 \otimes \text{Hom}(V, V))$. Then $\tilde{\nabla}^2 f = \nabla^2 f + \Gamma(f)(f') \cdot f'$, and inductively we get

$$\tilde{\nabla}^s f = \nabla^s f + P_s(f \, ; \, \nabla^1 f, \ldots, \nabla^{s-1} f)$$

where $P(x \, ; \, \xi_1, \ldots, \xi_{s-1})$ is a polynomial with holomorphic coefficients in $x \in U$ which is of weighted homogeneous degree s in $(\xi_1, \ldots, \xi_{s-1})$. In other words, the corresponding change in the parametrization of $J^k V_{|U}$ is given by a $\mathbb{C}^*$-homogeneous transformation

$$\tilde{\xi}_s = \xi_s + P_s(x \, ; \, \xi_1, \ldots, \xi_{s-1}).$$

Let us introduce the corresponding rescaled components

$$(\xi_{1,\varepsilon}, \ldots, \xi_{k,\varepsilon}) = (\varepsilon_1^1 \xi_1, \ldots, \varepsilon_k^k \xi_k), \qquad (\tilde{\xi}_{1,\varepsilon}, \ldots, \tilde{\xi}_{k,\varepsilon}) = (\varepsilon_1^1 \tilde{\xi}_1, \ldots, \varepsilon_k^k \tilde{\xi}_k).$$

Then

$$\tilde{\xi}_{s,\varepsilon} = \xi_{s,\varepsilon} + \varepsilon_s^s \, P_s(x \, ; \, \varepsilon_1^{-1} \xi_{1,\varepsilon}, \ldots, \varepsilon_{s-1}^{-(s-1)} \xi_{s-1,\varepsilon})$$
$$= \xi_{s,\varepsilon} + O(\varepsilon_s/\varepsilon_{s-1})^s \, O(\|\xi_{1,\varepsilon}\| + \ldots + \|\xi_{s-1,\varepsilon}\|^{1/(s-1)})^s$$

and the error terms are thus polynomials of fixed degree with arbitrarily small coefficients as $\varepsilon_s/\varepsilon_{s-1} \to 0$. Now, the definition of $\Psi_{h,p,\varepsilon}$ consists of gluing the sums

$$\sum_{1 \le s \le k} \varepsilon_s^{2p} \|\xi_k\|_h^{2p/s} = \sum_{1 \le s \le k} \|\xi_{k,\varepsilon}\|_h^{2p/s}$$

corresponding to $\xi_k = \nabla_\alpha^s f(0)$ by means of the partition of unity $\sum \theta_\alpha(x) = 1$. We see that by using the rescaled variables $\xi_{s,\varepsilon}$ the changes occurring when replacing a connection ∇_α by an alternative one ∇_β are arbitrary small in C^∞ topology, with error terms uniformly controlled in terms of the ratios $\varepsilon_s/\varepsilon_{s-1}$ on all compact subsets of $V^k \smallsetminus \{0\}$. This shows that in C^∞ topology, $\Psi_{h,p,\varepsilon} \circ \rho_{\nabla,\varepsilon}^{-1}(\xi_1, \ldots, \xi_k)$ converges uniformly towards $(\sum_{1 \le s \le k} \|\xi_k\|_h^{2p/s})^{1/p}$, whatever the trivializing open set U and the holomorphic connection ∇ used to evaluate the components and perform the rescaling are. □

Now, we fix a point $x_0 \in X$ and a local holomorphic frame $(e_\alpha(z))_{1 \le \alpha \le r}$ satisfying (111) on a neighborhood U of x_0. We introduce the rescaled components $\xi_s = \varepsilon_s^s \nabla^s f(0)$ on $J^k V_{|U}$ and compute the curvature of

$$\Psi_{h,p,\varepsilon} \circ \rho_{\nabla,\varepsilon}^{-1}(z; \xi_1, \ldots, \xi_k) \simeq \left(\sum_{1 \le s \le k} \|\xi_s\|_h^{2p/s} \right)^{1/p}$$

(by Lemma 108, the errors can be taken arbitrary small in C^∞ topology). We write $\xi_s = \sum_{1 \le \alpha \le r} \xi_{s\alpha} e_\alpha$. By (111) we have

$$\|\xi_s\|_h^2 = \sum_\alpha |\xi_{s\alpha}|^2 + \sum_{i,j,\alpha,\beta} c_{ij\alpha\beta} z_i \bar{z}_j \xi_{s\alpha} \bar{\xi}_{s\beta} + O(|z|^3 |\xi|^2).$$

The question is to evaluate the curvature of the weighted metric defined by

$$\Psi(z; \xi_1, \ldots, \xi_k) = \left(\sum_{1 \le s \le k} \|\xi_s\|_h^{2p/s} \right)^{1/p}$$

$$= \left(\sum_{1 \le s \le k} \left(\sum_\alpha |\xi_{s\alpha}|^2 + \sum_{i,j,\alpha,\beta} c_{ij\alpha\beta} z_i \bar{z}_j \xi_{s\alpha} \bar{\xi}_{s\beta} \right)^{p/s} \right)^{1/p} + O(|z|^3).$$

We set $|\xi_s|^2 = \sum_\alpha |\xi_{s\alpha}|^2$. A straightforward calculation yields

$$\log \Psi(z; \xi_1, \ldots, \xi_k) =$$

$$= \frac{1}{p} \log \sum_{1 \le s \le k} |\xi_s|^{2p/s} + \sum_{1 \le s \le k} \frac{1}{s} \frac{|\xi_s|^{2p/s}}{\sum_t |\xi_t|^{2p/t}} \sum_{i,j,\alpha,\beta} c_{ij\alpha\beta} z_i \bar{z}_j \frac{\xi_{s\alpha} \bar{\xi}_{s\beta}}{|\xi_s|^2} + O(|z|^3).$$

By (115), the curvature form of $L_k = \mathcal{O}_{X_k^{\mathrm{GG}}}(1)$ is given at the central point x_0 by the following formula.

Proposition 109. *With the above choice of coordinates and with respect to the rescaled components $\xi_s = \varepsilon_s^s \nabla^s f(0)$ at $x_0 \in X$, we have the approximate expression*

$$\Theta_{L_k, \Psi_{h,p,\varepsilon}^*}(x_0, [\xi]) \simeq \omega_{a,r,p}(\xi) + \frac{i}{2\pi} \sum_{1 \le s \le k} \frac{1}{s} \frac{|\xi_s|^{2p/s}}{\sum_t |\xi_t|^{2p/t}} \sum_{i,j,\alpha,\beta} c_{ij\alpha\beta} \frac{\xi_{s\alpha} \bar{\xi}_{s\beta}}{|\xi_s|^2} dz_i \wedge d\bar{z}_j$$

where the error terms are $O(\max_{2 \le s \le k}(\varepsilon_s/\varepsilon_{s-1})^s)$ uniformly on the compact variety X_k^{GG}. Here $\omega_{a,r,p}$ is the (degenerate) Kähler metric associated with the weight $a = (1^{[r]}, 2^{[r]}, \ldots, k^{[r]})$ of the canonical $\mathbb{C}^*$ action on $J^k V$.

Thanks to the uniform approximation, we can (and will) neglect the error terms in the calculations below. Since $\omega_{a,r,p}$ is positive definite on the fibers of $X_k^{GG} \to X$ (at least outside of the axes $\xi_s = 0$), the index of the $(1,1)$ curvature form $\Theta_{L_k, \Psi_{h,p,\varepsilon}^*}(z, [\xi])$ is equal to the index of the $(1,1)$-form

$$\gamma_k(z, \xi) := \frac{i}{2\pi} \sum_{1 \le s \le k} \frac{1}{s} \frac{|\xi_s|^{2p/s}}{\sum_t |\xi_t|^{2p/t}} \sum_{i,j,\alpha,\beta} c_{ij\alpha\beta}(z) \frac{\xi_{s\alpha}\overline{\xi}_{s\beta}}{|\xi_s|^2} dz_i \wedge d\overline{z}_j \qquad (116)$$

depending only on the differentials $(dz_j)_{1 \le j \le n}$ on X. The q-index integral of $(L_k, \Psi_{h,p,\varepsilon}^*)$ on X_k^{GG} is therefore equal to

$$\int_{X_k^{GG}(L_k, q)} \Theta_{L_k, \Psi_{h,p,\varepsilon}^*}^{n+kr-1} =$$

$$= \frac{(n+kr-1)!}{n!(kr-1)!} \int_{z \in X} \int_{\xi \in P(1^{[r]}, \ldots, k^{[r]})} \omega_{a,r,p}^{kr-1}(\xi) \mathbb{1}_{\gamma_k, q}(z, \xi) \gamma_k(z, \xi)^n$$

where $\mathbb{1}_{\gamma_k, q}(z, \xi)$ is the characteristic function of the open set of points where $\gamma_k(z, \xi)$ has signature $(n-q, q)$ in terms of the dz_j's. Notice that since $\gamma_k(z, \xi)^n$ is a determinant, the product $\mathbb{1}_{\gamma_k, q}(z, \xi) \gamma_k(z, \xi)^n$ gives rise to a continuous function on X_k^{GG}. Formula (104) with $r_1 = \ldots = r_k = r$ and $a_s = s$ yields the slightly more explicit integral

$$\int_{X_k^{GG}(L_k, q)} \Theta_{L_k, \Psi_{h,p,\varepsilon}^*}^{n+kr-1} = \frac{(n+kr-1)!}{n!(k!)^r} \times$$

$$\times \int_{z \in X} \int_{(x,u) \in \Delta_{k-1} \times (S^{2r-1})^k} \mathbb{1}_{g_k, q}(z, x, u) g_k(z, x, u)^n \frac{(x_1 \ldots x_k)^{r-1}}{(r-1)!^k} dx\, d\mu(u),$$

where $g_k(z, x, u) = \gamma_k(z, x_1^{1/2p} u_1, \ldots, x_k^{k/2p} u_k)$ is given by

$$g_k(z, x, u) = \frac{i}{2\pi} \sum_{1 \le s \le k} \frac{1}{s} x_s \sum_{i,j,\alpha,\beta} c_{ij\alpha\beta}(z) u_{s\alpha} \overline{u}_{s\beta} dz_i \wedge d\overline{z}_j \qquad (117)$$

and $\mathbb{1}_{g_k, q}(z, x, u)$ is the characteristic function of its q-index set. Here

$$d\nu_{k,r}(x) = (kr-1)! \frac{(x_1 \ldots x_k)^{r-1}}{(r-1)!^k} dx \qquad (118)$$

is a probability measure on Δ_{k-1}, and we can rewrite

$$\int_{X_k^{GG}(L_k, q)} \Theta_{L_k, \Psi_{h,p,\varepsilon}^*}^{n+kr-1} = \frac{(n+kr-1)!}{n!(k!)^r (kr-1)!} \times$$

$$\times \int_{z \in X} \int_{(x,u) \in \Delta_{k-1} \times (S^{2r-1})^k} \mathbb{1}_{g_k, q}(z, x, u) g_k(z, x, u)^n\, d\nu_{k,r}(x)\, d\mu(u). \qquad (119)$$

Now, formula (117) shows that $g_k(z, x, u)$ is a "Monte Carlo" evaluation of the curvature tensor, obtained by averaging the curvature at random points $u_s \in S^{2r-1}$ with certain positive weights x_s/s; we should then think of the k-jet f as some sort of random parameter such that the derivatives $\nabla^k f(0)$ are uniformly distributed in all directions. Let us compute the expected value of $(x, u) \mapsto g_k(z, x, u)$ with respect to the probability measure $dv_{k,r}(x)\, d\mu(u)$. Since $\int_{S^{2r-1}} u_{s\alpha} \bar{u}_{s\beta} d\mu(u_s) = \frac{1}{r}\delta_{\alpha\beta}$ and $\int_{\Delta_{k-1}} x_s\, dv_{k,r}(x) = \frac{1}{k}$, we find

$$\mathbf{E}(g_k(z, \bullet, \bullet)) = \frac{1}{kr} \sum_{1 \le s \le k} \frac{1}{s} \cdot \frac{i}{2\pi} \sum_{i,j,\alpha} c_{ij\alpha\alpha}(z)\, dz_i \wedge d\bar{z}_j.$$

In other words, we get the normalized trace of the curvature, i.e.

$$\mathbf{E}(g_k(z, \bullet, \bullet)) = \frac{1}{kr}\left(1 + \frac{1}{2} + \ldots + \frac{1}{k}\right) \Theta_{\det(V^*),\det h^*}, \qquad (120)$$

where $\Theta_{\det(V^*),\det h^*}$ is the $(1,1)$-curvature form of $\det(V^*)$ with the metric induced by h. It is natural to guess that $g_k(z, x, u)$ behaves asymptotically as its expected value $\mathbf{E}(g_k(z, \bullet, \bullet))$ when k tends to infinity. If we replace brutally g_k by its expected value in (119), we get the integral

$$\frac{(n + kr - 1)!}{n!(k!)^r(kr-1)!} \frac{1}{(kr)^n} \left(1 + \frac{1}{2} + \ldots + \frac{1}{k}\right)^n \int_X \mathbb{1}_{\eta,q}\, \eta^n,$$

where $\eta := \Theta_{\det(V^*),\det h^*}$ and $\mathbb{1}_{\eta,q}$ is the characteristic function of its q-index set in X. The leading constant is equivalent to $(\log k)^n / n!(k!)^r$ modulo a multiplicative factor $1 + O(1/\log k)$. By working out a more precise analysis of the deviation, we will prove the following result.

Probabilistic estimate 110. *Fix smooth Hermitian metrics h on V and $\omega = \frac{i}{2\pi}\sum \omega_{ij} dz_i \wedge d\bar{z}_j$ on X. Denote by $\Theta_{V,h} = -\frac{i}{2\pi}\sum c_{ij\alpha\beta} dz_i \wedge d\bar{z}_j \otimes e_\alpha^* \otimes e_\beta$ the curvature tensor of V with respect to an h-orthonormal frame (e_α), and put*

$$\eta(z) = \Theta_{\det(V^*),\det h^*} = \frac{i}{2\pi} \sum_{1 \le i,j \le n} \eta_{ij} dz_i \wedge d\bar{z}_j, \qquad \eta_{ij} = \sum_{1 \le \alpha \le r} c_{ij\alpha\alpha}.$$

Finally consider the k-jet line bundle $L_k = \mathscr{O}_{X_k^{\mathrm{GG}}}(1) \to X_k^{\mathrm{GG}}$ equipped with the induced metric $\Psi_{h,p,\varepsilon}^*$ (as defined above, with $1 = \varepsilon_1 \gg \varepsilon_2 \gg \ldots \gg \varepsilon_k > 0$). When k tends to infinity, the integral of the top power of the curvature of L_k on its q-index set $X_k^{\mathrm{GG}}(L_k, q)$ is given by

$$\int_{X_k^{\mathrm{GG}}(L_k,q)} \Theta^{n+kr-1}_{L_k,\Psi_{h,p,\varepsilon}^*} = \frac{(\log k)^n}{n!\,(k!)^r}\left(\int_X \mathbb{1}_{\eta,q}\, \eta^n + O((\log k)^{-1})\right)$$

for all $q = 0, 1, \ldots, n$, and the error term $O((\log k)^{-1})$ can be bounded explicitly in terms of Θ_V, η and ω. Moreover, the left hand side is identically zero for $q > n$.

The final statement follows from the observation that the curvature of L_k is positive along the fibers of $X_k^{GG} \to X$, by the plurisubharmonicity of the weight (this is true even when the partition of unity terms are taken into account, since they depend only on the base); therefore the q-index sets are empty for $q > n$. We start with three elementary lemmas.

Lemma 111. *The integral*

$$I_{k,r,n} = \int_{\Delta_{k-1}} \left(\sum_{1 \le s \le k} \frac{x_s}{s} \right)^n d\nu_{k,r}(x)$$

is given by the expansion

(a) $I_{k,r,n} = \displaystyle\sum_{1 \le s_1, s_2, \ldots, s_n \le k} \frac{1}{s_1 s_2 \ldots s_n} \cdot \frac{(kr-1)!}{(r-1)!^k} \cdot \frac{\prod_{1 \le i \le k}(r-1+\beta_i)!}{(kr+n-1)!}.$

where $\beta_i = \beta_i(s) = \text{card}\{j \,;\, s_j = i\}$, $\sum \beta_i = n$, $1 \le i \le k$. *The quotient*

$$I_{k,r,n} \Big/ \frac{r^n}{kr(kr+1)\ldots(kr+n-1)} \left(1 + \frac{1}{2} + \ldots + \frac{1}{k}\right)^n$$

is bounded below by 1 and bounded above by

(b) $1 + \dfrac{1}{3} \displaystyle\sum_{m=2}^{n} \dfrac{2^m n!}{(n-m)!} \left(1 + \dfrac{1}{2} + \ldots + \dfrac{1}{k}\right)^{-m} = 1 + O((\log k)^{-2}).$

As a consequence

(c)

$$I_{k,r,n} = \frac{1}{k^n}\left(\left(1 + \frac{1}{2} + \ldots + \frac{1}{k}\right)^n + O((\log k)^{n-2})\right)$$

$$= \frac{(\log k + \gamma)^n + O((\log k)^{n-2})}{k^n}$$

where γ is the Euler–Mascheroni constant.

Proof. Let us expand the n-th power $\left(\sum_{1 \le s \le k} \frac{x_s}{s}\right)^n$. This gives

$$I_{k,r,n} = \sum_{1 \le s_1, s_2, \ldots, s_n \le k} \frac{1}{s_1 s_2 \ldots s_n} \int_{\Delta_{k-1}} x_1^{\beta_1} \ldots x_k^{\beta_k} d\nu_{k,r}(x)$$

and by definition of the measure $\nu_{k,r}$ we have

$$\int_{\Delta_{k-1}} x_1^{\beta_1} \ldots x_k^{\beta_k} d\nu_{k,r}(x) = \frac{(kr-1)!}{(r-1)!^k} \int_{\Delta_{k-1}} x_1^{r+\beta_1-1} \ldots x_k^{r+\beta_k-1} dx_1 \ldots dx_k.$$

Applications of Pluripotential Theory to Algebraic Geometry

By formula (33), we find

$$\int_{\Delta_{k-1}} x_1^{\beta_1} \ldots x_k^{\beta_k} \, dv_{k,r}(x) = \frac{(kr-1)!}{(r-1)!^k} \frac{\prod_{1 \leq i \leq k}(r+\beta_i-1)!}{(kr+n-1)!}$$

$$= \frac{r^n \prod_{i,\beta_i \geq 1}(1+\frac{1}{r})(1+\frac{2}{r})\ldots(1+\frac{\beta_i-1}{r})}{kr(kr+1)\ldots(kr+n-1)},$$

and Lemma 111(a) follows from the first equality. The final product is minimal when $r = 1$, thus

$$\frac{r^n}{kr(kr+1)\ldots(kr+n-1)} \leq \int_{\Delta_{k-1}} x_1^{\beta_1} \ldots x_k^{\beta_k} \, dv_{k,r}(x)$$

$$\leq \frac{r^n \prod_{1 \leq i \leq k} \beta_i!}{kr(kr+1)\ldots(kr+n-1)}. \quad (121)$$

Also, the integral is maximal when all β_i vanish except one, in which case one gets

$$\int_{\Delta_{k-1}} x_j^n \, dv_{k,r}(x) = \frac{r(r+1)\ldots(r+n-1)}{kr(kr+1)\ldots(kr+n-1)}. \quad (122)$$

By (121), we find the lower and upper bounds

$$I_{k,r,n} \geq \frac{r^n}{kr(kr+1)\ldots(kr+n-1)} \left(1 + \frac{1}{2} + \ldots + \frac{1}{k}\right)^n, \quad (123)$$

$$I_{k,r,n} \leq \frac{r^n}{kr(kr+1)\ldots(kr+n-1)} \sum_{1 \leq s_1,\ldots,s_n \leq k} \frac{\beta_1! \ldots \beta_k!}{s_1 \ldots s_n}. \quad (124)$$

In order to make the upper bound more explicit, we reorganize the n-tuple $(s_1, \ldots, s_n)$ into those indices $t_1 < \ldots < t_\ell$ which appear a certain number of times $\alpha_i = \beta_{t_i} \geq 2$, and those, say $t_{\ell+1} < \ldots < t_{\ell+m}$, which appear only once. We have of course $\sum \beta_i = n - m$, and each choice of the t_i's corresponds to $n!/\alpha_1! \ldots \alpha_\ell!$ possibilities for the n-tuple $(s_1, \ldots, s_n)$. Therefore we get

$$\sum_{1 \leq s_1,\ldots,s_n \leq k} \frac{\beta_1! \ldots \beta_k!}{s_1 \ldots s_n} \leq n! \sum_{m=0}^n \sum_{\ell, \Sigma \alpha_i = n-m} \sum_{(t_i)} \frac{1}{t_1^{\alpha_1} \ldots t_\ell^{\alpha_\ell}} \frac{1}{t_{\ell+1} \ldots t_{\ell+m}}.$$

A trivial comparison series vs. integral yields

$$\sum_{s < t < +\infty} \frac{1}{t^\alpha} \leq \frac{1}{\alpha - 1} \frac{1}{s^{\alpha-1}}$$

and in this way, using successive integrations in $t_\ell, t_{\ell-1}, \ldots$, we get inductively

$$\sum_{1 \leq t_1 < \ldots < t_\ell < +\infty} \frac{1}{t_1^{\alpha_1} \ldots t_\ell^{\alpha_\ell}} \leq \frac{1}{\prod_{1 \leq i \leq \ell}(\alpha_{\ell-i+1} + \ldots + \alpha_\ell - i)} \leq \frac{1}{\ell!},$$

since $\alpha_i \geq 2$ implies $\alpha_{\ell-i+1} + \ldots + \alpha_\ell - i \geq i$. On the other hand

$$\sum_{1 \leq t_{\ell+1} < \ldots < t_{\ell+m} \leq k} \frac{1}{t_{\ell+1} \ldots t_{\ell+m}} \leq \frac{1}{m!} \sum_{1 \leq s_1, \ldots, s_m \leq k} \frac{1}{s_1 \ldots s_m} = \frac{1}{m!}\left(1 + \frac{1}{2} + \ldots + \frac{1}{k}\right)^m.$$

Since partitions $\alpha_1 + \ldots + \alpha_\ell = n - m$ satisfying the additional restriction $\alpha_i \geq 2$ correspond to $\alpha'_i = \alpha_i - 2$ satisfying $\sum \alpha'_i = n - m - 2\ell$, their number is equal to

$$\binom{n - m - 2\ell + \ell - 1}{\ell - 1} = \binom{n - m - \ell - 1}{\ell - 1} \leq 2^{n-m-\ell-1}$$

and we infer from this

$$\sum_{1 \leq s_1, \ldots, s_n \leq k} \frac{\beta_1! \ldots \beta_k!}{s_1 \ldots s_n} \leq \sum_{\substack{\ell \geq 1 \\ 2\ell + m \leq n}} \frac{2^{n-m-\ell-1} n!}{\ell! \, m!} \left(1 + \frac{1}{2} + \ldots + \frac{1}{k}\right)^m + \left(1 + \frac{1}{2} + \ldots + \frac{1}{k}\right)^n$$

where the last term corresponds to the special case $\ell = 0, m = n$. Therefore

$$\sum_{1 \leq s_i \leq k} \frac{\beta_1! \ldots \beta_k!}{s_1 \ldots s_n} \leq \frac{e^{1/2} - 1}{2} \sum_{m=0}^{n-2} \frac{2^{n-m} n!}{m!} \left(1 + \frac{1}{2} + \ldots + \frac{1}{k}\right)^m + \left(1 + \frac{1}{2} + \ldots + \frac{1}{k}\right)^n$$

$$\leq \frac{1}{3} \sum_{m=2}^{n} \frac{2^m n!}{(n-m)!} \left(1 + \frac{1}{2} + \ldots + \frac{1}{k}\right)^{n-m} + \left(1 + \frac{1}{2} + \ldots + \frac{1}{k}\right)^n.$$

This estimate combined with (123), (124) implies the upper bound Lemma 111(b) (the lower bound 1 being now obvious). The asymptotic estimate Lemma 111(c) follows immediately. □

Lemma 112. *If A is a Hermitian $n \times n$ matrix, set $\mathbb{K}_{A,q}$ to be equal to 1 if A has signature $(n - q, q)$ and 0 otherwise. Then for all $n \times n$ Hermitian matrices A, B we have the estimate*

$$\left|\mathbb{K}_{A,q} \det A - \mathbb{K}_{B,q} \det B\right| \leq \|A - B\| \sum_{0 \leq i \leq n-1} \|A\|^i \|B\|^{n-1-i},$$

where $\|A\|$, $\|B\|$ are the Hermitian operator norms of the matrices.

Proof. We first check that the estimate holds for $|\det A - \det B|$. Let $\lambda_1 \leq \ldots \leq \lambda_n$ be the eigenvalues of A and $\lambda'_1 \leq \ldots \leq \lambda'_n$ be the eigenvalues of B. We have

$|\lambda_i| \le \|A\|, |\lambda'_i| \le \|B\|$ and the minimax principle implies that $|\lambda_i - \lambda'_i| \le \|A-B\|$. We then get the desired estimate by writing

$$\det A - \det B = \lambda_1 \ldots \lambda_n - \lambda'_1 \ldots \lambda'_n = \sum_{1 \le i \le n} \lambda_1 \ldots \lambda_{i-1}(\lambda_i - \lambda'_i)\lambda'_{i+1} \ldots \lambda'_n.$$

This already implies Lemma 112 if A or B is degenerate. If A and B are non degenerate we only have to prove the result when one of them (say A) has signature $(n-q, q)$ and the other one (say B) has a different signature. If we put $M(t) = (1-t)A + tB$, the already established estimate for the determinant yields

$$\left|\frac{d}{dt}\det M(t)\right| \le n\|A-B\|\,\|M(t)\| \le n\|A-B\|\big((1-t)\|A\| + t\|B\|\big)^{n-1}.$$

However, since the signature of $M(t)$ is not the same for $t=0$ and $t=1$, there must exist $t_0 \in\,]0,1[$ such that $(1-t_0)A + t_0 B$ is degenerate. Our claim follows by integrating the differential estimate on the smallest such interval $[0, t_0]$, after observing that $M(0) = A$, $\det M(t_0) = 0$, and that the integral of the right hand side on $[0, 1]$ is the announced bound. □

Lemma 113. *Let Q_A be the Hermitian quadratic form associated with the Hermitian operator A on $\mathbb{C}^n$. If μ is the rotation invariant probability measure on the unit sphere S^{2n-1} of $\mathbb{C}^n$ and λ_i are the eigenvalues of A, we have*

$$\int_{|\zeta|=1} |Q_A(\zeta)|^2 d\mu(\zeta) = \frac{1}{n(n+1)}\Big(\sum \lambda_i^2 + \Big(\sum \lambda_i\Big)^2\Big).$$

The norm $\|A\| = \max|\lambda_i|$ satisfies the estimate

$$\frac{1}{n^2}\|A\|^2 \le \int_{|\zeta|=1}|Q_A(\zeta)|^2 d\mu(\zeta) \le \|A\|^2.$$

Proof. The first identity is an easy calculation, and the inequalities follow by computing the eigenvalues of the quadratic form $\sum \lambda_i^2 + (\sum \lambda_i)^2 - c\lambda_{i_0}^2$, $c > 0$. The lower bound is attained e.g. for $Q_A(\zeta) = |\zeta_1|^2 - \frac{1}{n}(|\zeta_2|^2 + \ldots + |\zeta_n|^2)$ when we take $i_0 = 1$ and $c = 1 + \frac{1}{n}$. □

Proof of the Probabilistic estimate 110. Take a vector $\zeta \in T_{X,z}$, $\zeta = \sum \zeta_i \frac{\partial}{\partial z_i}$, with $\|\zeta\|_\omega = 1$, and introduce the trace free sesquilinear quadratic form

$$Q_{z,\zeta}(u) = \sum_{i,j,\alpha,\beta} \tilde{c}_{ij\alpha\beta}(z)\,\zeta_i\overline{\zeta}_j\,u_\alpha\overline{u}_\beta, \qquad \tilde{c}_{ij\alpha\beta} = c_{ij\alpha\beta} - \frac{1}{r}\eta_{ij}\delta_{\alpha\beta}, \qquad u \in \mathbb{C}^r$$

where $\eta_{ij} = \sum_{1\le\alpha\le r} c_{ij\alpha\alpha}$. We consider the corresponding trace free curvature tensor

$$\widetilde{\Theta}_V = \frac{i}{2\pi} \sum_{i,j,\alpha,\beta} \tilde{c}_{ij\alpha\beta}\, dz_i \wedge d\bar{z}_j \otimes e_\alpha^* \otimes e_\beta. \tag{125}$$

As a general matter of notation, we adopt here the convention that the canonical correspondence between Hermitian forms and $(1,1)$-forms is normalized as $\sum a_{ij} dz_i \otimes d\bar{z}_j \leftrightarrow \frac{i}{2\pi}\sum a_{ij} dz_i \wedge d\bar{z}_j$, and we take the liberty of using the same symbols for both types of objects; we do so especially for $g_k(z,x,u)$ and $\eta(z) = \frac{i}{2\pi}\sum \eta_{ij}(z) dz_i \wedge d\bar{z}_j = \operatorname{Tr}\Theta_V(z)$. First observe that for all k-tuples of unit vectors $u = (u_1,\ldots,u_k) \in (S^{2r-1})^k$, $u_s = (u_{s\alpha})_{1\le\alpha\le r}$, we have

$$\int_{(S^{2r-1})^k} \left|\sum_{1\le s\le k} \frac{1}{s} x_s \sum_{i,j,\alpha,\beta} \tilde{c}_{ij\alpha\beta}(z)\, \zeta_i \bar{\zeta}_j u_{s\alpha}\bar{u}_{s\beta}\right|^2 d\mu(u) = \sum_{1\le s\le k} \frac{x_s^2}{s^2} \mathbf{V}(Q_{z,\zeta})$$

where $\mathbf{V}(Q_{z,\zeta})$ is the variance of $Q_{z,\zeta}$ on S^{2r-1}. This is so because we have a sum over s of independent random variables on $(S^{2r-1})^k$, all of which have zero mean value (Lemma 113 shows that the variance $\mathbf{V}(Q)$ of a trace free Hermitian quadratic form $Q(u) = \sum_{1\le\alpha\le r} \lambda_\alpha |u_\alpha|^2$ on the unit sphere S^{2r-1} is equal to $\frac{1}{r(r+1)}\sum \lambda_\alpha^2$, but we only give the formula to fix the ideas). Formula (122) yields

$$\int_{\Delta_{k-1}} x_s^2 dv_{k,r}(x) = \frac{r+1}{k(kr+1)}.$$

Therefore, according to notation (117), we obtain the partial variance formula

$$\int_{\Delta_{k-1}\times(S^{2r-1})^k} |g_k(z,x,u)(\zeta) - \bar{g}_k(z,x)(\zeta)|^2 dv_{k,r}(x) d\mu(u)$$

$$= \frac{(r+1)}{k(kr+1)}\left(\sum_{1\le s\le k} \frac{1}{s^2}\right) \sigma_h(\widetilde{\Theta}_V(\zeta,\zeta))^2$$

in which

$$\bar{g}_k(z,x)(\zeta) = \sum_{1\le s\le k} \frac{1}{s} x_s \frac{1}{r}\sum_{ij\alpha} c_{ij\alpha\alpha}\zeta_i \bar{\zeta}_j = \left(\sum_{1\le s\le k} \frac{1}{s} x_s\right)\frac{1}{r}\eta(z)(\zeta),$$

$$\sigma_h(\widetilde{\Theta}_V(\zeta,\zeta))^2 = \mathbf{V}\big(u \mapsto \langle \widetilde{\Theta}_V(\zeta,\zeta)u, u\rangle_h\big) = \int_{u\in S^{2r-1}} |\langle \widetilde{\Theta}_V(\zeta,\zeta)u, u\rangle_h|^2 d\mu(u).$$

By integrating over $\zeta \in S^{2n-1} \subset \mathbb{C}^n$ and applying the left hand inequality in Lemma 113 we infer

Applications of Pluripotential Theory to Algebraic Geometry

$$\int_{\Delta_{k-1}\times(S^{2r-1})^k} \|g_k(z,x,u) - \overline{g}_k(z,x)\|_\omega^2 dv_{k,r}(x)d\mu(u)$$
$$\leq \frac{n^2(r+1)}{k(kr+1)}\left(\sum_{1\leq s\leq k}\frac{1}{s^2}\right)\sigma_{\omega,h}(\tilde{\Theta}_V)^2 \qquad (126)$$

where $\sigma_{\omega,h}(\tilde{\Theta}_V)$ is the standard deviation of $\tilde{\Theta}_V$ on $S^{2n-1}\times S^{2r-1}$:

$$\sigma_{\omega,h}(\tilde{\Theta}_V)^2 = \int_{|\zeta|_\omega=1,|u|_h=1}\left|\langle\tilde{\Theta}_V(\zeta,\zeta)u,u\rangle_h\right|^2 d\mu(\zeta)\,d\mu(u).$$

On the other hand, brutal estimates give the Hermitian operator norm estimates

$$\|\overline{g}_k(z,x)\|_\omega \leq \left(\sum_{1\leq s\leq k}\frac{1}{s}x_s\right)\frac{1}{r}\|\eta(z)\|_\omega, \qquad (127)$$

$$\|g_k(z,x,u)\|_\omega \leq \left(\sum_{1\leq s\leq k}\frac{1}{s}x_s\right)\|\Theta_V\|_{\omega,h} \qquad (128)$$

where

$$\|\Theta_V\|_{\omega,h} = \sup_{|\zeta|_\omega=1,|u|_h=1}\left|\langle\Theta_V(\zeta,\zeta)u,u\rangle_h\right|.$$

We use these estimates to evaluate the q-index integrals. The integral associated with $\overline{g}_k(z,x)$ is much easier to deal with than $g_k(z,x,u)$ since the characteristic function of the q-index set depends only on z. By Lemma 112 we find

$$\left|1\!\!\!1_{g_k,q}(z,x,u)\det g_k(z,x,u) - 1\!\!\!1_{\eta,q}(z)\det\overline{g}_k(z,x)\right|$$
$$\leq \|g_k(z,x,u) - \overline{g}_k(z,x)\|_\omega \sum_{0\leq i\leq n-1}\|g_k(z,x,u)\|_\omega^i\|\overline{g}_k(z,x)\|_\omega^{n-1-i}.$$

The Cauchy–Schwarz inequality combined with (126)–(128) implies

$$\int_{\Delta_{k-1}\times(S^{2r-1})^k}\left|1\!\!\!1_{g_k,q}(z,x,u)\det g_k(z,x,u) - 1\!\!\!1_{\eta,q}(z)\det\overline{g}_k(z,x)\right|dv_{k,r}(x)d\mu(u)$$
$$\leq \left(\int_{\Delta_{k-1}\times(S^{2r-1})^k}\|g_k(z,x,u)-\overline{g}_k(z,x)\|_\omega^2 dv_{k,r}(x)d\mu(u)\right)^{1/2}\times$$
$$\times\left(\int_{\Delta_{k-1}\times(S^{2r-1})^k}\left(\sum_{0\leq i\leq n-1}\|g_k(z,x,u)\|_\omega^i\|\overline{g}_k(z,x)\|_\omega^{n-1-i}\right)^2 dv_{k,r}(x)d\mu(u)\right)^{1/2}$$

$$\leq \frac{n(1+1/r)^{1/2}}{(k(k+1/r))^{1/2}} \left(\sum_{1 \leq s \leq k} \frac{1}{s^2} \right)^{1/2} \sigma_{\omega,h}(\tilde{\Theta}_V) \sum_{1 \leq i \leq n-1} \|\Theta_V\|_{\omega,h}^i \left(\frac{1}{r} \|\eta(z)\|_\omega \right)^{n-1-i}$$

$$\times \left(\int_{\Delta_{k-1}} \left(\sum_{1 \leq s \leq k} \frac{x_s}{s} \right)^{2n-2} dv_{k,r}(x) \right)^{1/2} = O\left(\frac{(\log k)^{n-1}}{k^n} \right)$$

by Lemma 111 with n replaced by $2n - 2$. This is the essential error estimate. As one can see, the growth of the error mainly depends on the final integral factor, since the initial multiplicative factor is uniformly bounded over X. In order to get the principal term, we compute

$$\int_{\Delta_{k-1}} \det \overline{g}_k(z,x)\, dv_{k,r}(x) = \frac{1}{r^n} \det \eta(z) \int_{\Delta_{k-1}} \left(\sum_{1 \leq s \leq k} \frac{x_s}{s} \right)^n dv_{k,r}(x)$$

$$\sim \frac{(\log k)^n}{r^n k^n} \det \eta(z).$$

From there we conclude that

$$\int_{z \in X} \int_{(x,u) \in \Delta_{k-1} \times (S^{2r-1})^k} \mathcal{K}_{g_k,q}(z,x,u) g_k(z,x,u)^n\, dv_{k,r}(x) d\mu(u)$$

$$= \frac{(\log k)^n}{r^n k^n} \int_X \mathcal{K}_{\eta,q} \eta^n + O\left(\frac{(\log k)^{n-1}}{k^n} \right)$$

The probabilistic estimate 110 follows by (119). □

Remark 114. If we take care of the precise bounds obtained above, the proof gives in fact the explicit estimate

$$\int_{X_k^{GG}(L_k,q)} \Theta_{L_k,\Psi_{h,p,\varepsilon}^*}^{n+kr-1} = \frac{(n+kr-1)!\, I_{k,r,n}}{n!(k!)^r(kr-1)!} \left(\int_X \mathcal{K}_{\eta,q} \eta^n + \varepsilon_{k,r,n} J \right)$$

where

$$J = n(1+1/r)^{1/2} \left(\sum_{s=1}^k \frac{1}{s^2} \right)^{1/2} \int_X \sigma_{\omega,h}(\tilde{\Theta}_V) \sum_{i=1}^{n-1} r^{i+1} \|\Theta_V\|_{\omega,h}^i \|\eta(z)\|_\omega^{n-1-i} \omega^n$$

and

$$|\varepsilon_{k,r,n}| \leq \frac{\left(\int_{\Delta_{k-1}} \left(\sum_{s=1}^k \frac{x_s}{s} \right)^{2n-2} dv_{k,r}(x) \right)^{1/2}}{(k(k+1/r))^{1/2} \int_{\Delta_{k-1}} \left(\sum_{s=1}^k \frac{x_s}{s} \right)^n dv_{k,r}(x)}$$

Applications of Pluripotential Theory to Algebraic Geometry

$$\leq \frac{\left(1+\frac{1}{3}\sum_{m=2}^{2n-2}\frac{2^m(2n-2)!}{(2n-2-m)!}\left(1+\frac{1}{2}+\ldots+\frac{1}{k}\right)^{-m}\right)^{1/2}}{1+\frac{1}{2}+\ldots+\frac{1}{k}} \sim \frac{1}{\log k}$$

by the lower and upper bounds of $I_{k,r,n}$, $I_{k,r,2n-2}$ obtained in Lemma 111. As $(2n-2)!/(2n-2-m)! \leq (2n-2)^m$, one easily shows that

$$|\varepsilon_{k,r,n}| \leq \frac{(31/15)^{1/2}}{\log k} \quad \text{for } k \geq e^{5n-5}. \tag{129}$$

Also, we see that the error terms vanish if $\tilde{\Theta}_V$ is identically zero, but this is of course a rather unexpected circumstance. In general, since the form $\tilde{\Theta}_V$ is trace free, Lemma 2.23 applied to the quadratic form $u \mapsto \langle \tilde{\Theta}_V(\zeta,\zeta)u,u\rangle$ on $\mathbb{C}^r$ implies $\sigma_{\omega,h}(\tilde{\Theta}_V) \leq (r+1)^{-1/2}\|\tilde{\Theta}_V\|_{\omega,h}$. This yields the simpler bound

$$J \leq n\, r^{1/2}\left(\sum_{s=1}^{k}\frac{1}{s^2}\right)^{1/2}\int_X \|\tilde{\Theta}_V\|_{\omega,h}\sum_{i=1}^{n-1}r^i\|\tilde{\Theta}_V\|_{\omega,h}^i\|\eta(z)\|_{\omega}^{n-1-i}\omega^n. \tag{130}$$

□

It will be useful to extend the above estimates to the case of sections of

$$L_k = \mathcal{O}_{X_k^{\mathrm{GG}}}(1) \otimes \pi_k^*\mathcal{O}\left(\frac{1}{kr}\left(1+\frac{1}{2}+\ldots+\frac{1}{k}\right)F\right) \tag{131}$$

where $F \in \mathrm{Pic}_\mathbb{Q}(X)$ is an arbitrary $\mathbb{Q}$-line bundle on X and $\pi_k : X_k^{\mathrm{GG}} \to X$ is the natural projection. We assume here that F is also equipped with a smooth Hermitian metric h_F. In formula (2.20), the renormalized metric $\eta_k(z,x,u)$ of L_k takes the form

$$\eta_k(z,x,u) = \frac{1}{\frac{1}{kr}(1+\frac{1}{2}+\ldots+\frac{1}{k})}g_k(z,x,u) + \Theta_{F,h_F}(z), \tag{132}$$

and by the same calculations its expected value is

$$\eta(z) := \mathbf{E}(\eta_k(z,\bullet,\bullet)) = \Theta_{\det V^*,\det h^*}(z) + \Theta_{F,h_F}(z). \tag{133}$$

Then the variance estimate for $\eta_k - \eta$ is unchanged, and the L^p bounds for η_k are still valid, since our forms are just shifted by adding the constant smooth term $\Theta_{F,h_F}(z)$. The probabilistic estimate (120) is therefore still true in exactly the same form, provided we use (131)–(133) instead of the previously defined L_k, η_k and η. An application of holomorphic Morse inequalities gives the desired cohomology estimates for

$$h^q\left(X, E_{k,m}^{\mathrm{GG}}V^* \otimes \mathcal{O}\left(\frac{m}{kr}\left(1+\frac{1}{2}+\ldots+\frac{1}{k}\right)F\right)\right)$$
$$= h^q\left(X_k^{\mathrm{GG}}, \mathcal{O}_{X_k^{\mathrm{GG}}}(m) \otimes \pi_k^*\mathcal{O}\left(\frac{m}{kr}\left(1+\frac{1}{2}+\ldots+\frac{1}{k}\right)F\right)\right),$$

provided m is sufficiently divisible to give a multiple of F which is a $\mathbb{Z}$-line bundle.

Theorem 115. *Let (X, V) be a directed manifold, $F \to X$ a $\mathbb{Q}$-line bundle, (V, h) and (F, h_F) smooth Hermitian structure on V and F respectively. We define*

$$L_k = \mathcal{O}_{X_k^{GG}}(1) \otimes \pi_k^* \mathcal{O}\Big(\frac{1}{kr}\Big(1 + \frac{1}{2} + \ldots + \frac{1}{k}\Big)F\Big),$$

$$\eta = \Theta_{\det V^*, \det h^*} + \Theta_{F, h_F}.$$

Then for all $q \geq 0$ and all $m \gg k \gg 1$ such that m is sufficiently divisible, we have

(a) $h^q(X_k^{GG}, \mathcal{O}(L_k^{\otimes m})) \leq \dfrac{m^{n+kr-1}}{(n+kr-1)!} \dfrac{(\log k)^n}{n!\,(k!)^r} \Big(\int_{X(\eta, q)} (-1)^q \eta^n + O((\log k)^{-1})\Big),$

(b) $h^0(X_k^{GG}, \mathcal{O}(L_k^{\otimes m})) \geq \dfrac{m^{n+kr-1}}{(n+kr-1)!} \dfrac{(\log k)^n}{n!\,(k!)^r} \Big(\int_{X(\eta, \leq 1)} \eta^n - O((\log k)^{-1})\Big),$

(c) $\chi(X_k^{GG}, \mathcal{O}(L_k^{\otimes m})) = \dfrac{m^{n+kr-1}}{(n+kr-1)!} \dfrac{(\log k)^n}{n!\,(k!)^r} (c_1(V^* \otimes F)^n + O((\log k)^{-1})).$

Green and Griffiths [51] already checked the Riemann–Roch calculation Theorem 115(c) in the special case $V = T_X^*$ and $F = \mathcal{O}_X$. Their proof is much simpler since it relies only on Chern class calculations, but it cannot provide any information on the individual cohomology groups, except in very special cases where vanishing theorems can be applied; in fact in dimension 2, the Euler characteristic satisfies $\chi = h^0 - h^1 + h^2 \leq h^0 + h^2$, hence it is enough to get the vanishing of the top cohomology group H^2 to infer $h^0 \geq \chi$; this works for surfaces by means of a well-known vanishing theorem of Bogomolov which implies in general

$$H^n\Big(X, E_{k,m}^{GG} T_X^* \otimes \mathcal{O}\Big(\frac{m}{kr}\Big(1 + \frac{1}{2} + \ldots + \frac{1}{k}\Big)F\Big)\Big) = 0$$

as soon as $K_X \otimes F$ is big and $m \gg 1$.

In fact, thanks to Bonavero's singular holomorphic Morse inequalities [13], everything works almost unchanged in the case where $V \subset T_X$ has singularities and h is an admissible metric on V (see (8)). We only have to find a blow-up $\mu: \tilde{X}_k \to X_k$ so that the resulting pull-backs $\mu^* L_k$ and $\mu^* V$ are locally free, and $\mu^* \det h^*$, $\mu^* \Psi_{h, p, \varepsilon}$ only have divisorial singularities. Then η is a $(1, 1)$-current with logarithmic poles, and we have to deal with smooth metrics on $\mu^* L_k^{\otimes m} \otimes \mathcal{O}(-mE_k)$ where E_k is a certain effective divisor on X_k (which, by our assumption (8), does not project onto X). The cohomology groups involved are then the twisted cohomology groups

$$H^q(X_k^{GG}, \mathcal{O}(L_k^{\otimes m}) \otimes \mathcal{J}_{k,m})$$

where $\mathcal{J}_{k,m} = \mu_*(\mathcal{O}(-mE_k))$ is the corresponding multiplier ideal sheaf, and the Morse integrals need only be evaluated in the complement of the poles, that is on $X(\eta, q) \smallsetminus S$ where $S = \text{Sing}(V) \cup \text{Sing}(h)$. Since

Applications of Pluripotential Theory to Algebraic Geometry

$$(\pi_k)_*\big(\mathcal{O}(L_k^{\otimes m}) \otimes \mathcal{J}_{k,m}\big) \subset E_{k,m}^{GG} V^* \otimes \mathcal{O}\Big(\frac{m}{kr}\Big(1 + \frac{1}{2} + \ldots + \frac{1}{k}\Big)F\Big)$$

we still get a lower bound for the H^0 of the latter sheaf (or for the H^0 of the un-twisted line bundle $\mathcal{O}(L_k^{\otimes m})$ on X_k^{GG}). If we assume that $K_V \otimes F$ is big, these considerations also allow us to obtain a strong estimate in terms of the volume, by using an approximate Zariski decomposition on a suitable blow-up of (X, V). The following corollary implies in particular Theorem 104.

Corollary 116. *If F is an arbitrary $\mathbb{Q}$-line bundle over X, one has*

$$h^0\Big(X_k^{GG}, \mathcal{O}_{X_k^{GG}}(m) \otimes \pi_k^* \mathcal{O}\Big(\frac{m}{kr}\Big(1 + \frac{1}{2} + \ldots + \frac{1}{k}\Big)F\Big)\Big)$$
$$\geq \frac{m^{n+kr-1}}{(n+kr-1)!} \frac{(\log k)^n}{n!\,(k!)^r} \Big(\mathrm{Vol}(K_V \otimes F) - O((\log k)^{-1})\Big) - o(m^{n+kr-1}),$$

when $m \gg k \gg 1$, in particular there are many sections of the k-jet differentials of degree m twisted by the appropriate power of F if $K_V \otimes F$ is big.

Proof. The volume is computed here as usual, i.e. after performing a suitable modification $\mu : \tilde{X} \to X$ which converts K_V into an invertible sheaf. There is of course nothing to prove if $K_V \otimes F$ is not big, so we can assume $\mathrm{Vol}(K_V \otimes F) > 0$. Let us fix smooth Hermitian metrics h_0 on T_X and h_F on F. They induce a metric $\mu^*(\det h_0^{-1} \otimes h_F)$ on $\mu^*(K_V \otimes F)$ which, by our definition of K_V, is a smooth metric. By the result of Fujita [47] on approximate Zariski decomposition, for every $\delta > 0$, one can find a modification $\mu_\delta : \tilde{X}_\delta \to X$ dominating μ such that

$$\mu_\delta^*(K_V \otimes F) = \mathcal{O}_{\tilde{X}_\delta}(A + E)$$

where A and E are $\mathbb{Q}$-divisors, A ample and E effective, with

$$\mathrm{Vol}(A) = A^n \geq \mathrm{Vol}(K_V \otimes F) - \delta.$$

If we take a smooth metric h_A with positive definite curvature form Θ_{A,h_A}, then we get a singular Hermitian metric $h_A h_E$ on $\mu_\delta^*(K_V \otimes F)$ with poles along E, i.e. the quotient $h_A h_E / \mu^*(\det h_0^{-1} \otimes h_F)$ is of the form $e^{-\varphi}$ where φ is quasi-psh with log poles $\log|\sigma_E|^2$ (mod $C^\infty(\tilde{X}_\delta)$) precisely given by the divisor E. We then only need to take the singular metric h on T_X defined by

$$h = h_0 e^{\frac{1}{r}(\mu_\delta)_*\varphi}$$

(the choice of the factor $\frac{1}{r}$ is there to correct adequately the metric on $\det V$). By construction h induces an admissible metric on V and the resulting curvature current $\eta = \Theta_{K_V,\det h^*} + \Theta_{F,h_F}$ is such that

$$\mu_\delta^* \eta = \Theta_{A,h_A} + [E], \qquad [E] = \text{current of integration on } E.$$

Then the 0-index Morse integral in the complement of the poles is given by

$$\int_{X(\eta,0)\smallsetminus S} \eta^n = \int_{\widetilde{X}_\delta} \Theta_{A,h_A}^n = A^n \geq \mathrm{Vol}(K_V \otimes F) - \delta$$

and Corollary 116 follows from the fact that δ can be taken arbitrary small. $\square$

Example 117. In some simple cases, the above estimates can lead to very explicit results. Take for instance X to be a smooth complete intersection of multidegree $(d_1, d_2, \ldots, d_s)$ in $\mathbb{P}_\mathbb{C}^{n+s}$ and consider the absolute case $V = T_X$. Then

$$K_X = \mathcal{O}_X(d_1 + \ldots + d_s - n - s - 1).$$

Assume that X is of general type, i.e. $\sum d_j > n + s + 1$. Let us equip $V = T_X$ with the restriction of the Fubini–Study metric $h = \Theta_{\mathcal{O}(1)}$; a better choice might be the Kähler–Einstein metric but we want to keep the calculations as elementary as possible. The standard formula for the curvature tensor of a submanifold gives

$$\Theta_{T_X,h} = (\Theta_{T_{\mathbb{P}^{n+s}},h})_{|X} + \beta^* \wedge \beta$$

where $\beta \in C^\infty(\Lambda^{1,0}T_X^* \otimes \mathrm{Hom}(T_X, \bigoplus \mathcal{O}(d_j)))$ is the second fundamental form. In other words, by the well known formula for the curvature of projective space, we have

$$\langle \Theta_{T_X,h}(\zeta,\zeta)u, u \rangle = |\zeta|^2|u|^2 + |\langle \zeta, u \rangle|^2 - |\beta(\zeta) \cdot u|^2.$$

The curvature ρ of $(K_X, \det h^*)$ (i.e. the opposite of the Ricci form $\mathrm{Tr}\,\Theta_{T_X,h}$) is given by

$$\rho = -\mathrm{Tr}\,\Theta_{T_X,h} = \mathrm{Tr}(\beta \wedge \beta^*) - (n+1)h \geq -(n+1)h. \qquad (134)$$

We take here $F = \mathcal{O}_X(-a)$, $a \in \mathbb{Q}_+$, and we want to determine conditions for the existence of sections

$$H^0\!\left(X, E_{k,m}^{\mathrm{GG}} T_X^* \otimes \mathcal{O}\!\left(-a\frac{m}{kr}\Big(1 + \frac{1}{2} + \ldots + \frac{1}{k}\Big)\right)\right), \qquad m \gg 1. \qquad (135)$$

We have to choose $K_X \otimes \mathcal{O}_X(-a)$ ample, i.e. $\sum d_j > n + s + a + 1$, and then (by an appropriate choice of the metric of $F = \mathcal{O}_X(-a)$), the form $\eta = \Theta_{K_X \otimes \mathcal{O}_X(-a)}$ can be taken to be any positive form cohomologous to $(\sum d_j - (n + s + a + 1))h$. We use Remark 114 and estimate the error terms by considering the Kähler metric

$$\omega = \rho + (n + s + 2)h \equiv \Big(\sum d_j + 1\Big)h.$$

Inequality (134) shows that $\omega \geq 2h$ and also that $\omega \geq \mathrm{Tr}(\beta \wedge \beta^*)$. From this, one easily concludes that $\|\eta\|_\omega \leq 1$ by an appropriate choice of η, as well as

$\|\Theta_{T_X,h}\|_{\omega,h} \leq 1$ and $\|\tilde{\Theta}_{T_X,h}\|_{\omega,h} \leq 2$. By (130), we obtain for $n \geq 2$

$$J \leq n^{3/2} \frac{\pi}{\sqrt{6}} \times 2 \frac{n^n - 1}{n - 1} \int_X \omega^n < \frac{4\pi}{\sqrt{6}} n^{n+1/2} \int_X \omega^n$$

where $\int_X \omega^n = \left(\sum d_j + 1\right)^n \deg(X)$. On the other hand, the leading term $\int_X \eta^n$ equals $\left(\sum d_j - n - s - a - 1\right)^n \deg(X)$ with $\deg(X) = d_1 \ldots d_s$. By the bound (129) on the error term $\varepsilon_{k,r,n}$, we find that the leading coefficient of the growth of our spaces of sections is strictly controlled below by a multiple of

$$\left(\sum d_j - n - s - a - 1\right)^n - 4\pi \left(\frac{31}{90}\right)^{1/2} \frac{n^{n+1/2}}{\log k} \left(\sum d_j + 1\right)^n$$

if $k \geq e^{5n-5}$. A sufficient condition for the existence of sections in (135) is thus

$$k \geq \exp\left(7.38 \, n^{n+1/2} \left(\frac{\sum d_j + 1}{\sum d_j - n - s - a - 1}\right)^n\right). \tag{136}$$

This is good in view of the fact that we can cover arbitrary smooth complete intersections of general type. On the other hand, even when the degrees d_j tend to $+\infty$, we still get a large lower bound $k \sim \exp(7.38 \, n^{n+1/2})$ on the order of jets, and this is far from being optimal: Diverio [39, 40] has shown e.g. that one can take $k = n$ for smooth hypersurfaces of high degree. It is however not unlikely that one could improve estimate (136) with more careful choices of ω, h. □

Acknowledgements The author expresses his warm thanks to the organizers of the CIME School in Pluripotential Theory held in Cetraro in July 2011, Filippo Bracci and John Erik Fornæss, for their invitation and the opportunity to deliver these lectures to an audience of young researchers. The author is also grateful to the referee for his (her) suggestions, and for a very careful reading of the manuscript.

References

1. M.F. Atiyah, R. Bott, V.K. Patodi, On the heat equation and the index theorem. Invent. Math. **19**, 279–330 (1973)
2. Y. Akizuki, S. Nakano, Note on Kodaira-Spencer's proof of Lefschetz theorems. Proc. Jpn. Acad. **30**, 266–272 (1954)
3. A. Andreotti, H. Grauert, Théorèmes de finitude pour la cohomologie des espaces complexes. Bull. Soc. Math. Fr. **90**, 193–259 (1962)
4. A. Andreotti, E. Vesentini, Carleman estimates for the Laplace-Beltrami equation in complex manifolds. Publ. Math. I.H.E.S. **25**, 81–130 (1965)
5. F. Angelini, An algebraic version of Demailly's asymptotic Morse inequalities. Proc. Am. Math. Soc. **124**, 3265–3269 (1996)

6. G. Bérczi, *Thom Polynomials and the Green-Griffiths Conjecture*. arXiv: 1011.4710, Contributions to Algebraic Geometry (EMS Series of Congress Reports – European Mathematical Society, Zürich, 2012), pp. 141–167
7. R. Berman, J.-P. Demailly, Regularity of plurisubharmonic upper envelopes in big cohomology classes|arXiv:math.CV/0905.1246v1, in *Proceedings of the Symposium "Perspectives in Analysis, Geometry and Topology"* in honor of Oleg Viro (Stockholm University, May 2008), Progress in Math., **296** (Birkhäuser/Springer, New York, 2012), pp. 39–66
8. J.-M. Bismut, Demailly's asymptotic inequalities: a heat equation proof. J. Funct. Anal. **72**, 263–278 (1987)
9. A. Bloch, Sur les systèmes de fonctions uniformes satisfaisant à l'équation d'une variété algébrique dont l'irrégularité dépasse la dimension. J. Math. **5**, 19–66 (1926)
10. S. Bochner, Curvature and Betti numbers (I) and (II). Ann. Math. **49**, 379–390 (1948); **50**, 77–93 (1949)
11. F.A. Bogomolov, Holomorphic tensors and vector bundles on projective varieties. Math. USSR Izv. **13/3**, 499–555 (1979)
12. E. Bombieri, Algebraic values of meromorphic maps. Invent. Math. **10**, 267–287 (1970); Addendum, Invent. Math. **11**, 163–166 (1970)
13. L. Bonavero, Inégalités de Morse holomorphes singulières. C. R. Acad. Sci. Paris Sér. I Math. **317**, 1163–1166 (1993)
14. L. Bonavero, Inégalités de Morse holomorphes singulières. J. Geom. Anal. **8**, 409–425 (1998)
15. D. Borthwick, A. Uribe, Nearly Kählerian embeddings of symplectic manifolds. Asian J. Math. **4**, 599–620 (2000)
16. Th. Bouche, Inégalités de Morse pour la d''-cohomologie sur une variété holomorphe non compacte. Ann. Sci. École Norm. Sup. **22**, 501–513 (1989)
17. Th. Bouche, Convergence de la métrique de Fubini-Study d'un fibré linéaire positif. Ann. Inst. Fourier (Grenoble) **40**, 117–130 (1990)
18. S. Boucksom, Cônes positifs des variétés complexes compactes. Thesis, Grenoble, 2002
19. S. Boucksom, J.-P. Demailly, M. Păun, Th. Peternell, The pseudo-effective cone of a compact Kähler manifold and varieties of negative Kodaira dimension, manuscript May 2004 [math.AG/0405285]
20. R. Brody, Compact manifolds and hyperbolicity. Trans. Amer. Math. Soc. **235**, 213–219 (1978)
21. H. Clemens, Curves on generic hypersurfaces. Ann. Sci. Éc. Norm. Sup. **19**, 629–636 (1986); Erratum: Ann. Sci. Éc. Norm. Sup. **20**, 281 (1987)
22. M. Cowen, P. Griffiths, Holomorphic curves and metrics of negative curvature. J. Anal. Math. **29**, 93–153 (1976)
23. F. Campana, Th. Peternell, Geometric stability of the cotangent bundle and the universal cover of a projective manifold [arXiv:math.AG/0405093]
24. J.-P. Demailly, Estimations L^2 pour l'opérateur $\overline{\partial}$ d'un fibré vectoriel holomorphe semi-positif au dessus d'une variété kählérienne complète. Ann. Sci. Ec. Norm. Sup. **15**, 457–511 (1982)
25. J.-P. Demailly, Champs magnétiques et inégalités de Morse pour la d''-cohomologie. Ann. Inst. Fourier (Grenoble) **35**, 189–229 (1985)
26. J.-P. Demailly, in *Singular Hermitian Metrics on Positive Line Bundles*, ed. by K. Hulek, T. Peternell, M. Schneider, F. Schreyer. Proceedings of the Conference on Complex Algebraic Varieties, Bayreuth, April 2–6, 1990. Lecture Notes in Mathematics, vol. 1507 (Springer, Berlin, 1992)
27. J.-P. Demailly, in *Holomorphic Morse Inequalities*. Lectures given at the AMS Summer Institute on Complex Analysis held in Santa Cruz, July 1989. Proceedings of Symposia in Pure Mathematics, vol. 52, Part 2 (1991), pp. 93–114
28. J.-P. Demailly, Regularization of closed positive currents and Intersection Theory. J. Algebr. Geom. **1**, 361–409 (1992)
29. J.-P. Demailly, A numerical criterion for very ample line bundles. J. Differ. Geom. **37**, 323–374 (1993)
30. J.-P. Demailly, in L^2 *Vanishing Theorems for Positive Line Bundles and Adjunction Theory*, ed. by F. Catanese, C. Ciliberto. Lecture Notes of the CIME Session "Transcendental methods

in Algebraic Geometry", Cetraro, Italy, July 1994. Lecture Notes in Mathematics, vol. 1646, pp. 1–97
31. J.-P. Demailly, in *Algebraic Criteria for Kobayashi Hyperbolic Projective Varieties and Jet Differentials*, ed. by J. Kollár, R. Lazarsfeld. AMS Summer School on Algebraic Geometry, Santa Cruz 1995. Proceedings of Symposia in Pure Mathematics, 76 p.
32. J.-P. Demailly, Variétés hyperboliques et équations différentielles algébriques. Gaz. Math. **73**, 3–23 (juillet 1997)
33. J.-P. Demailly, in *Multiplier Ideal Sheaves and Analytic Methods in Algebraic Geometry*. Lecture Notes, School on "Vanishing theorems and effective results in Algebraic Geometry, ICTP Trieste, April 2000 (Publications of ICTP, 2001)
34. J.-P. Demailly, Holomorphic Morse inequalities and asymptotic cohomology groups: a tribute to Bernhard Riemann. Milan J. Math. **78**, 265–277 (2010) [arXiv: math.CV/1003.5067]
35. J.-P. Demailly, A converse to the Andreotti-Grauert theorem. Ann. Faculté des Sciences de Toulouse, Volume spécial en l'honneur de Nguyen Thanh Van. **20**, 123–135 (2011)
36. J.-P. Demailly, J. El Goul, Hyperbolicity of generic surfaces of high degree in projective 3-space. Am. J. Math. **122**, 515–546 (2000)
37. J.-P. Demailly, L. Ein, R. Lazarsfeld, A subadditivity property of multiplier ideals math. AG/0002035; Michigan Math. J. (special volume in honor of William Fulton) **48**, 137–156 (2000)
38. J.-P. Demailly, J. Kollár, Semicontinuity of complex singularity exponents and Kähler-Einstein metrics on Fano orbifolds. Ann. Ec. Norm. Sup. **34**, 525–556 (2001) [math.AG/9910118]
39. S. Diverio, Differential equations on complex projective hypersurfaces of low dimension. Compos. Math. **144**, 920–932 (2008)
40. S. Diverio, Existence of global invariant jet differentials on projective hypersurfaces of high degree. Math. Ann. **344**, 293–315 (2009)
41. S. Diverio, J. Merker, E. Rousseau, Effective algebraic degeneracy. Invent. Math. **180**, 161–223 (2010)
42. J.-P. Demailly, M. Păun, Numerical characterization of the Kähler cone of a compact Kähler manifold. Ann. Math. **159**, 1247–1274 (2004) [arXiv: math.AG/0105176]
43. J.-P. Demailly, Th. Peternell, M. Schneider, Compact complex manifolds with numerically effective tangent bundles. J. Algebr. Geometry **3**, 295–345 (1994)
44. J.-P. Demailly, Th. Peternell, M. Schneider, Pseudo-effective line bundles on compact Kähler manifolds. Int. J. Math. **12**, 689–741 (2001)
45. S. Diverio, S. Trapani, A remark on the codimension of the Green-Griffiths locus of generic projective hypersurfaces of high degree. J. Reine Angew. Math. **649**, 55–61 (2010)
46. T. de Fernex, A. Küronya, R. Lazarsfeld, Higher cohomology of divisors on a projective variety. Math. Ann. **337**, 443–455 (2007)
47. T. Fujita, Approximating Zariski decomposition of big line bundles. Kodai Math. J. **17**, 1–3 (1994)
48. E. Getzler, Pseudodifferential operators on supermanifolds and the Atiyah-Singer index theorem. Commun. Math. Phys. **92**, 167–178 (1983)
49. E. Getzler, An analogue of Demailly's inequality for strictly pseudoconvex CR manifolds. J. Differ. Geom. **29**, 231–244 (1989)
50. H. Grauert, O. Riemenschneider, Verschwindungssätze für analytische Kohomologiegruppen auf komplexen Räumen. Invent. Math. **11**, 263–292 (1970)
51. M. Green, P. Griffiths, in *Two Applications of Algebraic Geometry to Entire Holomorphic Mappings*. The Chern Symposium 1979, Proceedings of the International Symposium, Berkeley, CA, 1979 (Springer, New York, 1980), pp. 41–74
52. P.A. Griffiths, in *Hermitian Differential Geometry, Chern Classes and Positive Vector Bundles*. Global Analysis, papers in honor of K. Kodaira (Princeton University Press, Princeton, 1969), pp. 181–251
53. R. Hartshorne, in *Ample Subvarieties of Algebraic Varieties*. Lecture Notes in Mathematics, vol. 156 (Springer, Berlin 1970)

54. H. Hess, R. Schrader, D.A. Uhlenbock, Kato's inequality and the spectral distribution of Laplacians on compact Riemannian manifolds. J. Differ. Geom. **15** (1980), 27–38
55. H. Hironaka, Resolution of singularities of an algebraic variety over a field of characteristic zero. Ann. Math. **79**, 109–326 (1964)
56. L. Hörmander, L^2 estimates and existence theorems for the $\bar{\partial}$ operator. Acta Math. **113**, 89–152 (1965)
57. A.G. Hovanski, Geometry of convex bodies and algebraic geometry. Uspehi Mat. Nau **34**(4), 160–161 (1979)
58. S. Ji, Inequality for distortion function of invertible sheaves on abelian varieties. Duke Math. J. **58**, 657–667 (1989)
59. S. Ji, B. Shiffman, Properties of compact complex manifolds carrying closed positive currents. J. Geom. Anal. **3**, 37–61 (1993)
60. G. Kempf, in *Metrics on Invertible Sheaves on Abelian Varieties*. Topics in Algebraic Geometry, Guanajuato, 1989. Aportaciones Mat. Notas Investigación, vol. 5 (Soc. Mat. Mexicana, México 1992), pp. 107–108
61. K. Kodaira, On a differential geometric method in the theory of analytic stacks. Proc. Nat. Acad. Sci. USA **39**, 1268–1273 (1953)
62. K. Kodaira, On Kähler varieties of restricted type. Ann. Math. **60**, 28–48 (1954)
63. A. Küronya, Asymptotic cohomological functions on projective varieties. Am. J. Math. **128**, 1475–1519 (2006)
64. L. Laeng, Estimations spectrales asymptotiques en géométrie hermitienne. Thèse de Doctorat de l'Université de Grenoble I, October 2002, http://www-fourier.ujf-grenoble.fr/THESE/ps/laeng.ps.gz and http://tel.archives-ouvertes.fr/tel-00002098/en/
65. S. Lang, Hyperbolic and diophantine analysis. Bull. Amer. Math. Soc. **14**, 159–205 (1986)
66. S. Lang, *Introduction to Complex Hyperbolic Spaces* (Springer, New York, 1987)
67. R. Lazarsfeld, in*Positivity in Algebraic Geometry I.-II*. Ergebnisse der Mathematik und ihrer Grenzgebiete, vols. 48–49 (Springer, Berlin, 2004)
68. P. Lelong, Intégration sur un ensemble analytique complexe. Bull. Soc. Math. Fr. **85**, 239–262 (1957)
69. P. Lelong, *Plurisubharmonic Functions and Positive Differential Forms* (Gordon and Breach, New York/Dunod, Paris, 1969)
70. L. Manivel, Un théorème de prolongement L^2 de sections holomorphes d'un fibré vectoriel. Math. Z. **212**, 107–122 (1993)
71. M. McQuillan, Diophantine approximation and foliations. Inst. Hautes Études Sci. Publ. Math. **87**, 121–174 (1998)
72. M. McQuillan, Holomorphic curves on hyperplane sections of 3-folds. Geom. Funct. Anal. **9**, 370–392 (1999)
73. J. Merker, Algebraic differential equations for entire holomorphic curves in projective hypersurfaces of general type, 89 p. arXiv:1005.0405
74. J. Milnor, in *Morse Theory*. Based on Lecture Notes by M. Spivak and R. Wells. Annals of Mathematics Studies, vol. 51 (Princeton University Press, Princeton, 1963) 153 pp.
75. A.M. Nadel, Multiplier ideal sheaves and Kähler-Einstein metrics of positive scalar curvature. Proc. Nat. Acad. Sci. USA **86**, 7299–7300, (1989); Ann. Math. **132**, 549–596 (1990)
76. S. Nakano, On complex analytic vector bundles. J. Math. Soc. Jpn. **7**, 1–12 (1955)
77. T. Ohsawa, On the extension of L^2 holomorphic functions, II. Publ. RIMS, Kyoto Univ. **24**, 265–275 (1988)
78. T. Ohsawa, K. Takegoshi, On the extension of L^2 holomorphic functions. Math. Z. **195**, 197–204 (1987)
79. D. Popovici, Regularization of currents with mass control and singular Morse inequalities. J. Differ. Geom. **80**, 281–326 (2008)
80. J.-P. Serre, Fonctions automorphes: quelques majorations dans le cas où X/G est compact. Sém. Cartan (1953–1954), 2-1 à 2-9
81. J.-P. Serre, Un théorème de dualité. Comment. Math. **29**, 9–26 (1955)

82. B. Shiffman, S. Zelditch, Asymptotics of almost holomorphic sections of ample line bundles on symplectic manifolds. J. Reine Angew. Math. **544**, 181–222 (2002)
83. C.L. Siegel, Meromorphic Funktionen auf kompakten Mannigfaltigkeiten. Nachrichten der Akademie der Wissenschaften in Göttingen. Math. Phys. Klasse **4**, 71–77 (1955)
84. Y.T. Siu, Analyticity of sets associated to Lelong numbers and the extension of closed positive currents. Invent. Math. **27**, 53–156 (1974)
85. Y.T. Siu, A vanishing theorem for semi-positive line bundles over non-Kähler manifolds. J. Differ. Geom. **19**, 431–452 (1984)
86. Y.T. Siu, in *Some Recent Results in Complex Manifold Theory Related to Vanishing Theorems for the Semi-positive Case*. Proceedings of the Math. Arbeitstagung 1984. Lecture Notes in Mathematics, vol. 1111 (Springer, Berlin, 1985), pp. 169–192
87. Y.T. Siu, Calculus inequalities derived from holomorphic Morse inequalities. Math. Ann. **286**, 549–558 (1990)
88. Y.T. Siu, An effective Matsusaka big theorem. Ann. Inst. Fourier **43**, 1387–1405 (1993)
89. Y.T. Siu, A proof of the general schwarz lemma using the logarithmic derivative lemma. Communication personnelle, avril 1997
90. Y.T. Siu, in *Some Recent Transcendental Techniques in Algebraic and Complex Geometry*. Proceedings of the International Congress of Mathematicians, vol. I, Beijing, 2002 (Higher Ed. Press, Beijing, 2002), pp. 439–448
91. Y.T. Siu, in *Hyperbolicity in Complex Geometry*. The Legacy of Niels Henrik Abel (Springer, Berlin, 2004), pp. 543–566
92. Y.T. Siu, S.K. Yeung, Hyperbolicity of the complement of a generic smooth curve of high degree in the complex projective plane. Invent. Math. **124**, 573–618 (1996)
93. Y.T. Siu, S.K. Yeung, Defects for ample divisors of Abelian varieties, Schwarz lemma and hyperbolic surfaces of low degree. Am. J. Math. **119**, 1139–1172 (1997)
94. H. Skoda, Sous-ensembles analytiques d'ordre fini ou infini dans $\mathbb{C}^n$. Bull. Soc. Math. Fr. **100**, 353–408 (1972)
95. H. Skoda, in *Estimations L^2 Pour L'opérateur $\bar{\partial}$ et Applications Arithmétiques*. Séminaire P. Lelong (Analyse), année 1975/1976. Lecture Notes in Mathematics, vol. 538 (Springer, Berlin, 1977), pp. 314–323
96. B. Teissier, Du théorème de l'index de Hodge aux inégalités isopérimétriques. C. R. Acad. Sc. Paris, sér. A **288**, 287–289 (1979)
97. B. Teissier, Bonnesen-type inequalities in algebraic geometry, in *Seminar on Differential Geometry*, ed. by S. T. Yau (Princeton University Press, Princeton, 1982), pp. 85–105
98. G. Tian, On a set of polarized Kähler metrics on algebraic manifolds. J. Differ. Geom. **32**, 99–130 (1990)
99. B. Totaro, Line bundles with partially vanishing cohomology, July 2010 [arXiv: math.AG/1007.3955]
100. S. Trapani, Numerical criteria for the positivity of the difference of ample divisors. Math. Z. **219**, 387–401 (1995)
101. C. Voisin, On a conjecture of Clemens on rational curves on hypersurfaces. J. Differ. Geom. **44**, 200–213 (1996); Correction: J. Differ. Geom. **49**, 601–611 (1998)
102. E. Witten, Supersymmetry and Morse theory. J. Differ. Geom. **17**, 661–692 (1982)

Pluripotential Theory and Monge–Ampère Foliations

G. Patrizio and A. Spiro

Abstract A regular, rank one solution u of the complex homogeneous Monge–Ampère equation $(\partial\bar{\partial}u)^n = 0$ on a complex manifold is associated with the Monge–Ampère foliation, given by the complex curves along which u is harmonic. Monge–Ampère foliations find many applications in complex geometry and the selection of a good candidate for the associated Monge–Ampère foliation is always the first step in the construction of well behaved solutions of the complex homogeneous Monge–Ampère equation. Here, after reviewing some basic notions on Monge–Ampère foliations, we concentrate on two main topics. We discuss the construction of (complete) modular data for a large family of complex manifolds, which carry regular pluricomplex Green functions. This class of manifolds naturally includes all smoothly bounded, strictly linearly convex domains and all smoothly bounded, strongly pseudoconvex circular domains of $\mathbb{C}^n$. We then report on the problem of defining pluricomplex Green functions in the almost complex setting, providing sufficient conditions on almost complex structures, which ensure existence of almost complex Green pluripotentials and equality between the notions of stationary disks and of Kobayashi extremal disks, and allow extensions of known results to the case of non integrable complex structures.

1 Introduction

Pluripotential theory might be considered as the analogue in several complex variables of the potential theory associated with the Laplace operator. Indeed, it

---

G. Patrizio (✉)
Dipartimento di Matematica "U.DIni", Università di Firenze, Italy
e-mail: patrizio@math.unifi.it

A. Spiro
Scuola di Scienze e Tecnologie, Università di Camerino, Italy
e-mail: andrea.spiro@unicam.it

can be regarded as the potential theory in higher dimensions associated with the complex homogeneous Monge–Ampère equation.

For a function u of class $\mathcal{C}^2$ on an open set U of a complex manifold, the *complex (homogeneous) Monge–Ampère equation* is the equation on u of the form

$$(dd^c u)^n = (2i\partial\bar{\partial}u)^n = (2i)^n \underbrace{\partial\bar{\partial}u \wedge \ldots \wedge \partial\bar{\partial}u}_{n \text{ times}} = 0, \tag{1}$$

which, in local coordinates, is equivalent to

$$\det(u_{i\bar{k}})dz^1 \wedge \ldots \wedge dz^n \wedge d\bar{z}^1 \wedge \ldots \wedge d\bar{z}^n = 0 \iff \det(u_{j\bar{k}}) = 0.$$

It is immediate to realize that, in complex dimension one, this equation reduces to the Laplace equation and it is well known that the Monge–Ampère operator may be meaningfully extended to much larger classes of functions. As (1) is invariant under biholomorphic maps, it is natural to expect that its solutions play a role of great importance in several complex variables as much as harmonic functions do in complex dimension one.

A distinctive feature of classical potential theory is the fact that harmonic functions, which are very regular, may be constructed maximizing families of (non regular) subharmonic functions. In fact, on one hand subharmonic functions are abundant and easy to construct since they do not need be very regular, on the other hand envelopes of suitable families of subharmonic functions are very regular and in fact harmonic. This construction scheme, systematized as *Perron method*, is based on maximum principle and it is both a basic tool and a key aspect of classical potential theory.

In higher dimension the peculiar role of subharmonic functions is played by the class of plurisubharmonic functions. As the suitable maximum principle holds also for the complex Monge–Ampère operator, Perron method has been successfully applied to construct solutions for the complex homogeneous Monge–Ampère equation satisfying boundary conditions. It turns out that the appropriate notion of maximality among plurisubharmonic functions is equivalent to be solution of (1) at least in a generalized sense. Here the analogy with the complex one dimensional case breaks down. The highly non linearity and the non ellipticity nature of equation (1) forces its solutions to be not regular even for very regular initial data. For instance, while positive results regarding the existence of solutions to the Dirichlet problem for (1) have been known since long time (see [6] for instance) it has been soon realized that it has at most $\mathcal{C}^{1,1}$ solutions even for the unit ball in $\mathbb{C}^n$ with real analytic datum on the boundary (see [4, 19]).

Potential theory in one complex variable plays a fundamental role in many areas of function theory, in particular, in the uniformization theory of Riemann surfaces. For instance one singles out hyperbolic surfaces by the existence of Green functions, which are bounded above harmonic functions with a logarithmic singularity at one point and may be constructed by Perron method. It is natural to try and repeat the scheme in higher dimension replacing subharmonic functions with plurisubharmonic functions, their natural counterpart in higher dimension in order to define a natural generalization of Green function: the *pluricomplex Green function*. For instance for a domain $D \subset \mathbb{C}^n$, the pluricomplex Green function with logarithmic pole $z_0 \in D$ is defined by

$$G_D(z_0, z) = \sup\{u(z) \mid u \in PSH(D), u < 0, \limsup_{z \to z_0}[u(z) - \log \|z - z_0\|] < +\infty\}.$$

This is in complete analogy with the definition of Green function in complex dimension one and in fact $G_D(z_0, z)$ satisfies (1) on $D \setminus \{z_0\}$ (in the weak sense). It is known that pluricomplex Green functions exist for any hyperconvex domain in $\mathbb{C}^n$ (see [17]). On the other hand pluricomplex Green function does not satisfy basic properties which one may expect (and desire) to hold. For instance, even for a real analytic bounded strongly pseudoconvex domain D, the pluricomplex Green function $G_D(z_0, z)$ need not be of class $\mathcal{C}^2$ on $D \setminus \{z_0\}$, in general it fails to be symmetric, i.e., $G_D(w, z) \neq G_D(z, w)$, and one cannot even expect that $G_D(z_0, z)$ is subharmonic in z_0 [3].

There are several layers of understanding for the lack of regularity of solutions of complex homogeneous Monge–Ampère equation. The first is the most obvious one: non regularity as "defect of differentiability" and this motivates the need of understanding the equation in the weak sense. A second aspect is that non regularity is coupled with—and in many cases caused by—an excess or, rather, non constancy of degeneracy. More precisely, for a function u being solution of (1) is equivalent to the degeneracy of the form $dd^c u$, which is the same to ask that $dd^c u$ has non trivial annihilator. In general the rank of the annihilator of $dd^c u$ for a solution of (1) need not be the smallest possible (i.e., one) or constant. In some sense, this is a geometric aspect of the non regular behavior of (1).

The existence of regular solutions for the complex homogeneous Monge–Ampère equation with the least possible degeneracy defines a reach geometry: There exists a foliation in complex curves of the domain of existence of the solution such that the restriction of the solution to one of the leaf is harmonic. This geometric byproduct of existence of well behaved solutions, known as *Monge–Ampère foliation*, was explicitly studied for the first time by Bedford and Kalka [5]. Starting with the work of Stoll [52] (see [13, 56] for alternative proofs), many of these ideas were exploited in questions of classification and characterization of special complex manifolds: See, for instance, [27, 39, 40, 57] for applications to the classification of circular domains and their generalizations and [14, 23, 32, 45, 47, 53] for the study of complexifications of Riemannian manifolds.

In fact, whenever such regular solutions exist, its construction starts with the determination of a suitable foliations. This is a well known fact, used also to provide examples of solutions with bad behaviors. In this regard, we mention the early example [2] and the most recent work by Lempert and Vivas [33] on the non-existence of regular geodesics joining points in the space of Kähler metrics (see also [10], in this volume).

On the other hand, in the seminal work of Lempert on convex domains [30], the existence of regular pluricomplex Green functions is related to a very good behavior of the Kobayashi metric for such domains, namely to the existence of a smooth foliation by Kobayashi extremal disks through any given point (see also for similar construction for pluri-Poisson kernels [11, 12]). In this case, the existence of a foliation by extremal disks is based on the equivalence between the notions of extremal and stationary disks, the latter being characterized as solution of manageable differential problem. Such link between Monge–Ampère equation, Kobayashi metric and stationary disks determines connections between many different problems and ways to approach them from various points of view.

Within this framework, we will report on two lines of research.

Following and simplifying ideas that go back to papers of Lempert for strictly convex domains [31] and of Bland–Duchamp for domains that are small deformations of the unit ball [7–9], we discuss the construction of (complete) modular data for a large family of complex manifolds, which carry regular pluricomplex Green functions. This class of manifolds naturally includes all smoothly bounded, strictly linearly convex domains and all smoothly bounded, strongly pseudoconvex circular domains of $\mathbb{C}^n$.

The modular data for this class of manifolds and, even more, the methods used, naturally suggest to ask similar questions for almost complex manifolds and to investigate the possibility of defining a useful notion of almost complex pluricomplex Green function. The generality of this setting poses new difficulties. The abundance of J-holomorphic curves, which is an advantage in many geometrical considerations, turns into a drawback when considering objects such as the Kobayashi metric. In particular, the notions of stationary and extremal disks are in general different [20]. As for the construction of Green pluripotentials, it is necessary to cope with the behavior of plurisubharmonic functions in the non-integrable case, which may be is rather unexpected even for arbitrarily small deformations of the standard complex structure. Finally, the kernel distribution of (the natural candidate of) almost-complex Monge–Ampère operator, even if appropriate non-degenericity conditions are assumed, in principle are neither integrable, nor J-invariant. All this is in clear contrast with the classical setting and hence it cannot be expected that, for completely arbitrary non-integrable structures, one can reproduce the whole pattern of fruitful properties relating regular solutions of complex Monge–Ampère equations, foliations in disks and Kobayashi metric.

Nevertheless it is possible to determine sufficient conditions on the almost complex structure, which ensure the existence of almost complex Green pluripotential and the equality between the two notions of stationary disks and of extremal disks. The class of such structures is very large in many regards, in fact determined by a finite set of conditions (it is finite-codimensional) in an infinite dimensional space.

2 Domains of Circular Type and Monge–Ampère Foliations

2.1 Circular Domains and Domains of Circular Type

Let $D \subset \mathbb{C}^n$ be a *complete circular domain*, i.e., such that $z \in D$ if and only if $\lambda z \in D$ for any $\lambda \in \mathbb{C}$ with $|\lambda| \leq 1$. For simplicity, let us assume that it is smoothly bounded and strictly pseudoconvex. It is well known that any such domain D is completely determined by its *Minkowski functional* μ_D, which is the real-valued function

$$\mu_D : \mathbb{C}^n \longrightarrow \mathbb{R}_{\geq 0}, \qquad \mu_D(z) = \begin{cases} 0 & \text{if } z = 0 \\ \frac{1}{t_z} & \text{if } z \neq 0, \end{cases}$$

where $t_z = \sup\{ t \in \mathbb{R} : tz \in D \}$. The square of the Minkowski functional

$$\rho_D : D \longrightarrow \mathbb{R}_{\geq 0}, \qquad \rho_D(z) = \mu_D(z)^2$$

is called *Monge–Ampère exhaustion of D* and satisfies some crucial properties. It can be considered as the modeling example of the Monge–Ampère exhaustions of the class of domains that we are going to analyze in the sequel.

One of the very first properties that can be inferred just from the definitions is the fact that ρ_D is always a map of the form

$$\rho_D(z) = G_D(z)\|z\|^2,$$

where $G_D : D \setminus \{0\} \longrightarrow \mathbb{R}$ is a bounded function of class $\mathcal{C}^\infty$, which is constant on each complex line through the origin and hence identifiable with a $\mathcal{C}^\infty$ map $G_D : \mathbb{C}P^{n-1} \longrightarrow \mathbb{R}$, defined on the complex projective space $\mathbb{C}P^{n-1}$. It turns out that the exhaustion ρ_D provides full biholomorphic data for the moduli space of complete circular domains in the sense described in the following theorem.

Theorem 1 ([46]). *Two bounded circular domains $D_1, D_2 \subset \mathbb{C}^n$ are biholomorphic if and only if their Monge–Ampère exhaustions are such that*

$$\rho_{D_1} = \rho_{D_2} \circ A$$

for some $A \in \mathrm{GL}_n(\mathbb{C})$. Moreover, if we denote by $\mathcal{D}$ the set of biholomorphic classes of smoothly bounded complete circular domains, $\mathcal{D}^+ \subset \mathcal{D}$ the subset of biholomorphic classes of strictly pseudoconvex domains, by ω_{FS} the Fubini-Study 2-form of $\mathbb{C}P^{n-1}$, then

$$\mathcal{D} \simeq [\omega_{FS}]/\operatorname{Aut}(\mathbb{C}P^{n-1}) \qquad \text{and} \qquad \mathcal{D}^+ \simeq [\omega_{FS}]^+/\operatorname{Aut}(\mathbb{C}P^{n-1}),$$

where we denote by $[\omega_{FS}]$ the cohomology class of ω_{FS} in $H^{1,1}(\mathbb{C}P^{n-1})$ and $[\omega_{FS}]^+ = \{$1-forms $\omega \in [\omega_{FS}]$ that are positive definite$\}$.

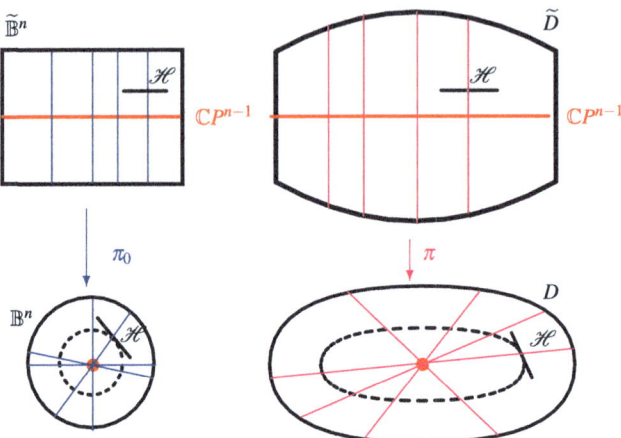

Fig. 1 Blowning up Unit Ball and Circular Domains

The description of the moduli spaces $\mathcal{D}$ and $\mathcal{D}^+$, given in this theorem, can be considered as a consequence of the following observation. Consider the unit ball $\mathbb{B}^n = \{ \|z\|^2 < 1 \}$ and a complete circular domain $D = \{ \rho_D(z) = \mu_D^2(z) < 1\}$ with Monge–Ampère exhaustion $\rho_D(z) = G_D(z)\|z\|^2$. Both domains are blow downs of diffeomorphic disk bundles over $\mathbb{CP}^{n-1}$ (Fig. 1), which have

- "Same" complex structure along fibers
- "Same" holomorphic bundle $\mathcal{H}$ normal to the fibers
- "Different" complex structures along $\mathcal{H}$

The differences between the complex structures on the normal holomorphic bundles $\mathcal{H}$ of $\mathbb{B}^n$ and D can be completely recovered from the map $G_D : \mathbb{CP}^{n-1} \longrightarrow \mathbb{R}$. This fact can be used to prove that the biholomorphic class of $[D]$ is completely determined (modulo actions of elements in $\mathrm{Aut}(\mathbb{CP}^{n-1})$) by the $(1,1)$-form $\omega = \omega_{FS} + \partial\bar\partial G_D$ and

$$[D] = [\mathbb{B}^n] \qquad \text{if and only if} \qquad \omega = \omega_{FS}$$

of course, up to actions of elements in $\mathrm{Aut}(\mathbb{CP}^{n-1})$.

Let us now consider the so-called *domains of circular type*, which are our main object of study for the first segment of these notes.

Before going into details, let us say a few words about notation. Since later we will have to deal with generalizations concerning almost complex manifolds it is useful to adopt notations that can be easily extended to the cases of non-integrable almost complex structures. With this purpose in mind, we recall that the familiar ∂- and $\bar\partial$-operators are related with the differential geometric operators d and $d^c = -J^* \circ d \circ J^*$ by the identities

$$d = \partial + \overline{\partial}, \qquad d^c = i(\overline{\partial} - \partial).$$

and that $dd^c = -d^c d$ and $dd^c u = 2i\partial\overline{\partial} u$ for any $\mathcal{C}^2$ function $u : \mathcal{U} \subset M \longrightarrow \mathbb{R}$.

Let us now begin introducing the notion of manifolds of circular type. Here we give a definition which is slightly different from the original one, but nonetheless equivalent, as it follows from the results in [40].

Definition 2 ([40]). A pair (M, τ), formed by a complex manifold (M, J) of dimension n and a real valued function $\tau : M \longrightarrow [0, 1)$, is called *(bounded) manifold of circular type with center x_o* if

(i) $\tau : M \longrightarrow [0, 1)$ is an exhaustion of M with $\{\tau = 0\} = \{x_o\}$ and satisfies the regularity conditions:
   (a) $\tau \in \mathcal{C}^0(M) \cap \mathcal{C}^\infty(\{\tau > 0\})$;
   (b) $\tau|_{\{\tau > 0\}}$ extends smoothly over the blow up $\tilde{M}$ at x_o of M.

(ii) $\begin{cases} 2i\partial\overline{\partial}\tau = dd^c \tau > 0, \\ 2i\partial\overline{\partial}\log\tau = dd^c \log\tau \geq 0, \\ (dd^c \log\tau)^n \equiv 0 \text{ (Monge–Ampère Equation)}; \end{cases}$

(iii) in some (hence *any*) system of complex coordinates $z = (z^i)$ centered at x_o, the function τ has a logarithmic singularity at x_o, i.e.

$$\log\tau(z) = \log\|z\| + O(1).$$

A *domain of circular type with center x_o* is a pair (D, τ), given a relatively compact domain $D \subset M$ of a complex manifold M with smooth boundary, and an exhaustion $\tau : D \longrightarrow [0, 1]$ smooth up to the boundary, such that (D, τ) is a manifold of circular type, i.e., satisfying the above conditions (i), (ii), (iii).

The simplest example of a domain of circular type is the unit ball $B^n \subset \mathbb{C}^n$, endowed with the standard exhaustion $\tau_o(z) = \|z\|^2$. In fact, *any* pair (D, ρ_D), formed by a strictly pseudoconvex, smoothly bounded, complete circular domain $D \subset \mathbb{C}^n$ and its Monge–Ampère exhaustion ρ_D, is a domain of circular type. To see this, it is enough to observe that ρ_D is strictly plurisubharmonic and that $\log \rho_D$ is plurisubharmonic, with harmonic restrictions on each (punctured) disk through the origin. The conditions (i) and (iii) are easily seen to be satisfied.

A much larger and interesting class of examples is given by the strictly (linearly) convex domains, whose properties, determined by the seminal work of Lempert, can be summarized as follows.

Theorem 3 ([30]). *Let $D \subset\subset \mathbb{C}^n$ be a smooth, bounded strictly (linearly) convex domain and denote by δ_D its Kobayashi distance and, for any given $x_o \in D$, by δ_{x_o} the function $\delta_{x_o} = \delta_D(x_o, \cdot)$. Then*

- $\delta_D \in \mathcal{C}^\infty(D \times D \setminus \mathrm{Diag})$, *where we denoted by* $\mathrm{Diag} = \{(z, z) \mid z \in D\}$;

- the function $u = 2\log(\tanh \delta_{x_o})$ is in $C^\infty(\overline{D} \setminus \{x_o\})$ and it is the unique solution of the problem

$$\begin{cases} \det(u_{\mu\bar{\nu}}) = 0 & \text{on } D \setminus \{x_o\}, \\ u_{|\partial D} = 0 & \text{and } u(z) = 2\log \|z - x_o\| + O(1) \quad \text{near } D \setminus \{x_o\}. \end{cases}$$

In fact, the pair (D, τ), with $\tau = (\tanh \delta_{x_o})^2$, is a domain of circular type with center x_0.

This theorem is indeed the result of a deep proof of geometric nature, which can be outlined as follows. Let x_o be a point of a smoothly bounded, strictly (linearly) convex domain $D \subset M$ and v a tangent vector in $T_{x_o}D$ of unit length w.r.t. to the infinitesimal Kobayashi metric κ_D. Lempert proved that there exists a unique complex geodesic

$$f_v : \Delta \longrightarrow D, \qquad \Delta = \{ |\zeta| < 1 \} \subset \mathbb{C},$$

(i.e., a holomorphic map which is also an isometry between Δ, with its standard hyperbolic metric, and $\Delta^{(v)} = f_v(\Delta) \subset D$, endowed with metric induced by the Kobayashi metric of D) such that

$$f_v(0) = x_o, \qquad f_v'(0) = v.$$

He also shows that the complex geodesic f_v depends smoothly on the vector v and that the images of the punctured disk $\Delta \setminus \{0\}$,

$$f_v(\Delta \setminus \{0\}), \qquad v \in T_{x_o}M,$$

determine (Fig. 2) a smooth foliation of the punctured domain $D \setminus \{x_o\}$.

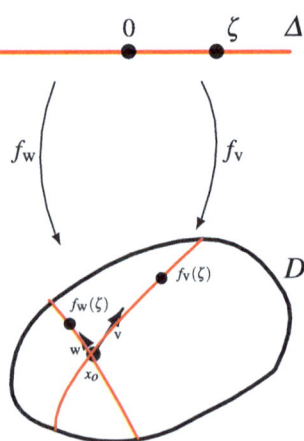

Fig. 2 Complex geodesics of strictly convex domains

Using this and the fact that a holomorphic disc f_v is an isometry between $(\Delta \setminus \{0\}, \delta_\Delta)$ and $(f_v(\Delta \setminus \{0\}), \delta_D) \subset (D, \delta_D)$, one gets that the map

$$u = 2\log(\tanh \delta_{x_o}) : \overline{D} \setminus \{x_o\} \longrightarrow \mathbb{R}$$

satisfies the equality

$$u(f_v(\zeta)) = \log|\zeta|$$

for any complex geodesic f_v. In particular, $\tau(f_v) = |\zeta|^2$. From these information, all other claims of the statement can be derived.

We remark that Lempert's result on existence of the foliation by extremal disks through a given point is based on the equivalence (for strictly convex domains) between the notions of extremal disks and of *stationary disks*, these being precisely the disks that realize the stationarity condition for the appropriate functional on holomorphic disks. We will come back to this point later on.

The problem of determining moduli for (pointed) strictly convex domains was addressed—and to a large extent solved—by Lempert, Bland and Duchamp in [7–9, 31].

The results in [31] can be summarized as follows. In that paper, it is proved that, for a given strictly convex domain $D \subset \mathbb{C}^n$ with a distinguished point $x_o \in D$ and for any given Kobayashi extremal disk $f_v : \overline{\Delta} \longrightarrow \overline{D}$ with $f_v(0) = x_o$ and $f_v'(0) = v$, there exists a special set of coordinates, defined on a neighborhood $\mathcal{U}$ of $f_v(\partial\Delta) \subset \partial D$, in which the boundary ∂D admits a defining function $r : \mathcal{U} \longrightarrow \mathbb{R}$ of a special kind, called "normal form". The lower order terms of such defining functions in normal form (which can be considered as functions of the vectors $v = f_v'(0) \in T_{x_o}^{10}D$) determine biholomorphic invariants, which completely characterize the pointed domain (D, x_o) up to biholomorphic equivalences.

Bland and Duchamp's approach is quite different. Roughly speaking, they succeeded in constructing a complete class of invariants for any pointed, strictly convex domain (D, x_o) (and also for any pointed domain which is a sufficiently small deformation of the unit ball) using the Kobayashi indicatrix at x_o and a suitable "deformation tensor", defined on the holomorphic tangent spaces, which are normal to the extremal disks through x_o.

Lempert, Bland and Duchamp's results provide an excellent description of the moduli space of strictly convex domains, but they also motivate the following problems.

Problem 1. The moduli space of pointed convex domains appears to be naturally sitting inside a larger space. *Find the "right" family of domains corresponding to such larger space.*

Problem 2. The singular foliation of a circular domain by its stationary disks through the origin is very similar to the singular foliation of a strictly convex domain by stationary disks through a fixed point x_o. But there are also some crucial differences between such two situation: In the latter case, *any* point is center of a singular foliation, while in the former there apparently is only one natural choice for the center, the origin. *Determine an appropriate framework for "understanding"* the possible differences between the "sets of centers" of the domains admitting (singular) stationary foliations.

In the following two sections, we are going to discuss a simplification and a generalization of Bland and Duchamp's invariants. The results that we will give may be roughly described as follows:

(a) *The manifolds of circular type determine a moduli space, which naturally includes the moduli of strictly convex domains and on which Bland and Duchamp's invariants are in bijective correspondence.*
(b) *This new construction of Bland and Duchamp's invariants determine a new setting, in which the sets of special points can be studied in a systematic way.*

2.2 Homogeneous Complex Monge–Ampère Equation and Monge–Ampère Foliations

Let M be a complex manifold of dimension n and $u : M \longrightarrow \mathbb{R}$ be a function of class C^∞. In complex coordinates $(z^1, \ldots, z^n)$, we have that $dd^c u = 2i \partial \bar{\partial} u = 2i \sum u_{j\bar{k}} dz^j \wedge d\bar{z}^k$ and the plurisubharmonicity of a function u is equivalent to require that $dd^c u = 2i \partial \bar{\partial} u \geq 0$ or, in local coordinates, that $(u_{j\bar{k}}) \geq 0$.

The *complex (homogeneous) Monge–Ampère equation* is the equation on u of the form
$$(dd^c u)^n = (2i \partial \bar{\partial} u)^n = (2i)^n \underbrace{\partial \bar{\partial} u \wedge \ldots \wedge \partial \bar{\partial} u}_{n \text{ times}} = 0,$$

i.e., in local coordinates,
$$\det(u_{j\bar{k}}) dz^1 \wedge \ldots \wedge dz^n \wedge d\bar{z}^1 \wedge \ldots \wedge d\bar{z}^n = 0 \quad \left(\Longleftrightarrow \quad \det(u_{j\bar{k}}) = 0 \right).$$

Assume now that:

(a) u is a smooth solution of a Monge–Ampère equation (that is $(dd^c u)^n = 0$).
(b) u is plurisubharmonic (that is $dd^c u \geq 0$).
(c) $\tau = e^u$ is *strictly* plurisubharmonic (that is $dd^c \tau > 0$).

We claim that from (a), (b), (c), it follows also that

$$(dd^c u)^{n-1} \neq 0, \tag{2}$$

i.e., that *the rank of $dd^c u$ is exactly $n - 1$ at all points*. In fact, at every point p, the 2-form $dd^c u|_p$ is positive along directions in the holomorphic tangent space to level sets of u through p and $dd^c u|_p$ has exactly $n - 1$ *positive eigenvalues and only one zero eigenvalue*.

Since it is useful for future developments, we give here some details of the proof. First of all, we observe that, by definitions,

$$e^{2u} dd^c u = \tau^2 dd^c \log \tau = \tau dd^c \tau - d\tau \wedge d^c \tau \tag{3}$$

and hence

$$(e^{2u})^n (dd^c u)^n = \tau^{2n}(dd^c \log \tau)^n = \tau^n(dd^c \tau)^n - n\tau^{n-1}(dd^c \tau)^{n-1} \wedge d\tau \wedge d^c\tau.$$

This implies that

$$(dd^c u)^n = 0 \quad \text{if and only if} \quad \tau(dd^c \tau)^n = n(dd^c \tau)^{n-1} \wedge d\tau \wedge d^c\tau$$

or, equivalently,

$$\det(u_{i\bar{k}}) = 0 \quad \text{if and only if} \quad \tau = -\sum_{\nu,\mu} \tau_{\bar{\nu}} \tau^{\bar{\nu}\mu} \tau_\mu, \quad (\tau^{\bar{\nu}\mu}) \stackrel{\text{def}}{=} (\tau_{\bar{\rho}\sigma})^{-1}. \quad (4)$$

Assume now that u satisfies the Monge–Ampère equation and consider the vector field Z in $T^{1,0}(M \setminus \{x_o\})$, determined by the condition

$$-\frac{i}{2} dd^c \tau(Z, \cdot) = \partial\bar{\partial}\tau(Z, \cdot) = \bar{\partial}\tau. \quad (5)$$

Such vector field necessarily exists and is unique, because by assumptions the 2-form $dd^c\tau$ is non-degenerate (in fact, a Kähler metric). In coordinates, the vector fields Z is of the form

$$Z = \sum_{\mu,\nu} \tau_{\bar{\nu}} \tau^{\bar{\nu}\mu} \frac{\partial}{\partial z^\mu}. \quad (6)$$

From (6), (5) and (4), it follows that

$$\partial\bar{\partial}\tau(Z, \overline{Z}) = \bar{\partial}\tau(\overline{Z}) = \tau \quad (\text{ and hence also } \partial\tau(Z) = \tau). \quad (7)$$

Moreover, decomposing an arbitrary $(1,0)$-vector field V into a sum of the form

$$V = \lambda Z + W,$$

where W is a $(1,0)$-vector field W tangent to the level sets $\{\tau = \text{const.}\}$ (and hence such that $d\tau(W) = 0$), using (3) and (7), we get

$$e^{2u} dd^c u(V, \overline{V}) = |\lambda|^2 \tau dd^c \tau(Z, \overline{Z}) + 2\operatorname{Re}(\lambda \tau dd^c \tau(Z, \overline{W})) + \tau dd^c \tau(W, \overline{W})$$
$$- (d\tau \wedge d^c\tau)(\lambda Z + W, \overline{\lambda}\overline{Z} + \overline{W})$$
$$= 2i|\lambda|^2 \tau^2 + 2\operatorname{Re}(\tau d\tau(\overline{W})) + \tau dd^c \tau(W, \overline{W}) - 2i|\lambda|^2 \tau^2$$
$$= \tau dd^c \tau(W, \overline{W}) \geq 0,$$

because the level sets of u coincide with the level sets of τ and these are strictly pseudoconvex. It follows that $dd^c u \geq 0$ is positively semi-definite and $\operatorname{Ann} dd^c u = \mathbb{C} Z$ on $M \setminus \{x_o\}$.

From these observations, we have that, for any $u : M \longrightarrow \mathbb{R}$ satisfying the above conditions (a), (b) and (c), the family of complex lines

$$\mathcal{Z} = \{ \mathcal{Z}_x \subset T_x M \, , \, x \neq x_o \, : \, \mathcal{Z}_x \text{ is the kernel of } dd^c u|_x \} \tag{8}$$

is actually a complex distribution of rank 1 with the following crucial properties:

- It is integrable (in fact, it coincides with $\operatorname{Ann} dd^c u$ and $dd^c u$ is a closed 2-form).
- Its integral leaves are holomorphic curves (in fact, $dd^c u$ is a $(1, 1)$-form).

The foliation $\mathcal{F}$ of the integral leaves of $\mathcal{Z}$ is called *Monge–Ampère foliation associated with u* (or $\tau = e^u$).

We point out that there exists a very simple criterion for determining whether a holomorphic curve is part of a leaf of $\mathcal{F}$: It suffices to observe that *the image $L(\Delta)$ of a holomorphic curve $L : \Delta \longrightarrow M$ is contained in an integral leaf of $\mathcal{Z}$ if and only if $u \circ L : \Delta \longrightarrow \mathbb{R}$ is a harmonic function*.

Various properties of the domains of circular type follows from the above conditions (a), (b), (c). We summarize them in the next theorem and we refer to [40, 41] for details and proofs.

Theorem 4. *Let (M, τ) be a manifold of circular type with center x_o and denote by $\pi_{x_o} : \tilde{M} \longrightarrow M$ and $\pi_0 : \tilde{\mathbb{B}}^n \longrightarrow \mathbb{B}^n$ the blow-ups of M and of the unit ball $\mathbb{B}^n \subset \mathbb{C}^n$ at x_o and 0, respectively.*

There exists a diffeomorphism $\Psi : \tilde{\mathbb{B}}^n \longrightarrow \tilde{M}$ such that, for any $v \in S^{2n-1} \subset T_0 \mathbb{C}^n$, the map

$$f_v : \Delta \longrightarrow M \, , \qquad f_v(\zeta) = \pi_{x_o} \left(\Psi(\pi_0^{-1}(\zeta v)) \right)$$

is such that

(a) *It is proper, one-to-one and holomorphic.*
(b) *Its image $f_v(\Delta)$ is (the closure of) a leaf of the Monge–Ampère foliation on $M \setminus \{x_o\}$ determined by τ.*
(c) *It is the unique complex geodesic for the Kobayashi metric of M, passing through x_o and tangent to the vector $v \in T_{x_o} M \simeq \mathbb{C}^n$.*

Moreover, the map Ψ satisfies the additional property

$$(\tau \circ \Psi)|_{\tilde{\mathbb{B}}^n \setminus \pi_0^{-1}(0)} = \| \cdot \|^2 \qquad (\| \cdot \| = \text{Euclidean norm of } \mathbb{C}^n) .$$

In Fig. 3, we try to schematize the properties of the map $\Psi : \tilde{\mathbb{B}}^n \longrightarrow \tilde{M}$ described in the above theorem

Rather than provide a complete argument, we sketch here the circle of ideas underlying these results. Under the assumptions of the theorem, there exist complex coordinates on a neighborhood $\mathcal{U} \subset M$ of x_o and centered at $x_o \simeq 0_{\mathbb{C}^n}$, in which τ assumes of the form

$$\tau(z) = h(z)\|z\|^2 + o(\|z\|^2) \tag{9}$$

for some $h : S^{2n-1} \subset T_{x_o} M \simeq \mathbb{C}^n \longrightarrow \mathbb{R}_*$ of class $\mathcal{C}^\infty$ and such that

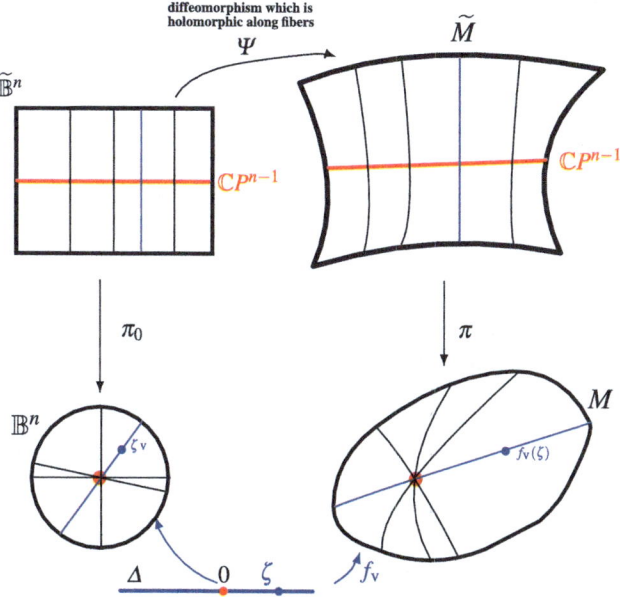

Fig. 3 The map Ψ

$$h(\lambda z) = h(z) \quad \text{for any } \lambda \in \mathbb{C} \text{ with } |\lambda| = 1.$$

Given the complex gradient $Z = \sum_{\mu,\nu} \tau_{\bar{\nu}} \tau^{\bar{\nu}\mu} \frac{\partial}{\partial z^\mu}$ of τ on $\mathcal{U} \setminus \{0\}$, consider the real vector fields

$$Y = \frac{1}{\sqrt{\tau}} \left(Z + \overline{Z} \right), \qquad W = i \left(Z - \overline{Z} \right),$$

which determine the real and imaginary parts of the flow of Z. By construction, Y and W are generators over $\mathbb{R}$ for the distribution (8), which we know it is integrable. Actually, using the Monge–Ampère equation satisfied by $u = \log \tau$, one can directly check that $[Y, W] = 0$.

Now, using (9) and once again the Monge–Ampère equation, one can show that near x_o the vector field Z is of the form

$$Z = \sum_{\mu,\nu} [z^\mu + G^\mu(z)] \frac{\partial}{\partial z^\mu} \quad \text{for some } G^\mu(z) = O(\|z\|^2), \tag{10}$$

so that the vector field $\tilde{Z}$ on $(\Delta_\varepsilon \setminus \{0\}) \times (\mathbb{B}_r^n \setminus \{0\})$ with $\Delta_\varepsilon = \{\zeta \in \mathbb{C} : |\zeta| < \varepsilon\}$, for a fixed $r > 1$ and $\varepsilon > 0$ sufficiently small, defined by

$$\tilde{Z}(\lambda, z) = \sum_{\mu,\nu} \tau_{\bar{\nu}}(\lambda z) \tau^{\bar{\nu}\mu}(\lambda z) \frac{\partial}{\partial z^\mu},$$

extends in fact of class $\mathcal{C}^\infty$ on the whole open set $\tilde{\mathcal{V}} = \Delta_\varepsilon \times \mathbb{B}_r^n$.

From previous remarks, the $\mathcal{C}^\infty$ vector fields

$$\tilde{Y} = \frac{1}{\sqrt{\tau}}\left(\tilde{Z} + \overline{\tilde{Z}}\right), \qquad \tilde{W} = i\left(\tilde{Z} - \overline{\tilde{Z}}\right)$$

satisfy

$$[\tilde{Y}, \tilde{W}] = 0$$

at all points of

$$\tilde{\mathcal{V}} \setminus \left(\{0\} \times \mathbb{B}_r^n \cup \Delta_\varepsilon \times \{0_{\mathbb{C}^n}\}\right)$$

and hence, by continuity, on the entire $\tilde{\mathcal{V}}$.

Due to this, one can integrate such vector fields and, for any $v \in \mathbb{B}_r^n$, construct a holomorphic map $\tilde{f}_v : \Delta_\varepsilon \longrightarrow \tilde{\mathcal{V}}$ with $\tilde{f}_v(0) = (0, v)$ and such that

$$\tilde{f}_{v*}\left(\frac{\partial}{\partial x}\bigg|_\zeta\right) = \tilde{Y}\big|_{\tilde{f}_v(\zeta)}$$

for any $\zeta \in \Delta_\varepsilon$. The collection of such holomorphic maps determine a map of class $\mathcal{C}^\infty$

$$\tilde{F} : \Delta_\varepsilon \times S^{2n-1} \longrightarrow \tilde{\mathcal{V}}, \qquad \tilde{F}(\zeta, v) = \tilde{f}_v(\zeta).$$

By restriction on $\Delta_\varepsilon \times S^{2n-1}$, where $S^{2n-1} = \{z \in \mathbb{C}^n : \|z\| = 1\} \subset \mathbb{B}_r^n$, and composing with the natural projection onto the blow up of $\mathcal{U} \simeq \mathbb{B}_r^n$,

$$\tilde{\pi} : \tilde{\mathcal{V}} = \Delta_\varepsilon \times \mathbb{B}_r^n \longrightarrow \tilde{\mathbb{B}}_r^n \simeq \tilde{\mathcal{U}}, \qquad \tilde{\pi}(\zeta, v) = ([v], \zeta v),$$

we get a smooth map

$$F : \Delta_\varepsilon \times S^{2n-1} \longrightarrow \mathcal{U} \subset M, \qquad F = \pi \circ \tilde{F}\big|_{\Delta_\varepsilon \times S^{2n-1}}.$$

This map extends uniquely to a smooth map $F : \Delta \times S^{2n-1} \longrightarrow M$ onto the whole complex manifold M, with the following properties:

- $\tau(F(\zeta, v)) = |\zeta|^2$ for any $\zeta \neq 0$, so that $F(0, v)$ coincides with the corresponding point $([v], \zeta v)$ of the singular set $E = \pi_0^{-1}(x_o) \simeq \mathbb{C}P^{n-1}$ for any $v \in S^{2n-1}$;
- $F(\lambda \zeta, v) = F(\zeta, \lambda v)$ for any λ with $|\lambda| = 1$;
- $f_v = F(\cdot, v) : \Delta \longrightarrow \tilde{M}$ is a biholomorphism between Δ and a (closure of) a leaf of the Monge–Ampère foliation determined by $u = \log \tau$ on $M \setminus \{x_o\}$;
- $\tilde{Z}\big|_{F(\zeta, v)} = \zeta f'_v(\zeta)$ for any $v \in S^{2n-1}$;
- $f'_v(0) = \sqrt{h(v)}\, v$, where h denotes the function in (9).

The map of Theorem 4 is precisely the map $\Psi : \tilde{\mathbb{B}}^n \longrightarrow \tilde{M}$ defined by

$$\Psi\big|_E = \mathrm{Id}_E \qquad \text{and} \qquad \Psi([v], \zeta v) = F(\zeta, v) = f_v(\zeta)$$

for any $([v], \zeta v) \in \tilde{\mathbb{B}}^n \setminus E$ with $v \in S^{2n-1}$. Using the above construction, one can directly check that Ψ is smooth.

It also turns out that the subset of $T_{x_o} M$ defined by

$$I_{x_o} = \{ v \in T_{x_o} M \simeq \mathbb{C}^n : h(v) \| v \|^2 < 1 \}$$

coincides with the *Kobayashi indicatrix of M at x_o*. In fact, this is a consequence of the following proposition [41].

Proposition 5. *For any $v \in S^{2n-1} \subset \mathbb{C}^n \simeq T_{x_o} M$, the holomorphic disk $f_v : \Delta \longrightarrow M$ is a Kobayashi extremal disk of M in the direction of v. This extremal disk is unique.*

Proof. To prove the first claim, we need to show that for any holomorphic map $g : \Delta \longrightarrow M$ with $g(0) = x_o$ and $g'(0) = t v$ for some $t \in \mathbb{R}_{>0}$, we have that

$$t = \| g'(0) \| \leq \| f'_v(0) \| = \sqrt{h(v)} .$$

For any such disk, consider the function

$$\ell : \Delta \longrightarrow \mathbb{R}, \qquad \ell(\zeta) = \log \tau(g(\zeta)) .$$

It is subharmonic with $|\ell(\zeta)| \leq 0$ at all points and

$$|\ell(\zeta) - \log(|\zeta|)^2| = o(|z|) .$$

This means that $\log(|\zeta|^2)$ is a harmonic majorant for $\ell(\zeta)$ and that

$$\tau(g(\zeta)) \leq |\zeta|^2 = \tau(f_v(\zeta)) . \tag{11}$$

One can also check that the map

$$\tilde{\tau} = \tau \circ \pi_{x_o} : \tilde{M} \setminus \pi_{x_o}^{-1}(x_o) \longrightarrow \mathbb{R}$$

extends smoothly to a function $\tilde{\tau} : \tilde{M} \longrightarrow \mathbb{R}$ defined over the whole blow up $\tilde{M}$ and that *the limit of $\tau_{\mu \bar{\nu}}(g(\zeta))$, for ζ tending to 0, exists and its value*

$$\lim_{\zeta \to 0} \tau(g(\zeta))_{\zeta \bar{\zeta}} = \tilde{\tau}_{\mu \bar{\nu}}([g'(0)], 0) v^\mu \overline{v}^\nu = \tilde{\tau}_{\mu \bar{\nu}}([v], 0) v^\mu \overline{v}^\nu$$

depends only on the element $[v] \in E = \mathbb{CP}^{n-1}$.

From (11), we have that $\tau(g(\zeta)) = r(\zeta) |\zeta|^2$ for some smooth function $0 \leq r \leq 1$ and that

$$\tau(g(\zeta))_{\zeta \bar{\zeta}} = \tau_{\mu \bar{\nu}}(g(\zeta)) g'^\mu(\zeta) \overline{g'^\nu(\zeta)} = r_{\zeta \bar{\zeta}}(\zeta) |\zeta|^2 + r_\zeta(\zeta) \zeta + r_{\bar{\zeta}}(\zeta) \bar{\zeta} + r(\zeta) .$$

All this implies that

$$t^2 \tilde{\tau}_{\mu\bar{\nu}}([v], 0) v^\mu \bar{v}^\nu = \lim_{\zeta \to 0} \tau_{\mu\bar{\nu}}(g(\zeta)) g'^\mu(\zeta) \overline{g'^\nu(\zeta)} = r(0) \leq 1.$$

On the other hand, recalling that $\tau(f_v(\zeta)) = |\zeta|^2$ and that $f'_v(0) = \sqrt{h(v)}\,v$, we have

$$1 = \tau(f_v(0))_{\zeta\bar\zeta} = \lim_{\zeta \to 0} \tau_{\mu\bar\nu}(f_v(\zeta)) f_v'^\mu(\zeta) \overline{f_v'^\nu(\zeta)} = h(v)\tilde{\tau}_{\mu\bar\nu}([v], 0) v^\mu \bar{v}^\nu,$$

from which it follows that

$$\tilde{\tau}_{\mu\bar\nu}([v], 0) v^\mu \bar{v}^\nu = \frac{1}{h(v)} = \frac{1}{\|f'_v(0)\|^2} \quad \text{so that} \quad \|g'(0)\| = t \leq \|f'_v(0)\|.$$

It now remains to check the uniqueness of such extremal disk. First of all, notice that if $g : \Delta \longrightarrow M$ is a holomorphic disk with $g(0) = x_o$ and $g'(0) = v = f'_v(0)$, and $r(\zeta) \leq 1$ is the function defined above, such that $\tau(g(\zeta)) = r(\zeta)|\zeta|^2$, then $r(0) = 1$. Moreover:

(a) The function $\log r$ is subharmonic. In fact

$$\Delta \log r = \Delta \log \tau \circ g - \Delta \log |\zeta|^2 = \Delta \log \tau \circ g \geq 0.$$

(b) The function $\log r$ is always less or equal to 0.

By Maximum Principle, conditions (a), (b) and the equality $r(0) = 1$ imply that $r(\zeta) = 1$ and $\tau(g(\zeta)) = |\zeta|^2$ at all points. A little additional computation shows that $g(\Delta)$ is necessarily included in a leaf of the Monge–Ampère foliation determined by $u = \log \tau$. Using the fact that $\log \tau|_{g(\Delta)}$ is harmonic, one concludes that there is only one possibility for g, namely $g = f_v$. □

This proposition concludes our outline of the ideas behind the proof of Theorem 4 and the various properties of the diffeomorphism $\Psi : \tilde{\mathbb{B}}^n \longrightarrow \tilde{M}$. There is however another very important information on the diffeomorphism Ψ which we want to point out.

Assume that M is a smoothly bounded, relatively compact domain in a larger complex manifold N and that the exhaustion τ extends smoothly up to the boundary ∂M in such a way that $dd^c \tau > 0$ also at the boundary points of M. In this case, one can check that the map Ψ extends smoothly up to the closures of the blow-ups

$$\Psi : \overline{\tilde{\mathbb{B}}^n} \longrightarrow \overline{\tilde{M}}$$

and, consequently, all holomorphic disk $f_v : \Delta \longrightarrow M$ extend to smooth maps $f_v : \overline{\Delta} \longrightarrow \overline{M}$. This extendability property will turn out to be important for the construction of the normal forms and Bland and Duchamp's invariants that we are going to present in the next section.

Furthermore, the fact that such extremal disks are smoothly attached to the boundary determine a crucial relation between these disks and the geometry of the boundary. It turns out that *for any* $v \in S^{2n-1} \subset T_{x_o}M$, *the corresponding holomorphic disk* $f_v : \overline{\Delta} \longrightarrow \overline{M}$ *is stationary*, i.e., (see [30] for the original definition) there exists a holomorphic map

$$\hat{f}_v : \overline{\Delta} \longrightarrow T^*M$$

such that

(i) $\hat{\pi} \circ \hat{f}_v(\zeta) = f_v(\zeta)$ for any $\zeta \in \Delta$, where $\hat{\pi} : T^*M \longrightarrow M$ is the standard projection.
(ii) For any $\zeta \in \partial \Delta$, the 1-form $\zeta^{-1} \cdot \hat{f}_v(\zeta) \in (T^*_{f_v(\zeta)}M)^{10}$ is non-zero and belongs to the conormal bundle of ∂M (i.e., vanishes on the tangent of ∂M).

In fact, since the restrictions of u on the leaves of the Monge–Ampère equation are harmonic, one can immediately check that, for any given disk $f_v : \overline{\Delta} \longrightarrow M$, the required map $\tilde{f}_v : \overline{\Delta} \longrightarrow (T^*_{f_v(\zeta)}M)^{10}$ is given by

$$\tilde{f}_v(\zeta) = \zeta \, \partial u|_{f_v(\zeta)} .$$

We conclude this section with the following result by Stoll ([52]; see also the alternative proofs in [13, 56]), which was essentially one of the starting points of the geometrical applications of the theory of Monge–Ampère foliation theory.

Theorem 6 ([52]). *Let M be a complex manifold complex manifold of dimension n. Then there exists a $\mathcal{C}^\infty$ exhaustion $\tau \colon M \to [0, 1)$ such that*

(1) $dd^c \tau > 0$ on M (2) $(dd^c \log \tau)^n = 0$ on $M \setminus \{\tau = 0\}$

if and only if there exists a biholomorphic map $F \colon M \to \mathbb{B}^n = \{Z \in \mathbb{C}^n \mid \|Z\|^2 < 1\}$ with $\tau(F(z)) = \|z\|^2$.

Two key remarks allow to prove Stoll's theorem using Theorem 4:

(i) The minimal set $\{\tau = 0\}$ of τ reduces to a singleton $\{x_o\}$ so that (M, τ) is a manifold of circular type with center x_o.
(ii) The map $\Phi : \mathbb{B}^n \longrightarrow M$, defined by requiring that the following diagram commutes (here π_0, π_{x_o} are blow down maps and $\Psi : \tilde{\mathbb{B}}^n \longrightarrow \tilde{M}$ is the map given in Theorem 4),

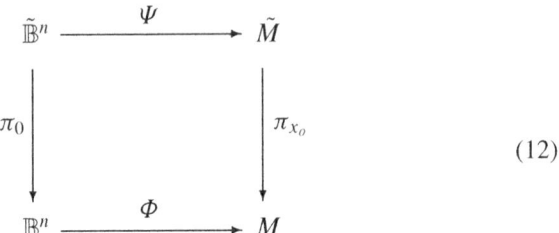 (12)

is a smooth diffeomorphism (even at the origin!).

To prove (i), one has first to observe that, as consequence of a result of Harvey and Wells [25], the level set $\{\tau = 0\}$ is totally real, compact and discrete (and hence finite). Then the conclusion follows by an argument of Morse theory: M is connected and retracts onto $\{\tau = 0\}$ along the flow of the vector field Y, which turns out to be the gradient of $\sqrt{\tau}$ with respect to the Kähler metric g determined by the Kähler form $dd^c \tau$.

Property (ii) follows from the fact that it is possible to show that Φ is a reparametrization of the exponential map at x_o of the metric g and it is therefore smooth at x_o.

By a classical result of Hartogs on series of homogeneous polynomials, from the fact that Φ is smooth and holomorphic along each disk through the origin, it follows that Φ is holomorphic. The fact that $\tau(\Phi(z)) = \|z\|^2$ is a consequence of Theorem 4 and the commutativity of the diagram (12).

3 Normal Forms and Deformations of CR Structures

3.1 *The Normal Forms of Domains of Circular Type*

In this section we constantly use the following notation:

- $\mathbb{B}^n$ is the unit ball of $\mathbb{C}^n$, centered at the origin.
- J_{st} is the standard complex structure of $\mathbb{C}^n$.
- $\pi : \tilde{\mathbb{B}}^n \longrightarrow \mathbb{B}^n$ is the blow up of $\mathbb{B}^n$ at the origin.
- $\tau_o : \mathbb{B}^n \longrightarrow \mathbb{R}_{\geq 0}$ is the standard Monge–Ampère exhaustion of $\mathbb{B}^n$, i.e., $\tau_o = \|\cdot\|^2$.
- $u_o : \mathbb{B}^n \longrightarrow \mathbb{R}_{\geq 0}$ is the function $u_o = \log \tau_o^2$.

We will also denote by $\mathcal{Z} = \bigcup_{x \in \mathbb{B}^n \setminus \{0\}} \mathcal{Z}_x$ and $\mathcal{H} = \bigcup_{x \in \mathbb{B}^n \setminus \{0\}} \mathcal{U}_x$ the distributions on $\mathbb{B}^n \setminus \{0\}$, determined by the following subspaces of tangent spaces:

$$\mathcal{Z}_x = \text{0-eigenspace of the matrix } (u_{o\,j\bar{k}})\Big|_x \subset T_x \mathbb{B}^n , \quad (13)$$

$$\mathcal{H}_x = \text{orthogonal complement of } \mathcal{Z}_x \text{ in } T_x \mathbb{B}^n \ (\text{ w.r.t Euclidean metric }) . \quad (14)$$

Notice that, for any point $x \in \mathbb{B}^n \setminus \{0\}$, the space $\mathcal{Z}_x$ is nothing else but the tangent space of the straight complex lines of $\mathbb{C}^n$, passing through x and 0 and that the (closures of) integral leaves of $\mathcal{Z}$ are the straight disks through the origin. The distributions $\mathcal{Z}$ and $\mathcal{H}$ will be called *radial* and *normal distributions*, respectively. They are both smoothly extendible at all points of the blow up $\tilde{\mathbb{B}}^n$.

The main ingredients of this section consist of the objects introduced in the following definition.

Definition 7. A complex structure J on $\tilde{\mathbb{B}}^n$ is called *L-complex structure* if and only if

  (i) The distributions $\mathcal{Z}$ and $\mathcal{H}$ are both J-invariant.
  (ii) $J|_\mathcal{Z} = J_{\text{st}}|_\mathcal{Z}$ (i.e., J and J_{st} differ only for their actions on $\mathcal{H}$!).
  (iii) There exists a smooth homotopy $J(t)$ of complex structures, all of them satisfying (i) and (ii), with $J(0) = J_{\text{st}}$ and $J(1) = J$.

A complex manifold $M = (\mathbb{B}^n, J)$, which is the blow-down at 0 of a complex manifold of the form $(\tilde{\mathbb{B}}^n, \tilde{J})$, for some L-complex structure $\tilde{J}$, is called *manifold of circular type in normal form*. Its complex structure J will be called *L-complex structure induced by $\tilde{J}$*.

Remark 8. The existence of a complex structure J on B^n that makes $M = (B^n, J)$ a blow-down at 0 of $(\tilde{B}^n, J)$ is a consequence of the fact that $(\tilde{B}^n, \tilde{J})$ is naturally endowed with a J-plurisubharmonic exhaustion $\tilde{\tau} : \tilde{B}^n \longrightarrow [0, 1)$, with $\tilde{\tau}^{-1}(0) = \mathbb{C}P^{n-1}$ which is strictly J-plurisubharmonic outside the exceptional set $\tilde{\tau}^{-1}(0)$. For this classical fact one may refer for instance to [36] (see also [42, Sect. 3.1], for more details about this point).

The crucial property of this class of manifolds is the following:

Proposition 9. *If $M = (\mathbb{B}^n, J)$ is a complex manifold, which is blow-down at 0 of a complex manifold of the form $(\tilde{\mathbb{B}}^n, \tilde{J})$, for some L-complex structure $\tilde{J}$, the pair $(M = (\mathbb{B}^n, J), \tau_o)$ is a manifold of circular type with center $x_o = 0$.*

The proof essentially consists of checking that the exhaustion $\tau_o : \mathbb{B}^n \longrightarrow \mathbb{R}_{\geq 0}$ is strictly plurisubharmonic w.r.t. the (non-standard) complex structure J, i.e., $dd^c_J \tau_o > 0$. Here, "d^c_J" is the operator $d^c_J = -J \circ d \circ J$ and is in general different from the usual operator $d^c = -J_{\text{st}} \circ d \circ J_{\text{st}} = i(\partial - \bar{\partial})$ determined by J_{st}.

Due to J-invariance, the radial and normal distributions $\mathcal{Z}, \mathcal{H}$ are not only orthogonal w.r.t. the Euclidean metric but also w.r.t. the J-invariant 2-form $dd^c_J \tau_o$. Moreover, since $J|_\mathcal{Z} = J_{\text{st}}|_\mathcal{Z}$, we have that

$$dd^c_J \tau_o|_{\mathcal{Z} \times \mathcal{Z}} = dd^c \tau_o|_{\mathcal{Z} \times \mathcal{Z}} > 0 \, .$$

Therefore, what one really needs to check is that $dd^c_J \tau_o|_{\mathcal{H} \times \mathcal{H}} > 0$. By construction, for any $x \in \mathbb{B}^n \setminus \{0\}$, the subspace $\mathcal{H}_x \subset T_x \mathbb{B}^n$ coincides with the J-holomorphic tangent space of the sphere $S_c = \{ \tau_o = c \}$ with $c = \tau_o(x)$. Indeed, it is also the $J(t)$-holomorphic tangent space of S^{2n-1} *for any complex structure $J(t)$ of the* *isotopy between* J_{st} *and* J. It follows that the restriction

$$dd^c_{J(t)}\tau_o\big|_{\mathcal{H}_x\times\mathcal{H}_x}$$

is the Levi forms of S_c at x for any complex structure $J(t)$. On the other hand, the distribution $\mathcal{H}|_{S_c} \subset TS_c$ is a contact distribution (it is the standard contact distribution of S_c) and, consequently, all such Levi forms are non-degenerate. Since $dd^c_{J(t)}\tau_o|_{\mathcal{H}_x\times\mathcal{H}_x}$ is positively when $t = 0$ and the complex structure is $J(0) = J_{st}$, by continuity, the Levi forms $dd^c_{J(t)}\tau_o\big|_{\mathcal{H}_x\times\mathcal{H}_x}$ are positively defined for any t and in particular when the complex structure is $J = J(1)$. This shows that $dd^c_J\tau_o|_{\mathcal{H}_x\times\mathcal{H}_x} > 0$ at all points as we needed.

By previous proposition, the manifolds "in normal form" constitute a very large family of examples manifolds of circular type, with an exhaustion $\tau_o = \|\cdot\|$ which is particularly simple. Moreover, the key result on such manifolds is the following.

Theorem 10 (Existence and uniqueness of normalizing maps). *For each manifold of circular type* (M, J), *with exhaustion* τ *and center* x_o, *there exists a biholomorphism* $\Phi : (M, J) \longrightarrow (\mathbb{B}^n, J')$ *to a manifold in normal form* $(\mathbb{B}^n, J')$ *with (Fig. 4)*

(a) $\Phi(x_o) = 0$ *and* $\tau = \tau_o \circ \Phi$;
(b) Φ *maps the leaves of the Monge–Ampère foliation of* M *into the straight disks through the origin of* $\mathbb{B}^n$.

Any biholomorphism $\Phi : (M, J) \longrightarrow (\mathbb{B}^n, J')$, satisfying the conditions (a) and (b), is called *normalizing map for the manifold* M. The lifted map $\tilde{\Phi} : \tilde{M} \longrightarrow \tilde{\mathbb{B}}^n$ between the blow ups at x_o and $\Phi(x_o) = 0$ is nothing but the inverse

$$\tilde{\Phi} = \Psi^{-1}$$

of the diffeomorphism $\Psi : \tilde{\mathbb{B}}^n \longrightarrow \tilde{M}$, described in Theorem 4. The L-complex structure J' on $\mathbb{B}^n$ is constructed in such a way that the corresponding complex structure $\tilde{J}'$ on the blow-up $\tilde{\mathbb{B}}^n$ coincides with the complex structure on $\hat{\mathbb{B}}^n$, obtained by push-forwarding the complex structure $\tilde{J}$ of $\tilde{M}$ onto $\tilde{\mathbb{B}}^n$, i.e.,

$$\tilde{J}' \stackrel{\text{def}}{=} \Phi_*(\tilde{J}).$$

Notice also that two normalizing maps, related with the same exhaustion τ and same center x_o, differ only by their action on the leaf space of the Monge–Ampère foliation. In fact, it turns out that the class $\mathcal{N}(M)$ of all normalizing maps, determined by all its exhaustions of M (which might correspond to distinct centers), is naturally parameterized by a suitable subset of $\text{Aut}(\mathbb{B}^n)$, which includes $\text{Aut}(\mathbb{B}^n)_0 = U_n$. We will discuss this point in more details in Sect. 3.3.

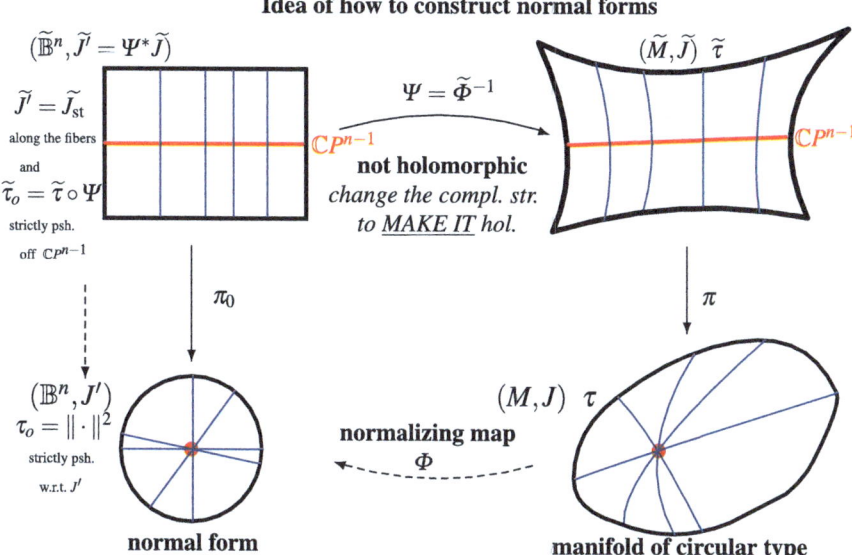

Fig. 4 Normalizing Map

3.2 Normal Forms and Deformations of CR Structures

Let $(\mathbb{B}^n, J)$ be a manifold of circular type in normal form, which is blow down of $(\tilde{\mathbb{B}}^n, J)$ for an L-complex structure J (for simplicity, we will use the same symbol for the two complex structures). By definitions, J is completely determined by the restriction $J|_{\mathcal{H}}$ and such restriction is uniquely determined by the corresponding J-anti-holomorphic subbundle $\mathcal{H}^{01}_J \subset \mathcal{H}^{\mathbb{C}}$, formed by the $(-i)$-eigenspaces $\mathcal{H}^{01}_{Jx} \subset \mathcal{H}^{\mathbb{C}}_x$, $x \in \tilde{\mathbb{B}}^n$ of the $\mathbb{C}$-linear maps $J_x : \mathcal{H}^{\mathbb{C}}_x \longrightarrow \mathcal{H}^{\mathbb{C}}_x$.

If we denote by $\mathcal{H}^{01} \subset \mathcal{H}^{\mathbb{C}}$ the J_{st}-anti-holomorphic subbundle, given by the standard complex structure J_{st}, in almost all cases the L-complex structure J can be recovered by the tensor field

$$\phi_J \in (\mathcal{H}^{01})^* \otimes \mathcal{H}^{10} = \bigcup_{x \in \tilde{\mathbb{B}}^n} \mathrm{Hom}(\mathcal{H}^{01}_x, \mathcal{H}^{10}_x) ,$$

defined by the condition (Fig. 5)

$$\mathcal{H}^{01}_{Jx} = \{ \, \mathrm{v} = \mathrm{w} + \phi_J(\mathrm{w}) , \; \mathrm{w} \in \mathcal{H}^{01} \, \} . \tag{15}$$

This tensor field ϕ_J is called *deformation tensor of J*.

We however point out that the existence of a deformation tensor ϕ_J, associated with a complex structure J, which satisfies only (i) and (ii) of Definition 7, is a priori not always granted: It exists whenever, at any $x \in \tilde{\mathbb{B}}^n$, the natural projection

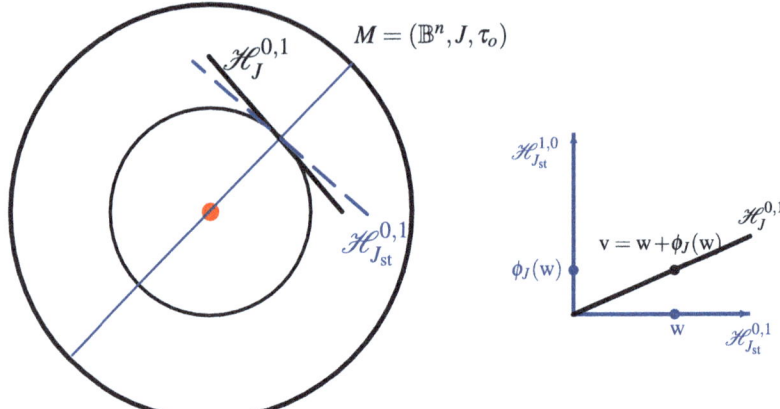

Fig. 5 Deformation Tensor

$$p : \mathcal{H}_x^{\mathbb{C}} = \mathcal{H}_x^{10} + \mathcal{H}_x^{01} \longrightarrow \mathcal{H}_x^{01}$$

determines a linear isomorphism $p|_{\mathcal{H}_{J_x}^{01}} : \mathcal{H}_{J_x}^{01} \xrightarrow{\sim} \mathcal{H}_x^{01}$. This is an "open" condition, meaning that if J can be represented by a deformation tensor ϕ_J, then also the sufficiently close complex structures J', satisfying (i) and (ii), are representable by deformation tensors.

However, as we will shortly see, the existence of a deformation tensor for L-complex structures is also a "closed" condition and hence <u>any L-complex structure is represented by a deformation tensor</u>.

By these remarks, we have that the biholomorphic classes of domains of circular type are in natural correspondence with the deformation tensors of L-complex structures on $\tilde{\mathbb{B}}^n$. It is therefore very important to find an efficient characterization of the tensor fields $\phi = (\mathcal{H}^{01})^* \otimes \mathcal{H}^{10}$ that correspond to L-complex structures, by a suitable set of intrinsic properties. This problem was solved by Bland and Duchamp in [7–9] via a suitable adaptation of the theory of deformations of complex structures (see [29] for a classical introduction).

In those papers, Bland and Duchamp were concerned with strictly linearly convex domains in $\mathbb{C}^n$ that are small deformations of the unit ball $\mathbb{B}^n$. To any such domain D, they associated a deformation tensor field ϕ_D on $\partial \mathbb{B}^n$, which, in our terminology, is the restriction to $\partial \mathbb{B}^n$ of the deformation tensor of the L-complex structure J of a normal form.

The arguments of Bland and Duchamp extend very naturally to all cases of our more general context and bring to the characterization of L-complex structures, which we are now going to describe.

First of all, notice that the holomorphic and anti-holomorphic distributions $\mathcal{H}^{10}$, $\mathcal{H}^{01}$ can be (locally) generated by vector fields $X^{1,0} \in H^{1,0}$, $Y^{0,1} \in \mathcal{H}^{0,1}$ such that

$$\hat{\pi}_*\left([X^{1,0}, Y^{0,1}]\right) = [\hat{\pi}_*(X^{1,0}), \hat{\pi}_*(Y^{0,1})] = 0 ,$$

where $\tilde{\pi} : \tilde{\mathbb{B}}^n \longrightarrow \mathbb{C}^{n-1}$ is the natural fibering over the exceptional set $\mathbb{C}P^{n-1}$ of the blow up $\tilde{\mathbb{B}}^n$. Let us call the vector fields of this kind *holomorphic (resp. anti-holomorphic) vector fields* of $\mathcal{H}^{\mathbb{C}}$.

Now, consider the following operators (see e.g. [29]) (here, we denote by $(\cdot)_{\mathcal{H}^{\mathbb{C}}} : T^{\mathbb{C}} \tilde{\mathbb{B}}^n \longrightarrow \mathcal{H}^{\mathbb{C}}$ the natural projection, determined by the decomposition $T^{\mathbb{C}} \tilde{\mathbb{B}}^n = \mathcal{Z}^{\mathbb{C}} + \mathcal{H}^{\mathbb{C}}$):

$$\bar{\partial}_b : H^{0,1*} \otimes H^{1,0} \to \Lambda^2 H^{0,1*} \otimes H^{1,0},$$

$$\bar{\partial}_b \alpha(X, Y) \stackrel{\text{def}}{=} [X, \alpha(Y)]_{\mathcal{H}^{\mathbb{C}}} - [Y, \alpha(X)]_{\mathcal{H}^{\mathbb{C}}} - \alpha([X, Y]), \qquad (16)$$

and

$$[\cdot, \cdot] : \left(H^{0,1*} \otimes H^{1,0}\right) \times \left(H^{0,1*} \otimes H^{1,0}\right) \longrightarrow \Lambda^2 H^{0,1*} \otimes H^{1,0},$$

$$[\alpha, \beta](X, Y) \stackrel{\text{def}}{=} \frac{1}{2} ([\alpha(X), \beta(Y)] - [\alpha(Y), \beta(X)]) \qquad (17)$$

for any pair of holomorphic and anti-holomorphic vector fields X, Y of $\mathcal{H}$. We then have the following:

Theorem 11 ([42]). *Let J be an L-complex structure on $\tilde{\mathbb{B}}^n$ that admits a deformation tensor ϕ (in fact, J is an arbitrary L-complex structure). Then:*

(i) $dd^c \tau_o(\phi(X), Y) + dd^c \tau_o(X, \phi(Y)) = 0$ *for any pair X, Y of vector fields in* $\mathcal{H}^{0,1}$;
(ii) $\bar{\partial}_b \phi + \frac{1}{2}[\phi, \phi] = 0$;
(iii) $\mathcal{L}_{Z^{0,1}}(\phi) = 0$.
Conversely, any tensor field $\phi \in H^{0,1*} \otimes H^{1,0}$ that satisfies (i)–(iii) is the deformation tensor of an L-complex structure.

In addition, an L-complex structure J, associated with a deformation tensor ϕ, is so that $(\mathbb{B}^n, J, \tau_o)$ is a manifold of circular type if and only if
(iv) $dd^c \tau_o(\phi(X), \overline{\phi(X)}) < dd^c \tau_o(\bar{X}, X)$ *for any $0 \neq X \in H^{0,1}$.*

For the proof we refer directly to [42]. Here, we only point out that the conditions (i)–(iii) comes out from the request of integrability for the almost complex structure J, coinciding with J_{st} on $\mathcal{Z}$ and with anti-holomorphic distribution $\mathcal{H}^{01}_J$ determined by (15).

Remark 12. Condition (iv) of previous theorem can be interpreted as an a-priori estimate for the deformation tensor ϕ: It gives an "upper bound" for the norm of ϕ w.r.t. to the Kähler metric $dd^c \tau_o$. It is this property that makes the representability of an L-complex structure by a deformation tensor a "closed" condition and that it implies the existence of a deformation tensor for *any* L-complex structure, as previously pointed out.

Consider now a local trivialization of the line bundle $\tilde{\pi} : \tilde{\mathbb{B}}^n \longrightarrow \mathbb{C}P^{n-1}$, which represents the points $z = ([v], v) \in \tilde{\pi}^{-1}(\mathcal{U})$ of some open subset $\mathcal{U} \subset \mathbb{C}P^{n-1}$ by pairs $(w, \zeta) \in S^{2n-1} \times \Delta$ with

$$w = \frac{v}{\|v\|} \in S^{2n-1} \quad \text{and} \quad v = \zeta w .$$

Condition (iii) of Theorem 11 implies that the restriction $\phi_J|_{\pi^{-1}(\mathcal{U})}$ of the deformation tensor of J is of the form

$$\phi_J = \sum_{k=0}^{\infty} \phi_J^{(k)}(w, \zeta) = \sum_{k=0}^{\infty} \phi_J^k(w)\zeta^k ,$$

where each $\phi_J^{(k)}(w, \zeta) \stackrel{\text{def}}{=} \phi_J^k(w)\zeta^k$ is a tensor in $(\mathcal{H}^{01*} \otimes \mathcal{H}^{10})|_{([w], \zeta w)}$.

One can check that the tensor fields $\phi^{(k)}$ do not depend on the trivializations and are well defined over $\tilde{\mathbb{B}}^n$. Indeed, one has a sequence $\{\phi_J^{(k)}\}$ of deformation tensors over $\tilde{\mathbb{B}}^n$ such that the series $\sum_{k=0}^{\infty} \phi_J^{(k)}$ converges uniformly on compact sets to ϕ_J. These observations bring directly to the following corollary.

Corollary 13. *A manifold $(\mathbb{B}^n, J)$ of circular type in normal form, given by the blow down at 0 of $(\tilde{\mathbb{B}}^n, J)$, is uniquely associated with a sequence of tensor fields $\phi_J^{(k)}$ in $(\mathcal{H}^{01})^* \otimes \mathcal{H}^{10}$, $0 \leq k < \infty$, each of them (locally) of the form*

$$\phi_J^{(k)}([w], \zeta w) = \phi_J^k([w])\zeta^k , \qquad w \in S^{2n-1} , \zeta \in \Delta ,$$

such that the series

$$\phi_J = \sum_{k \geq 0} \phi_J^{(k)} \tag{18}$$

converges uniformly on compacta and satisfies the following conditions:

(i) $dd^c \tau_o(\phi_J(X), Y) + dd^c \tau_o(X, \phi_J(Y)) = 0$ *for anti-holomorphic* $X, Y \in H^{0,1}$;
(ii) $\bar{\partial}_b \phi_J + \frac{1}{2}[\phi_J, \phi_J] = 0$;
(iii) $dd^c \tau_o(\phi_J(X), \overline{\phi_J(X)}) < dd^c \tau_o(\bar{X}, X)$ *for any* $0 \neq X \in H^{0,1}$.

Conversely, any sequence of tensor fields $\phi_J^{(k)} \in (\mathcal{H}^{01})^* \otimes \mathcal{H}^{10}$, $0 \leq k < \infty$, such that (18) converges uniformly on compacta and satisfies (i)–(iii), determines a manifold of circular type in normal form.

It is important to observe that the restriction $\tilde{\phi}_J = \phi_J|_{S^{2n-1}(r)}$ of a deformation tensor ϕ_J to a sphere

$$S^{2n-1}(r) = \{ ([v], v) , \|v\| = r \} \subset \tilde{\mathbb{B}}^n , \qquad 0 < r < 1$$

is the deformation tensor of the CR structure $(\mathcal{H}|_{S^{2n-1}(r)}, J)$, induced on $S^{2n-1}(r)$ by the complex structure J. Vice versa, any deformation tensor $\tilde{\phi}_J$ of a CR structure of the form $(\mathcal{H}|_{S^{2n-1}(r)}, J)$ on a sphere $S^{2n-1}(r)$ can be written as a Fourier series

$$\tilde{\phi}_J = \sum_{k \geq 0} \tilde{\phi}^{(k)}, \qquad (19)$$

whose terms are of the form

$$\tilde{\phi}_J^{(k)}([\mathrm{w}], re^{i\vartheta}) = \tilde{\phi}_J^k([\mathrm{w}]) r^k e^{ik\vartheta}, \qquad \mathrm{w} \in S^{2n-1}.$$

From this, one can see that all deformation tensors $\tilde{\phi}_J$ of such CR structures are exactly the restrictions of the deformation tensors ϕ_J as in (18). If such deformation tensor ϕ_J satisfies conditions (i)–(iii) of Corollary 13, it uniquely determines a complex structure J which makes $(\tilde{\mathbb{B}}^n, J)$ the blow up of a manifold of circular type in normal form.

Summing up all these observations, we see that there exists a natural bijection between the following two classes:

{Manifolds of circular type in normal formal
with the point 0 as distinguished center}

⇕

{Deformation tensors of CR-structures on $S^{2n-1}(r)$
satisfying suitable (explicit) conditions}

As we already mentioned, the Fourier developments of the tensors $\tilde{\phi}_J$, which characterize the CR structures of the form $(\mathcal{H}|_{S^{2n-1}(r)}, J)$ on S^{2n-1} were first considered by Bland and Duchamp in [8, 9] in case of *small* deformations of the standard CR structure. They managed to prove that they can be always realized as the CR structures of boundaries of bounded domains in $\mathbb{C}^n$.

3.3 *Parameterizations of Normalizing Maps*

For a given manifold of circular type (M, J) (considered without a distinguished exhaustion τ), there are in general many distinct normalizing map

$$\Phi : M \longrightarrow \mathbb{B}^n.$$

The class $\mathcal{N}(M)$ of all such normalizing maps is an important biholomorphic invariant of the manifolds of circular type. Let us see in more details how such class $\mathcal{N}(M)$ can be studied.

Let us first try to understand the structure of the subclass $\mathcal{N}(M)_{x_o} \subset \mathcal{N}(M)$, consisting of the normalizing maps that send a fixed center x_o into the center 0 of its normal form $(\mathbb{B}^n, J')$. For this we need to introduced the notion of "special frames".

Let $x_o \in M$ be the fixed center determined by a given Monge–Ampère exhaustion $\tau : M \longrightarrow \mathbb{R}_{\geq 0}$ and $\kappa = \kappa_{x_o} : T_{x_o} M = \mathbb{C}^n \longrightarrow \mathbb{R}_{\geq 0}$ the infinitesimal Kobayashi metric of M at x_o. Let also $I \subset T_{x_o} M$ be the Kobayashi indicatrix at x_o, that is

$$I = \{\, v \in T_{x_o} M \simeq \mathbb{C}^n \ : \ \kappa_{x_o}(v) < 1 \,\}.$$

Recall that I is a circular domain of $T_{x_o} M \simeq \mathbb{C}^n$ and that κ coincides with the Minkowski functional of I. We call *special frame at x_o* any linear frame $(e_0, e_1, \ldots, e_{n-1})$ for $T_{x_o} M$ such that:

(i) $e_0 \in \partial I$;
(ii) $(e_1, \ldots, e_{n-1})$ is a collection of vectors in the tangent space

$$T_{e_0}(\partial I) \subset T_{e_0}(T_{x_o} M) = T_{x_o} M \simeq \mathbb{C}^n,$$

that constitutes a linear frame for the holomorphic tangent space of ∂I at e_0, which is unitary w.r.t. the Levi form determined by the defining function $\rho = \kappa^2 - 1$ of I.

In [42], we proved the existence of a natural one-to-one correspondence between $\mathcal{N}(M)_{x_o}$ and the set P_{x_o} of all special frames at x_o. More precisely, given a fixed frame $u_o = (e_0^o, \ldots e_{n-1}^o) \in P_{x_o}$ and a normalizing map $\Phi_o \in \mathcal{N}(M)_{x_o}$, the new basis of $T_{x_o} M$ defined by

$$u^\Phi = (e_0, \ldots, e_{n-1}) \quad \text{with } e_i = (\Phi^{-1} \circ \Phi_o)_*(e_i^o), \quad \Phi \in \mathcal{N}(M)_{x_o},$$

is also a special frame and the correspondence

$$\iota_{x_o} : \mathcal{N}_{x_o}(M) \longrightarrow P_{x_o}, \qquad \iota(\Phi) = u^\Phi$$

is a bijection.

Therefore, for studying the whole class $\mathcal{N}(M)$ of normalizing maps, it is convenient to consider the collection of special frames

$$P(M) = \bigcup_{\substack{x \in M \text{ that} \\ \text{are centers w.r.t some } \tau}} P_x,$$

which we call *pseudo-bundle of special frames*. We stress the fact that $P(M)$ is not expected to be a manifold—its geometric properties strongly depend on the geometry of the set of centers of M. However, it turns out that when M is a strictly linearly convex domain of $\mathbb{C}^n$, the pseudo-bundle $P(M)$ coincides with the unitary frame bundle of the complex Finsler metric, given by the infinitesimal Kobayashi metric κ (for definitions and properties of unitary frame bundles of complex Finsler manifolds, see e.g. [50])

The previously defined correspondence between normalizing maps and special frames determines a natural bijection

$$\mathcal{N}(M) \xrightarrow{\sim} P(M).$$

Notice also that, if we identify M with one of its normal form $(\mathbb{B}^n, J)$, one can construct a diffeomorphism between the collection P_{x_o} of all special frames at a fixed center $x_o \in \mathbb{B}^n$ and the subgroup $U_n = \text{Aut}_0(\mathbb{B}^n, J_{st})$ of the automorphisms of $(\mathbb{B}^n, J_{st})$ fixing the origin. Such correspondence brings to an identification between $P(M)$ and a suitable subset of $\text{Aut}(\mathbb{B}^n, J_{st})$, which reveals to be a true diffeomorphism

$$\mathcal{N}(M) \xrightarrow{\sim} \text{Aut}(\mathbb{B}^n, J_{st})$$

when M is a strictly linearly convex domain of $\mathbb{C}^n$.

3.4 Some Geometrical Interpretations and Applications

Let $M = (\mathbb{B}^n, J, \tau_o)$ be a manifold of circular type in normal form, endowed with the standard exhaustion $\tau_o = \|\cdot\|^2$. Let also $\phi_J = \sum_{k=0}^{\infty} \phi_J^{(k)}$ be the corresponding deformation tensor and $I \subset T_0\mathbb{B}^n = \mathbb{C}^n$ the Kobayashi indicatrix at the center $x_o = 0$. Notice that, if we denote by $\mu = \kappa^2$ given by the square of the infinitesimal Kobayashi metric κ of $(\mathbb{B}^n, J)$ at 0, the pair (I, μ) is a domain of circular type in $T_0\mathbb{B}^n = \mathbb{C}^n$—in fact, I is a circular domain and its Minkowski functional is κ!

One can prove the following.

Theorem 14. *(i) The 0-th order component $\phi_J^{(0)}$ of ϕ_J coincides with the deformation tensor of the normal form of (I, μ).*
(ii) The difference $\phi_J - \phi_J^{(0)}$ vanishes identically if and only if M is biholomorphic to the circular domain I.

An application of this and all previous discussion is given by the following generalizations of results of Leung, Patrizio and P.M. Wong and for strictly convex domains and of Abate and Patrizio for Kähler–Finsler manifolds [1, 34]. In the next statement, given a manifold of circular type (M, τ) with center x_o and a real number $0 < r < 1$, we use the notation $M_{x_o,r} = \{x \in M : \tau(x) < r\}$ for any $0 < r < 1$.

Theorem 15. *(1) A manifold of circular type (M, τ) is biholomorphic to a circular domain if and only if the following condition holds:*

  ($\star$) there exists two distinct values $r_1, r_2 \in (0, 1)$ such that M_{x_o,r_1} is biholomorphic to M_{x_o,r_2}.

(2) A complex manifold (M, J) is biholomorphic to the standard unit ball $(\mathbb{B}^n, J_{st})$ if and only if it admits at least two distinct structures of manifold of circular type $(M, \tau), (M, \tau')$, relative to two distinct centers $x_o \neq x'_o$, for which condition $(\star)$ holds.

3.5 Remarks and Questions

Here are some open question, which we consider interesting and worth of investigations.

(a) Find geometric interpretations of (possibly all) terms of the expansion in Fourier series $\phi_J = \sum_{k=0}^{\infty} \phi_J^{(k)}$ of the deformation tensor of a manifold of circular type in normal form.
(b) Using modular data (in practice, using possible expression for the deformation tensors in normal forms), construct explicit examples of manifolds of circular type with prescribed properties, e.g.,

 - with exactly one center or with a discrete set of centers (if there are any);
 - with an open set of centers:
 - not embeddable in $\mathbb{C}^n$ (if there exist).

(c) Find conditions on the deformation tensor that characterize the domains of circular type, for which any point is a center.

We recall that Wong proved in [58] that any manifold of circular type admits non constant bounded holomorphic functions. In fact, such manifolds are hyperbolic and he proved that the Caratheodory metric of such manifolds is bounded below by a multiple of the Kobayashi metric. This stimulates further research towards the solution to the following basic question:

(d) Find conditions on modular data that characterize the manifolds of circular type biholomorphic to some strictly linearly convex domain or just to a bounded domain in $\mathbb{C}^n$.

4 The Definition of "Stationary Disk" in the Almost Complex Setting

4.1 First Definitions

From now on, our discussion will focus on the wider class of *almost* complex manifolds and we will be mainly concerned with generalizations of previous results in this larger context.

In what follows, M is always a $2n$-dimensional real manifold with an almost complex structure J, which is a tensor field of type $(1, 1)$ that gives a linear map at any $x \in M$

$$J_x : T_x M \longrightarrow T_x M \quad \text{such that} \quad J_x^2 = -\operatorname{Id}_{T_x M} .$$

We recall that an almost complex structure J is called *integrable* if the associated Nijenhuis tensor N_J vanishes identically. The definition of N_J is the following: It is the tensor field of type $(1,2)$ defined by the relation

$$N(X,Y) = \frac{1}{4}([X,Y] - [JX,JY] + J[JX,Y] + J[X,JY])$$

for any pair of vector fields X, Y of M.

By the celebrated Newlander–Nirenberg theorem, an almost complex structure J is integrable if and only if M admits a structure of complex manifold, i.e., if and only if there exists an atlas of complex charts for M

$$\xi = (z^1 = x^1 + iy^1, \ldots, z^n = x^n + iy^n) : \mathcal{U} \subset M \longrightarrow \mathbb{C}^n ,$$

such that

$$J\left(\frac{\partial}{\partial x^i}\right) = \frac{\partial}{\partial y^i} , \qquad J\left(\frac{\partial}{\partial y^i}\right) = -\frac{\partial}{\partial x^i}$$

and the changes of coordinates $\xi \circ \eta^{-1}$, $\eta \circ \xi^{-1}$ between any two overlapping charts of the atlas are holomorphic. In the following, the charts of such atlas will be called *systems of holomorphic coordinates*.

Given a pair of almost complex manifold (M,J), (M',J'), a map $f : M \longrightarrow M'$ is called (J,J')-*holomorphic* (or, simply, *holomorphic*) if

$$\overline{\partial}_{J,J'} f(v) = 0 \qquad \text{for any } v \in TM ,$$

where $\overline{\partial}_{J,J'} f$ is the map

$$\overline{\partial}_{J,J'} f : TM \longrightarrow TM' , \qquad \overline{\partial}_{J,J'} f(v) = f_*(Jv) - J'(f_*(v)) . \tag{20}$$

Notice that $\overline{\partial}_{J,J'}$ is a natural generalization of the usual $\overline{\partial}$-operator. In fact, when (M,J) and (M',J') are complex manifolds (i.e., when J, J' are both integrable) and

$$\xi = \left(x^i = \frac{z^i + \overline{z^i}}{2}, y^i = \frac{z^i - \overline{z^i}}{2i}\right) , \qquad \xi' = \left(x'^i = \frac{z'^i + \overline{z'^i}}{2}, y'^i = \frac{z'^i - \overline{z'^i}}{2i}\right)$$

are systems of holomorphic coordinates for M and M', respectively, the expression of an arbitrary smooth real map $f : M \longrightarrow M'$ is of the form

$$f(z^i, \overline{z^j}) = (f^m(z^i, \overline{z^j}), f^{\overline{m}}(z^i, \overline{z^n})) ,$$

where $f^m(z^i, \overline{z^j})$ and $f^{\overline{m}}(z^i, \overline{z^j})$ denote the values of $f(z^i, \overline{z^j})$ in the complex coordinates z'^m and $\overline{z'^m}$ of M'. In such coordinates the ($\mathbb{C}$-linear extended) map

$$\overline{\partial}_{J,J'} f : T^{\mathbb{C}} M \longrightarrow T^{\mathbb{C}} M$$

is

$$\overline{\partial}_{J,J'} f\left(\frac{\partial}{\partial \overline{z}^i}\right) = -i\frac{\partial f^j}{\partial \overline{z}^i}\frac{\partial}{\partial z'^j} - i\frac{\partial f^{\overline{j}}}{\partial \overline{z}^i}\frac{\partial}{\partial \overline{z'^j}} - i\frac{\partial f^j}{\partial \overline{z}^i}\frac{\partial}{\partial z'^j} + i\frac{\partial f^j}{\partial \overline{z}^i}\frac{\partial}{\partial \overline{z'^j}}$$

$$= -i2\frac{\partial f^j}{\partial \overline{z}^i}\frac{\partial}{\partial z'^j},$$

showing that $\overline{\partial}_{J,J'} f$ vanishes identically if and only if the f^ℓ's are holomorphic in the usual sense.

In the following, when $(M, J) \subset (\mathbb{C}^n, J_{\mathrm{st}})$, we will often use the simplified notation $\overline{\partial}_{J'} = \overline{\partial}_{J_{\mathrm{st}}, J'}$. Given an almost complex manifold (M, J), a map

$$f : \Delta = \{ |\zeta| < 1 \} \longrightarrow (M, J)$$

is called *J-holomorphic disk* if it is (J_{st}, J)-holomorphic or, equivalently, if $\overline{\partial}_J f = 0$. A simple argument shows that, for an arbitrary C^1-map $f : \Delta \longrightarrow (M, J)$, the J-holomorphicity condition $\overline{\partial}_J f = 0$ is equivalent to the differential equation

$$\partial_J f\left(\frac{\partial}{\partial x}\bigg|_\zeta\right) = 0 \qquad \text{for any } \zeta = x + iy \in \Delta$$

(see e.g. [26, 51]).

4.2 Canonical Lifts of Almost Complex Structures

When (M, J) is a (integrable) complex manifold, it is easy to define a corresponding pair of natural complex structures on the tangent bundle

$$\hat{\pi} : TM \longrightarrow M$$

and on the cotangent bundle

$$\tilde{\pi} : T^*M \longrightarrow M .$$

In fact, any system of holomorphic coordinates $\xi = (x^i)$ can be used to locally identify M, TM and T^*M with open subsets of $\mathbb{C}^n$, $T\mathbb{C}^n = \mathbb{C}^{2n}$ and $T^*\mathbb{C}^n = \mathbb{C}^{2n}$, respectively. These identifications determine integrable almost complex structures $\hat{\mathbb{J}}$ and $\tilde{\mathbb{J}}$ on TM and T^*M, which turn out to be independent on the choice of the considered system of holomorphic coordinates. They are therefore *naturally and globally defined complex structures on TM and T^*M*, respectively.

When (M, J) is a (non-integrable) almost complex manifold, the notion of "system of holomorphic coordinates" is meaningless and the above construction does not apply. However, it is still possible to define a pair of almost complex structures on TM and T^*M, which depend in a canonical way on the almost complex J of M. Such almost complex structures were introduced by Yano and Ishihara in the 1970's and are defined as follows (see [59]). Given a system of coordinates

$$\xi = (x^1, \ldots, x^{2n}) : \mathcal{U} \subset M \longrightarrow \mathbb{R}^n,$$

let us denote by

$$\hat{\xi} = (x^1, \ldots, x^{2n}, q^1, \ldots, q^{2n}) : \hat{\pi}^{-1}(\mathcal{U}) \subset TM \longrightarrow \mathbb{R}^{4n},$$

$$\tilde{\xi} = (x^1, \ldots, x^{2n}, p_1, \ldots, p_{2n}) : \tilde{\pi}^{-1}(\mathcal{U}) \subset T^*M \longrightarrow \mathbb{R}^{4n},$$

the associated coordinates on $TM|_\mathcal{U}$ and $T^*M|_\mathcal{U}$, determined by the components q^i of the vectors $v = q^i \frac{\partial}{\partial x^i}$ in the basis $\left(\frac{\partial}{\partial x^i}\right)$ and by the components p_j of the covectors $\alpha = p_j dx^j$ in the basis (dx^i). Let us also denote by $J^i_j = J^i_j(x)$ the components of the almost complex structure $J = J^i_j \frac{\partial}{\partial x^i} \otimes dx^j$.

The *canonical lifts of J on TM and T^*M* are the almost complex structures $\mathbb{J}$ on TM and $\tilde{\mathbb{J}}$ on T^*M, defined by

$$\mathbb{J} = J^a_i \frac{\partial}{\partial x^a} \otimes dx^i + J^a_i \frac{\partial}{\partial q^a} \otimes dq^i + q^b J^a_{i,b} \frac{\partial}{\partial q^a} \otimes dx^i, \tag{21}$$

$$\tilde{\mathbb{J}} = J^a_i \frac{\partial}{\partial x^a} \otimes dx^i + J^a_i \frac{\partial}{\partial p_i} \otimes dp_a +$$

$$+ \frac{1}{2} p_a \left(-J^a_{i,j} + J^a_{j,i} + J^a_\ell \left(J^\ell_{i,m} J^m_j - J^\ell_{j,m} J^m_i\right)\right) \frac{\partial}{\partial p_j} \otimes dx^i. \tag{22}$$

These tensor fields can be checked to be independent on the chart (x^i) and:

(i) The standard projections $\hat{\pi} : T^*M \longrightarrow M$, $\tilde{\pi} : T^*M \longrightarrow M$ are $(\mathbb{J}, J)$-holomorphic and $(\tilde{\mathbb{J}}, J)$-holomorphic, respectively.
(ii) Given a (J, J')-biholomorphism $f : (M, J) \longrightarrow (N, J')$ between almost complex manifolds, the tangent and cotangent maps

$$f_* : TM \longrightarrow TN \quad \text{and} \quad f^* : T^*N \longrightarrow T^*M$$

are $(\mathbb{J}, \mathbb{J}')$- and $(\tilde{\mathbb{J}}', \tilde{\mathbb{J}})$-holomorphic, respectively.
(iii) When J is integrable, $\mathbb{J}$ and $\tilde{\mathbb{J}}$ coincide with above described integrable complex structures of TM and T^*M, respectively.

In order to better understand the property (iii) and see the precise relation between the almost complex structures $\mathbb{J}$ and $\tilde{\mathbb{J}}$ with their analogues of the integrable case, it is convenient to rewrite them in (non-holomorphic) complex coordinates, i.e., in complex coordinates of the form

$$(z^A) = (z^a = x^a + ix^{a+n}, z^{\bar{a}} = \overline{z^a} = x^a - ix^{a+n}).$$

If $(q^A) = (q^a, \bar{q}^a)$ and $(p_A) = (p_a, p_{\bar{a}} \stackrel{\text{def}}{=} \bar{p}_a)$ are the complex components of *real* vector fields and *real* 1-forms

$$X = q^a \frac{\partial}{\partial z^a} + \bar{q}^a \frac{\partial}{\partial \bar{z}^a} \in TM, \qquad \omega = p_a dz^a + \bar{p}_a d\bar{z}^a \in T^*M,$$

the canonical lifts $\mathbb{J}$ and $\tilde{\mathbb{J}}$ can re-written in the following form:

$$\mathbb{J} = J^A_B \left(\frac{\partial}{\partial z^A} \otimes dz^B + \frac{\partial}{\partial q^A} \otimes dq^B \right) + q^C J^A_{B,C} \frac{\partial}{\partial q^A} \otimes dx^B,$$

$$\tilde{\mathbb{J}} = J^B_A \left(\frac{\partial}{\partial z^B} \otimes dz^A + \frac{\partial}{\partial p_A} \otimes dp_B \right)$$
$$+ \frac{1}{2} p_C \left(-J^C_{A,B} + J^C_{B,A} + J^C_L \left(J^L_{A,M} J^M_B - J^L_{B,M} J^M_A \right) \right) \frac{\partial}{\partial p_B} \otimes dz^A,$$

where the J^A_B's are the components of J w.r.t. the complex vector fields $\left(\frac{\partial}{\partial z^A} \right)$. When J is integrable and $(z^1, \ldots, z^n)$ are holomorphic coordinates, the J^A_B's are constant and equal to the entries of the matrix

$$(J^A_B) = \begin{pmatrix} i\delta^a_b & 0 \\ 0 & -i\delta^{\bar{a}}_{\bar{b}} \end{pmatrix}.$$

and $\mathbb{J}$ and $\tilde{\mathbb{J}}$ assume the familiar expressions

$$\mathbb{J} = J^A_B \left(\frac{\partial}{\partial z^A} \otimes dz^A + \frac{\partial}{\partial q^A} \otimes dq^B \right), \quad \tilde{\mathbb{J}} = J^B_A \left(\frac{\partial}{\partial z^B} \otimes dz^A + \frac{\partial}{\partial p_A} \otimes dp_B \right).$$

4.3 Strong Pseudoconvexity in the Almost Complex Setting

The classical notions of "CR structure", "Levi form", "(strong) pseudoconvexity" admit direct and simple generalizations in the context of almost complex manifolds. Let us recall them.

Let (M, J) be an almost complex manifold and $\Gamma \subset M$ a connected (smooth) hypersurface. The *(induced) CR structure of* Γ is the pair $(\mathcal{D}, J^\mathcal{D})$ formed by

(a) the distribution $\mathcal{D} \subset TM$ defined by

$$\mathcal{D} = \bigcup_{x \in M} \mathcal{D}_x, \qquad \mathcal{D}_x = \{ v \in T_x \Gamma : Jv \in T_x \Gamma \};$$

(b) the family $J^\mathcal{D}$ of complex structures

$$J^\mathcal{D}_x : \mathcal{D}_x \longrightarrow \mathcal{D}_x, \qquad J^\mathcal{D}_x(v) = J_x v.$$

A 1-form ϑ in $T^*\Gamma$ is called *defining form for* $\mathcal{D}$ if for any $x \in \Gamma$

$$\ker \vartheta_x = \mathcal{D}_x \ .$$

Notice that, for any pair of defining forms ϑ, ϑ' for $\mathcal{D}$, there exists a nowhere vanishing, smooth real function λ such that

$$\vartheta' = \lambda \cdot \vartheta \ . \tag{23}$$

In the following, we will assume that Γ is oriented, i.e., endowed with a fixed choice of a nowhere vanishing vector field $\xi \in T\Gamma \setminus \mathcal{D}$. Clearly, $N_x = J\xi_x$ is transversal to $T_x\Gamma$ at all points $x \in \Gamma$ and to be oriented in the previous sense coincide with the usual definition. If $\Gamma = \partial D$ is the boundary of a relatively compact domain $D \subset M$, we will always assume that the orientating field ξ is such that the vector $N_x = J\xi_x$ is pointing outwards D for any $x \in \Gamma$. A defining form ϑ such that $\vartheta(\xi) > 0$ (resp. < 0) will be called *positive* (resp. *negative*).

Given a fixed positive defining form ϑ, the *Levi form of* Γ *at* x is the quadratic form

$$\mathcal{L}_x : \mathcal{D}_x \longrightarrow \mathbb{R} \ , \qquad \mathcal{L}_x(\mathrm{v}) = d\vartheta_x(\mathrm{v}, J\mathrm{v}) = \vartheta_x([X^{(\mathrm{v})}, JX^{(\mathrm{v})}]) \ , \tag{24}$$

where $X^{(\mathrm{v})}$ is any smooth vector field with values in $\mathcal{D}$ such that $X_x^{(\mathrm{v})} = \mathrm{v}$. From the last expression in (24), it follows immediately that if ϑ is replaced by another positive defining form, the corresponding Levi form changes only by a positive factor.

Definition 16. An oriented smooth hypersurface $\Gamma \subset M$ is called *strongly pseudoconvex* if $\mathcal{L}_x > 0$ for any $x \in \Gamma$. A smooth, relatively compact domain $D \subset M$ is called *strongly pseudoconvex* if ∂D is strongly pseudoconvex.

Many properties of classical strongly pseudoconvex domains generalize to the case of strongly pseudoconvex domains in almost manifolds. For instance, it is known that $D \subset M$ *is strongly pseudoconvex if and only if it admits a strictly* J-*plurisubharmonic defining function* (for the definition of J-plurisubharmonicity, see later). For this and other basic properties of almost complex domains, we refer to the survey [16].

4.4 Stationary Disks

Let us now introduce the notion of stationary disks of almost complex strongly pseudoconvex domains. The original definition of stationary disk is due to Lempert [30] (see also [49]) and extended by Tumanov [54] to more general settings for submanifolds in (integrable) complex manifolds. Tumanov's definition extends directly to the almost complex environment (see [15,16,21]). As before, (M, J) is an almost complex manifold and $\Gamma = \partial D \subset M$ is the oriented, smooth hypersurface, which is the boundary of a relatively compact domain $D \subset M$.

Let us recall that the *conormal bundle of* Γ is the collection $\mathcal{N}$ of 1-forms at the points of Γ defined by

$$\mathcal{N} = \{\, \alpha \in T_x^* M \,:\, x \in \Gamma \text{ and } T_x \Gamma \subset \ker \alpha \,\} \subset T^* M|_\Gamma \,.$$

We denote $\mathcal{N}_* = \mathcal{N} \setminus \{\text{zero section}\}$.

Definition 17. Given $\alpha \geq 1$, $\varepsilon > 0$, a map $f : \overline{\Delta} \to M$ from the closed unit disk $\overline{\Delta} \subset \mathbb{C}$ into M is called $\mathcal{C}^{\alpha,\varepsilon}$-stationary disk of D if

(i) $f|_\Delta$ is a J-holomorphic embedding and $f(\partial \Delta) \subset \partial D$;
(ii) there exists a $\widetilde{J}$-holomorphic maps $\widetilde{f} : \overline{\Delta} \to T^* M$ with $\pi \circ \widetilde{f} = f$ and $\widetilde{\xi} \circ \widetilde{f}$ in $\mathcal{C}^{\alpha,\varepsilon}(\overline{\Delta}, \mathbb{C}^{2n})$ for some system of complex coordinates $\widetilde{\xi} = (z^i, w_j)$, such that

$$\zeta^{-1} \cdot \widetilde{f}(\zeta) \in \mathcal{N}_* \quad \text{for any } \zeta \in \partial \Delta \,. \tag{25}$$

If f is a stationary disk, the maps $\widetilde{f}$ that satisfy (ii) are called *stationary lifts of* f.

In (25), the product "$\cdot$" denotes the $\mathbb{C}$-action on $T^* M$ defined by

$$\zeta \cdot \alpha = \operatorname{Re}(\zeta) \alpha - \operatorname{Im}(\zeta) J^* \alpha \quad \text{for any } \alpha \in T^* M, \ \zeta \in \mathbb{C} \,. \tag{26}$$

We point out that, as a consequence of the maximum principle for subharmonic functions and of the fact that D is strongly pseudoconvex, for any disk $f : \overline{\Delta} \to M$ satisfying (i), one has that $f(\overline{\Delta}) \subset \overline{D}$ and $f(\zeta) \in \partial D$ if and only if $\zeta \in \partial \Delta$.

Remark 18. When J is an integrable complex structure, condition (ii) implies that the restriction along $f(\partial \Delta)$ of the CR distribution of ∂D extends to a $\widetilde{J}$-holomorphic bundle over Δ ($\simeq f(\Delta)$), this being a characterizing property of the usual stationary disks of the domains of $\mathbb{C}^n$ [30]. This is one of the reasons why the previous definition can be considered as the natural generalization of the concept of stationary disk in the almost complex setting.

5 Almost Complex Domains of Circular Type

5.1 Looking for the Stationary Disks of Almost Complex Domain

In this and the next sections, D is a smooth, relatively compact, strongly pseudoconvex domain in an almost complex manifold (M, J) with boundary $\Gamma = \partial D$ and

$$\mathcal{N}_* = \mathcal{N} \setminus \{\text{zero section}\} \,, \quad \text{where } \mathcal{N} \subset T^* M|_{\partial D} \ \text{conormal bundle}.$$

We will also assume that:

- $\overline{D} \subset M$ is contained in a globally coordinatizable open subset $\mathcal{U} \subset M$ or, equivalently, D is a domain of $M = \mathbb{R}^{2n} \simeq \mathbb{C}^n$ equipped with a non-standard complex structure J.
- D has a smooth global defining function $\rho : \mathcal{U} \subset M \longrightarrow \mathbb{R}$, i.e.,

$$D = \{\, x \in M \;:\; \rho(x) < 0 \,\}$$

with $d\rho_x \neq 0$ for any $x \in \Gamma = \partial D$.

Let us study the differential problem that characterizes the lifts $\tilde{f} : \overline{\Delta} \to T^*M$ of stationary disks of D. Consider the map

$$\tilde{\rho} : \mathbb{R}_* \times T^*M|_{\mathcal{U}} \longrightarrow \mathbb{R} \times T^*M|_{\mathcal{U}}, \quad \tilde{\rho}(t,\alpha) \stackrel{\text{def}}{=} (\rho(\tilde{\pi}(\alpha)), \alpha - t \cdot d\rho_{\tilde{\pi}(\alpha)}). \quad (27)$$

Notice that $\mathcal{N}_*$ is a $2n$-dimensional submanifold of T^*M and that it can be identified with the level set

$$\{(t,\alpha) : t \neq 0, \; \tilde{\rho}(t,\alpha) = (0_{\mathbb{R}}, 0_{T^*_{\tilde{\pi}(\alpha)}M})\} \subset \mathbb{R}_* \times T^*M|_{\mathcal{U}},$$

which is a $2n$-dimensional submanifold of $\mathbb{R}_* \times T^*M$. Therefore, using a system of coordinates $\tilde{\xi} = (x^i, p_j)$ on $T^*M|_{\mathcal{U}}$, associated with coordinates $\xi = (x^i)$, we may identify $\mathbb{R}_* \times T^*M|_{\mathcal{U}}$ with an open subset $\mathcal{V} \subset \mathbb{R}^{4n+1}$ and $\mathcal{N}_*$ with the level set in $\mathcal{V}$ defined by

$$\mathcal{N}_* \simeq \{\, (t,\alpha) \in \mathcal{V} \;:\; \tilde{\rho}^i(t,\alpha) = 0, \; 1 \leq i \leq 2n+1 \,\}.$$

By a direct check of the rank of the Jacobian, one can see that $\tilde{\rho} = (\tilde{\rho}^1, \ldots, \tilde{\rho}^{2n+1})$ *is a smooth defining function for $\mathcal{N}_*$.*

We now consider the map $r : \mathbb{C} \times \mathcal{V} \subset \mathbb{C} \times \mathbb{R}^{4n+1} \longrightarrow \mathbb{R}^{2n+1}$, defined by

$$r(\zeta, t, \alpha) \stackrel{\text{def}}{=} \left(\tilde{\rho}^1(t, \zeta^{-1} \cdot \alpha), \ldots, \tilde{\rho}^n(t, \zeta^{-1} \cdot \alpha)\right). \quad (28)$$

By the above identifications, we have that a disk $f : \overline{\Delta} \to \overline{D} \subset \mathbb{R}^{2n}$ is stationary if and only if there exists $\tilde{f} \in C^{\alpha,\epsilon}(\overline{\Delta}; \mathbb{C}^{2n})$ and $\lambda \in C^{\epsilon}(\partial\Delta; \mathbb{R})$ such that

$$\begin{cases} \overline{\partial}_J \tilde{f}(\zeta) = 0, & \zeta \in \Delta, \\ r(\zeta, \lambda(\zeta), \tilde{f}(\zeta)) = 0, & \zeta \in \partial\Delta, \end{cases} \quad (29)$$

where $\overline{\partial}_J = \overline{\partial}_{J_{\text{st}}, J} : C^{\alpha}(\overline{\Delta}; \mathbb{C}^{2n}) \longrightarrow C^{\alpha-1}(\overline{\Delta}; \mathbb{C}^{2n})$ is the operator (20). Problem (29) belongs to the class usually called *of generalized Riemann–Hilbert problems*, for which there exists a well developed theory (see e.g. [35, 55]).

In order to study the solution space of (29) and its stability w.r.t. small deformations of data, one has to find explicit coordinate expressions for the operators, which determines this problem. For this, let us fix an almost complex structure $J = J_o$, a point $x_o \in D(\subset \mathbb{R}^{2n})$ and a vector $v_o \in T_{x_o}D \simeq \mathbb{R}^{2n}$. Denote by $\mathcal{R}_1, \ldots, \mathcal{R}_5$ the operators

$$(\tilde{f}, \lambda, \mu) \in \mathcal{C}^{\alpha,\varepsilon}(\overline{\Delta}; \mathbb{C}^{2n}) \times \mathcal{C}^{\varepsilon}(\partial\Delta; \mathbb{R}) \times \mathbb{R}_*,$$

which correspond to the conditions of (29) plus some additional conditions, which are convenient to introduce in order to fully parameterize the solution space of the problem:

$$\mathcal{R}_1(\tilde{f}, \lambda, \mu) = \overline{\partial}_{\tilde{J}_o} \tilde{f} \qquad (\tilde{J}\text{-holomorphicity of } \tilde{f}),$$

$$\mathcal{R}_2(\tilde{f}, \lambda, \mu) = r(\zeta, \lambda(\zeta), \tilde{f}(\zeta)) \qquad (\text{boundary data for } \tilde{f}),$$

$$\mathcal{R}_3(\tilde{f}, \lambda, \mu) = \tilde{\pi}(\tilde{f})|_{\zeta=0} - x_o \qquad (\text{center of } f = \tilde{\pi} \circ \tilde{f}), \qquad (30)$$

$$\mathcal{R}_4(\tilde{f}, \lambda, \mu) = \tilde{\pi}(\tilde{f})_* \left(\tfrac{\partial}{\partial x}\big|_{\zeta=0} \right) - \mu v_o \qquad (\text{tangent vector at the center}),$$

$$\mathcal{R}_5(\tilde{f}, \lambda, \mu) = \tilde{f} \left(\tilde{\pi}(\tilde{f})_* \left(\tfrac{\partial}{\partial x}\big|_1 \right) \right) - 1 \qquad (\text{normalizing condition on } \tilde{f}).$$

Using Hopf's Lemma, one can check that, for any stationary disk, there exists exactly one stationary lift satisfying the condition

$$\tilde{f} \left(\tilde{\pi}(\tilde{f})_* \left(\frac{\partial}{\partial x}\bigg|_1 \right) \right) = 1.$$

Therefore, if we denote by $\mathcal{R}_{(J_o, x_o, v_o)} = (\mathcal{R}_1, \ldots, \mathcal{R}_5)$ the operator

$$\mathcal{R}_{(J_o, x_o, v_o)} = (\mathcal{R}_1, \ldots, \mathcal{R}_5) : \mathcal{C}^{\alpha,\varepsilon}(\overline{\Delta}; \mathbb{C}^{2n}) \times \mathcal{C}^{\varepsilon}(\partial\Delta; \mathbb{R}) \times \mathbb{R}_* \longrightarrow$$

$$\longrightarrow \mathcal{C}^{\alpha-1,\varepsilon}(\overline{\Delta}; \mathbb{C}^{2n}) \times \mathcal{C}^{\varepsilon}(\partial\Delta; \mathbb{R}^{2n+1}) \times \mathbb{C}^n \times \mathbb{C}^n \times \mathbb{R},$$

we see that there is a one-to-one correspondence between the stationary disks $f : \overline{\Delta} \longrightarrow \overline{D}$ with $f(0) = x_o$ and $f_*\left(\tfrac{\partial}{\partial x}\big|_{x_o} \right)$ and the solutions to the problem

$$\mathcal{R}_{(J_o, x_o, v_o)}(\tilde{f}, \lambda, \mu) = 0. \qquad (31)$$

The following is a well-known fact of Lempert's theory of stationary disks [30]: If $D \subset \mathbb{C}^n$ is a strictly (linearly) convex, smoothly bounded domain, endowed with the standard complex structure $J_o = J_{\text{st}}$, for any $x_o \in D$ and $v_o \in T_{x_o}D$, the problem (31) has a unique solution smoothly depending on data x_o and v_o.

Now, a similar existence and uniqueness result and a smooth dependence on the data for the stationary disks of almost complex domains (D, J) can be proved whenever J is small deformations of J_{st} in the following sense.

Consider a solution $(\tilde{f}_o, \lambda_o, \mu_o)$ of (31) and denote by

$$\mathfrak{R}_{(J_o, x_o, v_o; \tilde{f}_o, \lambda_o, \mu_o)} \stackrel{\text{def}}{=} \dot{\mathcal{R}}_{(J_o, x_o, v_o)}|_{(\tilde{f}_o, \lambda_o, \mu_o)}$$

the linearized operator at $(\tilde{f}_o, \lambda_o, \mu_o)$ determined by $\mathcal{R}_{(J_o, x_o, v_o)}$. By the Implicit Function Theorem (see e.g. [28]), when the linear operator $\mathfrak{R} = \mathfrak{R}_{(J_o, x_o, v_o; \tilde{f}_o, \lambda_o, \mu_o)}$ is invertible, there exists a solution to the problem $\mathcal{R}_{(J_t, x_t, v_t)}(\tilde{f}, \lambda, \mu) = 0$ for any smooth deformation (J_t, x_t, v_t) of (J_o, x_o, v_o) for a sufficiently small t. In this case $\dim_{\mathbb{R}} \ker \mathfrak{R}_{(J_o, x_o, v_o; \tilde{f}_o, \lambda_o, \mu_o)}$ is equal to the dimension of the solutions space. This fact motivates the following definition.

Definition 19. Let $f_o : \overline{\Delta} \to \overline{D}$ be a stationary disk of (D, J_o) with $x_o = f(0)$ and $v_o = f_*\left(\frac{\partial}{\partial x}\big|_{\zeta=0}\right)$. We say that ∂D is a good boundary for (J_o, f_o) if there exists a lift $\tilde{f}_o$ of f_o and a function λ_o such that $(\tilde{f}_o, \lambda_o, 1)$ is a solution to (31) and the linearized operator $\mathfrak{R} = \mathfrak{R}_{(J_o, x_o, v_o; \tilde{f}_o, \lambda_o, 1)}$ is invertible.

The Implicit Function Theorem and previous remarks bring immediately to the next proposition. In the statement, we denote by $g = g_{ij} dx^i \otimes dx^j$ a fixed Riemannian metric on a neighborhood of $\overline{D}$ and by $g^* = g_{ij} dx^i \otimes dx^j + g^{ij} dp_i \otimes dp_j$ the corresponding Riemannian metric on T^*M. We also set

$$\|J - J'\|_{\overline{D}}^{(1)} \stackrel{\text{def}}{=} \sup_{x \in \overline{D}, v \in T(T_x^*M)} \frac{\|\mathbb{J}(v) - \mathbb{J}'(v)\|_{g^*}}{\|v\|_g}, \qquad (32)$$

where $\|\cdot\|_g$ and $\|\cdot\|_{g^*}$ are the norm functions determined by g and g^*. The topology determined by the norm $\|\cdot\|_{\overline{D}}^{(1)}$ is clearly independent on the choice of g.

Proposition 20. Let $f_o : \overline{\Delta} \to \overline{D}$ be a stationary disk of $D \subset (M, J_o)$ with $x_o = f_o(0)$ and $v_o = f_{o*}\left(\frac{\partial}{\partial x}\big|_{\zeta=0}\right)$. Assume also that ∂D is a good boundary for (J_o, f_o).

Then, there exists neighborhoods $\mathcal{V} \subset D$, $\mathcal{W} \subset TD$ of x_o and v_o, with $\tilde{\pi}(\mathcal{W}) = \mathcal{V} \subset D$, and $\varepsilon > 0$ such that, for any

$$x \in \mathcal{V}, \qquad v \in \mathcal{W}, \qquad \|J - J_o\|_{\overline{D}}^{(1)} < \varepsilon,$$

there exists a unique stationary disk f with

$$f(0) = x \quad \text{and} \quad f_*\left(\frac{\partial}{\partial x}\bigg|_{\zeta=0}\right) = \mu v \quad \text{for some } \mu \neq 0. \qquad (33)$$

The disk f depends smoothly on x, v and J.

The previous result reduces the problem of finding domains with well-behaved families of stationary disks to the query for almost complex domains (D, J) with stationary disks and good boundaries for such disks—let us informally call this kind of domains "good". By these remarks, we can determined useful and efficient results if we are able to determine conditions that ensure that a domain is "good". Such conditions do exist and we will shortly discuss them.

For the moment, let us see what one can do if he has to deal with a "good" almost complex domain. As usual, let $D \subset (M, J)$ be a smooth, relatively compact, strongly pseudoconvex domain in almost complex manifold. For a fixed $x_o \in D$, let $\pi : \tilde{M} \longrightarrow M$ be the *blow up of M at x_o*. Here some care is needed: Keep in mind that the definition of "blow up at a point" is usually defined just for *complex* manifolds. Nonetheless there exists a generalization that makes sense also in case of almost complex domains and this is the notion we refer to (for details, see e.g. [44]).

Now, if $f : \overline{\Delta} \longrightarrow \overline{D}$ is a stationary disk with $f(0) = x_o$ and $f_* \left(\frac{\partial}{\partial x}\big|_0\right) = w$, we may define a J-holomorphic lift of f with image in the closure $\overline{\tilde{D}}$ in $\tilde{M}$ of the blow up $\tilde{D}$ at x_o:

$$\hat{f} : \overline{\Delta} \longrightarrow \overline{\tilde{D}} \subset \hat{M}, \qquad \hat{f}(\zeta) = \begin{cases} (f(\zeta), [f(\zeta)]) & \text{if } \zeta \neq 0, \\ (x_o, [w]) & \text{if } \zeta = 0. \end{cases}$$

This allows to consider the next definition.

Definition 21 ([43]). Let $x_o \in D$ and $\overline{\tilde{D}}$ as above and denote by $\mathcal{F}^{(x_o)}$ the family of all stationary disks of D with $f(0) = x_o$. We say that $\mathcal{F}^{(x_o)}$ is a *foliation of circular type of the pointed domain* (D, x_o) if the following conditions are satisfied:

(i) for any $v \in T_{x_o} D$, there exists a unique disk $f^{(v)} \in \mathcal{F}^{(x_o)}$ such that $f_*^{(v)} \left(\frac{\partial}{\partial x}\big|_0\right) = \mu \cdot v$ for some $0 \neq \mu \in \mathbb{R}$;
(ii) under a fixed identification $(T_{x_o} D, J_{x_o}) \simeq (\mathbb{C}^n, J_{\text{st}})$, the map

$$\tilde{E} : \tilde{\mathbb{B}}^n \subset \tilde{\mathbb{C}}^n \longrightarrow \tilde{D}, \qquad \tilde{E}(v, [v]) = \widetilde{f^{(v)}}(|v|), \qquad (34)$$

between the blow ups of $\mathbb{B}^n \subset \mathbb{C}^n$ and D at 0 and x_o, respectively, is smooth and extends smoothly up to the boundary, determining a diffeomorphism $\tilde{E}|_{\partial \mathbb{B}^n} : \partial \mathbb{B}^n \longrightarrow \partial D$.

If $\mathcal{F}^{(x_o)}$ is a foliation of circular type, we say that x_o is the *center of the foliation* and that D is a *domain of circular type with center x_o*.

Proposition 20 brings to the following stability result for foliation of circular type of "good" domains.

Proposition 22. *Let D be of circular type w.r.t. to an almost complex structure J_o and with center x_o, such that ∂D is a good boundary for (J_o, f) for any $f \in \mathcal{F}^{(x_o)}$.*

Then there exists $\varepsilon > 0$ and an open neighborhood $\mathcal{U} \subset D$ of x_o such that for all almost complex structures J (defined on a neighborhood of $\overline{D}$) with $\|J - J_o\|_{\overline{D}}^{(1)} < \varepsilon$, D is a domain of circular type w.r.t. J with center $x \in \mathcal{U}$ (i.e., for any such J and x, the corresponding collection of stationary disks $\mathcal{F}^{(x)}$ is a foliation of circular type).

Let us now come to the main results on existence of foliations by stationary disks.

Theorem 23 ([43]). *Let $D \subset M$ be a bounded, relatively compact, strongly pseudoconvex domain with smooth boundary in an almost complex manifold (M, J_o). If there exists a diffeomorphism*

$$\varphi : \mathcal{U} \subset M \longrightarrow \varphi(\mathcal{U}) \subset \mathbb{C}^n,$$

between an open neighborhood $\mathcal{U}$ of $\overline{D}$ and an open subset of $\mathbb{C}^n$, with the property that $D' = \varphi(D)$ is a strictly linearly convex domain $D' \subset \mathbb{C}^n$ and $\varphi_*(J_o)$ is sufficiently close to J_{st} in C^1-norm, then D is a domain of circular type w.r.t. J_o with center $x \in D$ (for any $x \in D$!).

This result is essentially a generalization of the result for the case $D' = \mathbb{B}^n$, proved by Coupet et al. in [15]. Its proof is also very close to a similar result, proved independently by Gaussier and Joo in [20].

The proof is technical and we just outline the key ingredients. The first thing to be done is to show that the boundary ∂D of a strictly (linearly) convex domain $D \subset (\mathbb{C}^n, J_{st})$ is good for any pair (J_{st}, f) formed by the standard complex structure J_{st} and a stationary disk f of D through any $x_o \in D$ (here, "stationary" is in the usual sense, in the context of integrable complex structures). In this case, existence and uniqueness results are determined by Lempert's theory.

Secondly, for any $v_o \in T_{x_o} D$, one considers the Riemann–Hilbert operator

$$\mathcal{R} = \mathcal{R}_{(J_{st}, x_o, v_o)} = (\mathcal{R}_1, \mathcal{R}_2, \mathcal{R}_3, \mathcal{R}_4, \mathcal{R}_5)$$

defined in (30) and the corresponding linearization

$$\mathfrak{R} = \mathfrak{R}_{(J_o, x_o, v_o; \tilde{f}_o, \lambda_o, \mu_o)} = (\mathfrak{R}_1, \mathfrak{R}_2, \mathfrak{R}_3, \mathfrak{R}_4, \mathfrak{R}_5)$$

at a given triple $(\tilde{f}_o, \lambda_o, \mu_o)$, corresponding to some lift $\tilde{f}_o$ of a stationary disk f_o. One of the key points of the whole proof consists in showing that the operator $\tilde{\mathfrak{R}} = (\mathfrak{R}_1, \mathfrak{R}_2)$, given by just the first two components of $\mathfrak{R}$, is surjective and with finite dimensional kernel. After this, if one consider the other three components $(\mathfrak{R}_3, \mathfrak{R}_4, \mathfrak{R}_5)$ (which correspond to the operators which fix the initial data and impose an additional normalizing condition), the resulting operator $\mathfrak{R}$ is "nailed down" to become a linear isomorphism.

The surjectivity of $\tilde{\mathfrak{R}} = (\mathfrak{R}_1, \mathfrak{R}_2)$ is in fact a direct consequence of a result by Globevnik [22] and of general facts of the theory of Riemann–Hilbert problems. Such results can be used only if one is able to compute explicitly (and hence check whether they satisfy or not certain conditions) the so-called *partial indices* and the

Maslov index of the conormal bundle along the boundaries of stationary disks of a strictly (linearly) convex domain $D \subset \mathbb{C}^n$. The computation of these indices is radically simplified by considering the so-called *flattening coordinates* of Lempert and Pang, in which a given stationary disk and the boundary nearby assume very simple expressions (see [38, Proposition 2.36 and Theorem 2.45]).

We conclude this section, mentioning that there exists also the following "boundary version" for Theorem 23. As before, let $D \subset M$ be a smoothly bounded, relatively compact, strongly pseudoconvex domain in an almost complex manifold (M, J_o). For a fixed point $x_o \in \partial D$, consider a Riemannian metric $< \cdot, \cdot >$ around x_o, which is Hermitian w.r.t. J_o, a normal vector v to ∂M at x_o, pointing inwards, and for any $a > 0$, let us denote by $\mathcal{C}^{(a)}$ the cone

$$\mathcal{C}^{(a)} = \{ \mathrm{v} \in T_{x_o} M \ : <\mathrm{v}, v>> a \, \}.$$

With the same techniques of Theorem 23, one can prove the following:

Theorem 24 ([43]). *Assume that there exists a diffeomorphism*

$$\varphi : \mathcal{U} \subset M \longrightarrow \varphi(\mathcal{U}) \subset \mathbb{C}^n \, ,$$

between an open neighborhood $\mathcal{U}$ of $\overline{D}$ and an open subset of $\mathbb{C}^n$, with the property that $D' = \varphi(D)$ is a strictly linearly convex domain $D' \subset \mathbb{C}^n$ and that $\varphi_*(J_o)$ is sufficiently close to J_{st} in C^1-norm. Then for any $x_o \in \partial D$ and $a > 0$, there exists a foliation by stationary disks of a subdomain $D^{(a)}_{(x_o)} \subset D$, in which all disks map $1 \in \partial \Delta$ into $x_o \in \partial D$ and have boundary tangent vector at x_o contained in the cone $\mathcal{C}^{(a)}$.

For the definition of *boundary tangent vector* and a more detailed description of the subdomain $D^{(a)}_{(x_o)} \subset D$, see [43].

5.2 Almost Complex Domains of Circular Type and Normal Forms

The project of this section can be roughly described as follows:

– Try to reproduce the steps, performed in the study of stationary disks and Monge–Ampère foliations of complex domains, in the new wider context of domains in *almost* complex manifolds.
– Show that such steps can in fact be performed for a very large class of almost complex domains and give useful information about pluripotential theory on almost complex domains.

As usual, let $D \subset (M, J)$ be a smooth, relatively compact domain in an almost complex manifold and, for any point x_o of an almost complex domain $D \subset (M, J)$, denote by $\mathcal{F}^{(x_o)}$ the family of stationary disks of D with $f(0) = x_o$. By analogy with foliations of circular domains in complex manifolds, one is naturally driven to consider the notion of *foliation of circular type* $\mathcal{F}^{(x_o)}$, given in Definition 21. For any such foliation, the point x_o is called *center of the foliation* and the corresponding map $\tilde{E} : \tilde{\mathbb{B}}^n \longrightarrow \tilde{D}$ is called *(generalized) Riemann map of (D, x_o)*. Any domain $D \subset (M, J)$ admitting a foliation of circular type centered at x_o is called *almost complex domain of circular type with center x_o*.

As in the integrable case, consider the blow-down map

$$\pi : \tilde{\mathbb{B}}^n \longrightarrow \mathbb{B}^n$$

and the (uniquely defined) map $E : \mathbb{B}^n \longrightarrow D$ such that the following diagram commutes

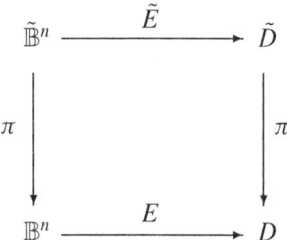

The map E defined in this way is $\mathcal{C}^\infty$ on $\mathbb{B}^n \setminus \{0\}$ and it is continuous at 0. However, in general, E is not smooth at 0. Nevertheless, one can consider a new differentiable structure on $\mathbb{B}^n$, formed by the atlas of coordinate charts of the form

$$\xi' : \mathcal{U} \longrightarrow \mathbb{R}^{2n}, \qquad \xi' = \xi \circ E,$$

where the $\xi = (x^i)$ are charts of the differentiable structure of $D \subset M$. By construction, these charts overlap smoothly with the standard coordinates of $\mathbb{R}^{2n}$ on any open subsets of $\mathbb{B}^n \setminus \{0\}$. Hence their restrictions on $\mathbb{B}^n \setminus \{0\}$ belong to the standard differentiable structure of $\mathbb{B}^n$. On the other hand, when E is not smooth at 0, they cannot smoothly overlap with standard coordinates in neighborhoods of 0. This means that *they give a non-standard differentiable structure on $\mathbb{B}^n$, which coincides with the standard one only on $\mathbb{B}^n \setminus \{0\}$*.

In the following, we always implicitly consider on $\mathbb{B}^n$ such new differentiable structure. Notice that, by construction, the map $E : \mathbb{B}^n \longrightarrow D$ is smooth also at 0 if $\mathbb{B}^n$ is endowed with such non-standard differentiable structure.

Now, let us consider on $\mathbb{B}^n$ the almost complex structure J' defined by

$$J' = E^*(J).$$

We stress once again that *the tensor field J' is smooth over the whole $\mathbb{B}^n$, provided that one considers the differentiable structure on $\mathbb{B}^n$ defined above; w.r.t. the standard differentiable structure, J' is smooth only on $\mathbb{B}^n \setminus \{0\}$ and it is possibly non well-defined at $0 \in \mathbb{B}^n$*.

The pair $(\mathbb{B}^n, J')$ is called *normal form of the almost complex domain of circular type (D, J)*. After all necessary verifications, one can conclude that the following perfect analogue of the situation in the integrable case, holds true: *Any almost complex domain $D \subset (M, J)$ of circular type is (J, J')-biholomorphic to its normal form $(\mathbb{B}^n, J')$ through a map that sends the J-stationary disks of D into the straight radial disks of $\mathbb{B}^n$, which are therefore J'-stationary disks.*

This shows that the analysis of almost complex domains can be reduced to the study of almost complex structures on $\mathbb{B}^n$ with the above properties, i.e., such that the straight radial disks are J'-stationary.

Consider the distribution $\mathcal{Z} \subset T\mathbb{B}^n$ defined in (13). We have the following.

Theorem 25. *A pair $(\mathbb{B}^n, J)$, formed by $\mathbb{B}^n$ endowed with a non-standard differentiable structure, coinciding with the standard one on $\mathbb{B}^n \setminus \{0\}$, and an almost complex structure J which is smooth w.r.t. such differentiable structure, is a domain of circular type in normal form if and only if*

(i) *$\mathcal{Z}$ is J-invariant (i.e., $J\mathcal{Z} = \mathcal{Z}$);*
(ii) *the straight radial disks of $\mathbb{B}^n$ are J-stationary;*
(iii) *the non-standard differentiable structure is such that the blow-up $\widetilde{\mathbb{B}}^n$ of $\mathbb{B}^n$ at 0, determined by J, is equivalent, as differentiable manifold, to the blow-up determined by the standard complex structure.*

In the following, the almost complex structures J on $\mathbb{B}^n$, satisfying (i)–(iii) of previous theorem, will be called *L-almost complex structure*.

It is important to observe that (ii) holds if and only if, in suitable systems of coordinates, the components of J belong to the range of a Fredholm operator, i.e., they belong to a space, which is finite codimensional in an appropriate Hilbert space, and can be characterized by a finite number of equations. This fact has a pair of interesting consequences, namely that:

(a) The class of L-almost complex structure is in practice a very large class.
(b) Such class naturally includes two smaller classes, characterized by very simple conditions, which are very useful to construct a number of interesting examples.

The definitions and the analysis of such smaller classes are the contents of next section.

5.3 "Nice" and "Very Nice" L-Almost Complex Structures

Consider the distributions $\mathcal{Z}$ and $\mathcal{H}$ on $\mathbb{B}^n$ defined in (13) and (14). We recall that:

– denoting, as usual, $\tau_o(z) = \|z\|^2$ and $d_{st}^c = J_{st}^* \circ d \circ J_{st}^*$, one has that

$$\begin{cases} J_{st}\mathcal{Z}_z = \mathcal{Z}_z \\ \mathcal{Z}_z = \ker dd_{st}^c \log \tau_o \end{cases} \text{for any } z \in \mathbb{B}^n \setminus \{0\} \ .$$

- $\mathcal{Z}$ and $\mathcal{H}$ are not only orthogonal w.r.t. the Euclidean metric but also w.r.t. the J_{st}-invariant 2-form $dd^c_{J_{\text{st}}}\tau_o$.
- For any $z \in \mathbb{B}^n \setminus \{0\}$, the subspace $\mathcal{H}_z \subset T_z\mathbb{B}^n$ coincides with the J_{st}-holomorphic tangent space of the sphere $S_c = \{\tau_o = c\}$, $c = \tau_o(z)$.
- For any tangent space $T_z\mathbb{B}^n$, $z \neq 0$, the complexification $T_z^\mathbb{C}\mathbb{B}^n$ decomposes into the direct sum

$$T_z^\mathbb{C}\mathbb{B}^n = \mathcal{Z}_z^\mathbb{C} \oplus \mathcal{H}_z^\mathbb{C} = \left(\mathcal{Z}_z^{10} \oplus \mathcal{Z}_z^{01}\right) \oplus \left(\mathcal{H}_z^{10} \oplus \mathcal{H}_z^{01}\right),$$

where we denoted by $\mathcal{Z}_z^{01}$, $\mathcal{H}_z^{01}$ the $(-i)$-eigenspaces of J_{st} in $\mathcal{Z}_z^\mathbb{C}$ and $\mathcal{H}_z^\mathbb{C}$, and by $\mathcal{Z}_z^{10}$, $\mathcal{H}_z^{10}$ their complex conjugates (which are the $(+i)$-eigenspaces).

Now, an arbitrary almost complex structure J on $\mathbb{B}^n$ is uniquely determined by the corresponding distribution of $(-i)$-eigenspaces $(T_z\mathbb{B}^n)_J^{01}$ in $T_x^\mathbb{C}\mathbb{B}^n$. Generically, these eigenspaces are determined by a tensor field $\varphi \in \text{Hom}(T^{01}\mathbb{B}^n, T^{10}\mathbb{B}^n)$ such that

$$(T_z\mathbb{B}^n)_J^{01} = T_z^{01}\mathbb{B}^n + \varphi(T_z^{01}\mathbb{B}^n),$$

where, as before, we denoted $T_z^{01}\mathbb{B}^n = (T_z\mathbb{B}^n)_{J_{\text{st}}}^{01}$ and $T_z^{10}\mathbb{B}^n = (T_z\mathbb{B}^n)_{J_{\text{st}}}^{10}$. If we consider the decomposition of ϕ as a sum of the form

$$\phi = \phi^\mathcal{Z} \oplus \phi^\mathcal{H} \oplus \phi^{\mathcal{Z},\mathcal{H}} \oplus \phi^{\mathcal{H},\mathcal{Z}}$$

with

$$\phi_z^\mathcal{Z} \in \text{Hom}(\mathcal{Z}^{01}, \mathcal{Z}^{10}), \quad \phi_z^\mathcal{H} \in \text{Hom}(\mathcal{H}^{01}, \mathcal{H}^{10}), \quad \phi_z^{\mathcal{Z},\mathcal{H}} \in \text{Hom}(\mathcal{Z}^{01}, \mathcal{H}^{10}),$$

$$\phi_z^{\mathcal{H},\mathcal{Z}} \in \text{Hom}(\mathcal{H}^{01}, \mathcal{Z}^{10}),$$

we have that the $(-i)$-eigenspaces $(T_z\mathbb{B}^n)_J^{01}$ can be written as

$$(T_z\mathbb{B}^n)_J^{01} = \left(\mathcal{Z}^{01} + \phi_z^\mathcal{Z}(\mathcal{Z}^{01}) + \phi_z^{\mathcal{Z},\mathcal{H}}(\mathcal{Z}^{01})\right) + \\ + \left(\mathcal{H}^{01} + \phi_z^\mathcal{H}(\mathcal{H}^{01}) + \phi_z^{\mathcal{H},\mathcal{Z}}(\mathcal{H}^{01})\right). \tag{35}$$

We observe that J satisfies condition (i) of Theorem 25 (i.e., $J(\mathcal{Z}) \subset \mathcal{Z}$ and $J|_\mathcal{Z} = J_{\text{st}}|_\mathcal{Z}$) if and only if the components $\varphi^\mathcal{Z}$ and $\varphi^{\mathcal{Z},\mathcal{H}}$ are identically equal to 0, that is

$$\varphi = \varphi^\mathcal{H} + \varphi^{\mathcal{H},\mathcal{Z}}.$$

In such class of almost complex structures J on $\mathbb{B}^n$, it is very convenient to consider the following conditions.

1. J is called *nice* if the corresponding deformation tensor φ is, in addition, of the form

$$\varphi = \varphi^\mathcal{H}.$$

This is equivalent to assume that the distribution $\mathcal{H}$ is J-invariant.

2. J is called *very nice* if

$$\varphi = \varphi^{\mathcal{H}} \quad \text{and} \quad \mathcal{L}_{Z^{0,1}}\varphi^{\mathcal{H}} = 0 \, .$$

This assumption corresponds to require that J is nice and that the deformation tensor φ depends J_{st}-holomorphically on the complex parameter that describe the straight radial disks of $\mathbb{B}^n$.

A geometric motivation for considering the notion of "very nice structures" comes from the following. It is well known that, in case of *integrable* complex structures, there exists a strict relation between stationary disks and Kobayashi extremal disks. This is a fact that goes back to the ideas of Poletski [49] and Lempert [30], which showed that, under appropriate regularity conditions, the stationary disks are the solutions to the Euler–Lagrange conditions for the extremal Kobayashi disks (they are the critical point of a suitable functional!). In fact, by the results of [30] we know that the two notions agree for the strictly linearly convex domains in $\mathbb{C}^n$.

Now, it is important to have in mind that, for generic almost complex domains, *these two notions—stationarity and extremality—are no longer related*. Counterexamples have been recently exhibited by Gaussier and Joo in [20]. The authors determined also some conditions, which are sufficient for a stationary disk to be also extremal and which can be described as follows.

Let us first recall a few concepts related with the geometry of the tangent bundle of a manifold M. We recall that the *vertical distribution in $T(TM)$* is the subbundle of $T(TM)$ defined by

$$T^V(TM) = \bigcup_{(x,v)\in TM} T^V_{(x,v)}M \, , \quad T^V_{(x,v)}M = \ker \pi_*|_{(x,v)} \, .$$

For any $x \in M$, let us denote by $(\cdot)^V : T_x M \longrightarrow T^V_{(x,v)}M$ the map

$$\left(\mathrm{w}^i \frac{\partial}{\partial x^i} \bigg|_x \right)^V \stackrel{\mathrm{def}}{=} \mathrm{w}^i \frac{\partial}{\partial q^i} \bigg|_{(x,v)} \, .$$

It is possible to check that this map does not depend on the choice of coordinates and that it determines a natural map from TM to $T(TM)$ (see [59]). For any $\mathrm{w} \in TM$, the corresponding vector $\mathrm{w}^V \in T(TM)$ is called *vertical lift of* w.

Definition 26 ([21,44]). Let $f : \overline{\Delta} \longrightarrow M$ be a $\mathcal{C}^{\alpha,\epsilon}$, J-holomorphic embedding with $f(\partial\Delta) \subset \partial D$. We call *infinitesimal variation of f* any $\mathbb{J}$-holomorphic map $W : \overline{\Delta} \longrightarrow TM$ of class $\mathcal{C}^{\alpha-1,\epsilon}$ with $\pi \circ W = f$ (here, $\pi : TM \longrightarrow M$ is the natural projection). An infinitesimal variation W is called *attached to ∂D and with fixed center* if

(a) $\alpha(W_\zeta) = 0$ for any $\alpha \in \mathcal{N}_{f(\zeta)}$, $\zeta \in \partial\Delta$,
(b) $W|_0 = 0$.

It is called *with fixed central direction* if in addition it satisfies

(c) $W_*\left(\frac{\partial}{\partial \operatorname{Re}\zeta}\Big|_0\right) \in T^V_{W_0}(TM)$ and it is equal to $\lambda \left(f_*\left(\frac{\partial}{\partial \operatorname{Re}\zeta}\Big|_0\right)\right)^V$ for some $\lambda \in \mathbb{R}$.

The disk f is called *Kobayashi critical* if for any infinitesimal variation W, attached to ∂D and with fixed central direction, one has $W_*\left(\frac{\partial}{\partial \operatorname{Re}\zeta}\Big|_0\right) = 0$.

Such definition is motivated by the fact that, when $f^{(t)} : \overline{\Delta} \longrightarrow M$, $t \in]-a, a[$, is a smooth 1-parameter family of J-holomorphic disks of class $\mathcal{C}^{\alpha,\epsilon}$ with $f^{(0)} = f$, then $W = \frac{df^{(t)}}{dt}\Big|_{t=0}$ is a variational field on f. Moreover, if $f^{(t)}$ is such that, for all $t \in]a, a[$

$$f^{(t)}(\partial \Delta) \subset \partial D, \quad f^{(t)}(0) = f(0), \quad f^{(t)}_*\left(\frac{\partial}{\partial \operatorname{Re}\zeta}\Big|_0\right) \in \mathbb{R} f_*\left(\frac{\partial}{\partial \operatorname{Re}\zeta}\Big|_0\right), \quad (36)$$

then W satisfies (a)–(c). On the other hand, a disk f is a *locally extremal disk* if for any J-holomorphic disk $g : \overline{\Delta} \longrightarrow M$ of class $\mathcal{C}^{\alpha,\epsilon}$, with image contained in some neighborhood of $f(\overline{\Delta})$ and such that, for some $\lambda \in \mathbb{R}$,

$$g(\partial \Delta) \subset \partial D, \quad g(0) = f(0) = x_o, \quad g_*\left(\frac{\partial}{\partial \operatorname{Re}\zeta}\Big|_0\right) = \lambda f_*\left(\frac{\partial}{\partial \operatorname{Re}\zeta}\Big|_0\right),$$

then $\lambda \leq 1$. It is known that the notions of "Kobayashi critical" and "locally extremal Kobayashi disk" are tightly related. In fact, *any locally extremal disk f, with $f(\partial \Delta) \subset \partial D$, is Kobayashi critical. Conversely, when $D \subset \mathbb{C}^n$ is strictly convex around $f(\overline{\Delta})$ and J is sufficiently close to J_{st}, any Kobayashi critical disk f is locally extremal* [20, 21, 44].

Next theorem gives conditions that imply the equality between stationary and critical disks and will be used in the sequel. It is a refinement of a result and arguments given in Gaussier and Joo (2009, Preprint of Extremal discs in almost complex spaces, unpublished) (see [44]). In this statement, $f : \overline{\Delta} \longrightarrow M$ is a J-holomorphic embedding, of class $\mathcal{C}^{\alpha,\epsilon}$ with $f(\partial \Delta) \subset \partial D$, and $\mathfrak{Var}_o(f)$ denotes the class of infinitesimal variations of f attached to ∂D and with fixed center.

Theorem 27. *Assume that $D \subset M$ is of the form $D = \{ \rho < 0 \}$ for some J-plurisubharmonic ρ and that $\mathfrak{Var}_o(f)$ contains a $(2n - 2)$-dimensional J-invariant vector space, generated by infinitesimal variations e_i, Je_i, $1 \leq i \leq n - 1$, such that the maps*

$$\zeta^{-1} \cdot e_i(\zeta), \quad \zeta^{-1} \cdot Je_i(\zeta) : \overline{\Delta} \longrightarrow TM$$

are of class $\mathcal{C}^{\alpha,\epsilon}$ on $\overline{\Delta}$. Assume also that, for any $\zeta \in \overline{\Delta}$, the set $\{e_i(\zeta), Je_i(\zeta)\} \subset T_{f(\zeta)}M$ span a subspace, which is complementary to $T_{f(\zeta)}f(\Delta) \subset T_{f(\zeta)}M$.

Then f is critical if and only if it is stationary.

This result has a direct application in our case. Assume that $(\mathbb{B}^n, J)$ is an almost complex domain of circular type in normal form (i.e., such that J is an L-almost complex structure).

One can construct variations of the straight radial disks of $\mathbb{B}^n$, deforming them through the directions of $\mathcal{H}$ and obtain a special subspace $\widetilde{\mathfrak{V}} \subset \mathfrak{Var}_o(f)$ of infinitesimal variations for any given straight radial disk f. It turns out that when J is nice, (i.e., $\mathcal{H}$ is J-invariant),

$$J\widetilde{\mathfrak{V}} \subset \mathfrak{Var}_o(f) \quad \Longleftrightarrow \quad \mathcal{L}_{Z^{01}} J = 0 ,$$

i.e., if and only if J is very nice. Combining this fact with the previous theorem, one is able to prove the following proposition which motivates the interest for "very nice structures".

Proposition 28. *Let $(\mathbb{B}^n, J)$ be an almost complex domain of circular type in normal form. If J is very nice, the straight radial disks of $\mathbb{B}^n$ are not only stationary disks but also Kobayashi critical.*

6 Plurisubharmonic Functions and Pseudoconvex Almost Complex Manifolds

Let (M, J) be an almost complex manifold and $\Omega^k(M)$, $k \geq 0$, the space of k-forms of M. We denote by $d^c : \Omega^k(M) \longrightarrow \Omega^{k+1}(M)$ the classical d^c-operator

$$d^c \alpha = (-1)^k (J^* \circ d \circ J^*)(\alpha) ,$$

where J^* denotes the usual action of J on k-forms, i.e.,

$$J^*\beta(v_1, \ldots, v_k) \stackrel{\text{def}}{=} (-1)^k \beta(J v_1, \ldots, J v_k)$$

Let us recall, once again, that when J is integrable,

$$d^c = i(\overline{\partial} - \partial) , \quad \partial\overline{\partial} = \frac{1}{2i} dd^c , \quad dd^c = -d^c d$$

and that $dd^c u$ is a J-Hermitian 2-form for any C^2-function u. We stress the fact that, when J is not integrable, $d^c d \neq -dd^c$ and the 2-forms $dd^c u$, determined by the functions $u \in C^2(M)$, are usually not J-Hermitian. In fact, one has that

$$dd^c u(JX_1, X_2) + dd^c u(X_1, JX_2) = 4N_{X_1 X_2}(u) , \tag{37}$$

where $N_{X_1 X_2}$ is the Nijenhuis tensor evaluated on X_1, X_2 and is—of course—in general non zero. This fact suggests the following definition.

Definition 29. Let $u : \mathcal{U} \subset M \longrightarrow \mathbb{R}$ be of class C^2. We call J-*Hessian of* u *at* x the symmetric form $Hess(u)_x \in S^2 T_x M$, whose associated quadratic form is

$\mathcal{L}(u)_x(v) = dd^c u(v, J v)_x$. By polarization formula and (37), one has that, for any $v, w \in T_x M$,

$$\mathrm{Hess}(u)_x(v, w) = \frac{1}{2}(dd^c u(v, J w) + dd^c u(w, J v))\Big|_x =$$
$$= dd^c u(v, J w)_x - 2N_{vw}(u) . \qquad (38)$$

We remark that $\mathrm{Hess}(u)_x$ is not only symmetric, but also J-Hermitian, i.e.,

$$\mathrm{Hess}(u)_x(J v, J w) = \mathrm{Hess}(u)_x(v, w) \qquad \text{for any } v, w$$

and it is associated with the Hermitian antisymmetric tensor

$$\mathrm{Hess}(u)(J \cdot, \cdot) = \frac{1}{2}(dd^c u(\cdot, \cdot) + dd^c u(J \cdot, J \cdot)) = \frac{1}{2}\left(dd^c u + J^* dd^c u\right). \qquad (39)$$

The *Levi form of u at x* is the quadratic form

$$\mathcal{L}(u)_x(v) = dd^c u(v, J v)|_x .$$

The operator dd^c defined above turns out to be suitable to study plurisubharmonicity on almost complex manifolds. It has been used for instance by Pali in [37] for his study of positivity questions and in a very recent work of Harvey and Lawson [24], using a completely different point of view involving viscosity approach, to provide a satisfactory *weak* pluripotential theory in the almost complex setting. We will further give evidence that it is appropriate to define the almost complex Monge–Ampère operator. Finally, we point out that Pliś [48] uses it to study the inhomogeneous almost complex Monge–Ampère equation.

An upper semicontinuous function $u : \mathcal{U} \subset M \longrightarrow \mathbb{R}$ is called J-plurisubharmonic if

$$u \circ f : \Delta \longrightarrow \mathbb{R}$$

is subharmonic for any J-holomorphic disk $f : \Delta \longrightarrow \mathcal{U} \subset M$. As for complex manifolds, for any $u \in \mathcal{C}^2(\mathcal{U})$ one has that

$$u \text{ is } J\text{-plurisubharmonic}$$

if and only if

$$\mathcal{L}(u)_x(v) = \mathrm{Hess}(u)_x(v, v) \geq 0 \qquad \text{for any } x \in \mathcal{U} \text{ and } v \in T_x M .$$

This motivates the following generalizations of classical notions. In the following, for any $\mathcal{U} \subset M$, the symbol $\mathrm{Psh}(\mathcal{U})$ denotes the class of J-plurisubharmonic functions on $\mathcal{U}$.

Definition 30. Let (M, J) be an almost complex manifold and $\mathcal{U} \subset M$ an open subset. We say that $u \in \text{Psh}(\mathcal{U})$ is *strictly J-plurisubharmonic* if:

(a) $u \in L^1_{\text{loc}}(\mathcal{U})$;
(b) for any $x_o \in \mathcal{U}$ there exists a neighborhood $\mathcal{V}$ of x_o and $v \in C^2(\mathcal{V}) \cap \text{Psh}(\mathcal{V})$ for which $Hess(v)_x$ is positive definite at all points and $u - v$ is in $\text{Psh}(\mathcal{V})$.

In particular, $u \in \text{Psh}(\mathcal{U}) \cap C^2(\mathcal{U})$ is strictly plurisubharmonic if and only if $Hess(u)_x$ is positive definite at any $x \in \mathcal{U}$.

The almost complex manifold (M, J) is called *strongly pseudoconvex* (or *Stein*) *manifold* if it admits a C^2-strictly plurisubharmonic exhaustion $\tau : M \longrightarrow]-\infty, \infty[$.

6.1 Maximal Plurisubharmonic Functions

The J-plurisubharmonic functions share most of the basic properties of classical plurisubharmonic functions. In particular, as it occurs for the domains in complex manifolds, for any open domain $\mathcal{U} \subset (M, J)$, the class $\text{Psh}(\mathcal{U})$ is a convex cone and for any given $u_i \in \text{Psh}(\mathcal{U})$ and $\lambda_i \in \mathbb{R}$, also

$$u = \sum_{i=1}^n \lambda_i u_i \quad \text{and} \quad u' = \max\{u_1, \ldots, u_n\}$$

are in $\text{Psh}(\mathcal{U})$. It is therefore natural to consider the following notion of "maximal" J-plurisubharmonic functions.

Definition 31. Let D be a domain in a strongly pseudoconvex almost complex manifold (M, J). A function $u \in \text{Psh}(D)$ is called *maximal* if for any open $\mathcal{U} \subset\subset D$ and $h \in \text{Psh}(\mathcal{U})$ satisfying the condition

$$\limsup_{z \to x} h(z) \leq u(x) \quad \text{for all } x \in \partial \mathcal{U}, \tag{40}$$

one has that $h \leq u|_{\mathcal{U}}$.

The following characterization of maximal plurisubharmonic functions "nails down" the right candidate for what should be considered as "almost complex Monge–Ampère operator".

Theorem 32. *Let $D \subset M$ be a domain of a strongly pseudoconvex almost complex manifold (M, J) of dimension $2n$. A function $u \in \text{Psh}(D) \cap C^2(D)$ is maximal if and only if it satisfies*

$$\left(dd^c u + J^*(dd^c u)\right)^n = 0. \tag{41}$$

Proof. Let $\tau : M \longrightarrow]-\infty, +\infty[$ be a C^2 strictly plurisubharmonic exhaustion for M and assume that u satisfies (41). We need to show that for any $h \in \text{Psh}(\mathcal{U})$ on an

$\mathcal{U} \subset\subset D$ that satisfies (40), one has that $h \leq u|_{\mathcal{U}}$. Suppose not and pick $\mathcal{U} \subset\subset D$ and $h \in \mathrm{Psh}(\mathcal{U})$, so that (40) is true but there exists $x_o \in \mathcal{U}$ with $u(x_o) < h(x_o)$. Let $\lambda > 0$ so small that

$$h(x_o) + \lambda(\tau(x_o) - M) > u(x_o), \qquad \text{where } M = \max_{y \in \overline{\mathcal{U}}} \tau(y),$$

and denote by $\hat{h}$ the function

$$\hat{h} \stackrel{\text{def}}{=} h + \lambda(\tau - M)|_{\mathcal{U}}. \tag{42}$$

By construction, $\hat{h} \in \mathrm{Psh}(\mathcal{U})$, satisfies (40) and $(\hat{h} - u)(x_o) > 0$. In particular, $\hat{h} - u$ achieves its maximum at some inner point $y_o \in \mathcal{U}$. Now, we remark that (41) is equivalent to say that, for any $x \in D$, there exists $0 \neq v \in T_x M$ so that

$$\left(dd^c u + J^*(dd^c u)\right)_x (v, J v) = \mathrm{Hess}_x(u)(v, v) = 0. \tag{43}$$

Let $0 \neq v_o \in T_{y_o} M$ be a vector for which (43) is true and let $f : \Delta \longrightarrow M$ be a J-holomorphic disk so that $f(0) = y_o$ and with

$$f_*\left(\frac{\partial}{\partial x}\bigg|_0\right) = v_o, \quad f_*\left(\frac{\partial}{\partial y}\bigg|_0\right) = f_*\left(J_{\mathrm{st}} \frac{\partial}{\partial x}\bigg|_0\right) = J v_o.$$

Then, consider the function $G : \Delta \longrightarrow \mathbb{R}$ defined by

$$G \stackrel{\text{def}}{=} \hat{h} \circ f - u \circ f = h \circ f + (\lambda \tau - \lambda M - u) \circ f. \tag{44}$$

We claim that there exists a disk $\Delta_r = \{|\zeta| < r\}$ such that $G|_{\Delta_r}$ is subharmonic. In fact, since τ is $\mathcal{C}^2$ and strictly plurisubharmonic and $\mathrm{Hess}(u)_{y_o}(v_o, v_o) = 0$, we have that

$$0 < \mathrm{Hess}((\lambda\tau - \lambda M - u))_{y_o}(v_o, v_o) = 2i \, \partial\bar{\partial}((\lambda\tau - \lambda M - u) \circ f)\big|_0.$$

Hence, by continuity, there exists $r > 0$ so that

$$0 < 2i \, \partial\bar{\partial}((\lambda\tau - \lambda M - u) \circ f)\big|_\zeta \qquad \text{for any } \zeta \in \overline{\Delta_r}.$$

It follows that $(\lambda\tau - \lambda M - u) \circ f|_{\Delta_r}$ is strictly subharmonic and that $G|_{\Delta_r}$ is subharmonic, being sum of subharmonic functions. At this point, it suffices to observe that, since y_o is a point of maximum for $\hat{h} - u$ on $f(\Delta) \subset \mathcal{U}$, then $0 = f^{-1}(y_o) \in \Delta_r$ is an inner point of maximum for $G|_{\Delta_r}$. In fact, from this and the maximum principle, we get that $G|_{\Delta_r}$ is constant and hence that $h \circ f|_{\Delta_r}$ is $\mathcal{C}^2$ with $2i \, \partial\bar{\partial}(h \circ f)\big|_{\Delta_r} < 0$, contradicting the hypothesis on subharmonicity of $h \circ f$.

Conversely, assume that $u \in C^2(D) \cap \text{Psh}(D)$ is maximal, but that there exists $y_o \in D$ for which $\text{Hess}_{y_o}(u)(v,v) > 0$ for any $0 \neq v \in T_{y_o}M$ and consider the following well known result (see e.g. [18]).

Lemma 33. *For any $\varepsilon > 0$, there exists a relatively compact neighborhood $\mathcal{U}$ of y_o, such that $(\mathcal{U}, J)$ is (J, J')-biholomorphic to $(\mathbb{B}^n, J')$ for some J' such that $\|J' - J_{\text{st}}\|_{\overline{\mathbb{B}^n}, C^2} < \varepsilon$.*

Due to this, we may assume that $\tau = \tau_o \circ \varphi$, with $\tau_o(z) = \|z\|^2$, is a C^2 strictly J-plurisubharmonic exhaustion on $\mathcal{U}$, tending to 1 at the points of $\partial \mathcal{U}$. Hence, there is a constant $c > 0$ such that

$$\text{Hess}_x(u + c(1-\tau))(v,v) = \text{Hess}_x(u)(v,v) - c\text{Hess}_x(\tau)(v,v) \geq 0,$$

for all $x \in \mathcal{U}$ and $v \in T_x M \simeq \mathbb{R}^{2n}$ with $\|v\| = 1$. This means that

$$\hat{h} \stackrel{\text{def}}{=} (u + c(1-\tau))|_{B_{y_o}(r)}$$

is in $C^2(\mathcal{U}) \cap \text{Psh}(\mathcal{U})$, satisfies (40) and, by maximality of u, satisfies $\hat{h} \leq u$ at all points of $\mathcal{U}$. But there is also an $\epsilon > 0$ such that $\emptyset \neq \tau^{-1}([0, 1-\epsilon[) \subsetneq \mathcal{U}$ and hence such that, on this subset, $\hat{h} \geq u + c\epsilon > u$, contradicting the maximality of u. $\square$

6.2 Green Functions of Nice Circular Domains

The results of previous section motivate the following generalized notion of Green functions.

Definition 34. Let D be a domain in a strongly pseudoconvex, almost complex manifold (M, J). We call *almost pluricomplex Green function with pole at $x_o \in D$* an exhaustion $u : \overline{D} \longrightarrow [-\infty, 0]$ such that

(i) $u|_{\partial D} = 0$ and $u(x) \simeq \log \|x - x_o\|$ when $x \to x_o$, for some Euclidean metric $\|\cdot\|$ on a neighborhood of x_o;
(ii) it is J-plurisubharmonic;
(iii) it is a solution of the generalized Monge–Ampere equation $(dd^c u + J^*(dd^c u))^n = 0$ on $D \setminus \{x_o\}$.

Notice that, if a Green function with pole x_o exists, by a direct consequence of property of maximality (Theorem 32), it is unique.

Consider an almost complex domain D of circular type in (M, J) with center x_o and denote by $\tilde{E} : \tilde{\mathbb{B}}^n \longrightarrow \tilde{D}$ the corresponding Riemann map. We call *standard exhaustion of D* the map

$$\tau_{(x_o)} : D \longrightarrow [0, 1[\, , \qquad \tau(x) = \begin{cases} |\tilde{E}^{-1}(x)|^2 & \text{if } x \neq 0 \, , \\ 0 & \text{if } x = x_o \, . \end{cases}$$

so that, when D is in normal form, i.e., when $D = (\mathbb{B}^n, J)$ with J almost L-complex structure, its standard exhaustion is just $\tau_o(z) = \|z\|^2$.

Proposition 35. *Let D be a domain of circular type in (M, J) with center x_o and standard exhaustion $\tau_{(x_o)}$. If $u = \log \tau_{(x_o)}$ is J-plurisubharmonic, then u is an almost pluricomplex Green function with pole at x_0.*

Proof. With no loss of generality, we may assume that the domain is in normal form, i.e., $D = (\mathbb{B}^n, J)$ and $\tau_{(x_o)}(z) = \tau_o(x) = \|x\|^2$. Since τ_o is smooth on $\mathbb{B}^n \setminus \{0\}$ and $u = \log \tau_o$ is J-plurisubharmonic, we have that $Hess(u)_x \geq 0$ for any $x \neq 0$. On the other hand, for any straight disk $f : \Delta \longrightarrow \mathbb{B}^n$ of the form $f(\zeta) = v \cdot \zeta$, we have that $u \circ f$ is harmonic and $Hess(u)_{f(\zeta)}(v, v) = 0$ for any $\zeta \neq 0$. This means that $Hess(u)_x \geq 0$ has at least one vanishing eigenvalue at any point of $\mathbb{B}^n \setminus \{0\}$ and that (41) is satisfied. The other conditions of Definition 34 can be checked directly from definitions. □

When J is integrable, the standard exhaustion $u = \log \tau_{(x_o)}$ of the normal form of a domain of circular type is automatically plurisubharmonic [42], but *in the almost complex case, this is no longer true*, even for small deformations of the standard complex structure. Though this fact is known, it is important to try and understand why it happens. In the next section, we will illustrate how it is easy to produce illuminating examples using our deformation arguments and we provide hints on how to avoid such pathologies.

6.3 A Counterexample and Pluricomplex Green Functions of Nice Domains

On the blow up $\pi : \tilde{B}^2 \longrightarrow B^2$ of the unit ball $B^2 \subset \mathbb{C}^2$ (defined in the usual way, via the standard complex structure J_{st}) consider the vector fields

$$Z \, , \quad J_{st} Z \, , \quad E \, , \quad J_{st} E \, ,$$

where Z is the lift on $\tilde{\mathbb{B}}^n$ of the real vector field $\operatorname{Re}\left(z^i \frac{\partial}{\partial z^i}\right)$ on $\mathbb{B}^n$ and E is any vector field in the distribution $\mathcal{H}$ that satisfies the conditions

$$[Z, E] = [J_{st} Z, E] = 0 \, , \qquad [E, J_{st} E] = -J_{st} Z \, . \tag{45}$$

The standard holomorphic bundle $T^{10} \tilde{B}^2$ is generated at all points by the complex vector field

$$Z^{10} = Z - i J_{st} Z \, ,$$

which determines the "radial" distribution, and by the complex vector field

$$E^{10} = E - iJ_{st}E,$$

which determines the holomorphic tangent bundles of the spheres $S_c = \{\tau_o(z) = \|z\| = c\}$.

Let us also denote by $(E^{10*}, E^{01*}, Z^{10*}, Z^{01*})$ the field of complex coframes, which is dual to the complex frame field $(E^{10}, E^{01} = \overline{E^{10}}, Z^{10}, Z^{01} = \overline{Z^{10}})$ at all points.

Consider now a smooth real valued function $h : \tilde{\mathbb{B}}^n \longrightarrow \mathbb{R}$ such that

- on each sphere S_c, the restriction $h|_{S_c}$ is constant,
- $h \equiv 0$ on an open neighborhood of $\pi^{-1}(0) = \mathbb{C}P^1$,

and let $\varphi \in \text{Hom}(\mathcal{H}^{01}, \mathcal{Z}^{10} + \mathcal{H}^{10})$ be the deformation tensor

$$\varphi_z = h(z)Z_z^{10} \otimes E_z^{01*}.$$

The almost complex structure J, corresponding to φ, is uniquely determined by the J-holomorphic spaces

$$T_{J_z}^{10}\tilde{\mathbb{B}}^n = \mathbb{C}Z_z^{10} \oplus \mathbb{C}\tilde{E}_z^{10} \quad \text{where} \quad \tilde{E}_z^{10} \stackrel{\text{def}}{=} E_z^{10} + h(z)Z_z^{01},$$

By direct inspection, it is not hard to check that J is an almost L-complex structure and that $(\tilde{\mathbb{B}}^n, J)$ is an almost complex domain of circular type in normal form [44]. But we also have the following crucial fact.

Fact. *If $h \not\equiv 0$, the function $u = \log \tau_o$ is not J-plurisubharmonic.*

Indeed, using the definition of J and (45), one computes

$$\text{Hess}(\tilde{E}^{10}, \tilde{E}^{10}) = 2(1 + 2hh_Z),$$

$$\text{Hess}(\tilde{E}^{10}, Z^{10}) = 2h_Z, \quad \text{Hess}(Z^{10}, Z^{10}) = 0$$

(here, we used the notation "$(\cdot)_Z$" to indicate the derivative $(\cdot)_Z = Z(\cdot)$ in the direction of Z) so that the matrix H of the components of $\text{Hess}(u)_z$ w.r.t. the frame $\{E^{10}, Z^{10}\}$ is

$$H = 2\begin{pmatrix} 1 + 2hh_Z & h_Z \\ h_Z & 0 \end{pmatrix}.$$

Since the eigenvalues of H are $\lambda_\pm = 2\frac{(1+2hh_Z)\pm\sqrt{(1+2hh_Z)^2+4h_Z^2}}{2}$, we conclude that u is J-plurisubharmonic if and only if $h_Z \equiv 0$ and hence if and only if $h \equiv 0$ at all points (recall that, by assumptions, h vanishes identically around 0).

Therefore, we may construct arbitrary examples in which $u = \log \tau_o$ is not J-plurisubharmonic for almost complex structures J arbitrarily close to the standard one. This shows that in order to avoid situations like this it is not sufficient to restrict to a class of sufficiently small deformations of integrable structures. One needs some additional assumptions. One of them is the condition that J is also *"nice"*. In fact, one has the following result.

Theorem 36. *Let D be a nice circular domain with standard exhaustion $\tau_{(x_o)}$ and normal form $(\mathbb{B}^n, J)$. If J is a sufficiently small $\mathcal{C}^1$-deformation of J_{st}, then $u = \log \tau_{(x_o)}$ is the Green function with pole at x_o.*

Proof. We only need to show that $u = \log \|z\|^2$ is J-plurisubharmonic on $\mathbb{B}^n \setminus \{0\}$. If $(\mathbb{B}^n, J)$ is nice, then $Hess(u)_z(\mathcal{Z}, \mathcal{H}) = 0$ at any $z \neq 0$. Since the spheres S_c are J-pseudoconvex for any J sufficiently close to the standard structure, the plurisubharmonicity of u follows directly by computing the Hessian along "orthogonal" directions. □

Putting all these facts together, one gets

Theorem 37. *Let D be an almost complex domain of circular type with center x_o in (M, J) strongly pseudoconvex. If the normal form $(\mathbb{B}^n, J')$ of (D, J) is very nice with J' sufficiently close to J_{st}, then*

(a) *The stationary foliation $\mathcal{F}^{(x_o)}$ consists of extremal disks w.r.t. Kobayashi metric.*
(b) *The function $u = \log \tau_{(x_o)}$ is the almost pluricomplex Green function of D with pole x_o.*
(c) *The distribution $\mathcal{Z}_z = \ker(Hess(u)_z)$ is integrable and the closures of its integral leaves are the disks in $\mathcal{F}^{(x_o)}$.*

References

1. M. Abate, G. Patrizio, in *Finsler Metrics – A Global Approach*. With Applications to Geometric Function Theory. Lecture Notes in Mathematics vol. 1591 (Springer, Berlin, 1994)
2. E. Bedford, D. Burns, Holomorphic mapping of annuli in Cn and the associated extremal function. Ann. Sc. Norm. Sup. Pisa Cl. Sci. **6**(4), 381–414 (1979)
3. E. Bedford, J.P. Demailly, Two counterexamples concerning the pluri-complex Green function in $\mathbb{C}^n$. Indiana Univ. Math. J. **37**, 865–867 (1988)
4. E. Bedford, J.E. Fornæss, Counterexamples to regularity for the complex Monge-Ampère equation. Invent. Math. **50**, 129–134 (1978/1979)
5. E. Bedford, M. Kalka, Foliations and complex Monge-Ampére equations. Commun. Pure Appl. Math. **30**, 543–571 (1977)
6. E. Bedford, B.A. Taylor, The Dirichlet problem for a complex Monge-Ampère equation. Invent. Math. **37**, 1–44 (1976)
7. J. Bland, Contact geometry and CR structures on S^3. Acta Math. **173**, 1–49 (1994)
8. J. Bland, T. Duchamp, Moduli for pointed convex domains. Invent. Math. **104**, 61–112 (1991)
9. J. Bland, T. Duchamp, Contact geometry and CR-structures on spheres, in *Topics in Complex Analysis*, Warsaw, 1992. Banach Center Publ., vol. 31, (Polish Acad. Sci., Warsaw, 1995), pp. 99–113
10. Z. Błocki, The complex Monge-Ampére equation in Kahler geometry, in *Pluripotential Theory, CIME*, Cetraro (CS), 11–16 July 2011. Springer Lecture Notes in Mathematics 2075 (Springer, Berlin 2013), pp. 95–141

11. F. Bracci, G. Patrizio, Monge-Ampére foliations with singularities at the boundary of strongly convex domains. Math. Ann. **332**, 499–522 (2005)
12. F. Bracci, G. Patrizio, S. Trapani, The pluricomplex Poisson kernel for strongly convex domains. Trans. Am. Math. Soc. **361**, 979–1005 (2009)
13. D. Burns, Curvature of the Monge-Ampére foliation and parabolic manifolds. Ann. Math. **115**, 349–373 (1982)
14. D. Burns, On the uniqueness and characterization of Grauert tubes, in *Complex Analysis and Geometry*. Dekker Lecture Notes in Pure and Applied Mathematics, vol. 173 (Dekker, New York, 1995), pp. 119–133
15. B. Coupet, H. Gaussier, A. Sukhov, Riemann maps in Almost Complex Manifolds. Ann. Sc. Norm. Sup. Pisa, Cl. Sci. Ser. V, **II** 761–785 (2003)
16. B. Coupet, H. Gaussier, A. Sukhov, Some aspects of analysis on almost complex manifolds with boundary. J. Math. Sci. **154**, 923–986 (2008)
17. J.P. Demailly, Mesures de Monge-Ampère et mesures pluriharmoniques. Math. Z. **194**, 519–564 (1987)
18. K. Diederich, A. Sukhov, Plurisubharmonic exhaustion functions and almost complex Stein Structures. Mich. Math. J. **56**, 331–355 (2008)
19. T.W. Gamelin, N. Sibony, Subharmonicity for uniform algebras. J. Funct. Anal. **35**, 64–108 (1980)
20. H. Gaussier, A.C. Joo, Extremal discs in almost complex spaces. Ann. Sc. Norm. Sup. Pisa, Cl. Sc. Ser. V **IX**, 759–783 (2010)
21. H. Gaussier, A. Sukhov, On the geometry of model almost complex manifolds with boundary. Math. Z. **254**, 567–589 (2006)
22. J. Globevnik, Perturbation by analytic disks along maximal real submanifolds of $\mathbb{C}^N$. Math. Z. **217**, 287–316 (1994)
23. V. Guillemin, M. Stenzel, Grauert Tubes and the Homogeneous Monge-Ampère Equation. J. Differ. Geom. **34**, 561–570 (1991)
24. F.R. Harvey, H.B. Lawson, Potential Theory on Almost Complex Manifolds. Preprint posted on arXiv:1107.2584 (2011)
25. F.R. Harvey, R.O. Jr. Wells, Zero sets of non-negative strictly plurisubharmonic functions. Math. Ann. **201**, 165–170 (1973)
26. S. Ivashkovich, V. Shevchisin, Reflection principle and J-complex curves with boundary on totally real immersions. Commun. Contemp. Math. **4**(1), 65–106 (2002)
27. M. Kalka, G. Patrizio, Monge-Ampére foliations for degenerate solutions. Ann. Mat. Pura Appl. (4) **18**, 381–393 (2010)
28. L.V. Kantorovich, G.P. Akilov, *Functional Analysis in Normed Spaces* (Pergamon Press, Oxford, 1964)
29. K. Kodaira, J. Morrow, *Complex Manifolds* (Holt, Rinehart and Winston, 1971)
30. L. Lempert, La métrique de Kobayashi et la représentation des domaines sur la boule. Bull. Soc. Math. Fr. **109**, 427–474 (1981)
31. L. Lempert, Holomorphic invariants, normal forms and the moduli space of convex domain. Ann. Math. **128**, 43–78 (1988)
32. L. Lempert, R. Szöke, Global solutions of the homogeneous complex Monge-Ampère equation and complex structures on the tangent bundle of Riemannian manifolds. Math. Ann. **290**, 689–712 (1991)
33. L. Lempert, L. Vivas, Geodesics in the space of Kähler metrics. Preprint posted on arXiv:1105.2188 (2011)
34. K. Leung, G. Patrizio, P.M. Wong, Isometries of intrinsic metrics on strictly convex domains. Math. Z. **196**, 343–353 (1987)
35. S. Mikhlin, S. Prosdorf, *Singular Integral Operators* (Springer, Berlin, 1986)
36. R. Narasimhan, The Levi problem for complex spaces II. Math. Ann. **146**, 195–216 (1962)
37. N. Pali, Fonctions plurisousharmoniques et courants positifs de type (1,1) sur une varit presque complexe. Manuscripta Math. **118**, 311–337 (2005)

38. M.Y. Pang, Smoothness of the Kobayashi metric of non-convex domains. Int. J. Math. **4**(6), 953–987 (1993)
39. G. Patrizio, Parabolic Exhaustions for Strictly Convex Domains. Manuscripta Math. **47**, 271–309 (1984)
40. G. Patrizio, A characterization of complex manifolds biholomorphic to a circular domain. Math. Z. **189**, 343–363 (1985)
41. G. Patrizio, Disques extrémaux de Kobayashi et équation de Monge-Ampère complex. C. R. Acad. Sci. Paris, Sér. I **305**, 721–724 (1987)
42. G. Patrizio, A. Spiro, Monge-Ampère equations and moduli spaces of manifolds of circular type. Adv. Math. **223**, 174–197 (2010)
43. G. Patrizio, A. Spiro, Foliations by stationary disks of almost complex domains. Bull. Sci. Math. **134**, 215–234 (2010)
44. G. Patrizio, A. Spiro, Stationary disks and Green functions in almost complex domains. Ann. Sc. Nor. Sup. Pisa (Preprint posted on arXiv:1103.5383) (to appear)
45. G. Patrizio, A. Tomassini, A characterization of Affine hyperquadrics. Ann. Mat. Pura Appl. **184**(4), 555–564 (2005)
46. G. Patrizio, P.M. Wong, Stability of the Monge-Ampère foliation. Math. Ann. **263**, 13–19 (1983)
47. G. Patrizio, P.M. Wong, Stein manifolds with compact symmetric center. Math. Ann. **289**, 355–382 (1991)
48. S. Pliś, The Monge-Ampère equation on almost complex manifolds. Preprint posted on ArXiv:1106.3356 (2011)
49. E.A. Poletski, The Euler-Lagrange equations for extremal holomorphic mappings of the unit disk. Mich. Math. J. **30**, 317–333 (1983)
50. A. Spiro, The structure equation of a complex Finsler space. Asian J. Math. **5**(2), 291–396 (2001)
51. A. Spiro, A. Sukhov, An existence theorem for stationary disks in almost complex manifolds. J. Math. Anal. Appl. **327**, 269–286 (2007)
52. W. Stoll, The characterization of strictly parabolic manifolds. Ann. Sc. Nor. Sup. Pisa Ser. IV **VII**, 87–154 (1980)
53. R. Szöke, Complex structures on the tangent bundle of Riemannian manifolds. Math. Ann. **291**, 409–428 (1991)
54. A. Tumanov, Extremal disks and the regularity of CR mappings in higher codimension. Am. J. Math. **123**, 445–473 (2001)
55. W. Wendland, *Elliptic Systems in the Plane* (Pitman Publishing, Boston-London, 1979)
56. P.M. Wong, Geometry of the complex homogeneous Monge-Ampére equation. Invent. Math. **67**, 261–274 (1982)
57. P.M. Wong, *On Umbilical Hypersurfaces and Uniformization of Circular Domains*, in Complex Analysis of Several Variables. Proceedings of Symposium on Pure Mathematics, vol. 41 (American Mathematical Society, Providence, 1984), pp. 225–252
58. P.M. Wong, Complex Monge-Ampère equation and related problems, in *Complex Analysis, III*, College Park, Md., 1985–1986. Lecture Notes in Mathematics, vol. 1277 (Springer, Berlin, 1987)
59. K. Yano, S. Ishihara, *Tangent and Cotangent Bundles: Differential Geometry* (Marcel Dekker, New York, 1973)

LECTURE NOTES IN MATHEMATICS

Edited by J.-M. Morel, B. Teissier; P.K. Maini

Editorial Policy (for Multi-Author Publications: Summer Schools / Intensive Courses)

1. Lecture Notes aim to report new developments in all areas of mathematics and their applications - quickly, informally and at a high level. Mathematical texts analysing new developments in modelling and numerical simulation are welcome. Manuscripts should be reasonably selfcontained and rounded off. Thus they may, and often will, present not only results of the author but also related work by other people. They should provide sufficient motivation, examples and applications. There should also be an introduction making the text comprehensible to a wider audience. This clearly distinguishes Lecture Notes from journal articles or technical reports which normally are very concise. Articles intended for a journal but too long to be accepted by most journals, usually do not have this "lecture notes" character.

2. In general SUMMER SCHOOLS and other similar INTENSIVE COURSES are held to present mathematical topics that are close to the frontiers of recent research to an audience at the beginning or intermediate graduate level, who may want to continue with this area of work, for a thesis or later. This makes demands on the didactic aspects of the presentation. Because the subjects of such schools are advanced, there often exists no textbook, and so ideally, the publication resulting from such a school could be a first approximation to such a textbook. Usually several authors are involved in the writing, so it is not always simple to obtain a unified approach to the presentation.

   For prospective publication in LNM, the resulting manuscript should not be just a collection of course notes, each of which has been developed by an individual author with little or no coordination with the others, and with little or no common concept. The subject matter should dictate the structure of the book, and the authorship of each part or chapter should take secondary importance. Of course the choice of authors is crucial to the quality of the material at the school and in the book, and the intention here is not to belittle their impact, but simply to say that the book should be planned to be written by these authors jointly, and not just assembled as a result of what these authors happen to submit.

   This represents considerable preparatory work (as it is imperative to ensure that the authors know these criteria before they invest work on a manuscript), and also considerable editing work afterwards, to get the book into final shape. Still it is the form that holds the most promise of a successful book that will be used by its intended audience, rather than yet another volume of proceedings for the library shelf.

3. Manuscripts should be submitted either online at www.editorialmanager.com/lnm/ to Springer's mathematics editorial, or to one of the series editors. Volume editors are expected to arrange for the refereeing, to the usual scientific standards, of the individual contributions. If the resulting reports can be forwarded to us (series editors or Springer) this is very helpful. If no reports are forwarded or if other questions remain unclear in respect of homogeneity etc, the series editors may wish to consult external referees for an overall evaluation of the volume. A final decision to publish can be made only on the basis of the complete manuscript; however a preliminary decision can be based on a pre-final or incomplete manuscript. The strict minimum amount of material that will be considered should include a detailed outline describing the planned contents of each chapter.

   Volume editors and authors should be aware that incomplete or insufficiently close to final manuscripts almost always result in longer evaluation times. They should also be aware that parallel submission of their manuscript to another publisher while under consideration for LNM will in general lead to immediate rejection.

4. Manuscripts should in general be submitted in English. Final manuscripts should contain at least 100 pages of mathematical text and should always include
   - a general table of contents;
   - an informative introduction, with adequate motivation and perhaps some historical remarks: it should be accessible to a reader not intimately familiar with the topic treated;
   - a global subject index: as a rule this is genuinely helpful for the reader.

   Lecture Notes volumes are, as a rule, printed digitally from the authors' files. We strongly recommend that all contributions in a volume be written in the same LaTeX version, preferably LaTeX2e. To ensure best results, authors are asked to use the LaTeX2e style files available from Springer's web-server at
   ftp://ftp.springer.de/pub/tex/latex/svmonot1/ (for monographs) and
   ftp://ftp.springer.de/pub/tex/latex/svmultt1/ (for summer schools/tutorials).
   Additional technical instructions, if necessary, are available on request from:
   lnm@springer.com.

5. Careful preparation of the manuscripts will help keep production time short besides ensuring satisfactory appearance of the finished book in print and online. After acceptance of the manuscript authors will be asked to prepare the final LaTeX source files and also the corresponding dvi-, pdf- or zipped ps-file. The LaTeX source files are essential for producing the full-text online version of the book. For the existing online volumes of LNM see:
   http://www.springerlink.com/openurl.asp?genre=journal&issn=0075-8434.
   The actual production of a Lecture Notes volume takes approximately 12 weeks.

6. Volume editors receive a total of 50 free copies of their volume to be shared with the authors, but no royalties. They and the authors are entitled to a discount of 33.3 % on the price of Springer books purchased for their personal use, if ordering directly from Springer.

7. Commitment to publish is made by letter of intent rather than by signing a formal contract. Springer-Verlag secures the copyright for each volume. Authors are free to reuse material contained in their LNM volumes in later publications: a brief written (or e-mail) request for formal permission is sufficient.

Addresses:
Professor J.-M. Morel, CMLA,
École Normale Supérieure de Cachan,
61 Avenue du Président Wilson, 94235 Cachan Cedex, France
E-mail: morel@cmla.ens-cachan.fr

Professor B. Teissier, Institut Mathématique de Jussieu,
UMR 7586 du CNRS, Équipe "Géométrie et Dynamique",
175 rue du Chevaleret,
75013 Paris, France
E-mail: teissier@math.jussieu.fr

For the "Mathematical Biosciences Subseries" of LNM:

Professor P. K. Maini, Center for Mathematical Biology,
Mathematical Institute, 24-29 St Giles,
Oxford OX1 3LP, UK
E-mail : maini@maths.ox.ac.uk

Springer, Mathematics Editorial I,
Tiergartenstr. 17,
69121 Heidelberg, Germany,
Tel.: +49 (6221) 4876-8259
Fax: +49 (6221) 4876-8259
E-mail: lnm@springer.com

The manufacturer's authorised representative in the EU is Springer Nature Customer Service Centre GmbH, Europaplatz 3, 69115 Heidelberg, Germany. If you have any concerns regarding our products, please contact ProductSafety@springernature.com

Printed and bound by CPI Group (UK) Ltd, Croydon, CR0 4YY

25/03/2026

02078193-0014